Benefits of Silicon in the Nutrition of Plants

Renato de Mello Prado
Editor

Benefits of Silicon in the Nutrition of Plants

FAPESP
FUNDAÇÃO DE AMPARO À PESQUISA DO ESTADO DE SÃO PAULO

Springer

Editor
Renato de Mello Prado
São Paulo State University
Jaboticabal, São Paulo, Brazil

ISBN 978-3-031-26672-0 ISBN 978-3-031-26673-7 (eBook)
https://doi.org/10.1007/978-3-031-26673-7

This Springer imprint is published by the registered company Springer Nature Switzerland AG
The registered company address is: Gewerbestrasse 11, 6330 Cham, Switzerland

Foreword

Silicon (Si) is the second most abundant element in the Earth's crust after oxygen. It is therefore not unexpected that all plant species rooting in soil contain Si in their tissues to a varying extent. Silicon is of benefit for example in the form of amorphous silica or phytoliths in providing structure and rigidity to the upper plant parts. Si is not considered as an essential element according to the classic well-established definition of essentiality as met by the macronutrients (N, P, K, S, Ca, Mg) and micronutrients (Fe, Mn, Zn, B, Cu, Mo, Cl, Ni).

Nevertheless, it has been contended that for plants growing in solution cultures which rarely contain Si, that its absence amounts to the imposition of an atypical environmental stress which accords with the overwhelming evidence that Si exerts a positive influence on plant growth and development. In most texts, in plant nutrition Si has thus been described as a "beneficial element".

The possibility that Si may play a role in plant growth and development has intrigued plant scientists from the mid nineteenth century. It is only relatively recently, however, that this has been confirmed by extensive phytogenetic studies and evidence for example that in grass-like monocotylendenous plants such as sugar cane (*Saccharum officinarum*) and rice (*Oryza sativa*) as well as a number of other plant families, that Si may be the predominant mineral constituent. The similarity between essential and beneficial elements supports the current approach to describe both groups as plant nutrients, defining a plant nutrient as "an element which is essential or beneficial for plant growth and development or for quality attributes of the harvested product of a given plant species in its natural or cultivated environment". This definition is in keeping with the agronomic aims of high yield quality and sustainable crop production and recognition that Si is extensively involved in processes which bring this about.

This book in 19 chapters provides an up-to-date account of the indispensible nature of Si in plant and crop nutrition. Beginning with the role of Si in the environment from the biogeochemical cycle of terrestrial ecosystems, uptake to cellular and tissue accumulation is discussed as well as its influence in mitigating numerous abiotic and biotic stresses. Si is unique in being the only multi-stress mitigator in plant nutrition. The book focuses on the effects of Si on the mineral nutrients and for the first time provides convincing evidence of the mitigating effect of Si on deficiencies of all the macronutrients and a number of the micronutrients (Fe, Mn, Zn, B). Chapters are also devoted to its mitigating effect on Al toxicity and potentially toxic heavy metals such as Cd as well as the role of Si in the activity of microorganisms and in plant diseases and pests. The beneficial effect of Si on salinity, water deficit and high temperature is likely to take on increasing significance in relation to the threat of global warming.

Renato de Mello Prado is to be congratulated and thanked for his careful selection of eminent contributors to the book as well as his own major contribution to the text in addition to carrying out the extremely onerous task of editing the various versions of the chapters of the book. In looking through the names of those who have contributed, it is heartening to see the large number of young career scientists worldwide who have played a part in the production of this valuable text. *Benefits of Silicon in the Nutrition of Plants* fills a major gap in the literature of plant nutrition. It will be warmly welcomed by all those interested in Si as a plant nutrient, be they students, academics or agronomists working with farmers in the field with the aim of increasing the yields and quality of harvested crops.

Faculty of Biological Sciences
University of Leeds, Leeds, UK

Ernest A. Kirkby

Acknowledgment

We would like to thank the support of the São Paulo State Research Support Foundation FAPESP) (Process 22/10092-9) for helping with the publication and research carried out. FAPESP (fapesp.br/en) is a public foundation, funded by the taxpayer in the State of São Paulo, with the mission to support research projects in higher education and research institutions in all fields of knowledge. The book content express the authors opinions and not necessarily those of FAPESP. To Priscila Porchat for translating the book.

Contents

Chapter 1
Silicon Biogeochemistry in Terrestrial Ecosystems

Jörg Schaller and Daniel Puppe

1.1 Introduction

Uptake of silicon (Si) by plants can enhance their resistance to stress such as drought or fungal infections, which is why the International Plant Nutrition Institute (IPNI) categorized Si as a "beneficial substance" for plants (www.ipni.net). Despite the fact that the majority (>90%) of the earth crust consists of silicate minerals, only a small amount of Si is plant or bio-available and reactive. Mineral weathering is the ultimate source of bio-available Si, i.e., monomeric silicic acid (H_4SiO_4), on a geological time scale. Bio-available Si in soils follows different pathways, including (i) immobilization by adsorption and complexation processes, (ii) leaching as a function of rainfall and irrigation, and (iii) incorporation into living organisms (Sommer et al. 2006; Haynes 2019; Schaller et al. 2021a). The use of inorganic bio-available Si for the formation of biogenic silica, i.e., amorphous hydrated silica ($SiO_2 \bullet nH_2O$), by plants, protists (diatoms, testate amoebae), and animals (sponges) is called biosilicification (Ehrlich et al. 2010; Puppe 2020). Biosilicification and the release of biogenic silica from dead organic matter established a biological cycle that controls bio-available Si in soils on shorter time scales (Dürr et al. 2011; Struyf and Conley 2012; Schaller and Struyf 2013). This is because biogenic silica is much more soluble compared to silicate minerals (Fraysse et al. 2009).

Si cycling is actively influenced by humans, particularly in agricultural landscapes (Struyf et al. 2010). Cereal crops, which can have relatively high concentrations of Si, are key factors of biological Si cycling in agricultural plant-soil systems. The amount of plant-available Si decreases in agricultural soils as a result of silica exports by harvesting year by year (anthropogenic desilication; Guntzer et al. (2012a), Keller et al. (2012), Vandevenne et al. (2012)). As these Si losses are often

J. Schaller (✉) · D. Puppe
Leibniz Centre for Agricultural Landscape Research, Müncheberg, Germany
e-mail: Joerg.Schaller@zalf.de

R. de Mello Prado (ed.), *Benefits of Silicon in the Nutrition of Plants*,
https://doi.org/10.1007/978-3-031-26673-7_1

not compensated, especially in agricultural plant-soil systems of the temperate zone, negative consequences for crop resilience might be the result. In this context, crop straw recycling has been shown to replenish bio-available Si in temperate agricultural soils in the long term (Puppe et al. 2021).

In several biome types, such as forests, steppes, and cultivated lands, biogenic silica is accumulated to a large extent in the vegetation (Carey and Fulweiler 2012). In this context, phytoliths, i.e., biogenic silica formed in plants, are of special interest for biological Si cycling, especially in agricultural landscapes (Haynes 2017a, b). Moreover, phytoliths might offer a promising method for sequestering carbon to mitigate climate change, since phytoliths are capable of persisting in soils for centuries to millennia (Song et al. 2016). The distribution and size of biomes are directly affected by humans through intensified land use (forestry and agriculture) with consequences for soil properties and vegetation (Struyf et al. 2010; Vandevenne et al. 2015a, b). Moreover, increased greenhouse gas emissions and consequential changes in climate conditions might negatively affect Si cycling (Struyf et al. 2009).

For a better understanding of the underlying mechanisms of Si cycling in terrestrial ecosystems detailed interdisciplinary research is needed. In this chapter, we provide a comprehensive overview of current knowledge on Si chemistry in soils (Sect. 1.2), Si cycling in natural and agricultural plant-soil systems (Sect. 1.3), Si mitigating drought (Sect. 1.4), and Si controlling nutrient availability and carbon turnover (Sect. 1.5). The chapter ends with concluding remarks, where future directions are outlined (Sect. 1.6). We see our chapter as an encouragement for future research on biogeochemical Si cycling, especially in agricultural landscapes. In this context, Si might be a silver bullet for both the production of stress-resistant crops and for overcoming the negative effects of climate change in a modern, sustainable agriculture.

1.2 Silicon Chemistry in Soils

The determination of the different Si species in soil solution is very important to estimate the predominance of certain Si-related effects in soils. The speciation of Si in solution is comprising monomers to oligomers or even polymers, or precipitation after gel formation (Fig. 1.1) (Schaller et al. 2021a). Monomers are silicic acid with only one Si unit. Silicic acids with more than one Si unit (dimers, trimers, tetramers, oligomers/polymers) are called polysilicic acids (Dietzel 2002). Silicic acid solutions may become unstable by condensation (from Q^0, Q^1, Q^2, Q^3, to Q^4 groups, 0–4 denoting the number of Si units bound to silicon atoms via oxygen) tending to gel formation followed by precipitation (Fig. 1.1) (Greenwood and Earnshaw 2012). Factors controlling the condensation of silicic acid in solution are concentration, temperature, and pH, as well as the concentration of other solutes (Belton et al. 2012). Silicic acid is mobilized as polysilicic acid during the dissolution of Si-rich solids. This polysilicic acid is depolymerizing to monosilicic acid until equilibrium is reached (Belton et al. 2012). At early stages of dissolution, polymeric Si species

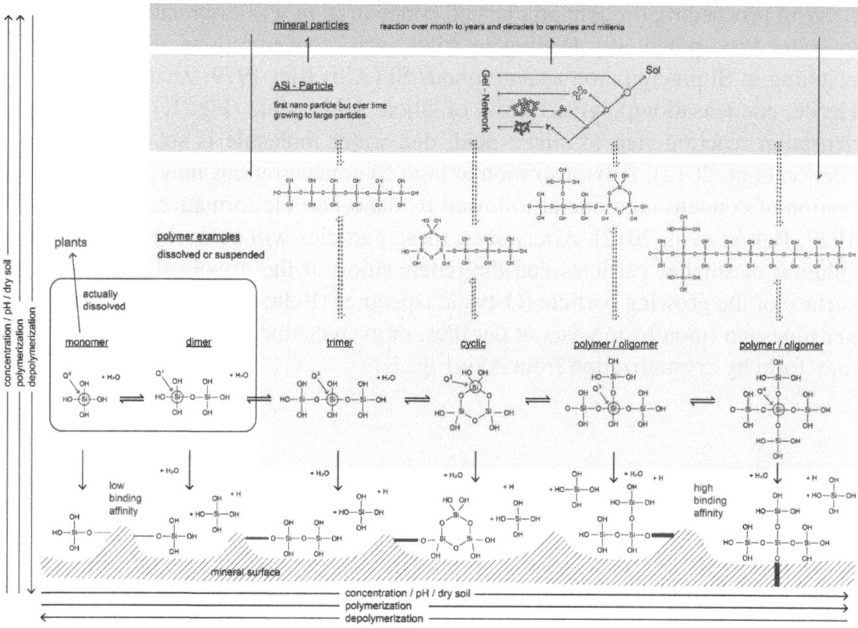

Fig. 1.1 The soil Si cycle modified after Schaller et al. (2021a) showing (i) polymerization/depolymerization of different species of silicic acid; (ii) precipitation to particulate ASi (first nano-scale particles and later larger particles), crystallization and mineralization; (iii) dissolution of particulate to dissolved Si species; as well as (iv) the different condensation states

are dominant (Dietzel 2000). This depolymerization may need between ~1 day and ~1 year, depending on concentration and pH (Dietzel 2000). However, most depolymerization experiments were conducted at concentrations far below saturation (Dietzel and Usdowski 1995; Christl et al. 2012).

If the concentration of silicic acid is higher, the polymerization to polysilicic acid will proceed (Fig. 1.1) (Schaller et al. 2021a). Polymerization and subsequent precipitation will increase if the pH is increasing (Icopini et al. 2005). This polymerization may require several months under low pH (pH ~4) and is much faster under higher pH (Christl et al. 2012). Elevated concentrations of other elements, such as lead, copper, or cadmium in soil solution increase polymerization and precipitation (Stein et al. 2020). In summary: The presence and concentration of the different species of silicic acid in solution is controlled by (i) polymerization/depolymerization and (ii) complexation of silicic acid with inorganic and organic ligands. At short reaction times, the binding strength of monosilicic acid to goethite is low (Klotzbücher et al. 2020). Compared to monosilicic acid, the adsorption of polysilicic acid to mineral surfaces is much faster (some minutes) and stronger because of the higher binding affinity of polysilicic acid compared to monosilicic acid (Dietzel 2002).

With proceeding drought silicic acid concentration will eventually increase due to water loss, potentially leading to silicic acid concentrations above saturation resulting in Si precipitation as amorphous Si (ASi) (Iler 1979; Zhang et al. 2014). Hence, condensation/polymerization of silicic acid occurs (Fig. 1.1). In each condensation reaction step of silicic acid, one water molecule is set free (Fig. 1.1) (Belton et al. 2012). Polymerization at high Si concentrations may lead to the formation of condensation nuclei followed by nano-particle formation (Fig. 1.1) (Iler 1979; Belton et al. 2012). Afterwards, these particles will grow driven by the dissolution of smaller particles and the redeposition of the dissolved species on the surface of the growing particle (Ostwald ripening) (Belton et al. 2012). Over a longer timespan (months to years or decades, or even centuries or millennia) minerals may form by crystallization from ASi (Fig. 1.1).

1.3 Silicon Cycling in Natural and Agricultural Plant-Soil Systems

1.3.1 Si Bioavailability

On geological timescales, mineral weathering represents the ultimate source of bioavailable Si in terrestrial ecosystems (Fig. 1.2). Weathering in soil-plant systems in turn is controlled by climate (precipitation, temperature), specific soil conditions (e.g., mineral composition, soil pH), and vegetation (Si uptake and recycling) (Sommer et al. 2006; Street-Perrott and Barker 2008; Schaller et al. 2021a). Mineral dissolution is much slower than that of amorphous, biogenic silica like phytoliths, which are 10^2 to 10^4 times more reactive than clay minerals and primary silicates at common soil pH (about 4–8) (Fraysse et al. 2006, 2009). Consequently, the cycling of Si by organisms, especially plants, has gained much attention as it strongly influences global Si cycling on a shorter timescale (Dürr et al. 2011; Struyf and Conley 2012). In addition to plants, diatoms, testate amoebae, and sponges have been found to contribute significantly to the formation of biogenic silica in the form of diatom frustules, testate amoeba shells, and sponge spicules, respectively. In this context, especially the significance of testate amoebae for Si cycling in some soils has been revealed and their significance for Si cycling in terrestrial ecosystems might be comparable to the role of protists (i.e., marine diatoms) for Si cycling in the oceans (see the review by Puppe (2020)). Studies indicate that protozoic (testate amoeba) Si pools are strongly affected by land use (Qin et al. 2020, 2021), but we do not yet know which consequences protozoic Si pool changes have on ecosystem scales (e.g., impacts on Si bioavailability).

Si bioavailability in soils is controlled by at least three key factors: (i) the Si concentration in soil solution, (ii) the reserve of Si in the solid phase as Si source (minerogenic/pedogenic, biogenic, adsorbed, or fertilizer Si), and (iii) the Si adsorption capacity or retention capability of the soil (Fig. 1.2) (Haynes 2014; Schaller

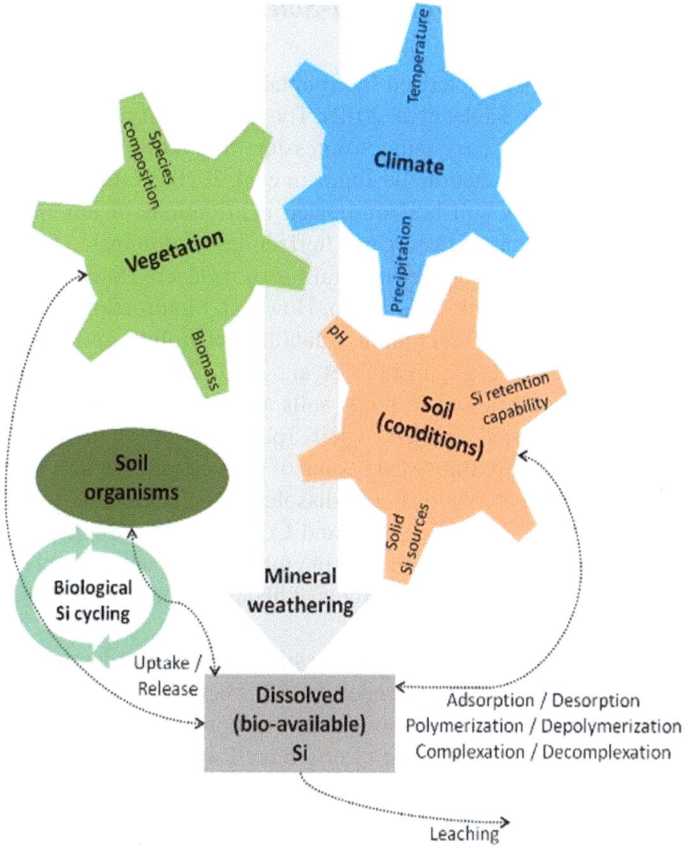

Fig. 1.2 Schematic overview of Si bioavailability in plant-soil systems. Mineral weathering as ultimate source of dissolved or bio-available Si in soils is mainly controlled by climate, vegetation, and soil conditions. Pathways of dissolved Si are indicated by arrows. Biological Si cycling is driven by vegetation and soil organisms

et al. 2021a). Because all of these factors are the result of complex biogeochemical interactions and differ from soil to soil, understanding Si availability in different soils and its uptake by plants and other organisms represents a challenging task. In fact, there is no standard procedure for evaluating plant-available Si in soils because these methods were developed for plants grown in specific climates, most notably sugarcane and rice in (sub)tropical zones (Sauer et al. 2006; Schaller et al. 2021a). Thus, different studies often show inconsistent results. Crusciol et al. (2018), for example, showed that correlations between plant-available Si in soils and Si concentrations in sugarcane depended not only on soil texture but also on the type of extractant.

1.3.2 Si Cycling in Natural Plant-Soil Systems

Natural ecosystems are characterized by Si uptakes of about 2–130 kg Si ha^{-1} y^{-1} depending on vegetation (Keller et al. 2012). The depletion of pedogenic Si pools in the long term, i.e., during ecosystem retrogression, might even increase the importance of Si recycling by vegetation (de Tombeur et al. 2020). Natural soils like forest soils are characterized by soil horizon-related distributions of amorphous silica, with highest concentrations in the organic horizon/layer (dominated by phytogenic silica) and a decrease in the deeper mineral horizons/layers (Saccone et al. 2008; Sommer et al. 2013; Kaczorek et al. 2019). However, bioturbation and percolation can affect the distribution of amorphous silica like phytoliths and testate amoeba shells (Alexandre et al. 1997; Fishkis et al. 2009, 2010; Puppe et al. 2015). Phytogenic silica concentrations in forest soils are primarily influenced by (i) the amount of organic material supplied to soils (phytolith input) and (ii) the loss of phytoliths (phytolith output) due to harvesting of trees, erosion (wind, water), translocation (bioturbation, percolation), and dissolution (e.g., Sommer et al. (2006), Street-Perrott and Barker (2008), Struyf and Conley (2009), Song et al. (2016)). However, it should be kept in mind that phytolith inputs are not only driven by aboveground vegetation but also by the roots of plants (Maguire et al. 2017; Turpault et al. 2018).

Physicochemical properties of phytogenic silica (phytoliths) determine their vulnerability to dissolution, and these properties differ between fresh and aged phytoliths, with implications for soil Si availability (Puppe and Leue 2018). Next to physicochemical properties, the place of origin (cell wall, cell lumen, and intercellular spaces), the morphology, and the specific surface area of phytoliths is assumed to influence phytolith dissolution kinetics (Puppe et al. 2017; Hodson 2019; Schaller et al. 2021a). In general, biological Si cycling in natural plant-soil systems can be assumed mostly undisturbed. However, GHG emissions, wildfires, and wet and dry depositions of pollutants can affect Si cycling in these systems (Fig. 1.3a). Moreover, direct human-induced changes in the vegetation (deforestation) can lead to large Si exports declining the concentration of amorphous Si in soils (Struyf et al. 2010). In this context, increased erosion (Struyf et al. 2010) or human set fires (Unzué-Belmonte et al. 2016; Schaller and Puppe 2021) have the potential to influence Si availability in soils, and thus Si cycling.

1.3.3 Si Cycling in Agricultural Plant-Soil Systems

Agricultural soils contain much less amorphous silica than natural soils. This is because Si exports through harvested crops generally lead to a Si loss in agricultural plant-soil systems, which is called anthropogenic desilication (Desplanques et al. 2006; Meunier et al. 2008; Guntzer et al. 2012b; Keller et al. 2012; Vandevenne et al. 2012). However, some agricultural practices potentially increase soil Si

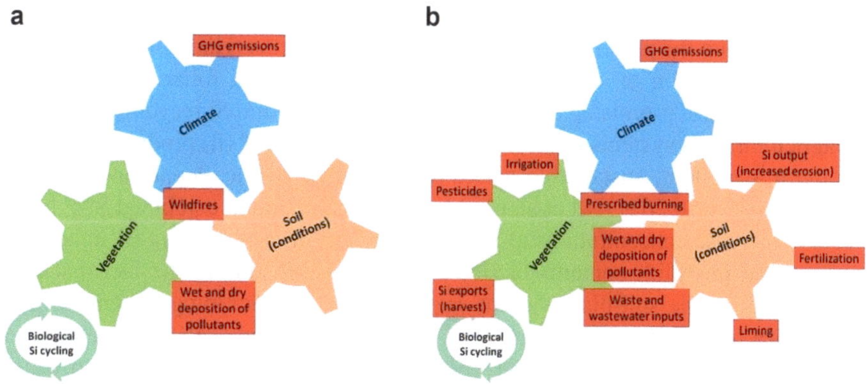

Fig. 1.3 Influencing factors (highlighted in red; assigned to climate, vegetation, soil conditions, and biological Si cycling, see Fig. 1.2) of Si cycling in (**a**) natural and (**b**) agricultural plant-soil systems. While Si cycling in natural plant-soil systems is affected only slightly, Si cycling in agricultural plant-soil systems is subject to diverse influences. Especially Si exports by harvesting largely influence biological Si cycling in agricultural plant-soil systems

availability such as prescribed fires (Schaller and Puppe 2021), the application of Si rich fertilizers (Schaller et al. 2021b), or liming (pH effect, see Haynes (2019). However, further research is necessary to explore to which extent the released Si is taken up by plants, immobilized by adsorption and complexation processes, or leached. Approximately 35% of Si accumulated in vegetation on a global scale is performed in field crops, and this proportion is going to increase with increased agricultural production within the next decades (Carey and Fulweiler 2016). In fact, Si uptakes of cereal crops (which can be assumed to equal Si exports by harvesting) are quite high and reach up to several 100 kilograms per hectare in a year [e.g., rice: 270–500 kg Si ha^{-1} y^{-1}, (Keller et al. 2012), 230–470 kg Si ha^{-1} y^{-1}, (Savant et al. 1996); sugarcane: 379 kg Si ha^{-1} y^{-1}, (Savant et al. 1999); wheat in the temperate zone: 20–113 kg Si ha^{-1} y^{-1}, (Keller et al. 2012)].

In contrast to natural ecosystems, where large amounts of Si are recycled year by year (e.g., Sommer et al. (2013), the annual Si exports in agricultural soil-plant systems are mostly not compensated (Fig. 1.3b). The targeted manipulation of Si cycling (e.g., Si fertilization, straw recycling) might be a promising strategy to both (i) prevent desilication of agricultural plant-soil systems and enhance crop resistance to abiotic and biotic stress (Puppe and Sommer 2018; Puppe et al. 2021) and (ii) potentially enhance carbon sequestration in agricultural biogeosystems to mitigate climate change (Song et al. 2014; Berhane et al. 2020). Carbon sequestration in agricultural systems might be enhanced by regulating (i) weathering (e.g., silicate rock powder amendment), (ii) organic C stabilization (e.g., crop straw recycling, biochar fertilization), and (iii) phytolith-occluded carbon (e.g., partial straw retention after harvest) (Song et al. 2014). It should be noted, however, that the role of phytoliths in C sequestration is still under debate (see the review of Hodson (2019) and references therein). The occluded carbon is potentially not more than the carbon remaining from the protein template shaping the phytoliths (Harrison 1996).

The combination of straw recycling as well as soil and foliar Si fertilization could be the most promising strategy for (i) restoring natural Si recycling processes in agricultural ecosystems to the highest possible extent and (ii) producing resilient crops in a modern, sustainable agriculture (Haynes 2017a; Puppe and Sommer 2018; Puppe et al. 2021). Li et al. (2020), for example, found combined Si-P fertilizers to increase concentrations of plant-available Si in soils, leading to higher biomasses and phytolith contents of rice plants. However, due to the fact that Si-P interactions in the soil-plant system are driven by complex biogeochemical processes that are still not fully understood (e.g., Schaller et al. 2019), further studies are needed to enlighten this aspect. In this context, it is of great interest to which degree Si uptake by cultured plants is determined by environmental factors (e.g., climate and Si availability) and/or their phylogenetic position [e.g., Prychid et al. (2003), Hodson et al. (2005), Cooke and Leishman (2012)].

1.4 Silicon Mitigating Drought

Lots of studies showed a better plant performance after Si fertilization under drought conditions compared to non-fertilized plants. This better plant performance after Si fertilization found expression in a higher biomass production (Gong et al. 2003; Hattori et al. 2005; Chen et al. 2011; Ibrahim et al. 2018). However, no clear picture of the underlying mechanisms emerged as most of these studies found different plant physiological adaptations. For example, some authors found a larger leaf area after Si fertilization (Gong et al. 2003; Alzahrani et al. 2018). An increase of root length and root to shoot (R/S) ratio showed that especially belowground biomass was enhanced after Si fertilization under drought (Hattori et al. 2005; Chen et al. 2011; Ibrahim et al. 2018). The response of nutrient uptake by plants has also been found ambivalent. Chen et al. (2011) found a decreased nutrient uptake under drought after Si fertilization, while Ibrahim et al. (2018) found an increase in plant nutrient uptake. Some studies clearly showed that after Si fertilization, the photosynthetic rate increased and, in most cases, the stomatal conductance of drought-stressed plants increased, too (Hattori et al. 2005; Chen et al. 2011; Ibrahim et al. 2018). It was also described that improved root formation after Si fertilization may be the basis for a better water supply for plants, which may also lead to enhanced nutrient uptake by the plants. Increased root length and surface may also be the reason for the higher water absorption after Si fertilization ameliorating the nutrient and water transport into the shoot, increasing growth and reducing the closure of stomata (Hattori et al. 2005; Chen et al. 2011; Ibrahim et al. 2018). Furthermore, leaf water status was shown to be improved by a decrease in transpirational water loss due to a formation of thicker leaves after Si fertilization (Gong et al. 2003). Another study claimed that Si was also found to prevent deterioration of cell membrane structures and functions during drought, maintaining an intact cell membrane

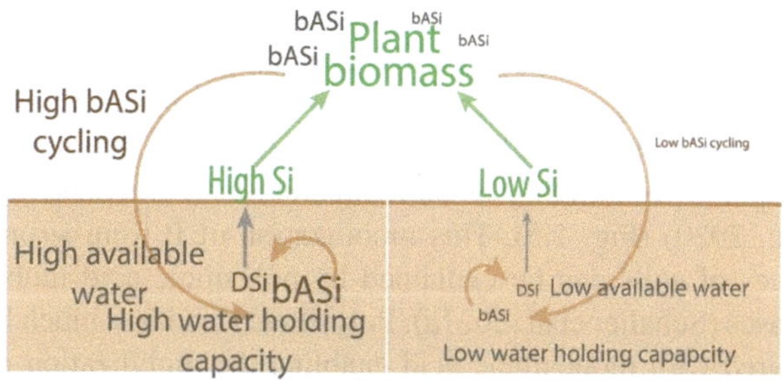

Fig. 1.4 Importance of biogenic ASi (bASi) cycling on soil water holding capacity. The scheme shows the role of bASi cycling via dissolved silicon (DSi) uptake and bASi cycling via litterfall for water holding capacity and available water of soils. (Figure modified after Schaller et al. (2020a))

in stressed plants, and thus promoting growth and productivity during drought (Agarie et al. 1998). Other studies found a larger leaf area due to increased growth combined with a reduced degradation of chlorophyll after Si fertilization to be responsible for a higher dry matter production (Agarie et al. 1992; Gong et al. 2003). However, all those studies showed positive effects of Si fertilization on plant performance during drought in systems with Si addition to soil. However, Si supply might change not only plant physiology but also the water availability in soils.

In a recent study, soil Si fertilization (using amorphous silica, ASi) was found to increase the water holding capacity of soils and strongly increased the plant-available water of soils (Schaller et al. 2020a) (Fig. 1.4). In particular, this study found an increased water content at any water potential and the plant-available water increased by up to >40% or >60% by ASi addition by 1% or 5% (weight), respectively. Such tremendous increase in soil water holding capacity was also found for a lysimeter study, where pure sand was fertilized with ASi (Schaller et al. 2020b). This positive effect of soil ASi fertilization on soil water relations was most recently found in a field plot experiment (Schaller et al. 2021b), showing higher soil moisture and both better plant performance during drought as well as later onset of senescence.

New data point out that the Si effect on soil properties may be the more important factor reducing drought stress for plants compared to the plant internal effects, which seem to be of less importance (Kuhla et al. 2021).

1.5 Si Controlling Nutrient Availability and Carbon Turnover

Silicic acid, especially polysilicic acid, has a strong binding affinity toward the surface of soil minerals (see above). It has been long known that Si fertilization increases plant P nutrition (Lemmermann and Wiessmann 1922; Ma and Takahashi 1990, 1991; Neu et al. 2017). As P nutrition is a very important factor for plant performance, biomass production, and yield (Elser et al. 2007; Elser 2012), increasing P nutrition of plants is also very important for food production in agriculture (Sharpley and Rekolainen 1997). The underlying mechanisms for this increase in plant P nutrition after soil Si fertilization is the binding competition between silicic acid and phosphate for binding at soil mineral surfaces (Schaller et al. 2019; Hömberg et al. 2020) (Fig. 1.5). This mobilization of P from strong bindings to mineral surfaces of soils can be explained by polysilicic acid mobilization from particulate phases (Schaller et al. 2021a). Polysilicic acid has a much higher binding affinity compared with monosilicic acid enabling the mobilization of P from soil increasing P availability for plants (see above).

Silicon, however, interferes not only with P but also with carbon (C) turnover. As Si is increasing P availability in soils and with this plants produce more biomass due to better nutrition the effects of Si on C can be really strong. There is currently a discussion about phytolith-occluded carbon (Song et al. 2014). It should be noted that the potential of phytoliths in C sequestration is low compared to the one of soil organic carbon and is still under controversial discussion (see the review by Hodson (2019) and references therein). The occluded carbon is potentially not more than the carbon remaining from the protein template shaping the phytoliths (Harrison 1996).

As mentioned before, Si is affecting the C turnover by increasing biomass production [e.g., Neu et al. (2017) or Li et al. (2020)]. This biomass will eventually be decomposed afterwards (Coûteaux et al. 1995). The Si effect on litter decomposition is not well understood yet. There are some articles showing an acceleration of decomposition by Si (Schaller and Struyf 2013; Schaller et al. 2014) but also some articles showing a slower decomposition (Nakamura et al. 2021). It was suggested that Si may also change the microbial decomposer community increasing bacteria and decreasing fungal abundance (Schaller et al. 2014, 2017). However, a more detailed picture on those effects on microbial community structure is still missing.

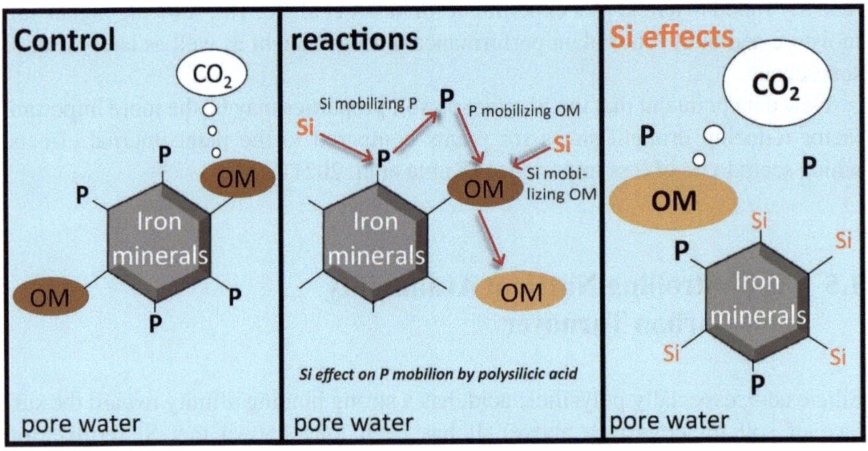

Fig. 1.5 Suggested Si effects on P and organic matter (OM) mobilization and soil respiration in terms of CO_2 release. (Modified after Schaller et al. (2019))

During organic matter decomposition, Si was suggested to increase greenhouse gas (GHG) production (Reithmaier et al. 2017). A few other articles also found increased GHG emissions from peatlands (Hömberg et al. 2021a), with the most pronounced increase for CH_4 emissions during peat rewetting (Hömberg et al. 2021b). Another important interdependency between Si and C is weathering of soil minerals (Si release) and CO_2 consumption (Hartmann et al. 2013). It is discussed if this potential of CO_2-binding by weathering (Schuiling and Krijgsman 2006) can be used to counteract for rising CO_2 concentrations in the atmosphere to mitigate global warming (Köhler et al. 2010). This may be a promising tool in agricultural or forest systems. In summary, there is promising evidence on Si interfering with the C turnover. However, the existing studies are in most cases pioneer studies, which should be elaborated to obtain a more solid overall picture on the underlying mechanisms for the interdependency between Si and C turnover.

1.6 Concluding Remarks

Biogeochemical Si cycling in terrestrial ecosystems is closely linked to ecosystem functioning and services, especially in agricultural plant-soil systems. What we need now are standardized methods for measuring Si availability in soils of natural and agricultural plant-soil systems. In this context, information on Si pools in soils (e.g., bio-available Si, solid biogenic, and pedogenic Si phases) and the findings of laboratory and (long-term) field experiments (e.g., identifying drivers of plant-available Si in different soils, balancing of Si cycling in plant-soil systems) will help us to enlighten the complex interactions in biogeochemical Si cycling. This knowledge is needed for a detailed understanding of Si cycling in natural terrestrial ecosystems and the implementation of corresponding conservation measures. Furthermore, it will allow us to use the recommended amount of Si in agricultural plant-soil systems to enhance plant health and thus ecosystem functioning and services. This is crucial for the production of resilient crops that are adapted to climate change in a modern, sustainable agriculture.

References

Agarie S, Agata W, Kubota F, Kaufman PB (1992) Physiological roles of silicon in photosynthesis and dry matter production in rice plants: i. effects of silicon and shading treatments. Jpn J Crop Sci 61:200–206. https://doi.org/10.1626/jcs.61.200

Agarie S, Uchida H, Agata W, Kubota F, Kaufman PB (1998) Effects of silicon on transpiration and leaf conductance in rice plants (*Oryza sativa* L.). Plant Prod Sci 1:89–95. https://doi.org/10.1626/pps.1.89

Alexandre A, Meunier J-D, Colin F, Koud J-M (1997) Plant impact on the biogeochemical cycle of silicon and related weathering processes. Geochim Cosmochim Acta 61:677–682. https://doi.org/10.1016/S0016-7037(97)00001-X

Alzahrani Y, Kuşvuran A, Alharby HF, Kuşvuran S, Rady MM (2018) The defensive role of silicon in wheat against stress conditions induced by drought, salinity or cadmium. Ecotoxicol Environ Saf 154:187–196. https://doi.org/10.1016/j.ecoenv.2018.02.057

Belton DJ, Deschaume O, Perry CC (2012) An overview of the fundamentals of the chemistry of silica with relevance to biosilicification and technological advances. FEBS J 279:1710–1720. https://doi.org/10.1111/j.1742-4658.2012.08531.x

Berhane M, Xu M, Liang Z, Shi J, Wei G, Tian X (2020) Effects of long-term straw return on soil organic carbon storage and sequestration rate in North China upland crops: a meta-analysis. Glob Change Biol 26:2686–2701. https://doi.org/10.1111/gcb.15018

Carey JC, Fulweiler RW (2012) The terrestrial silica pump. PLoS One 7:e52932. https://doi.org/10.1371/journal.pone.0052932

Carey JC, Fulweiler RW (2016) Human appropriation of biogenic silicon–the increasing role of agriculture. Funct Ecol 30:1331–1339. https://doi.org/10.1111/1365-2435.12544

Chen W, Yao X, Cai K, Chen J (2011) Silicon alleviates drought stress of rice plants by improving plant water status, photosynthesis and mineral nutrient absorption. Biol Trace Elem Res 142:67–76. https://doi.org/10.1007/s12011-010-8742-x

Christl I, Brechbühl Y, Graf M, Kretzschmar R (2012) Polymerization of silicate on hematite surfaces and its influence on arsenic sorption. Environ Sci Technol 46:13235–13243. https://doi.org/10.1021/es303297m

Cooke J, Leishman MR (2012) Tradeoffs between foliar silicon and carbon-based defences: evidence from vegetation communities of contrasting soil types. Oikos 121:2052–2060. https://doi.org/10.1111/j.1600-0706.2012.20057.x

Coûteaux M-M, Bottner P, Berg B (1995) Litter decomposition, climate and liter quality. Trends Ecol Evol 10:63–66. https://doi.org/10.1016/S0169-5347(00)88978-8

Crusciol CAC, De Arruda DP, Fernandes AM, Antonangelo JA, Alleoni LRF, Nascimento CACD, Rossato OB, McCray JM (2018) Methods and extractants to evaluate silicon availability for sugarcane. Sci Rep 8:1–14. https://doi.org/10.1038/s41598-018-19240-1

de Tombeur F, Turner B, Laliberté E, Lambers H, Mahy G, Faucon M-P, Zemunik G, Cornelis J-T (2020) Plants sustain the terrestrial silicon cycle during ecosystem retrogression. Science 369:1245–1248. https://doi.org/10.1126/science.abc039

Desplanques V, Cary L, Mouret JC, Trolard F, Bourrié G, Grauby O, Meunier JD (2006) Silicon transfers in a rice field in Camargue (France). J Geochem Explor 88:190–193. https://doi.org/10.1016/j.gexplo.2005.08.036

Dietzel M (2000) Dissolution of silicates and the stability of polysilicic acid. Geochim Cosmochim Acta 64:3275–3281. https://doi.org/10.1016/S0016-7037(00)00426-9

Dietzel M (2002) Interaction of polysilicic and monosilicic acid with mineral surfaces. In: Stober I, Bucher K (eds) Water-rock interaction. Springer, pp 207–235

Dietzel M, Usdowski E (1995) Depolymerization of soluble silicate in dilute aqueous solutions. Colloid Polymer Sci 273:590–597. https://doi.org/10.1007/BF00658690

Dürr H, Meybeck M, Hartmann J, Laruelle GG, Roubeix V (2011) Global spatial distribution of natural riverine silica inputs to the coastal zone. Biogeosciences 8:597–620. https://doi.org/10.5194/bg-8-597-2011

Ehrlich H, Demadis KD, Pokrovsky OS, Koutsoukos PG (2010) Modern views on desilicification: biosilica and abiotic silica dissolution in natural and artificial environments. Chem Rev 110:4656–4689. https://doi.org/10.1021/cr900334y

Elser JJ (2012) Phosphorus: a limiting nutrient for humanity? Curr Opin Biotechnol 23:833–838. https://doi.org/10.1016/j.copbio.2012.03.001

Elser JJ, Bracken MES, Cleland EE, Gruner DS, Harpole WS, Hillebrand H, Ngai JT, Seabloom EW, Shurin JB, Smith JE (2007) Global analysis of nitrogen and phosphorus limitation of primary producers in freshwater, marine and terrestrial ecosystems. Ecol Lett 10:1135–1142. https://doi.org/10.1111/j.1461-0248.2007.01113.x

Fishkis O, Ingwersen J, Lamers M, Denysenko D, Streck T (2010) Phytolith transport in soil: a field study using fluorescent labelling. Geoderma 157:27–36. https://doi.org/10.1016/j.geoderma.2010.03.012

Fishkis O, Ingwersen J, Streck T (2009) Phytolith transport in sandy sediment: experiments and modeling. Geoderma 151:168–178. https://doi.org/10.1016/j.geoderma.2009.04.003

Fraysse F, Pokrovsky OS, Schott J, Meunier J-D (2006) Surface properties, solubility and dissolution kinetics of bamboo phytoliths. Geochim Cosmochim Acta 70:1939–1951. https://doi.org/10.1016/j.gca.2005.12.025

Fraysse F, Pokrovsky OS, Schott J, Meunier JD (2009) Surface chemistry and reactivity of plant phytoliths in aqueous solutions. Chem Geol 258:197–206. https://doi.org/10.1016/j.chemgeo.2008.10.003

Gong HJ, Chen KM, Chen GC, Wang SM, Zhang CL (2003) Effects of silicon on growth of wheat under drought. J Plant Nutr 26:1055–1063. https://doi.org/10.1081/PLN-120020075

Greenwood NN, Earnshaw A (2012) Chemistry of the elements. Elsevier

Guntzer F, Keller C, Meunier J-D (2012a) Benefits of plant silicon for crops: a review. Agron Sustain Dev 32:201–213

Guntzer F, Keller C, Poulton PR, McGrath SP, Meunier J-D (2012b) Long-term removal of wheat straw decreases soil amorphous silica at Broadbalk, Rothamsted. Plant Soil 352:173–184. https://doi.org/10.1007/s11104-011-0987-4

Harrison CC (1996) Evidence for intramineral macromolecules containing protein from plant silicas. Phytochemistry 41:37–42. https://doi.org/10.1016/0031-9422(95)00576-5

Hartmann J, West AJ, Renforth P, Köhler P, De La Rocha CL, Wolf-Gladrow DA, Dürr HH, Scheffran J (2013) Enhanced chemical weathering as a geoengineering strategy to reduce atmospheric carbon dioxide, supply nutrients, and mitigate ocean acidification. Rev Geophys 51:113–149. https://doi.org/10.1002/rog.20004

Hattori T, Inanaga S, Araki H, An P, Morita S, Luxová M, Lux A (2005) Application of silicon enhanced drought tolerance in *Sorghum bicolor*. Physiol Plant 123:459–466. https://doi.org/10.1111/j.1399-3054.2005.00481.x

Haynes RJ (2014) A contemporary overview of silicon availability in agricultural soils. J Plant Nutr Soil Sci 177:831–844. https://doi.org/10.1002/jpln.201400202

Haynes RJ (2017a) The nature of biogenic Si and its potential role in Si supply in agricultural soils. Agric Ecosy Environ 245:100–111. https://doi.org/10.1016/j.agee.2017.04.021

Haynes RJ (2017b) Significance and role of Si in crop production. Adv Agron 146:83–166. https://doi.org/10.1016/bs.agron.2017.06.001

Haynes RJ (2019) What effect does liming have on silicon availability in agricultural soils? Geoderma 337:375–383. https://doi.org/10.1016/j.geoderma.2018.09.026

Hodson MJ (2019) The relative importance of cell wall and lumen phytoliths in carbon sequestration in soil: a hypothesis. Front Earth Sci 7:167. https://doi.org/10.3389/feart.2019.00167

Hodson MJ, White PJ, Mead A, Broadley MR (2005) Phylogenetic variation in the silicon composition of plants. Ann Bot 96:1027–1046. https://doi.org/10.1093/aob/mci255

Hömberg A, Broder T, Knorr K-H, Schaller J (2021a) Divergent effect of silicon on greenhouse gas production from reduced and oxidized peat organic matter. Geoderma 386:114916. https://doi.org/10.1016/j.geoderma.2020.114916

Hömberg A, Broder T, Schaller J, Knorr K-H (2021b) Methane fluxes but not respiratory carbon dioxide fluxes altered under Si amendment during drying–rewetting cycles in fen peat mesocosms. Geoderma 404:115338. https://doi.org/10.1016/j.geoderma.2021.115338

Hömberg A, Obst M, Knorr K-H, Kalbitz K, Schaller J (2020) Increased silicon concentration in fen peat leads to a release of iron and phosphate and changes in the composition of dissolved organic matter. Geoderma 374:114422. https://doi.org/10.1016/j.geoderma.2020.114422

Ibrahim MA, Merwad A-RM, Elnaka EA (2018) Rice (*Oryza sativa* L.) tolerance to drought can be improved by silicon application. Commun Soil Sci Plant Anal 49:945–957. https://doi.org/10.1080/00103624.2018.1448856

Icopini GA, Brantley SL, Heaney PJ (2005) Kinetics of silica oligomerization and nanocolloid formation as a function of pH and ionic strength at 25°C. Geochim Cosmochim Acta 69:293–303

Iler RK (1979) The chemistry of silica: solubility, polymerization, colloid and surface pro perties, and biochemistry. Wiley

Kaczorek D, Puppe D, Busse J, Sommer M (2019) Effects of phytolith distribution and characteristics on extractable silicon fractions in soils under different vegetation–An exploratory study on loess. Geoderma 356:113917. https://doi.org/10.1016/j.geoderma.2019.113917

Keller C, Guntzer F, Barboni D, Labreuche J, Meunier J-D (2012) Impact of agriculture on the Si biogeochemical cycle: input from phytolith studies. Comptes Rendus Geosci 344:739–746. https://doi.org/10.1016/j.crte.2012.10.004

Klotzbücher T, Treptow C, Kaiser K, Klotzbücher A, Mikutta R (2020) Sorption competition with natural organic matter as mechanism controlling silicon mobility in soil. Sci Rep 10:1–11. https://doi.org/10.1038/s41598-020-68042-x

Köhler P, Hartmann J, Wolf-Gladrow DA (2010) Geoengineering potential of artificially enhanced silicate weathering of olivine. Proc Natl Acad Sci U S A 107:20228–20233

Kuhla J, Pausch J, Schaller J (2021) Effect on soil water availability, rather than silicon uptake by plants, explains the beneficial effect of silicon on rice during drought. Plant Cell Environ 44:3336–3346. https://doi.org/10.1111/pce.14155

Lemmermann O, Wiessmann H (1922) Die ertragssteigernde Wirkung der Kieselsäure bei unzureichender Phosphorsäureernährung der Pflanzen. Zeitschrift für Pflanzenernährung und Düngung, A, Wissenschaftlicher Teil 1:185–246

Li Z, Guo F, Cornelis J-T, Song Z, Wang X, Delvaux B (2020) Combined silicon-phosphorus fertilization affects the biomass and phytolith stock of rice plants. Front Plant Sci 11:67. https://doi.org/10.3389/fpls.2020.00067

Ma JF, Takahashi E (1990) Effect of silicon on the growth and phosphorus uptake of rice. Plant Soil 126:115–119. https://doi.org/10.1007/BF00041376

Ma JF, Takahashi E (1991) Effect of silicate on phosphate availability for rice in a P-deficient soil. Plant Soil 133:151–155. https://doi.org/10.1007/BF0000918

Maguire TJ, Templer PH, Battles JJ, Fulweiler RW (2017) Winter climate change and fine root biogenic silica in sugar maple trees (*Acer saccharum*): implications for silica in the Anthropocene. J Geophyl Res Biogeosc 122:708–715. https://doi.org/10.1002/2016JG003755

Meunier J, Guntzer F, Kirman S, Keller C (2008) Terrestrial plant-Si and environmental changes. Mineral Magaz 72:263–267

Nakamura R, Amada G, Kajino H, Morisato K, Kanamori K, Hasegawa M (2021) Silicious trichomes as a trait that may slow down leaf decomposition by soil meso-and macrofauna. Plant Soil 1:11. https://doi.org/10.1007/s11104-021-05223-1

Neu S, Schaller J, Dudel EG (2017) Silicon availability modifies nutrient use efficiency and content, C:N:P stoichiometry, and productivity of winter wheat (*Triticum aestivum* L). Sci Rep 7:40829. https://doi.org/10.1038/srep40829

Prychid CJ, Rudall PJ, Gregory M (2003) Systematics and biology of silica bodies in monocotyledons. Bot Rev 69:377–440. https://doi.org/10.1663/0006-8101(2004)069[0377:SABOSB]2.0.CO;2

Puppe D (2020) Review on protozoic silica and its role in silicon cycling. Geoderma 365:114224. https://doi.org/10.1016/j.geoderma.2020.114224

Puppe D, Ehrmann O, Kaczorek D, Wanner M, Sommer M (2015) The protozoic Si pool in temperate forest ecosystems—quantification, abiotic controls and interactions with earthworms. Geoderma 243:196–204. https://doi.org/10.1016/j.geoderma.2014.12.018

Puppe D, Höhn A, Kaczorek D, Wanner M, Wehrhan M, Sommer M (2017) How big is the influence of biogenic silicon pools on short-term changes in water-soluble silicon in soils? Implications from a study of a 10-year-old soil-plant system. Biogeosciences 14:5239–5252. https://doi.org/10.5194/bg-14-5239-2017

Puppe D, Kaczorek D, Schaller J, Barkusky D, Sommer M (2021) Crop straw recycling prevents anthropogenic desilication of agricultural soil-plant systems in the temperate zone – results

from a long-term field experiment in NE Germany. Geoderma 403:115187. https://doi.org/10.1016/j.geoderma.2021.115187
Puppe D, Leue M (2018) Physicochemical surface properties of different biogenic silicon structures: results from spectroscopic and microscopic analyses of protistic and phytogenic silica. Geoderma 330:212–220. https://doi.org/10.1016/j.geoderma.2018.06.001
Puppe D, Sommer M (2018) Experiments, uptake mechanisms, and functioning of silicon foliar fertilization—a review focusing on maize, rice, and wheat. In: Advances in agronomy. Elsevier, pp 1–49
Qin Y, Puppe D, Payne R, Li L, Li J, Zhang Z, Xie S (2020) Land-use change effects on protozoic silicon pools in the Dajiuhu National Wetland Park, China. Geoderma 368:114305. https://doi.org/10.1016/j.geoderma.2020.114305
Qin Y, Puppe D, Zhang L, Sun R, Li P, Xie S (2021) How does Sphagnum growing affect testate Amoeba communities and corresponding protozoic Si pools? Results from field analyses in SW China. Microb Ecol 82:459–469. https://doi.org/10.1007/s00248-020-01668-6
Reithmaier GMS, Knorr KH, Arnhold S, Planer-Friedrich B, Schaller J (2017) Enhanced silicon availability leads to increased methane production, nutrient and toxicant mobility in peatlands. Sci Rep 7:8728. https://doi.org/10.1038/s41598-017-09130-3
Saccone L, Conley DJ, Likens GE, Bailey SW, Buso DC, Johnson CE (2008) Factors that control the range and variability of amorphous silica in soils in the Hubbard brook Exp Forest. Soil Sci Soc Am J 72:1637–1644. https://doi.org/10.2136/sssaj2007.0117
Sauer D, Saccone L, Conley DJ, Herrmann L, Sommer M (2006) Review of methodologies for extracting plant-available and amorphous Si from soils and aquatic sediments. Biogeochemistry 80:89–108. https://doi.org/10.1007/s10533-005-5879-3
Savant N, Snyder G, Datnoff L (1996) Silicon management and sustainable rice production. Adv Agron 58:151–199. https://doi.org/10.1016/S0065-2113(08)60255-2
Savant NK, Korndörfer GH, Datnoff LE, Snyder GH (1999) Silicon nutrition and sugarcane production: a review. J Plant Nutr 22:1853–1903. https://doi.org/10.1080/01904169909365761
Schaller J, Cramer A, Carminati A, Zarebanadkouki M (2020a) Biogenic amorphous silica as main driver for plant available water in soils. Sci Rep 10:2424. https://doi.org/10.1038/s41598-020-59437-x
Schaller J, Fauchere S, Joss H, Obst M, Goeckede M, Planer-Friedrich B, Peiffer S, Gilfedder B, Elberling B (2019) Silicon increases the phosphorus availability of Arctic soils. Sci Rep 9:449. https://doi.org/10.1038/s41598-018-37104-6
Schaller J, Frei S, Rohn L, Gilfedder BS (2020b) Amorphous silica controls water storage capacity and phosphorus mobility in soils. Front Environ Sci 8:94. https://doi.org/10.3389/fenvs.2020.00094
Schaller J, Hines J, Brackhage C, Bäucker E, Gessner MO (2014) Silica decouples fungal growth and litter decomposition without changing responses to climate warming and N enrichment. Ecology 95:3181–3189. https://doi.org/10.1890/13-2104.1
Schaller J, Hodson MJ, Struyf E (2017) Is relative Si/Ca availability crucial to the performance of grassland ecosystems? Ecosphere 8:e01726. https://doi.org/10.1002/ecs2.1726
Schaller J, Puppe D (2021) Heat improves silicon availability in mineral soils. Geoderma 386:114909. https://doi.org/10.1016/j.geoderma.2020.114909
Schaller J, Puppe D, Kaczorek D, Ellerbrock R, Sommer M (2021a) Silicon cycling in soils revisited. Plan Theory 10:295. https://doi.org/10.3390/plants10020295
Schaller J, Scherwietes E, Gerber L, Vaidya S, Kaczorek D, Pausch J, Barkusky D, Sommer M, Hoffmann M (2021b) Silica fertilization improved wheat performance and increased phosphorus concentrations during drought at the field scale. Sci Rep 11:1–12. https://doi.org/10.1038/s41598-021-00464-7
Schaller J, Struyf E (2013) Silicon controls microbial decay and nutrient release of grass litter during aquatic decomposition. Hydrobiologia 709:201–212. https://doi.org/10.1007/s10750-013-1449-1

Schuiling R, Krijgsman P (2006) Enhanced weathering: an effective and cheap tool to sequester CO_2. Clim Chang 74:349–354. https://doi.org/10.1007/s10584-005-3485-y
Sharpley A, Rekolainen S (1997) Phosphorus in agriculture and its environmental implications. Phosphorus loss from soil to water. Proceedings of a workshop, Irish Republic
Sommer M, Jochheim H, Hoehn A, Breuer J, Zagorski Z, Busse J, Barkusky D, Meier K, Puppe D, Wanner M, Kaczorek D (2013) Si cycling in a forest biogeosystem - the importance of transient state biogenic Si pools. Biogeosciences 10:4991–5007
Sommer M, Kaczorek D, Kuzyakov Y, Breuer J (2006) Silicon pools and fluxes in soils and landscapes - a review. J Plant Nutr Soil Sci 169:310–329. https://doi.org/10.1002/jpln.200521981
Song Z, McGrouther K, Wang H (2016) Occurrence, turnover and carbon sequestration potential of phytoliths in terrestrial ecosystems. Earth-Sci Rev 158:19–30. https://doi.org/10.1016/j.earscirev.2016.04.007
Song Z, Müller K, Wang H (2014) Biogeochemical silicon cycle and carbon sequestration in agricultural ecosystems. Earth-Sci Rev 139:268–278. https://doi.org/10.1016/j.earscirev.2014.09.009
Stein M, Georgiadis A, Gudat D, Rennert T (2020) Formation and properties of inorganic Si-contaminant compounds. Environ Pollut 265:115032. https://doi.org/10.1016/j.envpol.2020.115032
Street-Perrott FA, Barker PA (2008) Biogenic silica: a neglected component of the coupled global continental biogeochemical cycles of carbon and silicon. Earth Surf Process Landf 33:1436–1457
Struyf E, Conley DJ (2009) Silica: an essential nutrient in wetland biogeochemistry. Front Ecol Environ 7:88–94. https://doi.org/10.1890/070126
Struyf E, Conley DJ (2012) Emerging understanding of the ecosystem silica filter. Biogeochemistry 107:9–18. https://doi.org/10.1007/s10533-011-9590-2
Struyf E, Smis A, Van Damme S, Garnier J, Govers G, Van Wesemael B, Conley DJ, Batelaan O, Frot E, Clymans W, Vandevenne F, Lancelot C, Goos P, Meire P (2010) Historical land use change has lowered terrestrial silica mobilization. Nat Commun 1:129. https://doi.org/10.1038/ncomms1128
Struyf E, Smis A, Van Damme S, Meire P, Conley DJ (2009) The global biogeochemical silicon cycle. SILICON 1:207–213. https://doi.org/10.1007/s12633-010-9035-x
Turpault M-P, Calvaruso C, Kirchen G, Redon P-O, Cochet C (2018) Contribution of fine tree roots to the silicon cycle in a temperate forest ecosystem developed on three soil types. Biogeosciences 15:2231–2249. https://doi.org/10.5194/bg-15-2231-2018
Unzué-Belmonte D, Struyf E, Clymans W, Tischer A, Potthast K, Bremer M, Meire P, Schaller J (2016) Fire enhances solubility of biogenic silica. Sci Total Environ 572:1289–1296. https://doi.org/10.1016/j.scitotenv.2015.12.085
Vandevenne F, Barão L, Ronchi B, Govers G, Meire P, Kelly E, Struyf E (2015a) Silicon pools in human impacted soils of temperate zones. Glob Biogeochem Cycle 29:1439–1450. https://doi.org/10.1002/2014GB005049
Vandevenne F, Struyf E, Clymans W, Meire P (2012) Agricultural silica harvest: have humans created a new loop in the global silica cycle? Front Ecol Environ 10:243–248. https://doi.org/10.1890/110046
Vandevenne FI, Delvaux C, Hughes HJ, André L, Ronchi B, Clymans W, Barao L, Cornelis J-T, Govers G, Meire P (2015b) Landscape cultivation alters δ 30 Si signature in terrestrial ecosystems. Sci Rep 5:7732. https://doi.org/10.1038/srep07732
Zhang C, Li L, Lockington D (2014) Numerical study of evaporation-induced salt accumulation and precipitation in bare saline soils: mechanism and feedback. Water Resour Res 50:8084–8106. https://doi.org/10.1002/2013WR015127

Chapter 2
Silicon: Transcellular and Apoplastic Absorption and Transport in the Xylem

Rafael Ferreira Barreto and Lúcia Barão

2.1 Introduction

Before the roots absorb chemical elements, there is ion-root contact, either by the ion movement in the rhizosphere soil solution (diffusion or mass flow) or by the root growth itself encountering the ion (root interception) (Mengel et al. 2001; Prado 2021). However, the predominant contact form between silicon (Si) and roots has not yet been determined.

By either of these processes and upon contact with roots, an ion may immediately enter the symplast by crossing the plasma membrane of an epidermal cell. Moreover, it may penetrate the apoplast and diffuse between the epidermal cells through the cell walls. From the apoplast of the cortical parenchyma, the ion can either be transported across the plasma membrane of a cortical cell, thus entering the symplast, or diffuse radially into the endoderm via the apoplast. The apoplast forms a continuous phase from the root surface through the cortical parenchyma. However, the ions must enter the symplast before entering the stelae due to the presence of Casparian strips. This lignified or suberized layer that forms rings around endoderm cells and blocks the entry of water and solutes into the endoderm of the stele via apoplast (Assmann 2017).

The roots absorbed the Si from the soil solution or nutrient solution in monosilicic acid (H_4SiO_4) form, a neutrally charged molecule. The maximum solubility of H_4SiO_4 in solution is about 2 mmol L^{-1}, and most soils usually contain H_4SiO_4 in solution between 0.1–0.6 mmol L^{-1} (Ma and Takahashi 2002). After its absorption,

---

R. F. Barreto (✉)
Federal University of Mato Grosso do Sul, Chapadão do Sul, Brazil
e-mail: rafael.barreto@ufms.br

L. Barão
Center for Ecology, Evolution and Environmental Changes & CHANGE - Global Change and Sustainability Institute, University of Lisbon, Lisbon, Portugal

R. de Mello Prado (ed.), *Benefits of Silicon in the Nutrition of Plants*,
https://doi.org/10.1007/978-3-031-26673-7_2

the xylem transported it to the shoot in the same chemical form (Bauer et al. 2011; Takahashi and Hino 1979).

There are four main research areas on Si in plants: its absorption (or absence), its classification as beneficial for the plants, its application as a fertilizer, and the mechanisms by which it alleviates biotic and abiotic stresses. Among these, plants' differential absorption is one of the most intriguing Si properties. Under similar conditions, plant species have different abilities to accumulate Si. Considering that the benefits of Si are generally related to the amount absorbed, some plant species benefit more from Si fertilization than others (Coskun et al. 2019).

According to the Si content in the shoot, plants are classified into three categories. Non-accumulators, such as tomatoes, accumulate less than 5 g kg^{-1}. In contrast, intermediate ones, such as cucumbers, show contents between 5 and 10 g kg^{-1}. In turn, accumulators, such as rice, accumulate more than 10 g kg^{-1} of this element (Ma and Takahashi 2002). The difference in Si accumulation between species has been attributed to differences in the Si absorption capacity by the roots (Ma and Yamaji 2006). Thus, three modes of Si uptake were proposed: active, passive, and rejection (Mitani-Ueno and Ma 2021).

2.2 Active Uptake of Si

This absorption system is normally represented by Si-accumulating plants, such as rice, barley, maize, and wheat. Transporters must move Si from the soil solution to the plant organs. For example, in an experiment with wheat plants grown in a nutrient solution with 0.02 mmol L^{-1} of Si, the Si concentration in the xylem exudate reached values 400 times higher than those of the initial nutrient solution (8 mmol L^{-1}) in just 10 minutes. Thus, it reveals active mechanisms that absorb Si from the nutrient solution or soil solution and transport it to the xylem (Casey et al. 2004).

Monosilicic acid (H_4SiO_4) transporters were firstly described in rice. In this species, two transporters, Lsi1 and Lsi2 (Low silicon 1 and 2, so named due to the low Si content observed in the respective mutants with loss of function of these transporters), are expressed mainly in the mature zone of the roots and not in the root hairs, transporting Si from the soil solution to the xylem. These conveyors are of two types, i.e., channel type and transport type (Ma and Yamaji 2015).

Low silicon 1 is a bidirectional channel-type Si transporter that passively transports Si by concentration differences (Mitani-Ueno and Ma 2021). According to Coskun et al. (2021) and articles cited by them, Lsi1 transporters homologous to rice were characterized in other plants such as maize (*Zea mays*), barley (*Hordeum vulgare*), pumpkin (*Curcurbita moschata*), wheat (*Triticum aestivum*), horsetail (*Equisetum arvense*), soybean (*Glycine max*), poplar (*Populus trichocarpa*), cucumber (*Cucumis sativus*), tobacco (*Nicotiana sylvestris*), date palm (*Phoenix dactylifera*), grape (*Vitis vinifera*), and tomato (*Solanum lycopersicum*).

Low silicon 1 is a member of the Nodulin26-like intrinsic protein III (NIP-III) subgroup of Major Intrinsic Proteins (MIPs; also known as aquaporins) that passively transport water and/or small uncharged solutes such as H_4SiO_4 (Ma and Yamaji 2015). Thus, at the molecular level, aquaporins from plants belonging to the NIP-III subgroup with a GSGR (glycine-serine-glycine-arginine) selectivity filter and two NPA (asparagine-proline-alanine) domains separated by 108 amino acids are permeable to H_4SiO_4 (Coskun et al. 2019).

Low silicon 2 is an energy-dependent, antiport-type H_4SiO_4 efflux transporter. Monosilicic acid (H_4SiO_4) crosses the plasma membrane along with H^+ but in opposite directions. However, unlike Lsi1, the relationship between the structure and function of Lsi2 remains unknown (Coskun et al. 2021).

In rice, roots have distinct anatomy, characterized by two Casparian strips, one in the exoderm and the other in the endoderm, and the formation of aerenchyma (spaces without cells) in certain regions of the cortex. Therefore, H_4SiO_4 from the soil solution or nutrient solution first crosses the epidermis via the apoplast. Then, it is transported into the exodermis by Lsi1 located polarly on the distal side (farther from the root center), followed by efflux to the cortex by Lsi2 located polarly on the proximal side (closer to the root center) of the same cell's membrane. Thus, these transporters are on opposite sides of the same cell. Next, monosilicic acid (H_4SiO_4) flows via the apoplast to the endoderm, with low silicon 1 and 2 transporting it in and out, respectively, towards the xylem (Fig. 2.1a) (Mitani-Ueno and Ma 2021).

In addition to the polar location of Lsi1 and Lsi2 in rice roots, the exoderm with the Casparian strips increases Si accumulation. It prevents its reflux, as shown by the Si concentration between the exoderm and endoderm, which is duplicated due to the Casparian strips in the exodermis (Sakurai et al. 2015).

The expression of Si influx and efflux transporters is higher during the day than at night. Therefore, the xylem's transpiration rate and sap flow are higher during the day, and the Si absorbed in the root is efficiently transported to the upper tissues. On the other hand, the xylem's transpiration rate and sap flow are lower at night, and Si is not transported efficiently (Sakurai et al. 2017). According to the same authors, by analogy, when a conveyor belt is moving quickly, much luggage can be carried, but carrying much luggage on a slow conveyor is not a good strategy.

Recently, an H_4SiO_4 efflux transporter in the pericycle cells (layer of cells after the endoderm) of rice roots was identified and named Lsi3 (Fig. 2.1a). The low silicon 3 expression was much higher in the mature root region (>10 mm from the root tip) than in the root tip region. In the vegetative stage, OsLsi3 was mainly expressed in the roots, and in the reproductive stage, Lsi3 was expressed in the roots and nodes (Huang et al. 2022). The same authors identified a significant negative correlation between Si accumulation in shoots and Lsi3 expression, indicating that Si accumulation in shoots suppressed Lsi3 expression in roots. The OsLsi3 in the pericycle was responsible for 30% of the total Si loading to the xylem at low Si concentrations. However, it did not affect the absorption at high Si concentrations.

In maize roots, the expression of the ZmLsi1 transporter depends on the type of root. In the seminal (seed) roots, ZmLsi1 was located only in the epidermis and hypodermis cells (corresponding to the rice exoderm). In the lateral seminal roots,

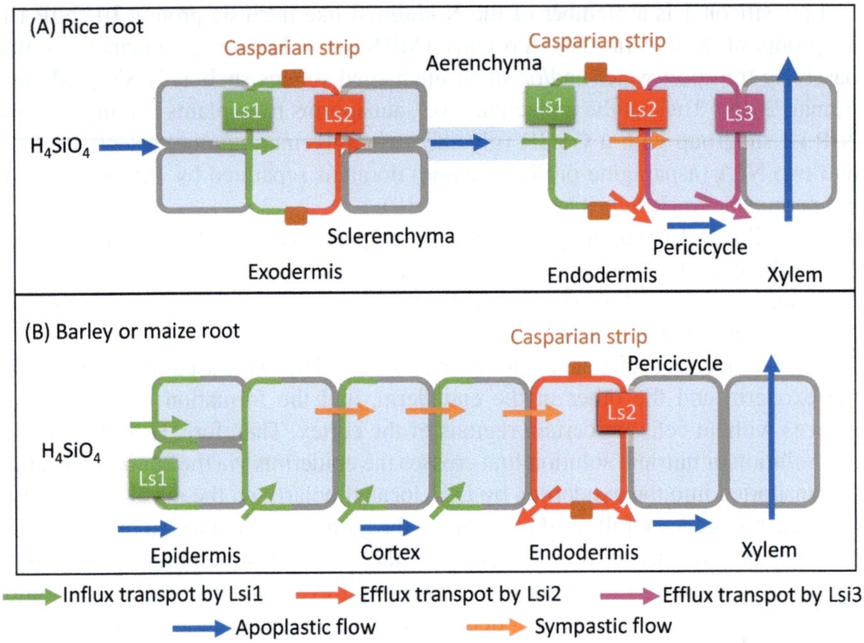

Fig. 2.1 Schematic presentation of Si uptake system in different plant species. (**a**) Active Si uptake system in rice. Low silicon 1 and 2 cooperatively mediate Si uptake. They are polarly localized at the distal and proximal side, respectively, of both exodermis and endodermis. In addition, Si loading to the xylem is facilitated by Lsi3, localized to the pericycle. (**b**) Active Si uptake system in other Si-accumulating species. Silicon uptake in these species, such as barley, is also cooperatively mediated by Lsi1 and Lsi2 but localized at different cell layers. (Redrawn from Mitani-Ueno and Ma (2021))

ZmLsi1 was expressed from the epidermis to the endodermis. In the crown roots, the location of ZmLsi1 was observed in some of the epidermal and hypodermic cells. In all root types, ZmLsi1 showed a polar location on the distal side of cells (Mitani et al. 2009a). To check the morphology of maize roots, see Hochholdinger (2009).

In barley (*Hordeum vulgare*) seminal roots, HvLsi1 is located in epidermal cells and all cortical cells. In the lateral roots, HvLsi1 is expressed in the hypodermic cells. Furthermore, HvLsi1 is more expressed in the mature region of the root than in its tip. This transporter has a polar location on the distal side of the epidermal and cortical cells of the seminal and lateral roots. (Chiba et al. 2009). However, HvLsi6 was highly expressed in the root tips compared to the mature root region. This transporter was located polarly in the epidermis and cortex of the root tip region (Yamaji et al. 2012). Therefore, HvLsi1 is responsible for Si absorption in the mature root region, while HvLsi6 is involved in Si absorption at root tips. On the other hand, Lsi2 from barley and maize is present only in the endoderm of the basal region of the roots and does not have a polar distribution. In other words, it is expressed throughout the plasma membrane of the endoderm cell (Mitani et al. 2009b).

In this context, the proposed model for the H_4SiO_4 uptake via roots in barley and maize indicates that this element is taken up by HvLsi1 or ZmLsi1, respectively, which is located polarly in epidermal, hypodermic, and cortical cells. Unlike rice roots, barley and maize generally lack aerenchyma in cortical cells. Thus, H_4SiO_4 is transported to the endodermis by the symplast pathway. The H_4SiO_4 transfer from the endoderm to the stele is performed by HvLsi2 or ZmLsi2 (Fig. 2.1b). This absorption system also results in high Si accumulation in the shoot but is not as efficient as the system present in rice (Mitani-Ueno and Ma 2021).

2.3 Passive Uptake of Si

This absorption system is more common in intermediate Si accumulator plants, such as cucumber (CsLsi1 and CsLsi2) and pumpkin (CmLsi1 and CmLsi2). Both influx transporters, CsLsi1 and CmLsi1, are expressed in almost all root cells around the entire plasma membrane. Except for CsLsi1, which shows a polar location in the endoderm. On the other hand, Lsi2 of both species is expressed only in endoderm cells without showing polar location (Fig. 2.2a). Thus, overall, the lack of polar location and the existence of the two transporters in the same cells result in low Si absorption efficiency (Mitani-Ueno and Ma 2021).

2.4 Rejection Uptake of Si

The rejection uptake occurs in plants that do not accumulate Si, such as tomatoes. However, as in rice, tomatoes have SlLsi1, a functional Si transporter. The SlLsi1 transporter was expressed at the roots' tips and basal regions and in the cells' plasma membrane but without polar distribution. In other words, without distribution on one side of the cell's plasma membrane but across the entire membrane. However, the Si efflux transporter, SlLsi2, is not expressed in tomatoes, which is attributed to the low accumulation of Si in this plant species. Based on this knowledge, the proposed model for the absorption of Si in tomatoes indicates that this element can be absorbed through the epidermis and cortex via apoplast, symplast, and SlLsi1. However, due to the lack of the efflux transporter SlLsi2 functional in the endodermis, Si is not actively exported to the stele, resulting in low accumulation in the shoot (Fig. 2.2b). Furthermore, SlLsi1 shows bidirectional transport, and part of the Si absorbed in the root cells can be released to the apoplast if there is a concentration gradient between the cytosol and the external solution. The SlLsi1 transporter showed 53–66% similarity with other Lsi1 from rice, cucumber, and squash (Sun et al. 2020).

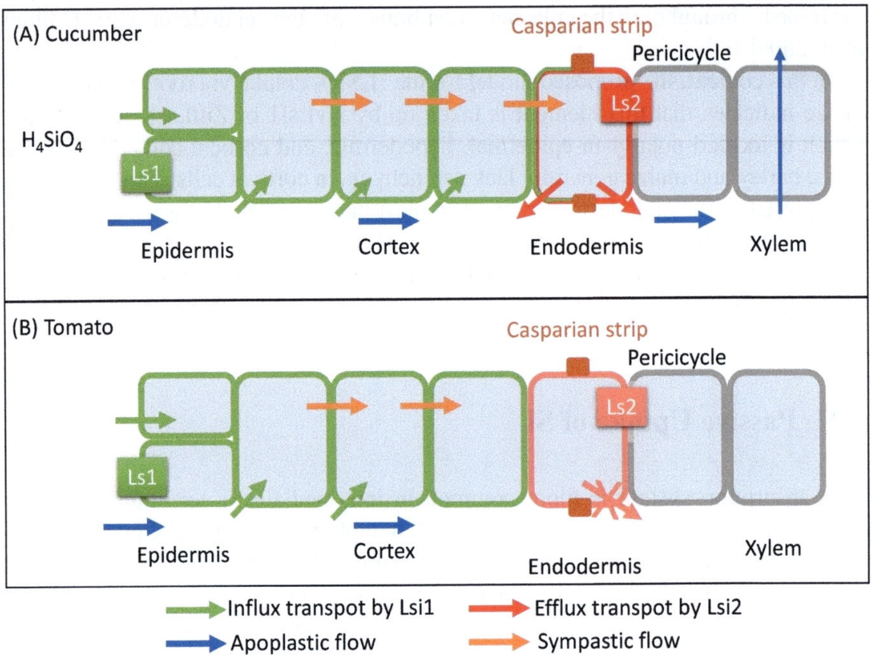

Fig. 2.2 Schematic presentation of Si uptake system in different plant species. (**a**) Passive uptake system. The CsLsi1 and CsLsi2 mediate Si uptake in these plant species, such as cucumber. The CsLsi1 transporter is expressed at all cell layers without polarity, while CsLsi2 is expressed at the endodermis without polarity. (**b**) Rejective uptake system. This system is employed by low Si-accumulators, such as tomatoes, which have functional Lsi1, but lack Lsi2. (Redrawn from Mitani-Ueno and Ma (2021))

2.5 Si Transport in the Xylem

The xylem is a vascular plant tissue responsible for transporting water and chemical elements from the roots to the other parts of plants. Therefore, xylem cells are set to function as water-conducting of tracheary elements. Each plant has its transporters to ensure the right ion movement to the outside of the cells through the xylem, and Si is no exception. Uptake and transport of Si in plants occur radially, from the cortical cells of roots to xylem vessels (Gaur et al. 2020).

More than 90% of the Si taken up by the roots is transferred into the shoot. Then, it gets distributed within the plant depending on the transpiration rate of the several organs (Ma 2010). Additionally, Si-accumulating plants, such as rice, have a Si efflux transporter (Lsi6) responsible for releasing silicic acid from the xylem and its subsequent distribution into the leaf sheath and midrib (Haynes 2017; Kaur and Greger 2019).

Low silicon 6 is polarly localized at the adaxial side of the xylem parenchyma cells in the leaf sheaths and leaf blades. Therefore, suppression of Lsi6 affects the silica deposition pattern in the leaf blades and sheaths. Other Si-accumulating crops, such as barley, have also shown this transporter to accumulate higher concentrations in leaves and grains. Low silicon 6 is also expressed in node 1 below the panicles during the reproductive stage. It transfers Si to vascular bundles connected to the panicles (Mitani-Ueno and Ma 2021).

Most Si will be distributed into the grains at the reproductive stage and mainly accumulated in the husk. Three Si transporters mediate this distribution; Lsi6, Lsi2, and Lsi3 in rice (Fig. 2.3), which are located at different cell layers of node I. The OsLsi6 transporters at the xylem transfer cells of the enlarged vascular bundles unload Si from the xylem. On the other hand, OsLsi2 and its homolog OsLsi3, located at the bundle sheath and parenchyma bridge cells, are responsible for the further transfer of Si to the diffuse vascular bundles (Mitani-Ueno and Ma 2021).

Once transported, Si reaches other parts of the plants and gets deposited. This deposition occurs in plant tissues where the transpiration rate is higher. Silicon precipitation is a chemical reaction named condensation. Because transpiration is the main driver for Si accumulation and deposition in plant cells, the growth stage duration of the plant is an important factor that determines that older cells will

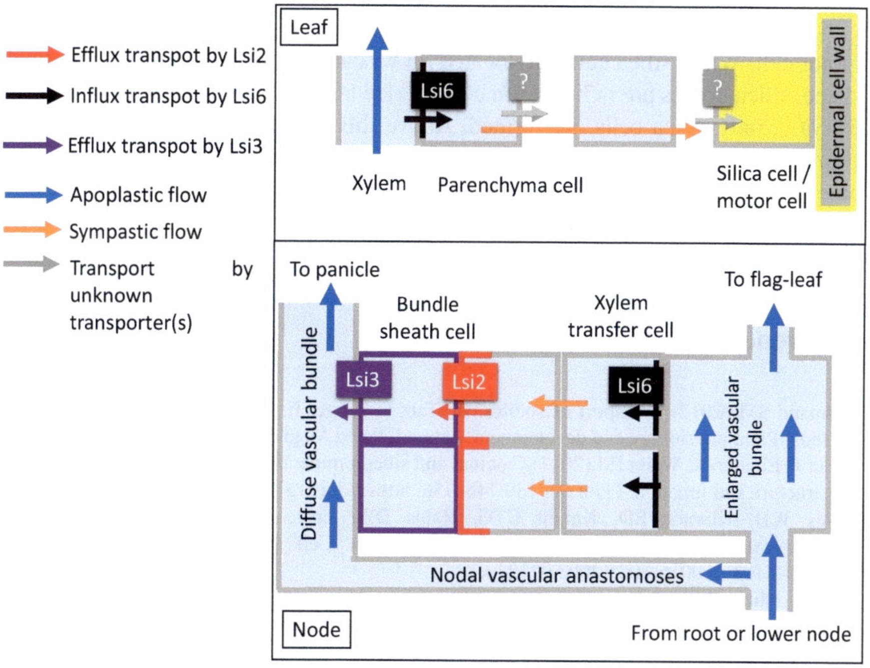

Fig. 2.3 Schematic presentation of Si distribution in rice shoots. In leaves, Si in the xylem sap is unloaded by Lsi6, but transporters for further deposition at specific cells are unidentified. In nodes, Si in the xylem of the enlarged vascular bundle is first unloaded by Lsi6 located at the xylem transfer cells. Then, Si is released toward diffuse vascular bundles by OsLsi2 polarly located at the bundle sheath cell layer and OsLsi3 located in the parenchyma tissues between enlarged and diffuse vascular bundles. Arrows with different colors indicate transport processes mediated by different transporters and the symplastic flow of Si (Mitani-Ueno and Ma 2021)

accumulate more Si than younger ones (Iler 1979). When the silicic acid concentration is higher than 100–200 mg kg^{-1}, monomers will form dimers and then oligomers with stable nuclei that eventually grow to form particles (Greenberg 1959). When these particles grow up to 1–3 nm, they begin carrying a negative surface charge, enabling interaction with the surrounding environment, such as the cell walls (Haynes 2017).

Deposition of silica can occur in a myriad of shapes and sizes. Often, silica precipitation in plants generates phytoliths, but that is not always the case. Phytoliths are amorphous microscopic opal structures produced in and between the cells of plants (Neethirajan et al. 2009). They are found in cells from the leaf epidermis and covering of seeds and fruits, the epidermis of bracts that surround and protect grass seeds, and in the subepidermal tissue of orchid and palm leaves. Interestingly, at the reproductive stage, most Si will be found in cells from grains, especially accumulated in the husk (Mitani-Ueno and Ma 2021).

Phytoliths range in size from 10 to 30 μm and are occasionally over 1000 μm in diameter. Depending on the carbon coating extension, colors can be as different as transparent or brown. Phytoliths' shapes and sizes differ as they assume the shapes and sizes of their host cells (Rashid et al. 2019).

Silicified cells are of two types: (i) the silica cell and (ii) the silica body or silica motor cell. While silica cells are located on vascular bundles, showing a dumbbell shape, silica bodies are in bulliform cells of rice leaves. Below the concentration of 5%SiO_2, only silica cells are formed. Above this threshold, silica bodies start to form, increasing with Si shoot concentration (Datnoff et al. 2001). In conclusion, among crop plants, Si concentrations generally increase in the order legumes < fruit crops < vegetable crops < grasses < grain crops (Haynes 2017).

References

Assmann SM (2017) Transport of solutes. In: Taiz L, Zeiger E, Moller IM, Murphy A, (Orgs.) (eds) Plant physiology and development. Artmed, Porto Alegre, pp 143–168

Bauer P, Elbaum R, Weiss IM (2011) Calcium and silicon mineralization in land plants: transport, structure and function. Plant Sci 180:746–756. https://doi.org/10.1016/j.plantsci.2011.01.019

Casey WH, Kinrade SD, Knight CTG, Rains DW, Epstein E (2004) Aqueous silicate complexes in wheat, *Triticum aestivum* L. Plant Cell Environ 27:51–54. https://doi.org/10.1046/j.0016-8025.2003.01124.x

Chiba Y, Mitani N, Yamaji N, Ma JF (2009) HvLsi1 is a silicon influx transporter in barley. Plant J 57:810–818. https://doi.org/10.1111/j.1365-313X.2008.03728.x

Coskun D, Deshmukh R, Sonah H, Menzies JG, Reynolds O, Ma JF, Kronzucker HJ, Bélanger RR (2019) The controversies of silicon's role in plant biology. New Phytol 221:67–85. https://doi.org/10.1111/nph.15343

Coskun D, Deshmukh R, Shivaraj SM, Isenring P, Bélanger RR (2021) Lsi2: a black box in plant silicon transport. Plant Soil 466:1–20. https://doi.org/10.1007/s11104-021-05061-1

Datnoff LE, Snyder GH, Korndörfer GH (2001) Silicon in Agriculture, 1st edn. Elsevier, New York

Gaur S, Kumar J, Kumar D, Chauhan DK, Prasad SM, Srivastava PK (2020) Fascinating impact of silicon and silicon transporters in plants: a review. Ecotoxicol Environ Saf 202:110885. https://doi.org/10.1016/j.ecoenv.2020.110885

Greenberg SA (1959) The chemistry of silicic acid. J Chem Educ 36:218–219
Haynes RJ (2017) The nature of biogenic Si and its potential role in Si supply in agricultural soils. Agric Ecosyst Environ 245:100–111. https://doi.org/10.1016/j.agee.2017.04.021
Hochholdinger F (2009) The maize root system: morphology, anatomy, and genetics. In: Handbook of maize: its biology. Springer, pp 145–160. https://doi.org/10.1007/978-0-387-79418-1_8
Huang S, Yamaji N, Sakurai G, Mitani-Ueno N, Konishi N, Ma JF (2022) A pericycle-localized silicon transporter for efficient xylem loading in rice. New Phytol 234:197–208. https://doi.org/10.1111/nph.17959
Iler RK (1979) The chemistry of silica: solubility, polymerization, colloid and surface properties and biochemistry of silica. Wiley, New York, Chichester
Kaur H, Greger M (2019) A review on Si uptake and transport system. Plan Theory 8:81. https://doi.org/10.3390/plants8040081
Ma JF (2010) Silicon transporters in higher plants. In: Jahn TP, Bienert GP, (Orgs.) (eds) MIPs and their role in the exchange of metalloids. Springer, New York, pp 99–106
Ma JF, Takahashi E (2002) Silicon-accumulating plants in the plant kingdom. In: Soil, fertilizer, and plant silicon research in Japan, pp 63–71p. https://doi.org/10.1016/b978-044451166-9/50005-1
Ma JF, Yamaji N (2006) Silicon uptake and accumulation in higher plants. Trends Plant Sci 11:392–397. https://doi.org/10.1016/j.tplants.2006.06.007
Ma JF, Yamaji N (2015) A cooperative system of silicon transport in plants. Trends Plant Sci 20:435–442. https://doi.org/10.1016/j.tplants.2015.04.007
Mengel K, Kirkby EA, Appel T (2001) Principles of plant nutrition. Springer. https://doi.org/10.1007/978-94-010-1009-2
Mitani N, Yamaji N, Ma JF (2009a) Identification of maize silicon influx transporters. Plant Cell Physiol 50:5–12. https://doi.org/10.1093/pcp/pcn110
Mitani N, Chiba Y, Yamaji N, Ma JF (2009b) Identification and characterization of maize and barley Lsi2-like silicon efflux transporters reveals a distinct silicon uptake system from that in rice. Plant Cell 21:2133–2142. https://doi.org/10.1105/tpc.109.067884
Mitani-Ueno N, Ma JF (2021) Linking transport system of silicon with its accumulation in different plant species. Soil Sci Plant Nutr 67:10–17. https://doi.org/10.1080/00380768.2020.1845972
Neethirajan S, Gordon R, Wang L (2009) Potential of silica bodies (phytoliths) for nanotechnology. Trends Biotechnol 27:461–467. https://doi.org/10.1016/j.tibtech.2009.05.002
Prado RM (2021) Mineral nutrition of tropical plants. Springer. https://doi.org/10.1007/978-3-030-71262-4
Rashid I, Mir SH, Zurro D, Dar RA, Reshi ZA (2019) Phytoliths as proxies of the past. Earth-Sci Rev 194:234–250. https://doi.org/10.1016/j.earscirev.2019.05.005
Sakurai G, Satake A, Yamaji N, Mitani-Ueno N, Yokozawa M, Feugier FG, Ma JF (2015) In silico simulation modeling reveals the importance of the casparian strip for efficient silicon uptake in rice roots. Plant Cell Physiol 56:631–639. https://doi.org/10.1093/pcp/pcv017
Sakurai G, Yamaji N, Mitani-Ueno N, Yokozawa M, Ono K, Ma JF (2017) A model of silicon dynamics in rice: an analysis of the investment efficiency of si transporters. Front Plant Sci 8:011187. https://doi.org/10.3389/fpls.2017.01187
Sun H, Duan Y, Mitani-Ueno N, Che J, Jia J, Liu J, Guo J, Ma JF, Gong H (2020) Tomato roots have a functional silicon influx transporter but not a functional silicon efflux transporter. Plant Cell Environ 43:732–744. https://doi.org/10.1111/pce.13679
Takahashi E, Hino K (1979) Silica uptake by rice plant with special reference to the forms of dissolved silica. J Sci Soil Manure 49:357–360
Yamaji N, Chiba Y, Mitani-Ueno N, Feng Ma J (2012) Functional characterization of a silicon transporter gene implicated in silicon distribution in barley. Plant Physiol 160:1491–1497. https://doi.org/10.1104/pp.112.204578

Chapter 3
Root Silicification and Plant Resistance to Stress

Zuzana Lukacova, Boris Bokor, Marek Vaculík, Jana Kohanová, and Alexander Lux

3.1 Introduction

The uptake of Si together with water and other elements and its translocation to other plant organs is the principal function of the root. Thus, most of the research focused on the transport processes of Si in roots, with important input on the role of Si transporters after their discovery in plants (Ma et al. 2006, 2007). Much less attention was paid to questions regarding Si deposition directly in the roots as well as to the function of these deposits. Most of the published data from the beginning of the detection of Si in plants was related to aboveground plant parts. The reason is not only easier work with these plant organs but also the simple identification of plant species according to their leaf, stem, or generative organs compared to the root

Z. Lukacova (✉) · J. Kohanová
Department of Plant Physiology, Faculty of Natural Sciences, Comenius University in Bratislava, Bratislava, Slovak Republic
e-mail: zuzana.lukacova@uniba.sk

B. Bokor
Department of Plant Physiology, Faculty of Natural Sciences, Comenius University in Bratislava, Bratislava, Slovak Republic

Comenius University Science Park, Bratislava, Slovak Republic

M. Vaculík
Department of Plant Physiology, Faculty of Natural Sciences, Comenius University in Bratislava, Bratislava, Slovak Republic

Institute of Botany, Plant Science and Biodiversity Centre, Slovak Academy of Sciences, Bratislava, Slovak Republic

A. Lux
Department of Plant Physiology, Faculty of Natural Sciences, Comenius University in Bratislava, Bratislava, Slovak Republic

Institute of Chemistry, Slovak Academy of Sciences, Bratislava, Slovak Republic

R. de Mello Prado (ed.), *Benefits of Silicon in the Nutrition of Plants*,
https://doi.org/10.1007/978-3-031-26673-7_3

if the research is performed in field conditions. Silicified special cells are and were used for plant classification, for detection of plant species for identification of plant evolution, in archaeology, palaeobotany, food identification in animal excrements, and some cases even in forensic sciences. In the case of plant evolution, Jones (1964) already detected phytoliths from 60-million-year-old sedimentary rocks. More recently, Stromberg (2002) identified phytoliths belonging to a variety of ferns, conifers, monocotyledons, and even some dicotyledons, from sediments ranging from the late Eocene to the late Miocene. Thus, it may be concluded that silica deposits in sediments date back to the early stages of vascular plant evolution. They were recognized as various types of phytoliths, and it was usually difficult to classify them for a particular plant organ. The term phytolith is composed of two Greek words: *phyton* = plant and *lithos* = stone. This term is used for all mineralized deposits in plants. Other terms used frequently in the literature are opaline silica depositions, plant opal, or silica aggregates. All these terms are used for Si deposits, however, with some specification in roots.

Bennett (1982) in his publication focused on the silicification of some cereals mentioned very few works dealing with root silicification. We may still agree with Sangster and Hodson (1992) concluding 30 years ago that: "very little information concerning silica deposition in roots of plants other than grasses and cereals. It seems either that little silica is deposited in these roots, or perhaps more likely that little research has been carried out."

Most of the work dealing with root silicification focused on silica deposits in the endodermis. This place of silicification is used almost entirely by grasses. Kumar et al. (2017), in their recent review, described three types of Si deposition in grasses: passive cell wall silicification, controlled cell wall silicification, and a special case of silica cells.

3.2 Sites of Si Deposition in Roots

A passive mode of silicification in roots suggests that Si condensation occurs as water is absorbed by the symplasm, resulting in the concentration of silicic acid in the apoplasm (Exley 2015). This possibility is supported by the observation of Si deposits in the epidermis and cortical cells of aerial roots of *Phalaris canariensis* (Hodson 1986). The silicification of inner tangential endodermal cell walls, i.e., areas located centripetally to the Casparian bands as observed in several grasses (Table 3.1), and the formation of Si aggregates in these parts of the walls (as in species of tribe Andropogonae) do not conform with the passive mode. Active uptake of Si takes place at the apical part of roots, bypassing the barrier of Casparian bands, and Si is transported through a central cylinder with the transpiration stream. Part of this Si transported to the shoot is transferred centrifugally and deposited into the maturing endodermis (Parry and Kelso 1975; Sangster and Parry 1976; Markovich et al. 2015). This way of transport of Si was demonstrated in adventitious roots of sorghum with silicified endodermis before reaching the growth medium (Sangster

Table 3.1 Published data about silica deposition in the roots of various plants

	Species	Place/form of deposition	Literature
Angiosperms **Monocots** **Clade**: Commelinids **Order**: Poales **Family**: Poaceae	*Oryza sativa* L.	Endodermis *Inner tangential walls*	Sangster (1978), Lux et al. (1999), Ma and Yamaji (2015), Parry and Soni (1972)
	Oryza sativa L.	Exodermis Sclerenchyma	Fleck et al. (2015), Shi et al. (2005), Gong et al. (2006)
	Sorghum bicolor L. Moench	Endodermis *Inner tangential walls*	Hodson and Sangster (1989b), Soukup et al. (2020)
	Digitaria sanguinalis L.	Endodermis *Inner tangential walls*	Sangster (1977b)
	Molinia caerulea (L.) Moench	Cortex Intercellular spaces	Montgomery and Parry (1979)
	Triticum aestivum L. *Hordeum vulgare* *Avena sativa*	Endodermis *Inner and outer tangential walls*	Bennett (1982)
	Triticum aestivum L.	Pericycle Outer parenchyma, metaxylem vessels ***Old root part**	Hodson and Sangster (1989a)
	Phalaris canariensis L.	Epidermis Endodermis Outer cortical cell layers ***Aerial roots**	Hodson (1986)

(continued)

Table 3.1 (continued)

	Species	Place/form of deposition	Literature
	Ammophila breviligulata *Vulpia octoflora* *Elymus canadensis* *Leptochloa fascicularis* *Hilaria jamesii* *Oryzopsis hymenoides* *Aristida wrightii* *Schizachyrium scoparium*	Not specified	Brown (1984)
	Genera: *Eustachys, Andropogon, Panicum* (the perennial grasses of the Serengeti National Park, Tanzania)	Not specified	McNaughton et al. (1985)
	Phyllostachys heterocycla Mitf.	Endodermis *Inner tangential walls*	Lux et al. (2003a)
	Sorghum bicolor L. Moench	****Silica aggregates/phytoliths** Endodermis *Inner tangential walls*	Borissow (1924), Sangster and Parry (1976), Sangster and Hodson (1992), Lux et al. (2002), Hattori et al. (2003), Lux et al. (2003b), Soukup et al. (2014, 2017, 2020), Zexer and Elbaum (2020)

	Species	Place/form of deposition	Literature
Angiosperms **Monocots** **Clade**: Commelinids **Order**: Poales **Family**: Poaceae **Tribe**: Andropogonae	*Andropogon gerardii* Vitman *A. scoparius* Michx.	****Silica aggregates/phytoliths** Endodermis *Inner tangential walls*	Sangster (1977a)
	Saccharum officinarum	****Silica aggregates/phytoliths** Endodermis *Inner tangential walls*	Parry and Kelso (1977)
	Cymbopogon	****Silica aggregates/phytoliths** Endodermis *Inner tangential walls*	Parry and Kelso (1977)
	Erianthus ravennae (Saccharum ravennae)	****Silica aggregates/phytoliths** Endodermis *Inner tangential walls*	Borissow (1925)
	Miscanthus	****Silica aggregates/phytoliths** Endodermis *Inner tangential walls*	Parry and Kelso (1977)
Angiosperms **Monocots** **Order**: Asparagales **Family**: Orchidaceae	*Maxillaria picta*	****Stegmata** *Associated with vascular sclerenchyma*	Holtzmeier et al. (1998)
	Cymbidium spp.	***Stegmata** *Replacing pericycle cells*	Prychid et al. (2003)

(continued)

Table 3.1 (continued)

	Species	Place/form of deposition	Literature
Angiosperms **Monocots** **Clade**: Commelinids **Order**: Zingiberales **Family**: Musaceae	The species from Musaceae family	***Stegmata**	Tomlinson (1959)
Angiosperms **Monocots** **Clade**: Commelinids **Order**: Arecales **Family**: Arecaceae	Some palm species	***Stegmata**	Tomlinson (1961)
	Phoenix dactylifera	***Stegmata**	Bokor et al. (2019)
	Lodoicea maldivica	***Stegmata**	Lux unpublished
	Elaeis guineensis	***Phytoliths**	Davamani et al. (2022)
Angiosperms **Eudicots** **Clade**: Asterids **Order**: Solanales **Family**: Solanaceae	*Lycopersicon esculentum*	Not specified	Zhou et al. (2022)
Angiosperms **Eudicots** **Clade**: Rosids **Order**: Cucurbitales **Family**: Cucurbitaceae	*Cucumis sativus*	Not specified	Zhou et al. (2022)
Angiosperms **Eudicots** **Clade**: Rosids **Order**: Fagales **Family**: Fagaceae	*Fagus sylvatica*	Cortical cell walls	Watteau and Villemin (2001)
Gymnosperms **Division**: Pinophyta **Class**: Pinopsida **Order**: Pinales **Family**: Pinaceae	*Picea abies* L. Karst.	Cortical cell walls	Hodson and Wilkins (1991)

Using some data from Sangster (1978) and Sangster and Hodson (1992) and Parry 1976). Other examples of this mode of Si deposition were shown by Lux et al. (2003b) and Soukup et al. (2017) using sorghum seminal roots supplied by Si to their apical parts and deposited in the basal parts.

Deposition of Si in older roots in central sclerified tissue of the stele, in vessel walls, epidermis, and subepidermal sclerenchyma was noted by Parry and Kelso (1975). At the beginning of silicification, silica deposits took the form of spheres and lined the cavities. Later, silica was found to fill the intercellular spaces in the mature roots of *Molinia caerulea*, and silica deposition was also found in the intercellular spaces of the root cortex near the endodermis (Montgomery and Parry, 1979) in intercellular spaces of the root cortex near the endodermis. In rice plants, in addition to endodermal silicification, accumulation of Si has also been observed in the root exodermis and sclerenchyma (Shi et al. 2005; Gong et al. 2006). Detailed subcellular analysis along the axis of the wheat roots showed the largest Si deposition in the endodermis and subsequently in peripheral early metaxylem vessels. In the oldest parts of the root, non-polymerized silicon was observed in the largest central late metaxylem vessel. This observation may partially explain the high concentration of Si detected in the xylem of some grasses. The soil solution contains variable concentrations of monosilicic acid (0.01–2.0 mM) depending on the pH and soil type. Above 2.0 mM concentration, Si starts polymerizing to form SiO_2. nH_2O. However, in Si-accumulating plants, the Si concentration in xylem sap can exceed 5.0 mM. It was found in wheat (Casey et al. 2004) and rice (Mitani and Ma 2005). It is not yet clear how it is possible. Organosilicate compounds were not detected in xylem sap, so the mode that prevents polymerization of such high concentrations of Si in the vascular system remains to be studied.

However, silica deposition in non-accumulating species is different. Fleck et al. (2015) comparing Si-accumulating and Si-excluding species found in the Si non-accumulator plant onion highest concentration of Si in root exodermis and gradually decreasing concentration in the mid-cortex and endodermis. Surprisingly, limited data are available on root silicification in herbaceous and woody eudicots. Most publications refer only to the presence or concentration of Si in roots (e.g., in cucumber and tomato – Zhou et al. 2022). In woody species, transmission electron microscopy and electron energy loss spectroscopy confirmed the presence of biogenic silica in beech roots (Watteau and Villemin 2001). The important role of fine beech roots and root decomposition in the silicon cycle between plants and soils in forest ecosystem was discussed by Turpault et al. (2018). Data on root silicification are also scarce. Considerable amounts of silicon were deposited in the root cortical walls of Norway spruce (*Picea abies* L. Karst.) (Hodson and Wilkins 1991). A completely different way of silica deposition was found in several taxonomically unrelated species in the form of Si deposits in the form of aggregates within specific stegmata cells. Stegmata are discussed in a special part of this manuscript.

As was shown in the older roots of some species of grasses, there may exist other places of Si deposition (Table 3.1). The heterogeneity of Si deposits in roots is summarized schematically in Fig. 3.1.

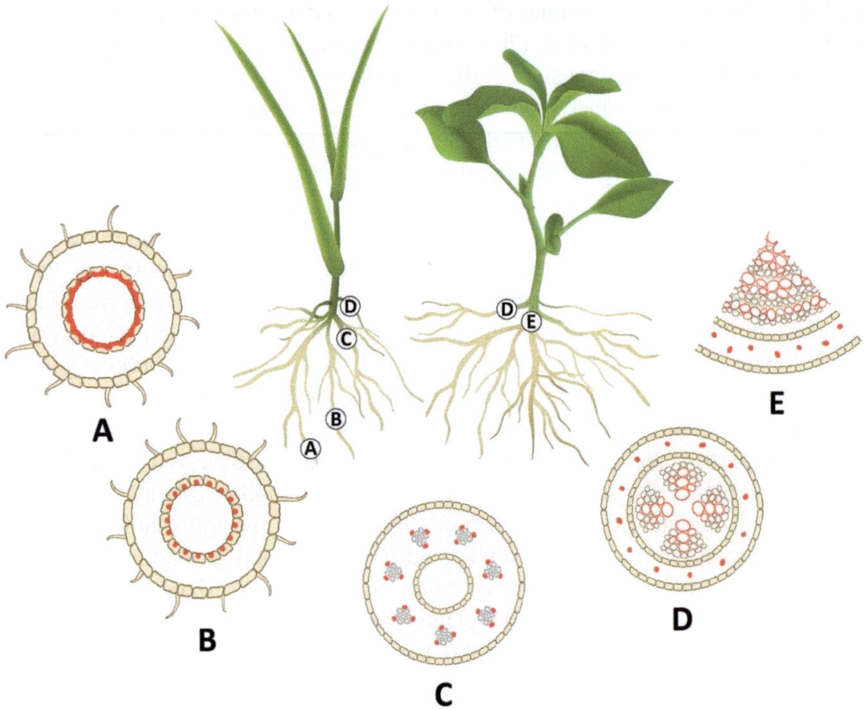

Fig. 3.1 Schematically summarized heterogeneity of Si deposits (red color) in roots. Two plant pictures represent variability known or expected in monocots (left plant) and in dicots (right plant). The best known and most intensively studied are deposits of Si in inner tangential walls of endodermis gradually developing during endodermal ontogenesis in some grasses (**a**) e.g., in rice and wheat. Well known are also silica aggregates on inner tangential endodermal walls in species from tribe Andropogonae (**b**) e.g., in sorghum and sugar cane. Less known are silica aggregates in specialized stegmata cells attached to the surface of sclerenchyma fibers or attached to sclerenchyma surrounding vascular bundles in cortical zone of some monocot species (**c**) e.g., in date palm and some orchids. Very limited evidence is about silica deposits both in monocots and dicots in other root cells and tissues, such as specialized idioblasts in cortical or epidermal layers, and cell wall impregnation of xylem vessels or tracheids (**d**, **e**)

3.3 Silicon Transport in Plants – From Chemistry to Cell Biology and Anatomy

Silicon as a beneficial element is bioavailable for plant roots in the form of orthosilicic acid $Si(OH)_4$, where silicon is tetrahedrally coordinated with four hydroxyl groups (Fig. 3.2). Roots absorb silicic acid from soil water, where it is soluble up to a certain limit (approximately 2 mmol/L concentration). Usually, the concentration of silicic acid is within the range of 0.1 to 0.6 mmol/L in soil solution (Epstein 1994). Because silicic acid is a very weak acid, it is not dissociated until the pH reaches a value of almost 10 and thus remains as an uncharged monomer molecule in most soil types. Below the solubility limit and pH of 10, this uncharged monomer

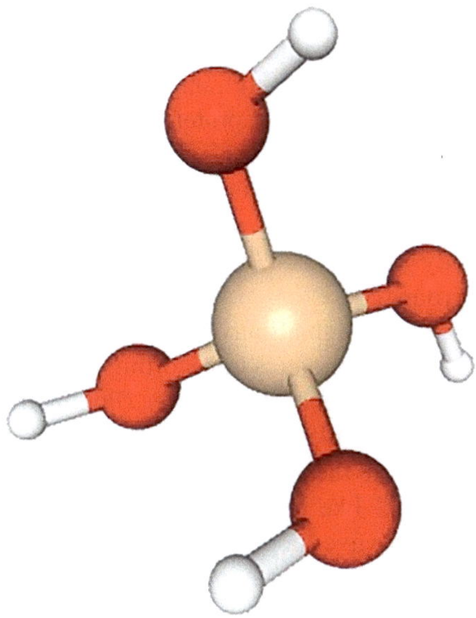

Fig. 3.2 The chemical 3D structure of silicic acid. Atoms are distinguished by colors: light brown – Si, red – O, and white – H. (The 3D model was retrieved from the PubChem chemistry database: https://pubchem.ncbi.nlm.nih.gov/compound/Orthosilicic-acid#section=3D-Conformer)

becomes inert and, according to Exley et al. (2019), does not react with any inorganic or organic compound.

Silicon may enter root tissue and continue radially through the cortex to the vasculature via (i) the apoplasmic (also known as apoplastic) pathway (ii) or be taken up by the initial root cell membrane and continue via the symplasmic (also known as symplastic) pathway (iii) and finally by the transcellular pathway. In the case of apoplasmic transport, Si is present in the intercellular spaces of cortical tissue, and radial transport is restricted by apoplasmic barriers, which are developed mostly in endodermal cells but also in exodermal cells in the form of Casparian strips. From intercellular spaces, Si can be transported to the cell lumen through specific membrane proteins. These membrane proteins play a key role in the symplasmic and transcellular transport of Si. In the symplasmic pathway, initial cell uptake is mediated by selective membrane proteins and, subsequently, Si is radially transported through the plasmodesmata of root cortical cells until the destination cell is reached. Then, Si might be released from this cell, again, by membrane proteins. Transcellular transport of Si requires selective membrane proteins, which are responsible for repeated influx and/or efflux of Si by cortical cells to achieve radial cell-to-cell transport (Barberon and Geldner 2014; Mitani-Ueno and Ma 2021).

Identification and functional annotation of selective membrane proteins involved in the influx and efflux of Si from root cells became crucial to understanding the molecular biology of Si transport in plants. Proteins responsible for Si transport are divided into passive channels that do not require energy supply and active transporters driven by a protonmotive force (Ma et al. 2006, 2007). However, both types of these proteins mediate the transport of molecules across the cytoplasmic membrane and thus could be considered transporters, so the term "transporter" is also widely used for passive channel-like proteins in the scientific literature (Mitani-Ueno and Ma 2021).

Lsi1 proteins are aquaporin-like proteins belonging to the Nodulin 26-like Intrinsic Protein III (NIP III) subfamily, which is part of a large and diverse Major Intrinsic Proteins family (Ma et al. 2006; Wallace and Roberts 2005). Proteins belonging to the NIP group (consisting of NIP I, NIP II, and NIP III) are also called metalloido-porins due to their high affinity for some metalloid transport across the cell membrane since many NIP proteins transporting metalloids are simultaneously poorly permeable or even impermeable to water (Pommerrenig et al. 2015). The Lsi1-like (also known as NIP2-like) channels are recognized as Si influx proteins responsible for Si uptake from the soil environment by roots (Ma et al. 2006). The amino acid structure of Lsi1-like proteins consists of six transmembrane helices (TM 1–TM 6) connected by five loops (A–E) (Fig. 3.3). A very important motif Asn-Pro-Ala (NPA motif), present twice in hydrophobic loops B and E, forms two short helices, which fold together to create a seventh pseudo-transmembrane helix forming a pore in the center of protein (Fig. 3.3) (Wallace and Roberts 2005; Kapilan et al. 2018). Another essential feature of Lsi1-like proteins responsible for selective transport of solutes, and especially silicic acid, is the ar/R (aromatic/arginine) selective filter (SF) that consists of glycine (G), serine (S), G, and arginine (R) amino acids (also called the GSGR filter) (Fig. 3.3) (Mitani-Ueno et al. 2011a).

Although some authors have expressed doubts about the Si transport activity of aquaporins, including the Lsi1 protein, due to the small pore size and the larger size of silicic acid (Exley et al. 2020), very recently, two independent scientific groups showed that the GSGR filter of Lsi1 is large and plays a key role in the transport of silicic acid in dehydrated form (Saitoh et al. 2021; van den Berg 2021). The structural analysis of Lsi1 protein showed that the SF includes the fifth extra residue threonine (T) (Fig. 3.3). This shows that the plant silicon channel SF consists of TGSGR amino acids, which makes the SF much wider and capable of the Si transport in comparison to the four residues of SF in common and/or conventional aquaporin channels (Figs. 3.3 and 3.4) (van den Berg et al. 2021). Saitoh et al. (2021) studied the crystal structure of Lsi1 at high resolution, and together with in silico calculations, these authors uncovered the principle of Si permeability. They showed that unique Lsi1 transmembrane helix orientations and five residues of SF play a key role in Si transport specificity. Probably, it is the T amino acid of SF that is essential for Si transport. It seems that during evolution, Lsi1 channels evolved to facilitate Si transport from canonical aquaporins without Si transport properties (Saitoh et al. 2021). In addition, transport mediated through the Lsi1 protein is bidirectional, and this protein shows tetrameric assembly, meaning four monomers per unit (Fig. 3.4) (Mitani-Ueno and Ma 2021; van den Berg et al. 2021).

Compared to Lsi1, the Lsi2 protein responsible for Si efflux is an active transporter that belongs to a putative anion transporter family (Ma et al. 2007). However, protein size, transmembrane domain organization, and amino acid composition vary greatly among plant species; thus, a general model of Lsi2 is missing (Coskun et al.

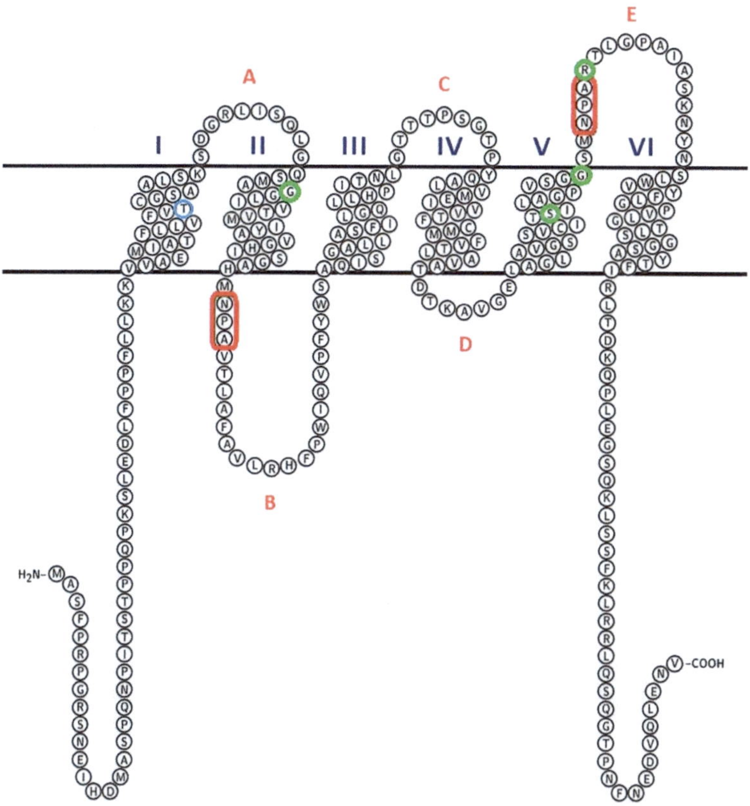

Fig. 3.3 Primary and predicted secondary structure of *Phoenix dactylifera* Lsi1 protein. Transmembrane helical domains are marked with Roman numerals (I–VI), and loops are labelled with letters (A–E). The N-terminus and the C-terminus of the protein are located in the cytosol. The NPA box and selective filter (GSGR or TGSGR) are highlighted by red and green-blue color, respectively. Amino acids (one-letter code) are circled. See the text for more details. The model was predicted by SOSUI algorithm (Hirokawa et al. 1998, http://harrier.nagahama-i-bio.ac.jp/sosui/sosui_submit.html)

2021). Nevertheless, the cooperation of the Lsi1 and Lsi2 proteins is essential for efficient Si transport in species with the highest Si accumulation among flowering plants (Sakurai et al. 2015; Ma and Yamaji 2015). In these Si-accumulating plant species, the Lsi1 and Lsi2 proteins show polar localization in cells of the root cortex. A great example is a root of rice (*Oryza sativa*), where Lsi1 is located distally and Lsi2 proximally in the plasma membrane of exodermal and endodermal cells, thus efficient influx and efflux of Si towards the xylem is achieved and results in high Si accumulation in shoots (Fig. 3.5) (Ma et al. 2006, 2007; Mitani-Ueno and Ma 2021).

The lower efficiency of the Si transport is found in some species, including maize (*Zea mays*) or barley (*Hordeum vulgare*); however, this system still results in high shoot Si accumulation, although not as high as in rice (Mitani-Ueno and Ma 2021). Plants within this group show polar localization of Lsi1 in all epidermal and cortical

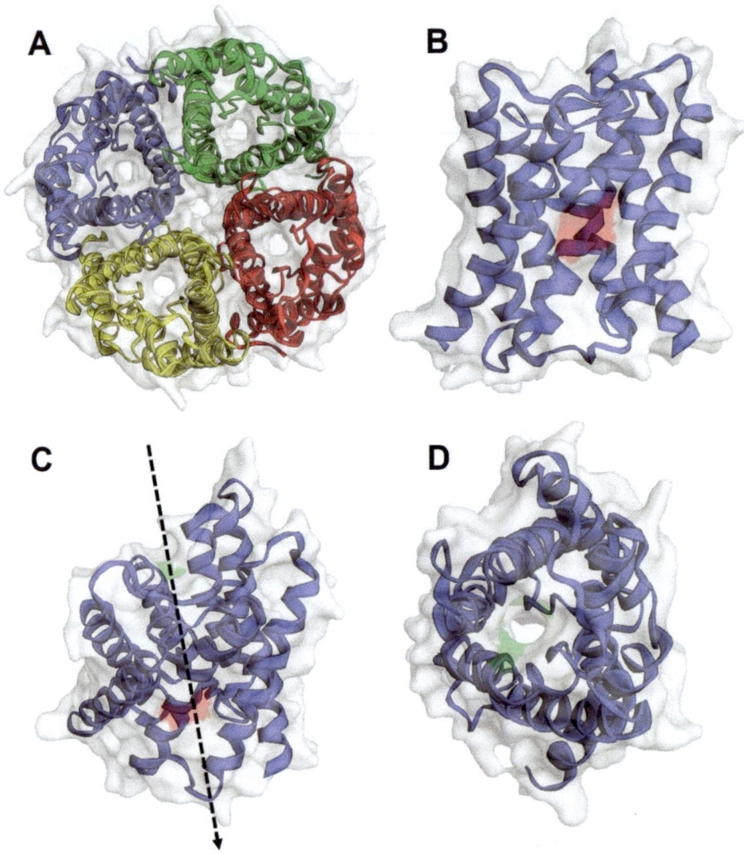

Fig. 3.4 The predicted structure of *Phoenix dactylifera* Lsi1 protein. (**a**) The overall structure of the tetramer viewed from the extracellular side. (**b**) The side view of Lsi1 monomer of Lsi1 with NPA motif highlighted in red color. (**c**) The side view of monomeric Lsi1 slightly rotated to show GSGR (or TGSGR) SF highlighted by green color. The dashed line schematically represents the transport of Si through the SF and NPA box. (**d**) The top view of monomeric Lsi1 from the extracellular side with GSGR (or TGSGR) SF highlighted by green color. See the text for more details. The predicted structure was modelled and retrieved from the SWISS-MODEL and EzMol web server (Reynolds et al. 2018; Waterhouse et al. 2018)

cells except the endodermis, where only the Lsi2 transporter is located with a nonpolar orientation (Mitani et al. 2009a, b; Chiba et al. 2009). Plants with intermediate accumulation of Si are characterized by the low efficiency of Si transport (Mitani-Ueno and Ma 2021). This Si transport system is found in dicots such as pumpkin (*Cucurbita moschata*) or cucumber (*Cucumis sativus*). Although the localization of these proteins was not fully characterized in the cucumber, some generalization might be suggested. The Lsi1 channels do not show polar localization in epidermal and all cortical cells in both species, except for the endodermis with polarly localized Lsi1 in cucumber. The Lsi2 transporter is found only in endodermis without

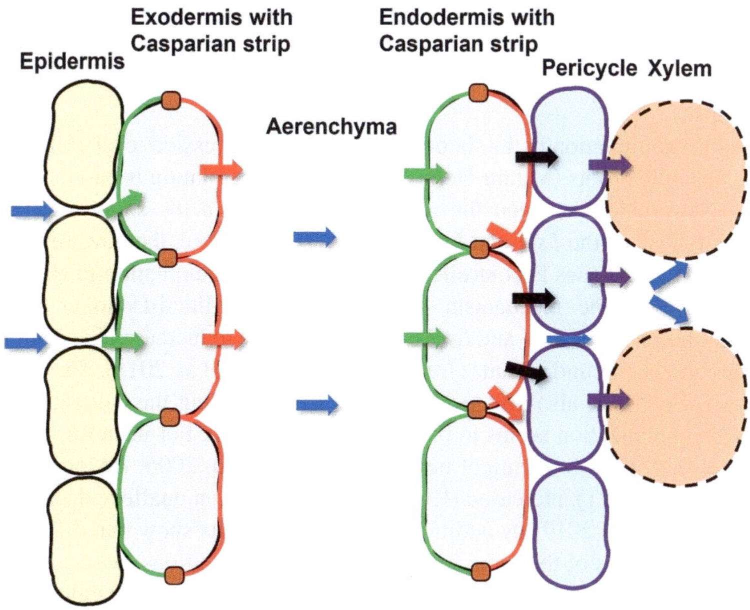

Fig. 3.5 General scheme of $Si(OH)_4$ radial transport in tissues of rice roots. Si is taken up by the Lsi1 influx channel (green arrow), localized distally in exodermal and endodermal cells (green part of the cell membrane), and subsequently transported by the Lsi2 efflux transporter (red arrow and membrane), localized proximally in the same tissues as Lsi1. The xylem loading of Si is mediated by the Lsi2 and also the Lsi3 transporter (purple arrow), which is located in the pericycle (purple membrane) without polar localization. Apoplasmic (blue arrows) and symplasmic (black arrow) transport of Si is shown. (Scheme modified according to Mitani-Ueno and Ma (2021) and Huang et al. (2022))

polar distribution (Mitani et al. 2011; Mitani-Ueno et al. 2011b; Sun et al. 2017, 2018; Mitani-Ueno and Ma 2021). The third group of plants based on the level of shoot Si accumulation is represented by species with a rejective uptake system. This system is well documented in tomato (*Lycopersicum esculentum*) (Sun et al. 2020). Despite the first study showing non-functional Lsi1 in tomato (Deshmukh et al. 2015), later it was shown that there is a functional Lsi1 channel, but the Lsi2 transporter is non-functional and thus unable to translocate Si from the root to aerial parts of the plant (Sun et al. 2020; Mitani-Ueno and Ma 2021).

Very recently, a new active efflux transporter Lsi3 was discovered in the rice root pericycle (Fig. 3.5) (Huang et al. 2022). This Si transporter shares 80% of amino acid sequence identity to the root Lsi2 transporter. In the pericycle, Lsi3 is not polarly localized and is responsible for the xylem loading of Si. The influx protein Lsi1 and the two efflux transporters Lsi2 and Lsi3 in roots act cooperatively, and their activities result in high Si accumulation in the shoot (Huang et al. 2022).

The list of identified Lsi1 and Lsi2 proteins, whether on the genomic, transcriptomic, or proteomic level is increasing; however, the cellular and tissue localization of many of these proteins has never been studied. These examples include Lsi1 in soybean (Deshmukh et al. 2013), Lsi2 in horsetail (Vivancos et al. 2016), Lsi1 and Lsi2 in potato (Vulavala et al. 2016), Lsi1 in palms (Bokor et al. 2019), Lsi1 in poinsettia (Hu et al. 2020), and many other plants within the dicot and monocot groups.

Silicon accumulation in the shoot is related to the expression level of the *Lsi1* and *Lsi2* genes in the roots (Mitani-Ueno et al. 2016). This relation is based on a strong negative correlation between the accumulation of Si in the shoot and the gene expression level of the *Lsi1* and *Lsi2* in the roots. Therefore, the downregulation of the *Lsi1* and *Lsi2* genes is controlled by increased Si accumulation in shoot of rice plants; however, the mechanism is not yet clear (Mitani-Ueno et al. 2016). Downregulation of the *Lsi1* and *Lsi2* genes in roots due to increased Si accumulation in leaves was also found in maize treated with Si (Bokor et al. 2015). The expression level of *Lsi* genes is also affected by silicon supply during the cultivation period, although generalization seems to be difficult at this point. For instance, the expression level of *Lsi1* in roots might be decreased (Ma et al. 2006; Bokor et al. 2015; Lukacova et al. 2021), increased (Kim et al. 2014), or even unaffected (Mitani et al. 2009a, Bokor et al. 2019) by addition of Si. These results show that different plant species (or cultivars of the same species), cultivation conditions, concentration, and source of Si in the growth substrate might be responsible for the variability of results.

3.4 Silicification in the Root Cell Walls

Silicon specifically interacts with different organic matter (polysaccharides and lignin) in the cell walls; on the other hand, there is evidence that cell wall proteins also act in templating SiO_2 deposition. However, the detailed functions in the shape-controlled mineralization process have not been elucidated in such a way as amorphous silicas (opals) (Nakamura et al. 2021). There is relatively lot of information about the deposition of SiO_2 within the cell walls of the aerial plant part, but less is known about this phenomenon in the roots. Research is underway in the Si-accumulating groups of plants, including monocots and pteridophytes, and is less documented in dicots. However, until now we know that the main effect is most probably the mechanical support and strengthening of the cell wall structure promoted by Si deposition (Guerriero et al. 2016).

Several decades back, some authors identified silica in the cell walls and showed different structures determined by the particle size and their orientation and also by surface functionality occurring in different regions of cell walls deposited at different developmental stages (e.g., Hodson et al. 1984; Parry et al. 1984; Perry and Fraser 1991). Especially Perry and Fraser (1991) achieved the detailed structural motifs of silica in Equisetum by using scanning electron microscopy which were principally fibrillar, globular, and sheet-like, and also tried to elucidate the importance of several relations between cell wall composition and silicified structures but, unfortunately, only in the aerial plant part. The authors concluded that the predominant function of cell wall silicification is the maintenance of rigidity: the fibrillar sheets of silica were intermeshed with the organic cell wall polymers, so the cellulose acted as the template for silica deposition (Perry and Fraser 1991; Kumar et al. 2017). Additionally, more recent studies have confirmed the interaction of cell wall polysaccharides with Si (Law and Exley 2011).

Several models have been proposed describing the process of cell wall silicification. The passive model explains silica deposition as a result of silicic acid condensation due to its dehydration during transpiration: silicification and transpiration are in verifiable positive correlation (Yoshida et al. 1962; Kumar et al. 2017). So, it is not surprising that the main places of silicification occurs in the aboveground plant parts. However, another mechanism not coupled with transpiration is also involved in silicification: there is evidence that biological factors contribute to this process and this mechanism is specific to the cell type (Kumar et al. 2017). This research confirmed that silica can be deposited directly on the cell wall matrix regardless of transpiration (Kumar et al. 2017). Also, in some plant species, silica was deposited directly in the roots cell walls impregnating inner tangential endodermal cell walls (Parry and Kelso 1975; Zexer and Elbaum 2020) (Fig. 3.6 and Table 3.1).

There are several components of the cell walls that can interact with silica, including cellulose (Guerriero et al. 2016; Perry and Lu 1992), hemicelluloses (He et al. 2015; Ma et al. 2015), callose (Brugiére and Exley 2017; Law and Exley 2011), lignin (Fang and Ma 2006; Montgomery and Parry 1979; Parry and Kelso 1975; Piperno et al. 2002; Watteau and Villemin 2001; Zhang et al. 2013), or ferulic acid (Soukup et al. 2017). Studies on rice cell wall mutants showed a positive correlation between the biomass of cellulose, hemicelluloses, and lignin and the Si content (Zhang et al. 2015). However, most of these studies focus on the shoot. A typical example of Si-accumulating plants, Commelinoid monocots (e.g., grasses), possess "type II" cell walls characterized by more phenylpropanoids (lignin) and fewer pectins and containing glucuronoarabinoxylan crosslinking cellulose and mixed-linkage glucans (Yokoyama et al. 2004) compared to the dicot cell wall

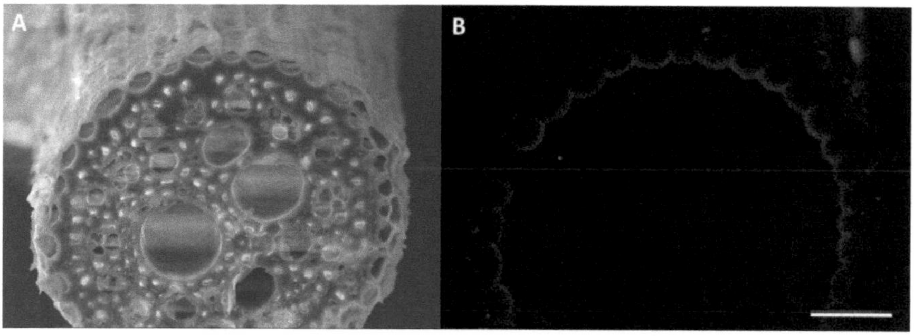

Fig. 3.6 Cross section of the base of older adventitious root of wheat visualized by (**a**) SEM and by (**b**) SEM-EDX for Si deposition. Peripheral root tissues degenerated, and the surface of vascular cylinder is covered and protected by heavily lignified endodermis impregnated with silica (purple color). Scale bar = 50 μm

representing type I. Interestingly, the complexation of Si with cell wall macromolecules occurs through the stabilization of sugars (He et al. 2013).

3.4.1 Cellulose and Polysaccharides

The main component of the plant cell wall is a polysaccharide cellulose. Since it forms tightly paced crystalline microfibrils, there were doubts about the possibility of binding Si (He et al. 2015). However, the artificial system of an aqueous solution of an octahedral silicon complex and insoluble cellulose favored the formation of ordered silica aggregates at neutral pH. In the absence of cellulose, the aggregates were disordered, suggesting a potential role for this polysaccharide in controlling SiO_2 deposition (Perry and Lu 1992). In addition, Si improved cell wall protoplast regeneration and the mechanical properties of new cell walls: cellulose filaments were longer and microfibrils were denser (He et al. 2015). The very recent study by Nakamura et al. (2021) describes the shape of opals controlled by intrafibrillar mineralization of silica nanoparticles on cellulose nanofibers that are stacked in the space between the epidermal leaf cell wall and the cuticle layer. The authors confirmed that cellulose nanofibers serve as a template for silicification.

Behind cellulosic microfibrils, the cell walls of land plants are based on a matrix of diverse polysaccharides (pectins and hemicelluloses) (Fry et al. 2008). The adjacent microfibrils of celluloses are tethered to each other by hemicelluloses bonding to them by hydrogen bonds (Fry 1989). The order Poales with the family Poaceae and some related families are unique among vascular plants in terms of hemicellulose species in cell walls. These species contain additional hemicellulose known as mixed-linkage (1→3),(1→4)-β-d-glucan (MLG) in cell walls (Labavitch and Ray 1978), and they can act as molecules in the immune reactions (Barghahn et al. 2021). However, Sørensen et al. (2008) provided evidence about the presence of mixed-linkage glucans also in the horsetail cell walls, which are phylogenetically distant from order Poales. Interestingly, Fry et al. (2008) suggested that these cell wall components also play role in silicification. Coming back to general hemicelluloses, He et al. (2015) also confirmed by the study on rice cell suspension cultures that Si occurs in the cell walls associated with hemicelluloses, and it is via Si-O-C bonds. Also, the cell walls of fern (*Adiantum raddianum*) was Si colocalized with epitopes of homogalacturonans (HG) and 1,5-arabinans (Leroux et al. 2013).

3.4.2 Lignin

There is an evidence showing that silica deposition in the cell walls is behind the polysaccharides, associated also with the processes of lignification (Montgomery and Parry 1979; Parry and Kelso 1975), because some studies confirmed associations between Si and lignin. For instance, Fang and Ma (2006) confirmed that only macromolecular lignin induced silica deposition in plants, and not its compounds (monolignols) themselves. When Na_2SiO_3 was in the presence of lignin in aqueous solution, the precipitation was more condensed than without silicon, and silica was induced to precipitate by the precipitated lignin. Even older studies confirmed these associations: Inanaga et al. (1995) showed that SiO_2 interacts with the aromatic ring of phenolic acid in the lignocarbohydrate complex, and Watteau and Villemin (2001) confirmed the association of SiO_2 with polyphenolic cell wall compounds. Soukup et al. (2017) and Zexer and Elbaum (2020) confirmed silica in roots aggregated on densely deposited cell walls enriched with phenolic compounds with blue autofluorescence. These places were confirmed by Raman microspectroscopy as loci for silica aggregation.

Silicon is also indirectly associated with cell wall lignification, by stimulating the process itself at the enzymatic or expression level. Zexer and Elbaum (2022) studied root cell wall lignification controlled by reactive oxygen species (ROS). Silicon was studied as a stress-related agent with an ameliorative role in the stress reaction. The oxidative state, especially the abundance of ROS in plants, also affected the presence of Si (Cooke and Leishman 2016) and, as a consequence, the cell wall is stimulated for lignification. This was confirmed, e.g., by Lukačová et al. (2013); however, the connection with suberization was not observed (Kreszies et al. 2020). The impact of the enhanced root lignification is obvious, especially in the root facing abiotic stress from soil pollutants, and can contribute to effective control of the apoplasmic movement of the ions.

3.4.3 *Callose*

Law and Exley (2011) were the first to provide the evidence for a connection between silicification and callose (β-1,3-glucan) formation and it was shown in aerial tissues of *Equisetum*. SiO_2 followed the deposition of callose, but also in vitro research confirmed that callose caused the precipitation of SiO_2. However, the role of callose in biosilicification is in initiation of this process rather than catalyzing silica formation (Brugiére and Exley 2017). Other studies also found that silica accumulation is dependent on callose deposition; however, this was investigated in the trichomes of *Arabidopsis* leaves (Kulich et al. 2018). The interaction between callose and silica precipitation in the root tissues must be elucidated in the future studies.

3.4.4 *Proteins*

Plant cell walls contain except of polysaccharide macromolecules also protein compounds, and in vitro studies have shown that their specific amino acid composition affects processes of biosilicification (Currie and Perry 2007). The side chains of amino acids are either positively charged or negatively charged, and those with positively charged groups can interact with the negatively charged silica species. These interactions are electrostatic, thus favoring biosilicification. In *the* hypocotyl cell wall of *Cucumis sativus*, proline-rich protein (PRP1) with clusters of positively charged amino acids (arginine and lysine) is involved in systemically resistant plants (Kauss et al. 2003). Synthetically prepared PRP1 induced the polymerization of $Si(OH)_4$ to SiO_2 that the authors evaluated so that the protein enhances the strengthening of the cell wall via silicification and thus induces acquired systemic resistance (Kauss et al. 2003). However, chemical binding of Si to any cell compound or biomolecule remains still obscure, because of the chemical properties of silicic acid, which is inert at physiological pH (Exley et al. 2019).

3.5 Phytoliths

Silica phytoliths are known as deposits of bioorganic silica found in wide spectra of plant species. These microscopic bodies form when monosilicic acid, the only form of Si that can be taken up and transported through the plant body, is polymerized, precipitated, and deposited in the form of hydrated amorphous silica (SiO_2 nH_2O), known as opal (Piperno 2006). Phytoliths are relatively widely distributed throughout the plant kingdom and can be found both in roots and shoots in various plant organs and tissues in intracellular and mostly extracellular space. The shape of phytoliths might copy the shape of cells that they are filling, might be regular but also irregular, and vary between plant species and taxa. The size of phytoliths varies between 0.1 and 200 μm, although larger structures up to 500 μm have also been reported (Nawaz et al. 2019). Moreover, their unique shape in particular plant species and their mechanical resistance to environmental factors preserved them for many years in the soil after the plant body was completely decomposed. Therefore, all these attributes predict them as a good determination tool, and together with other persistent plant components, such as pollen grains, phytoliths are an important research tool for the reconstruction of paleovegetation and are frequently used in paleoecology and paleobotany.

The mechanisms of phytoliths formation have been thoroughly studied; however, the exact ways of their formation and their importance for the plant organisms are not clear, and the data regarding these processes are still scarce. Although Si phytoliths are mostly known to form in aerial plant parts, there are some species, mostly belonging to the Andropogonae, such as *Sorghum bicolor* or *Saccharum officinarum*, commonly known for the formation of phytoliths especially in root endodermis (Fig. 3.7).

The formation of phytoliths in the root endodermis of sorghum is moreover closely related to the processes of lignification. Soukup et al. (2017) have described the specific pattern of phytoliths formation. They suggested that the movement of Si is driven by the basipetal water flow, and silica aggregates are formed only in the basal parts of the roots with developed suberin lamellae and

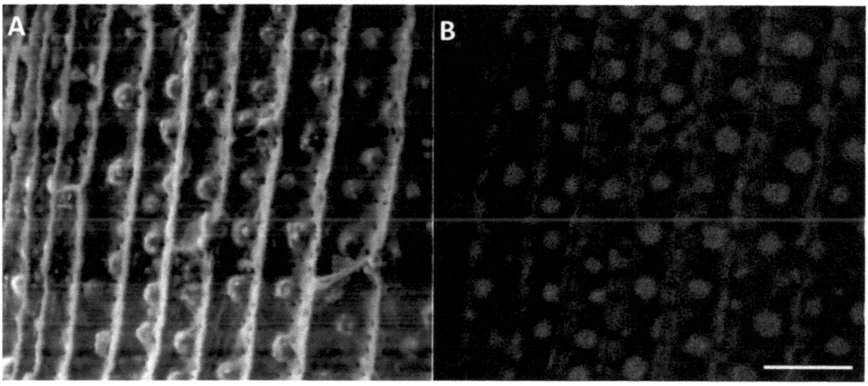

Fig. 3.7 Endodermal cells of sugar cane (*Saccharum officinarum*) covered by Si deposits (purple color) visualized by (**a**) SEM and by (**b**) SEM-EDX. Scale bar = 50 μm

lignified tertiary cell walls. More recently, Zexer and Elbaum (2020) suggested that polymerization of silicic acid occurs only at specific locations in the tangential endodermis cell walls where modification of lignin is present in the form of a lack of ferulic acid. However, recent studies contributed to the opinion that silicification is not only a passive and metabolically inactive process that is joined with plant transpiration. For example, Soukup et al. (2020) find that the process of Si phytolith deposition within the root endodermis is strictly associated with fully metabolic active living cells.

3.6 Stegmata

Stegmata are generally small cells filled out with amorphous silicon dioxide (SiO_2) (or solid silica) as silica aggregates/phytoliths adjacent to fibers in diverse monocotyledons such as orchids (Holtzmeier et al. 1998; Prychid et al. 2003), some species from the family Musaceae (Tomlinson 1959), and predominantly palm trees. Mettenius (1865) came up with the term "Deckzellen" (cover cells) or "stegmata"(from Greek *stegium* – a roof or covering) for specialized cells containing silica aggregates that lie over the sclerenchyma in ferns (Prichyd et al. 2003). A few years later, Kohl (1889) added more information about these cells (Prychid et al. 2003). They are associated with vascular and non-vascular fibers (Tomlinson 1961).

Kohl (1889) noted that the development of silica bodies is the main indicator for stegmata differentiation. Differentiation of stegmata starts in young tissue near to the apical meristem. At this developmental stage, stegmata with present cytoplasm can form silica bodies. In the later developmental stages, in older internodes, stegmata do not contain cytoplasm but contain silica bodies and become irregularly shaped (Prychid et al. 2003).

Stegmata cells are found in association with fibers in leaves, petioles, and leaf sheaths, and are rarely observed also in roots. In the following part, we mention the data about stegmata in various plants, mostly in leaves and stems, for the possibility of their presence in rarely investigated roots. Detail knowledge on the variability of shape and structure of the silica bodies in palm trees has been given by Tomlinson (1961). He has recognized mainly two types: hat-shaped and spherical (usually in rattan palms) silica bodies. Carlsward et al. (2006) analyzed a large group of epiphytic monopodial taxa Vandeae (Orchidaceae), which are characterized by the presence of spherical stegmata within the leaves but not in the roots. In Orchidaceae, stegmata are present in the form of conical silica bodies, spherical silica bodies (in Vandeae, in the genera *Apostasia*, *Eria,* and *Dendrobium*) or stegmata completely absent (Moller & Rasmussen, 1984). Stern and Carlsward (2008) investigated a small tribe of orchids Calypsoeae and found stegmata in leaves of only *Cremastra* and *Govenia* genera. The maturation process of stegmata within internodes in the rattan palm *Calamus axillaris* Becc. was described by Schmitt et al. (1995). This process is over when ten internodes are fully development. The result of the maturation are well developed silica-bodies, a thickened cell wall of stegmata at the fibre side, and absence of cytoplasm. Spherical stegmata were recently detected on the outer surface of the sclerenchyma bundles in the roots of date palm (Bokor et al. 2019) (Fig. 3.8). This arrangement is supposed to serve probably as mechanical protection strengthening the root tissues.

The shape and size of silica bodies are known to play a significant role in the taxonomic analysis of different plant groups (Shakoor et al. 2014). Sant'anna-Santos et al. (2015) used the characteristic feature of stegmata for the separation of the species *Butia capitata* and *B. odorata* (Arecaceae taxa) because the shape of the

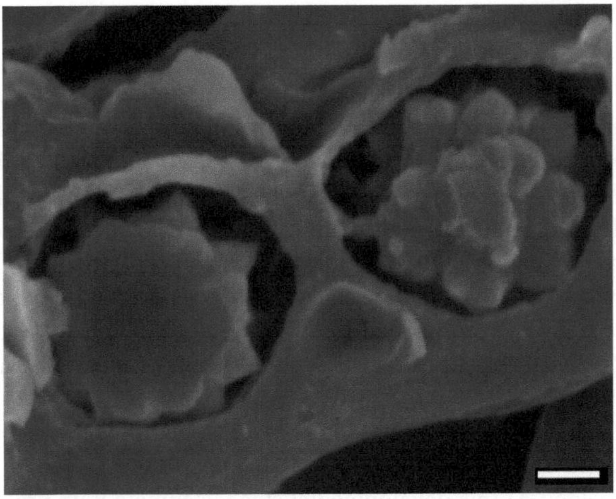

Fig. 3.8 Two adjacent stegmata with disrupted removed part of the cell walls exposing the Si phytoliths in date palm revealed by scanning electron microscopy. Si phytolith on the left is partially disrupted. Scale bar 2 μm

silica body is a useful taxonomy characteristic for this family (Uhl and Dransfield 1987). Stegmata in this family contain spherical and druse-like silica bodies. The pattern of silica bodies may be an important taxonomic characteristic for the genus *Syagrus*, as shown by Leite and Scatena (2001). An epiphytic orchid *Brasiliorchis picta (Maxillaria picta)* belonging to the tribe Cymbidieae (formerly Maxillarieae) contains stegmata with rough-surfaced, conical silica bodies that border the periphery of the thickened pericycle cells opposite the phloem in roots (Holtzmeier et al. 1998). Tomlinson (1961) reported that silica deposits in palm trees are common in aboveground plant parts, but in roots, they are present only in some species. In Musaceae, there are in all parts except the roots (Tomlinson 1959).

3.7 The Function of Silica Deposits in Roots

The functional role of silicon in plant organisms has been a topic of debate for a long time. After the first discovery of Si deposits in plant tissues, the questions about their functional role have emerged. The progress of Si research in plants has been considerable in recent years, although the importance is mostly paid to the aboveground plant parts. However, the root as the main organ responsible for uptake of available Si from the soil, has been discussed in several studies. In recent years, several evidence have shown that the deposition of Si within roots can contribute to improved mechanical strengthening, improved visco-elastic properties of cell walls, or that Si deposited within root tissues can serve as protection against herbivores, parasites, improve the stress from drought, metal(loid)s, or could help regulate nutrients uptake.

Several associations of Si with cell wall compounds may have a significant impact on plant survival under stress conditions. The association of hemicelluloses with Si may be important for Cd detoxification. The experiment was conducted in vitro on rice cell suspension culture. Cd^{2+} was naturally bound to cell walls, but once its concentration exceeded binding capacity, the Si-supplemented cells exhibited a smaller influx of toxic ions into cells (Ma et al. 2015). The results of this study showed that 64% of the total Si in cell walls was bound to hemicelluloses and consequently inhibited Cd uptake. This inhibition of Cd within cell walls by creating [Si-hemicellulose matrix] Cd complexes could be an effective mechanism of retaining heavy metal in roots, which was confirmed in rice seedlings (Shi et al. 2005). Enhanced binding of Cd to cell walls in root tips of *Kandelia obovata* also restricted its translocation (Ye et al. 2012). Similarly, Vaculik et al. (2009) showed that Si increased the elasticity and plasticity of cell walls of maize plants grown in the presence of Cd.

The ability of Si to co-precipitate with metals in the cell walls via modification of the cell wall properties has been proposed as one of the tolerance mechanisms how to cope with abiotic stress from metals (Neumann and zur Nieden 2001; Prabagar et al. 2011; for a literature review, see Vaculik et al. 2020). On the other hand, Si reduces the porosity of endodermal cell walls and thus decreases the apoplasmic movement of ions (da Cunha and do Nascimento 2009) or by changing the cell wall extensibility protects the stele (Hattori et al. 2003).

Several hypotheses about the functional role of Si phytoliths in plants have been postulated. In general, it is widely accepted that Si deposition in the form of phytoliths could contribute to better mechanical resistance and mechanically strengthen the plant body (Lux et al. 2020). For example, the formation of phytoliths in the endodermis of sorghum is probably related to the mechanical protection of the root central cylinder. Similarly, stegmata, specific cells filled with Si precipitates (phytoliths), are closely associated with additional mechanical tissue sclerenchyma and contribute to mechanical protection of root tissues in date palm (Bokor et al. 2019). Deposition of Si was also suggested as a metabolically less demanding alternative to lignin. Some studies suggested that Si was used as a cost-effective alternative to lignin deposition when land plants occur. However, the non-universality of silica biomineralization within land plants, which have inhabited the silica-rich terrestrial environment, suggests that active silica incorporation was not necessary for terrestrialization during the Paleozoic era (Stromberg et al. 2016). Deposition of Si, especially in the form of phytoliths around the cell walls of the endo- and exodermis, significantly improves water retention and consequently improves the maintenance of plant water balance. For example, the drought-resistant cultivar of sorghum contained more Si deposits in the endodermis than those sensitive to drought, suggesting that Si prevents water leakage from the central cylinder to the soil (Hattori et al. 2003). Similarly, an upland rice cultivar that showed better drought resistance showed a higher density of Si phytoliths in the endodermis than the lowland and less drought-resistant rice cultivar (Lux et al. 1999). Silicon root deposition has also been shown to be an efficient tool to decrease the infection pressure of fungal pathogens. Bathoova et al. (2018) found that in sorghum, the invasion of the necrotrophic fungus *Alternaria alternata* was restricted in the endodermis covered by Si phytoliths. Relatively broad spectra of evidence showed Si as an efficient tool in ameliorating the metal(loid) toxicity in plants. Although most of the findings are related to Si cell wall deposition, there are some suggestions that Si phytoliths might contribute to the amelioration of this kind of abiotic stress. Some of the published data showed the encapsulation and co-precipitation of various heavy metals and metalloids, such as lead, zinc, and arsenic, with Si phytoliths. Although most of the findings come from phytoliths originating from plant aboveground tissues (e.g., Tran et al. 2019; Delplace et al. 2020; Nguyen and Nguyen 2019; Nguyen et al. 2021; Liu et al. 2022), some findings also suggest root phytoliths as sites of Al-Si co-deposition (Liu et al. 2021). Therefore, it is highly probable that phytoliths in roots, which are in contact with an unfavorable metal environment from the beginning, could also act as one of the plant detoxification mechanisms.

In conclusion, it is possible to summarize that many data support the positive role of silica deposits in plants; moreover, there is some evidence supporting this beneficial functioning of Si at the root level. This positive role is also supported by the presence of passive and active proteins responsible for Si uptake and transport within roots and translocation to the shoots. Moreover, cooperation of these proteins seems to have an increased efficiency of Si precipitation and deposition. However, there are still many unanswered questions regarding the role of silica deposits in specific root cells or Si impregnated cell walls of root tissues. There is also a question related to biosilicification in almost the entire lumen of cortical cells that might be triggered by similar biochemistry known in silica cells of leaves, which opens the door to future research.

Acknowledgments This study was supported by the Operation Program of Integrated Infrastructure for the project, UpScale of Comenius University Capacities and Competence in Research, Development and Innovation, ITMS2014+: 313021BUZ3, IMTS: 313011V446 co-financed by the European Regional Development Fund and by the VEGA grant No. 1/0352/21.

References

Barberon M, Geldner N (2014) Radial transport of nutrients: the plant root as a polarized epithelium. Plant Physiol 166:528–537. https://doi.org/10.1104/pp.114.246124

Barghahn S, Arnal G, Jain N, Petutschnig E, Brumer H, Lipka V (2021) Mixed linkage β-1,3/1,4-glucan oligosaccharides induce defense responses in *Hordeum vulgare* and *Arabidopsis thaliana*. Front Plant Sci 12:682439. https://doi.org/10.3389/fpls.2021.682439

Bathoova M, Bokor B, Soukup M, Lux A, Martinka M (2018) Silicon-mediated cell wall modifications of sorghum root exodermis and suppression of invasion by fungus *Alternaria alternata*. Plant Pathol 67:1891–1900. https://doi.org/10.1111/ppa.12906

Bennett DM (1982) Silicon deposition in the roots of *Hordeum sativum* Jess, *Avena sativa* L. and *Triticum aestivum* L. Ann Bot 50:239–245. https://doi.org/10.1093/oxfordjournals.aob.a086361

Bokor B, Bokorová S, Ondoš S, Švubová R, Lukačová Z, Hýblová M, Szemes T, Lux A (2015) Ionome and expression level of Si transporter genes (Lsi1, Lsi2, and Lsi6) affected by Zn and Si interaction in maize. Environ Sci Pollut Res Int 22:6800–6811. https://doi.org/10.1007/s11356-014-3876-6

Bokor B, Soukup M, Vaculík M, Vd'ačný P, Weidinger M, Lichtscheidl I, Vávrová S, Šoltys K, Sonah H, Deshmukh R, Belanger RR, White PJ, El-Serehy HA, Lux A (2019) Silicon uptake and localisation in date palm (*Phoenix dactylifera*) — a unique association with sclerenchyma. Front Plant Sci 10:988. https://doi.org/10.3389/fpls.2019.00988

Borissow G (1924) Über die eigenartigen Kieselkörper in der Wurzelendodermis bei Andropogon-Arten. Berichte der Deutschen Botanischen Gesellschaft 42:366–380

Borissow G (1925) Rasdorskys Körpchen bei Ravenna-Gras. Berichte der Deutschen Botanischen Gesellschaft 43:178–184

Brown DA (1984) Prospects and limits of a phytolith key for grasses in the Central United States. J Archaeol Sci 11:345–368. https://doi.org/10.1016/0305-4403(84)90016-5

Brugiére T, Exley C (2017) Callose-associated silica deposition in *Arabidopsis*. J Trace Elem Med Biol 39:86–90. https://doi.org/10.1016/j.jtemb.2016.08.005

Carlsward BS, Stern WL, Bytebier B (2006) Comparative vegetative anatomy and systematics of the angraecoids (Vandeae, Orchidaceae) with an emphasis on the leafless habit. Bot J Linn Soc 151:165–218

Casey W, Kinrade S, Knight C, Rains D, Epstein E (2004) Aqueous silicate complexes in wheat, *Triticum aestivum* L. Plant Cell Environ 27:51–54

Chiba Y, Mitani N, Yamaji N, Ma JF (2009) HvLsi1 is a silicon influx transporter in barley. Plant J 57:810–818

Cooke J, Leishman MR (2016) Consistent alleviation of abiotic stress with silicon addition: a meta-analysis. Funct Ecol 30:1340–1357. https://doi.org/10.1111/1365-2435.12713

Coskun D, Deshmukh R, Shivaraj SM, Isenring P, Bélanger RR (2021) Lsi2: a black box in plant silicon transport. Plant Soil 466:1–20. https://doi.org/10.1007/s11104-021-05061-1

Currie HA, Perry CC (2007) Silica in plants: biological, biochemical and chemical studies. Ann Bot 100:1383–1389. https://doi.org/10.1093/aob/mcm247

da Cunha KPV, do Nascimento CWA (2009) Silicon effects on metal tolerance and structural changes in maize (*Zea mays* L.) gown on a cadmium and zinc enriched soil. Water Air Soil Pollut 197:323–330. https://doi.org/10.1007/s11270-008-9814-9

Davamani V, Sangeetha Piriya R, Rakesh SS, Parameswari E, Paul Sebastian S, Kalaiselvi P, Maheswari M, Santhi R (2022) Phytolith-occluded carbon sequestration potential of oil palm plantation in Tamil Nadu. ACS Omega 7:2809–2820. https://doi.org/10.1021/acsomega.1c05592

Deshmukh RK, Vivancos J, Guérin V, Sonah H, Labbé C, Belzile F, Bélanger RR (2013) Identification and functional characterization of silicon transporters in soybean using comparative genomics of major intrinsic proteins in arabidopsis and rice. Plant Mol Biol 83:303–315. https://doi.org/10.1007/s11103-013-0087-3

Deshmukh RK, Vivancos J, Ramakrishnan G, Guérin V, Carpentier G, Sonah H, Labbé C, Isenring P, Belzile FJ, Bélanger RR (2015) A precise spacing between the NPA domains of aquaporins is essential for silicon permeability in plants. Plant J 83:489–500. https://doi.org/10.1111/tpj.12904

Delplace G, Schreck E, Pokrovsky OS, Zouiten C, Blondet I, Darrozes J, Viers J (2020) Accumulation of heavy metals in phytoliths from reeds growing on mining environments in Southern Europe. Sci Total Environ 712:135595. https://doi.org/10.1016/j.scitotenv.2019.135595

Epstein E (1994) The anomaly of silicon in plant biology. PNAS 91:11–17

Exley C (2015) A possible mechanism of biological silicification in plants. Front Plant Sci 6:853. https://doi.org/10.3389/fpls.2015.00853

Exley C, Guerriero G, Lopez X (2019) Silicic acid: the omniscient molecule. Sci Total Environ 665:432–437

Exley C, Guerriero G, Lopez X (2020) How is silicic acid transported in plants? SILICON 12:2641–2645. https://doi.org/10.1007/s12633-019-00360-w

Fang JY, Ma XL (2006) *In vitro* simulation studies of silica deposition induced by lignin from rice. J Zhejiang Univ Sci B 7:267–271. https://doi.org/10.1631/jzus.2006.B0267

Fleck AT, Schulze S, Hinrichs M, Specht A, Waßmann F, Schreiber L, Schenk MK (2015) Silicon promotes exodermal casparian band formation in Si-accumulating and Si-excluding species by forming phenol complexes. PLoS One 10:e0138555. https://doi.org/10.1371/journal.pone.0138555

Fry SC (1989) Cellulases, hemicelluloses and auxin-stimulated growth: a possible relationship. Physiol Plant 75:532–536. https://doi.org/10.1111/j.1399-3054.1989.tb05620.x

Fry SC, Nesselrode BHWA, Miller JG, Mewburn BR (2008) Mixed-linkage (1→3,1→4)-β-d-glucan is a major hemicellulose of *Equisetum* (horsetail) cell walls. New Phytol 179:104–115. https://doi.org/10.1111/j.1469-8137.2008.02435.x

Gong HJ, Randall DP, Flowers TJ (2006) Silicon deposition in the root reduces sodium uptake in rice (*Oryza sativa* L.) seedlings by reducing bypass flow. Plant Cell Environ 29:1970–1979. https://doi.org/10.1111/j.1365-3040.2006.01572.x

Guerriero G, Hausman JF, Legay S (2016) Silicon and the plant extracellular matrix. Front Plant Sci 7:463. https://doi.org/10.3389/fpls.2016.00463

Hattori T, Inanaga S, Tanimoto E, Lux A, Luxova M, Sugimoto Y (2003) Silicon-induced changes in viscoelastic properties of sorghum root cell walls. Plant Cell Physiol 44:743–749. https://doi.org/10.1093/pcp/pcg090

He C, Wang L, Liu J, Liu X, Li X, Ma J, Lin Y, Xu F (2013) Evidence for 'silicon' within the cell walls of suspension-cultured rice cells. New Phytol 200:700–709. https://doi.org/10.1111/nph.12401

He C, Ma J, Wang L (2015) A hemicellulose-bound form of silicon with potential to improve the mechanical properties and regeneration of the cell wall of rice. New Phytol 206:1051–1062. https://doi.org/10.1111/nph.13282

Hirokawa T, Boon-Chieng S, Mitaku S (1998) SOSUI: classification and secondary structure prediction system for membrane proteins. Bioinformatics 14:378–379. https://doi.org/10.1093/bioinformatics/14.4.378

Hodson MJ (1986) Silicon deposition in the roots, culm and leaf of *Phalaris canariensis* L. Ann Bot 58:167–177. https://doi.org/10.1093/oxfordjournals.aob.a087194

Hodson MJ, Sangster AG (1989a) Subcellular localization of mineral deposits in the roots of wheat (*Triticum aestivum* L.). Protoplasma 151:19–32. https://doi.org/10.1007/BF01403298

Hodson MJ, Sangster AG (1989b) X-ray microanalysis of the seminal root of *Sorghum bicolor* with particular reference to silicon. Ann Bot 64:659–667. https://doi.org/10.1093/oxfordjournals.aob.a087891

Hodson MJ, Wilkins DA (1991) Localization of aluminium in the roots of Norway spruce (*Picea abies* (L.) Karst.) inoculated with *Paxillus involutus* Fr. New Phytol 118:273–278. https://doi.org/10.1111/j.1469-8137.1991.tb00977.x

Hodson MJ, Sangster AG, Parry DW (1984) An ultrastructural study on the development of silicified tissues in the lemma of *Phalaris canariensis* L. Philos Trans R Soc Lond B Biol Sci 222:413–425. https://doi.org/10.1098/rspb.1984.0074

Holtzmeier MA, Stern WL, Judd WS (1998) Comparative anatomy and systematics of Senghas's cushion species of Maxillaria (Orchidaceae). Bot J Linn Soc 127:43–82. https://doi.org/10.1111/j.1095-8339.1998.tb02087.x

Hu J, Li Y, Jeong BR (2020) Putative silicon transporters and effect of temperature stresses and silicon supplementation on their expressions and tissue silicon content in Poinsettia. Plan Theory 9:569. https://doi.org/10.3390/plants9050569

Huang S, Yamaji N, Sakurai G, Mitani-Ueno N, Konishi N, Ma JF (2022) A pericycle-localized silicon transporter for efficient xylem loading in rice. New Phytol 234:197–208. https://doi.org/10.1111/nph.17959

Inanaga S, Okasaka A, Tanaka S (1995) Does silicon exist in association with organic compounds in rice (*Oryza sativa*) plant? Soil Sci Plant Nutr 41:111–117. https://doi.org/10.1080/00380768.1995.10419564

Jones RL (1964) Note on occurrence of opal phytoliths in some Cenozoic sedimentary rocks. J Paleontol 38:773–775. https://doi.org/10.2136/sssaj1964.03615995002800050039x

Kapilan R, Vaziri M, Zwiazek JJ (2018) Regulation of aquaporins in plants under stress. Biol Res 51:4. https://doi.org/10.1186/s40659-018-0152-0

Kauss H, Seehaus K, Franke R, Gilbert S, Dietrich RA, Kröger N (2003) Silica deposition by a strongly cationic proline-rich protein from systemically resistant cucumber plants. Plant J 33:87–95. https://doi.org/10.1046/j.1365-313X.2003.01606.x

Kim Y-H, Khan AL, Kim D-H, Lee S-Y, Kim K-M, Waqas M, Jung H-Y, Shin J-H, Kim J-G, Lee I-J (2014) Silicon mitigates heavy metal stress by regulating P-type heavy metal ATPases, *Oryza sativa* low silicon genes, and endogenous phytohormones. BMC Plant Biol 14:1–13. https://doi.org/10.1186/1471-2229-14-13

Kohl FG (1889) Kalksalze und Kieselsäure in der Pflanze. Elwert'sche Verlagsbuchhandlung, Marburg

Kreszies T, Kreszies V, Ly F, Thangamani PD, Shellakkutti N, Schreiber L (2020) Suberized transport barriers in plant roots: the effect of silicon. J Ex Bot 71:6799–6806. https://doi.org/10.1093/jxb/eraa203

Kulich I, Vojtíková Z, Sabol P, Ortmannová J, Neděla V, Tihlaříková E, Žárský V (2018) Exocyst subunit EXO70H4 has a specific role in callose synthase secretion and silica accumulation. Plant Physiol 176:2040–2051. https://doi.org/10.1104/pp.17.01693

Kumar S, Soukup M, Elbaum R (2017) Silicification in grasses: variation between different cell types. Front Plant Sci 8:438. https://doi.org/10.3389/fpls.2017.00438

Labavitch JM, Ray PM (1978) Structure of hemicellulosic polysaccharides of *Avena sativa* coleoptile cell walls. Phytochemistry 17:933–937. https://doi.org/10.1016/S0031-9422(00)88649-5

Law C, Exley C (2011) New insight into silica deposition in horsetail (*Equisetum arvense*). BMC Plant Biol 11:112. https://doi.org/10.1186/1471-2229-11-112

Leite KRB, Scatena VL (2001) Anatomy of the leaf segment of species of *Syagrus* Mart. (Arecaceae) da Chapada Diamantina, Bahia, Brasil. Sitientibus Ser Cien Biol 1:3–14

Leroux O, Leroux F, Mastroberti AA, Santos-Silva F, Van Loo D, Bagniewska-Zadworna A, Van Hoorebeke L, Bals S, Popper ZA, de Araujo Mariath JE (2013) Heterogeneity of silica and glycan-epitope distribution in epidermal idioblast cell walls in *Adiantum raddianum* laminae. Planta 237:1453–1464. https://doi.org/10.1007/s00425-013-1856-6

Liu WS, Laird JS, Ryan CG, Tang J-T, Qiu R-L, Echevarria G, Morel J-L, van der Ent A (2021) Rare earth elements, aluminium and silicon distribution in the fern *Dicranopteris linearis* revealed by μPIXE Maia analysis. Ann Bot 128:17–30. https://doi.org/10.1093/aob/mcab026

Liu L, Song Z, Li Q, Ellam RM, Tang J, Wang Y, Sarkar B, Wang H (2022) Accumulation and partitioning of toxic trace metal(loid)s in phytoliths of wheat grown in a multi-element contaminated soil. Environ Pollut 294:118645. https://doi.org/10.1016/j.envpol.2021.118645

Lukačová Z, Švubová R, Kohanová J, Lux A (2013) Silicon mitigates the Cd toxicity in maize in relation to cadmium translocation, cell distribution, antioxidant enzymes stimulation and enhanced endodermal apoplasmic barrier development. Plant Growth Regul 70:89–103. https://doi.org/10.1007/s10725-012-9781-4

Lukacova Z, Liska D, Bokor B, Svubova R, Lux A (2021) Silicon and cadmium interaction of maize plants cultivated *in vitro*. Biologia 76:2721–2733. https://doi.org/10.1007/s11756-021-00799-6

Lux A, Luxová M, Morita S, Abe J, Inanaga S (1999) Endodermal silicification in developing seminal roots of lowland and upland cultivars of rice (*Oryza sativa* L.). Can J Bot 77:955–960. https://doi.org/10.1139/b99-043

Lux A, Luxová M, Hattori T, Inanaga S, Sugimoto Y (2002) Silicification in sorghum (*Sorghum bicolor*) cultivars with different drought tolerance. Physiol Plant 115:87–92. https://doi.org/10.1034/j.1399-3054.2002.1150110.x

Lux A, Luxová M, Abe J, Morita S, Inanaga S (2003a) Silicification of bamboo (*Phyllostachys heterocycla* Mitf.) root and leaf. In: Abe J (ed) Roots: the dynamic interface between plants and the earth, pp 85–91. https://doi.org/10.1007/978-94-017-2923-9_9

Lux A, Luxová M, Abe J, Tanimoto E, Hattori T, Inanaga S (2003b) The dynamics of silicon deposition in the sorghum root endodermis. New Phytol 158:437–441. https://doi.org/10.1046/j.1469-8137.2003.00764.x

Lux A, Lukačová Z, Vaculík M, Švubová R, Kohanová J, Soukup M, Martinka M, Bokor B. (2020) Silicification of Root Tissues Plants 9(1):111. https://doi.org/10.3390/plants9010111

Ma JF, Yamaji N (2015) A cooperative system of silicon transport in plants. Trends Plant Sci 20:435–442. https://doi.org/10.1016/j.tplants.2015.04.007

Ma JF, Tamai K, Yamaji N, Mitani N, Konishi S, Katsuhara M, Ishiguro M, Murata Y, Yano M (2006) A silicon transporter in rice. Nature 440:688–691. https://doi.org/10.1038/nature04590

Ma JF, Yamaji N, Mitani N, Tamai K, Konishi S, Fujiwara T, Katsuhara M, Yano M (2007) An efflux transporter of silicon in rice. Nature 448:209–212. https://doi.org/10.1038/nature05964

Ma J, Cai H, He C, Zhang W, Wang L (2015) A hemicellulose-bound form of silicon inhibits cadmium ion uptake in rice (*Oryza sativa*) cells. New Phytol 206:1063–1074. https://doi.org/10.1111/nph.13276

Markovich O, Kumar S, Cohen D, Addadi S, Fridman E, Elbaum R (2015) Silicification in leaves of sorghum mutant with low silicon accumulation. SILICON 11:2385–2391. https://doi.org/10.1007/s12633-015-9348-x

McNaughton SJ, Tarrants JL, McNaughton MM, Davis RD (1985) Silica as a defense against herbivory and a growth promotor in african grasses. Ecology 66:528–535. https://doi.org/10.2307/1940401

Mettenius G (1865) Ueber die Hymenophyllaceae. Abh d K s Gesell d Wiss 11:403–504

Mitani M, Ma JF (2005) Uptake system of silicon in different plant species. J Exp Bot 56:1255–1261. https://doi.org/10.1093/jxb/eri121

Mitani N, Yamaji N, Ma JF (2009a) Identification of maize silicon influx transporters. Plant Cell Physiol 50:5–12. https://doi.org/10.1093/pcp/pcn110

Mitani N, Chiba Y, Yamaji N, Ma JF (2009b) Identification and characterization of maize and barley Lsi2-like silicon efflux transporters reveals a distinct silicon uptake system from that in rice. Plant Cell 21:2133–2142. https://doi.org/10.1105/tpc.109.067884

Mitani N, Yamaji N, Ago Y, Iwasaki K, Ma JF (2011) Isolation and functional characterization of an influx silicon transporter in two pumpkin cultivars contrasting in silicon accumulation. Plant J 66:231–240. https://doi.org/10.1111/j.1365-313X.2011.04483.x

Mitani-Ueno N, Ma JF (2021) Linking transport system of silicon with its accumulation in different plant species. Soil Sci Plant Nutr 67:10–17. https://doi.org/10.1080/00380768.2020.1845972

Mitani-Ueno N, Yamaji N, Zhao FJ, Ma JF (2011a) The aromatic/arginine selectivity filter of NIP aquaporins plays a critical role in substrate selectivity for silicon, boron, and arsenic. J Exp Bot 62:4391–4398. https://doi.org/10.1093/jxb/err158

Mitani-Ueno N, Yamaji N, Ma JF (2011b) Silicon efflux transporters isolated from two pumpkin cultivars contrasting in Si uptake. Plant Signal Behav 6:991–994. https://doi.org/10.4161/psb.6.7.15462

Mitani-Ueno N, Yamaji N, Ma JF (2016) High silicon accumulation in the shoot is required for down-regulating the expression of Si transporter genes in rice. Plant Cell Physiol 57:2510–2518. https://doi.org/10.1093/pcp/pcw163

Moller JD, Rasmussen H (1984) Stegmata in orchidales: character state distribution and polarity. Bot J Lin Soc 89:53–76. https://doi.org/10.1111/j.1095-8339.1984.tb01000.x

Montgomery DJ, Parry DW (1979) The ultrastructure and analytical microscopy of silicon deposition in the intercellular spaces of the roots of *Molinia caerulea* (L.) Moench. Ann Bot 44:79–84. http://www.jstor.org/stable/42753868

Nakamura E, Ozaki N, Oaki Y, Imai H (2021) Cellulose intrafibrillar mineralization of biological silica in a rice plant. Sci Rep 11:7886. https://doi.org/10.1038/s41598-021-87144-8

Nawaz MA, Zakharenko AM, Zemchenko IV, Haider MS, Ali MA, Imtiaz M, Chung G, Tsatsakis A, Sun S, Golokhvast KS (2019) Phytolith formation in plants: from soil to cell. Plan Theory 8:249. https://doi.org/10.3390/plants8080249

Neumann D, zur Nieden U (2001) Silicon and heavy metal tolerance of higher plants. Phytochemistry 56:685–692. https://doi.org/10.1016/S0031-9422(00)00472-6

Nguyen ATQ, Nguyen MN (2019) Straw phytolith for less hazardous open burning of paddy straw. Sci Rep 9:20043.

Nguyen MN, Dam TTN, Nguyen ATQ, Nguyen AM, Nguyen LM, Duong LT, Dang QT, Tran TT (2021) Arsenic in rice straw phytoliths: Encapsulation and release properties. Appl Geochem 127:104907.

Parry DW, Kelso M (1975) The distribution of silicon deposits in the roots of *Molinia caerulea* (L.) Moench. and *Sorghum bicolor* (L.) Moench. Ann Bot 39:995–1001. https://www.jstor.org/stable/42756333

Parry DW, Kelso M (1977) The ultrastructure and analytical microscopy of silicon deposits in the roots of *Saccharum officinarum* (L.). Ann Bot 41:855–862. https://doi.org/10.1093/oxfordjournals.aob.a085361

Parry DW, Soni SL (1972) Electron-probe microanalysis of silicon in the roots of *Oryza sativa* L. Ann Bot 36:781–783. https://www.jstor.org/stable/42752089

Parry DW, Hodson MJ, Sangster AG, Jones WC, O'Neill CH (1984) Some recent advances in studies of silicon in higher plants. Philos Trans R Soc Lond Ser B Biol Sci 304:537–549. https://doi.org/10.1098/rstb.1984.0045

Perry CC, Fraser MA (1991) Silica deposition and ultrastructure in the cell wall of *Equisetum arvense*: the importance of cell wall structures and flow control in biosilicification? Phil Trans R Soc Lond B 334:149–157. https://www.jstor.org/stable/55565

Perry CC, Lu Y (1992) Preparation of silicas from silicon complexes: role of cellulose in polymerisation and aggregation control. J Chem Soc Faraday Trans 88:2915–2921. https://doi.org/10.1039/FT9928802915

Piperno DR (2006) Phytoliths: a comprehensive guide for archaeologists and paleoecologists. Rowman Altamira, Lanham

Piperno DR, Holst I, Wessel-Beaver L, Andres TC (2002) Evidence for the control of phytolith formation in *Cucurbita* fruits by the hard rind (Hr) genetic locus: archaeological and ecological implications. Proc Natl Acad Sci U S A 99:10923–10928

Pommerrenig B, Diehn TA, Bienert GP (2015) Metalloido-porins: essentiality of nodulin 26-like intrinsic proteins in metalloid transport. Plant Sci 238:212–227. https://doi.org/10.1016/j.plantsci.2015.06.002

Prabagar S, Hodson MJ, Evans (2011) Silicon amelioration of aluminium toxicity and cell death in suspension cultures of Norway spruce (*Picea abies* (L.) Karst.). Environ Exp Bot 70:266–276. https://doi.org/10.1016/j.envexpbot.2010.10.001

Prychid CJ, Rudall PJ, Gregory M (2003) Systematics and biology of silica bodies in monocotyledons. Bot Rev 69:377–440. https://www.jstor.org/stable/4354467

Reynolds CR, Islam SA, Sternberg MJE (2018) EzMol: a web server wizard for the rapid visualisation and image production of protein and nucleic acid structures. J Mol Bot 430:2244–2248. https://doi.org/10.1016/j.jmb.2018.01.013

Saitoh Y, Mitani-Ueno N, Saito K, Matsuki K, Huang S, Yang L, Yamaji N, Ishikita H, Shen JR, Ma JF, Suga M (2021) Structural basis for high selectivity of a rice silicon channel Lsi1. Nat Commun 12:6236. https://doi.org/10.1038/s41467-021-26535-x

Sakurai G, Satake A, Yamaji N, Mitani-Ueno N, Yokozawa M, Feugier FG, Ma JF (2015) *In silico* simulation modeling reveals the importance the of the Casparian strip for efficient silicon uptake in rice roots. Plant Cell Physiol 56:631–639. https://doi.org/10.1093/pcp/pcv017

Sangster AG (1977a) Electron-probe microassay studies of silicon deposits in the roots of two species of Andropogon. C J Bot 55:880–887

Sangster AG (1977b) Characteristics of silica deposition in *Digitaria sanguinalis* (L.) Scop. (Crabgrass). Ann Bot 41:341–350. https://doi.org/10.1093/oxfordjournals.aob.a085296

Sangster AG (1978) Silicon in the roots of higher plants. Am J Bot 65:929–935. https://doi.org/10.2307/2442679

Sangster AG, Hodson MJ (1992) Silica deposition in subterranean organs. In: Rapp G, Mulholland SC (eds) Phytolith systematics, advances in archaeological and museum science. Springer, pp 239–251

Sangster AG, Parry DW (1976) Endodermal silicification in mature, nodal roots of *Sorghum bicolor* (L.) Moench. Ann Bot 40:373–379. https://www.jstor.org/stable/42756380

Sant'Anna-Santos BF, Junior WC, Amaral VB (2015) *Butia capitata* (Mart.) Becc. lamina anatomy as a tool for taxonomic distinction from *B. odorata* (Barb.Rodr.) Noblick comb. nov. (Arecaceae). An Acad Bras Cienc 87:71–81. https://doi.org/10.1590/0001-3765201520130457

Schmitt U, Weiner G, Liese W (1995) The fine structure of the stegmata in *Calamus axillaris* during maturation. IAWA J 16:61–68. https://doi.org/10.1163/22941932-90001390

Shakoor SA, Bhat MA, Mir SH (2014) Phytoliths in plants: a review. J Bot Sci 3:10–24

Shi X, Zhang C, Wang H, Zhang F (2005) Effect of Si on the distribution of Cd in rice seedlings. Plant Soil 272:53–60. https://doi.org/10.1007/s11104-004-3920-2

Sørensen I, Pettolino FA, Wilson SM, Doblin MS, Johansen B, Bacic A, Willats WGT (2008) Mixed-linkage (1→3), (1→4)-β-d-glucan is not unique to the Poales and is an abundant component of *Equisetum arvense* cell walls. Plant 54:510–521. https://doi.org/10.1111/j.1365-313X.2008.03453.x

Soukup M, Martinka M, Cigáň M, Ravaszová F, Lux A (2014) New method for visualization of silica phytoliths in Sorghumbicolor roots by fluorescence microscopy revealed silicate concentration-dependent phytolith formation. Planta 240(6):1365–1372. https://doi.org/10.1007/s00425-014-2179-y

Soukup M, Martinka M, Bosnić D, Čaplovičová M, Elbaum R, Lux A (2017) Formation of silica aggregates in sorghum root endodermis is predetermined by cell wall architecture and development. Ann Bot 120:739–753. https://doi.org/10.1093/aob/mcx060
Soukup M, Zancajo VMR, Kneipp J, Elbaum R (2020) Formation of root silica aggregates in sorghum is an active process of the endodermis. J Exp Bot 71:6807–6817. https://doi.org/10.1093/jxb/erz387
Stern WL, Carlsward BS (2008) Vegetative anatomy of Calypsoeae (Orchidaceae). Lankesteriana 8:105–112
Stromberg CAE (2002) The origin and spread of grass-dominated ecosystems in the late Tertiary of North America: preliminary results concerning the evolution of hypsodonty. Palaeogeogr Palaeoclimatol Palaeoecol 77:5975
Stromberg CAE, Di Stilio VS, Song Z (2016) Functions of phytoliths in vascular plants: an evolutionary perspective. Funct Ecol 30:1286–1297. https://doi.org/10.1111/1365-2435.12692
Sun H, Guo J, Duan Y, Zhang T, Huo H, Gong H (2017) Isolation and functional characterization of CsLsi1, a silicon transporter gene in *Cucumis sativus*. Physiol Plant 159:201–214. https://doi.org/10.1111/ppl.12515
Sun H, Duan Y, Qi X, Zhang L, Huo H, Gong H (2018) Isolation and functional characterization of *CsLsi2*, a cucumber silicon efflux transporter gene. Ann Bot 122:641–648. https://doi.org/10.1093/aob/mcy103
Sun H, Duan Y, Mitani-Ueno N, Che J, Jia J, Liu J, Guo J, Ma JF, Gong H (2020) Tomato roots have a functional silicon influx transporter but not a functional silicon efflux transporter. Plant Cell Environ 43:732–744. https://doi.org/10.1111/pce.13679
Tomlinson PB (1959) An anatomical approach to the classification of the Musaceae. J Linn Soc Bot 55:779–809
Tomlinson PB (1961) Anatomy of the Monocotyledons, II. Palmae. Oxford at the Clarendon Press, London, UK, p 477
Tran TTT, Nguyen TT, Nguyen VT, Huynh HTH, Nguyen TTH, Nguyen MN (2019) Copper encapsulated in grass-derived phytoliths: characterization, dissolution properties and the relation of content to soil properties. J Environ Manag 249:109423. https://doi.org/10.1016/j.jenvman.2019.109423
Turpault MP, Calvaruso C, Kirchen G, Redon PO, Cochet C (2018) Contribution of fine tree roots to the silicon cycle in a temperate forest ecosystem developed on three soil types. Biogeosciences 15:2231–2249. https://doi.org/10.5194/bg-15-2231-2018
Uhl NW, Dransfield J (1987) Genera Palmarum – a classification of palms based on the work of Harold E. Moore Jr. Lawrence. Hortorium and the International Palm Society, p 610
van den Berg B, Pedebos C, Bolla JR, Robinson CV, Basle A, Khalid S (2021) Structural basis for sicic acid uptake by higher plants. J Mol Biol 433:167226. https://doi.org/10.1016/j.jmb.2021.167226
Vaculík M, Lux A, Luxová M, Tanimoto E, Lichtscheidl I (2009) Silicon mitigates cadmium inhibitory effects in young maize plants. Environ Exp Bot 67:52–58. https://doi.org/10.1016/j.envexpbot.2009.06.012
Vaculik M, Lukacova Z, Bokor B, Martinka M, Tripathi DK, Lux A (2020) Alleviation mechanisms of metal(loid) stress in plants by silicon: a review. J Exp Bot 71:6744–6757. https://doi.org/10.1093/jxb/eraa288
Vivancos J, Deshmukh R, Gregoire C, Remus-Borel W, Belzile F, Belanger RR (2016) Identification and characterization of silicon efflux transporters in horsetail (*Equisetum arvense*). J Plant Physiol 200:82–89. https://doi.org/10.1016/j.jplph.2016.06.011
Vulavala VK, Elbaum R, Yermiyahu U, Fogelman E, Kumar A, Ginzberg I (2016) Silicon fertilization of potato: expression of putative transporters and tuber skin quality. Planta 243:217–229. https://doi.org/10.1007/s00425-015-2401-6
Wallace IS, Roberts DM (2005) Distinct transport selectivity of two structural subclasses of the nodulin-like intrinsic protein family of plant aquaglyceroporin channels. Biochemist 44:16826–16834. https://doi.org/10.1021/bi0511888

Waterhouse A, Bertoni M, Bienert S, Studer G, Tauriello G, Gumienny R, Heer FT, de Beer TAP, Rempfer C, Bordoli L, Lepore R, Schwede T (2018) SWISS-MODEL: homology modelling of protein structures and complexes. Nucleic Acids Res 46:W296–W303

Watteau F, Villemin G (2001) Ultrastructural study of the biogeochemical cycle of silicon in the soil and litter of a temperate forest. E J Soil Sci 52:385–396. https://doi.org/10.1046/j.1365-2389.2001.00391.x

Ye J, Yan C, Liu J, Haoliang L, Liu T, Song Z (2012) Effects of silicon on the distribution of cadmium compartmentation in root tips of *Kandelia obovata* (S., L.) Yong. Environ Pollut 162:369–373. https://doi.org/10.1016/j.envpol.2011.12.002

Yokoyama R, Rose JKC, Nishitani K (2004) A surprising diversity and abundance of xyloglucan endotransglucosylase/hydrolases in rice. Classification and expression analysis. Plant Physiol 134:1088–1099. https://doi.org/10.1104/pp.103.035261

Yoshida S, Ohnishi Y, Kitagishi K (1962) Chemical forms, mobility and deposition of silicon in rice pant. Soil Sci Plant Nutr 8:15–21. https://doi.org/10.1080/00380768.1962.10430992

Zexer N, Elbaum R (2020) Unique lignin modifications pattern the nucleation of silica in sorghum endodermis. J Exp Bot 71:6818–6829. https://doi.org/10.1093/jxb/eraa127

Zexer N, Elbaum R (2022) Hydrogen peroxide modulates silica deposits in sorghum roots. J Exp Bot 73:1450–1463. https://doi.org/10.1093/jxb/erab497

Zhang C, Wang L, Zhang W, Zhang F (2013) Do lignification and silicification of the cell wall precede silicon deposition in the silica cell of the rice (*Oryza sativa* L.) leaf epidermis? Plant Soil 372:137–149. https://doi.org/10.1007/s11104-013-1723-z

Zhang J, Zou W, Li Y, Feng Y, Zhang H, Wu Z, Tu Y, Wang Y, Cai X, Peng L (2015) Silica distinctively affects cell wall features and lignocellulosic saccharification with large enhancement on biomass production in rice. Plant Sci 239:84–91. https://doi.org/10.1016/j.plantsci.2015.07.014

Zhou J, Tang S, Pan W, Pan W, Xiao H, Sun QMY, Xu M, Liu M, Wu L (2022) Silicon isotope fractionation dynamics during uptake and translocation by various crop species under three soil types. Plant Soil 477:41–55. https://doi.org/10.1007/s11104-021-05264-6

Chapter 4
Dynamics of Silicon in Soil and Plant to Establish Silicate Fertilization

Brenda S. Tubana

4.1 Introduction

The abundance of silicon in the earth's crust is one the bases to dispute the need to fertilize crops with silicon. However, silicon fertilization has long before an accepted agronomic practice in regions of the world with soils that are inherently low in plant-available silicon and where high silicon accumulator crops are cultivated. High-silicon accumulators are plant species which take up silicon at levels that greatly exceed other plant-essential nutrients. Sugarcane (*Saccharum* spp.) and rice (*Oryza sativa*) silicon removal rates are estimated at 300 and 500 kg ha^{-1}, respectively (Meyer and Keeping 2001; Makabe et al. 2009). For rice, it is estimated that at this rate, the reserve of plant-available silicon in an average soil would be exhausted after five years of continuous cultivation (Desplanques et al. 2006). While silicon research in agriculture progresses, there are areas that have been understudied which particularly are contributing to slow adoption of silicon fertilization in crop production, the most relevant to this chapter are (a) the absence of universally accepted soil test for estimating plant-available silicon for agronomic crops and (b) the lack of fertilization guidelines (Tubana et al. 2016; Zellner et al. 2021). The need to address these gaps in soil silicon research became more apparent after silicon was designated as a plant beneficial substance by the Association of American Plant Food Control Officials and upon the establishment of the official method for quantifying soluble silicon in dry silicon fertilizers (Sebastian 2012; Sebastian et al. 2013). These also prompted the marketing of silicon as fertilizer in the agricultural industry. Organizations which promote responsible management of plant nutrition like the International Plant Nutrition Institute also began highlighting the

B. S. Tubana (✉)
Louisiana State University, Agricultural Center, School of Plant, Environmental, and Soil Sciences, Baton Rouge, LA, USA
e-mail: btubana@agcenter.lsu.edu

R. de Mello Prado (ed.), *Benefits of Silicon in the Nutrition of Plants*,
https://doi.org/10.1007/978-3-031-26673-7_4

importance of silicon in plant's resilience to environmental stresses (http://www.ipni.net/nutrifacts-northamerican). This chapter elucidates the dynamics of silicon in soil and plant and how they influence two vital information for estimating silicon fertilizer application rates, i.e., soil silicon content and plant silicon requirement. In addition to this, a review and summary of published silicon fertilizer guidelines and recommendations, both as material (e.g., calcium silicate) and elemental (silicon), are provided.

4.2 Silicon in Soils

Silicon is ubiquitous in most soils existing primarily as silica and silicate bearing minerals such as basalt (23% Si) and orthoquartzite (46.5% Si) (Monger and Kelly 2002). Some silica and silicate bearing minerals are uniquely associated with certain soils. For instance, significant amounts of silicon are found in most arid soils from silcretes contribution (Singh et al. 1992). These are surface or near-surface soil derivatives or indurated mass cemented by various silica polymorphs with silicon content as high as 46% (Milnes and Twidale 1983; Singh et al. 1992; Nash and Ullyot 2007). The carbonates-indurated mass or petrocalcic horizon is also common in arid soils but contain lesser amount of silicon than silcretes. There are soils that are inherently low in silicon content, particularly the plant-available form (monosilicic acid, H_4SiO_4). The minerals mostly present in tropical or highly weathered soils such as bauxites and ferricretes contain low amount of silicon (Monger and Kelly 2002). These soils which include Oxisols and Ultisols are low in soluble (plant-available) silicon and generally highly leached, acidic, and with low base saturation (Foy 1992). The organic soils or Histosols contain very high level of organic matter and very low amount of the mineral component and is among the group of soils that carry low levels of plant-available silicon (Snyder et al. 1986). Crops that are cultivated on these low-silicon containing soils can significantly benefit from silicon fertilization. Additionally, crops' response to silicon addition is likely in soils that composed largely of quartz sand and have been under long-term crop production (Datnoff et al. 1997; Meunier 2003; Meunier et al. 2008). Soils with low plant-available silicon are found in many crop-producing areas of Africa, Asia, the Americas, and Europe (Zellner et al. 2021).

4.3 Components of Silicon Cycle in Soil

Silicon in soils is grouped into liquid, adsorbed, and solid phase with the solid being the largest fraction (Matichencov and Bocharnikova 2001; Sauer et al. 2006). Unlike the liquid and adsorbed phase, silicon in solid phase has little contribution for quick replenishment of plant-available silicon in soil solution. The silicon in liquid phase or also known as the silicon in soil solution exist primarily as monomeric (H_4SiO_4), oligomeric, and polymeric acid (Iler 1979). Oligomeric and polymeric forms chains, branch, or sphere of several units of H_4SiO_4. The degree of polymerization of polymeric acid is higher than oligomeric acid which holds only up to ten silicon atoms. Thus, the molecular weight of polymeric is higher than that of the oligomeric. Some of the dissolved silicic acid can form complexes with organic and inorganic compounds or adsorbed onto clay particles and hydroxides of iron and aluminum rendering them temporarily unavailable for plant uptake (Hansen et al. 1994; Dietzel 2002). The monomeric silicic acid, whether in solution or adsorbed onto solid phase, is the most important form for plant absorption and nutrition. In fact, the solutions that have been developed for extracting and estimating plant-available silicon in soil were formulated to specifically remove the monomeric form of silicic acid from the soil.

Monosilicic acid is slowly released to the soil solution from silicon dioxide, silicate minerals, and plant residues through microorganism-assisted and chemical-weathering processes. Once in the soil solution, the H_4SiO_4 concentration changes with soil pH, temperature, water and organic matter contents, and redox potential (Savant et al. 1997). These factors along with soil physical properties (e.g., texture, structure) can directly affect the H_4SiO_4 concentration and its release rate from silicon-containing minerals, clay particles, and iron and aluminum hydroxides. Generally, the concentration of H_4SiO_4 peaks during spring and summer when temperature favors biological activity as those observed in grassland ecosystems (Volkova 1980). The changes in H_4SiO_4 concentrations in agricultural soils can be more dynamic due to routine field practices employed during the cropping season. Liming and application of acid-producing fertilizers can abruptly impact soil pH and hence H_4SiO_4 concentration (Panov et al. 1982; Allmaras et al. 1991).

The specific dynamics of dissolved silicon (silicic acid) in soil solution along with the adsorbed and solid phase are illustrated in Fig. 4.1. Silicon, just like any nutrients, is acquired mainly through root absorption from the soil solution in its uncharged monomeric silicic form. In most soils, this monomeric silicic form remains uncharged at pH values <8 but can dissociate into $H^+ + H_3SiO_4^-$ at pH values above 9 and further into $2H^+ + H_2SiO_4^{2-}$ at pH values above 11 (McKeague and Cline 1963; Dietzel 2000). The concentration of silicon in H_4SiO_4 form in the soil solution ranges from 0.09 to 23.4 mg L^{-1} but can be higher in high pH soils and those that contain sodium carbonate (Volkova 1980; Kovda 1985). High concentrations of H_4SiO_4 in soil solution triggers polymerization, a process resulting in the formation of stable chains of several units of H_4SiO_4 (Knight and Kinrade 2001). Under most natural conditions, the stability of polymerized silicic acid is short-term and can last for hours or days before they breakdown into H_4SiO_4 unit (Dietzel 2000). However, even in its liquid phase, polysilicic acid is not plant-available; thus, polymerization is a process that does not favor silicon uptake by plant.

In the realm of nutrient management, the concentration of silicon in soil solution is the most relevant information. This is the form of silicon that estimates the amount available for plant consumption and can therefore be used to derive fertilizer recommendation. Silicon in soil solution (Fig. 4.1) is regulated by several processes mainly grouped into four (Tubana and Heckman 2015).

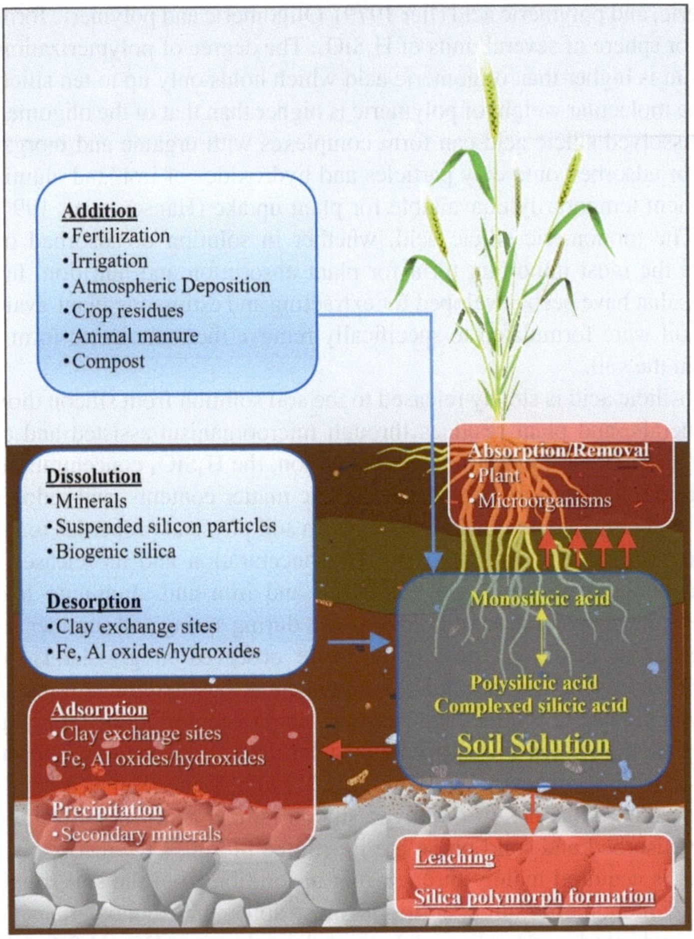

Fig. 4.1 Soil components and processes that affect the concentration of silicic acid in soil solution. Blue arrows pertain to those which increase silicic acid in soil solution whereas red arrows represent those that decrease silicic acid in soil solution. Double-headed yellow arrow indicates transformation of silicic acid

The first group include transformations or processes that raise silicon concentration in soil solution. This includes dissolution of primary and secondary silicon-containing minerals, desorption of silicon from iron, aluminum oxides/hydroxides and clay exchange sites, release from biogenic silica and suspended silicon particles, and addition through atmospheric dust, irrigation, and fertilization. Irrigation waters may contain silicon in ionic, molecular, and aggregate forms. Silicon contribution from wind-blown dust and ash containing-phytolith particles are very low compared to fertilization and plant litter/animal manure addition (Kurtz et al. 1987; Street-Perrott and Barker 2008; Opfergelt et al. 2010). The second group consists of processes that reduce silicon concentration in soil solutions, which include precipitation of silicon into secondary minerals, adsorption onto solid phases, and plant and microorganism removal. The third group includes leaching of silicic acid and conversion of H_4SiO_4 to stable, plant-unavailable silica polymorphs, making these classified as loss pathways of silicon from the soil system. The last (fourth) group includes the transformation processes for building up silica pools (secondary minerals, biogenic silica, and suspended silicon particles) that contribute to the concentration of dissolved silicon into the soil solution. In agricultural soils, the removal of silicon by plants and addition of silicon through fertilization are actively regulating the concentration of silicon in soil solution. Also, the reduction of silicon in the soil solution is mainly triggered by plant absorption causing the liquid and solid phases of silicon to be outside the equilibrium state. For soils with high buffering capacity, silicon in the soil solution can be easily replenished through mineral dissolution, depolymerization of polysilicic acid, and release of adsorbed silicic acid. However, the removal rate often can exceed the replenishment rate especially in highly weathered, organic, and intensively cropped soils.

Silicon can compete against other plant nutrients for adsorption sites in the soil. The result can be beneficial to plants such as those that results in precipitation of heavy metal like cadmium subsequently preventing toxicity in plants (Liang et al. 2005). A similar interaction is involved for nutrients that is taken by plant in trace amounts such as aluminum and iron. The adsorption of H_4SiO_4 on aluminum hydroxides transforms aluminum into a less soluble compound, which diminishes the activity of aluminum in the soil solution (Panov et al. 1982; Baylis et al. 1994). Aluminum ions are strongly adsorbed onto silica surfaces as well (Schulthess and Tokunaga 1996).

The reactivity of silicon also presents negative outcomes such as reduction in availability of several plant-essential nutrients and, in extreme cases, can induce deficiencies in plants. In addition, the precipitations of heavy metals by silicon reduces the concentration of H_4SiO_4 for plant uptake (Jones and Handreck 1967; Lindsay 1979; Snyder et al. 2007). The reactivity of silicon is attributed to the high adsorptive capacity of its forms that are naturally occurring in the soil and those that added to the soil. Silica surface is negatively charged and can thus attract positively charged nutrients. These adsorbed nutrients remained available for plant uptake and can reduce nutrient loss from the soil surface horizon via leaching (Tokunaga 1991). In fact, this has become the basis for controlled release fertilizer technology (Volker et al. 1985; Komisarov and Panfilova 1987).

The reaction of silicon and phosphorus is antagonistic in nature. Brown and Mahler (1987) attributed this to the strong competition between H_4SiO_4 and phosphate ions for specific sorption sites in the soil. Others reported that silicon fertilization increases phosphate availability (Gladkova 1982; Singh and Sarkar 1992; O'Reilly and Sims 1995; Matichenkov and Ammosova 1996). This happens because when silicon fertilization increases the H_4SiO_4 concentration in the soil solution, the adsorption of H_4SiO_4 onto the slight soluble aluminum, calcium, ferric, and magnesium phosphates starts taking place which subsequently releases phosphate ions into the soil solution (Matichenkov and Ammosova 1996; Lindsay 1979). Jones and Handreck (1967) also indicated that H_4SiO_4 lowers the activity of aluminum ion in soil solution hence its (aluminum) potential to precipitate phosphate ions. Overall, the notion established from these studies was that H_4SiO_4 having a lower negative surface charge than phosphate ions is more preferentially adsorbed on surfaces over phosphate ions. On the other hand, silicon fertilizer (depending on the source) itself can adsorb phosphate ions consequently reducing phosphorus supply for plant uptake. Matichencov and Bocharnikova (2001) specifically pointed out that steel slag has higher capacity to adsorb phosphate ions compared with amorphous silica and calcium silicate suggesting that the source of silicon fertilizer is a vital consideration for making fertilizer decision.

Most silicon fertilizer sources are agricultural- and industrial-based byproducts that have high liming potential. This property of silicon fertilizers indirectly affects the activity of the elements and nutrients in the soil, and the outcomes are mostly beneficial such as prevention of excessive uptake of heavy metals and contaminants such as arsenic. Several studies demonstrated that the application of fly ash and steel slag increased soil pH that subsequently reduced the availability of cadmium, copper, lead, and zinc in the soil and uptake by rice (Okuda and Takahashi 1962a; Chen et al. 2000; Gu et al. 2011). Similar action on aluminum in rice was reported by Lindsay (1979) and later by Liang et al. (2005) on cadmium. Reduction in iron intake by rice was also attributed to reduced iron solubility due to increase in soil pH (Wallace 1993) whereas the reduction in ferrous ion (a more preferred form of iron by plants) content in flooded soil was a specific action by silicon (Ma and Takahashi 2002). For nutrients that are preferably taken up by plants in their reduced forms such as iron and manganese, the silicon-induced oxidizing capacity of roots can prevent excessive accumulation and plant uptake of these nutrients in flooded soil (Ma and Takahashi 2002; Okuda and Takahashi 1962b).

4.4 Silicon in Plants

Plant takes up silicon as H_4SiO_4 primarily through root absorption. Earlier works strongly supported this mechanism, specifically in rice, following the identification of silicon transporters which facilitate the active transport of H_4SiO_4 across cell membranes (Mitani and Ma 2005; Ma et al. 2006, 2007). The high density of transporters, Lsi1 and Lsi2, in roots of high silicon accumulator plants like rice further support the notion that the major pathway of silicon acquisition by plants is via root absorption (Mitani and Ma 2005). The H_4SiO_4 can be also absorbed by roots through passive uptake and is likely driven by mass flow, a nutrient transport mechanism governed by the uptake of water by plant root and transpiration. The established knowledge on plant acquisition of silicon via root absorption stipulated that soil silicon fertilization is the most effective, if not the only way, to supply silicon to plants. Despite this, many studies were conducted to develop an alternative approach to soil silicon fertilization such as the foliar application of silicon-containing solution, mainly to address issues on handling of high application rates and accessibility to bulk sources of dry, high liming potential silicon fertilizers. Recent studies demonstrated the feasibility of supplying foliar application of silicon using the right sources and combining this method with nanotechnology (Laane 2018; Dos Santos Sarah et al. 2021; Verma et al. 2021; Hajihashemi and Kazemi 2022) while several studies reported otherwise and attributed positive responses of plants to foliar silicon due to reasons (e.g., pH change in apoplast that drives nutrient transport across cell membrane) other than enhanced silicon nutrition (Tebow et al. 2021). In addition, the lack of silicon transport mechanisms in plant organ does not support the idea of foliar absorption of silicon (Rodrigues et al. 2015).

The interaction of silicon is limited to several elements in the soil whereas it involved more nutrients in the plants through transcriptional regulation, acquisition, uptake, and translocation (Pavlovic et al. 2021). These are specifically presented in several chapters of this book. Overall, a vast amount of literature elucidates many silicon-mediated mechanisms in plants and how they contribute to plants' overall resilience to different forms of stresses. These mechanisms are categorized into two specific actions: one that aims to reinforce plant's protective layer and mechanical structure and the other to activate or prime plant's defense mechanism (Tubana et al. 2016).

One vital information for estimating fertilizer application rate is the plant's nutrient requirement or demand. Takahashi et al. (1990) categorized plants into high-, intermediate-, and low-silicon accumulators. The high accumulators contain 10 to 100 g Si kg^{-1} plant dry mass, and most are monocotyledons, such as barley, rice, sugarcane, and wheat (Liang et al. 2007; Ma et al. 2001; Ma and Takahashi 2002). With efficient uptake systems, high accumulators can take up silicon several times higher than essential nutrients like nitrogen and potassium. The intermediate accumulators have 5 to 10 g Si kg^{-1} dry weight and mainly monocotyledons plants whereas the low accumulators consist of dicot plants with silicon content less than 5 g kg^{-1}. These categories were based solely on shoot silicon content. However, some low silicon accumulator plants take up and store more silicon in their roots than in their shoot (Lewin and Reimann 1969; Carré-Missio et al. 2009; French-Monar et al. 2010; Huang et al. 2011) and with roots being the site of several silicon-induced mechanisms for alleviating stresses in plants, perhaps the plants' silicon requirement should be based on total (root + shoot) silicon uptake.

The root-absorbed H_4SiO_4 is transported through the xylem and deposited in leaf epidermal surfaces preferentially in abaxial side (Hodson and Sangster 1988). Following dehydration, the deposited H_4SiO_4 forms a hard, polymerized silica gel also known as phytolith (Yoshida et al. 1962; Jones and Handreck 1965, 1967; Raven 1983). The phytolith can also be found in specific cells in the vascular bundles. This is collectively called silica cell and exists as bulliform cells in rice, fusoid cells in wheat, and prickle hairs in bamboo (*Bambusa vulgaris*) (Dietrich et al. 2003; Motomura et al. 2004; Ma and Yamaji 2006). The deposited H_4SiO_4 is immobile thus cannot be translocated to supply silicon to new growing tissues (Elawad and Green 1979; Ma et al. 1989; Epstein 1999). Considering the immobile nature of silicon in plants, transpirational pull is the main driving force for silicon absorption, transport, and deposition in plants, especially those that rely heavily on the passive uptake mechanism. In such scenario, leaf age or duration of plant growth can significantly impact silicon concentration in plants and (De Saussure 1804; Henriet et al. 2006) therefore has implications on leaf silicon tissue testing and interpretation.

4.5 Bases of Silicon Fertilization

The two parameters that are required to estimate fertilizer application rates are the available silicon in the soil and the plant silicon requirement or demand. Soil silicon testing remains a major limitation for full adoption of silicon fertilization to manage plant health in crop production (Tubana et al. 2016; Zellner et al. 2021). Specifically, this is mainly about the lack of a standard procedure for extracting silicon across soil types. This is further complicated by the fact that many procedures that were established in the past 50 years are for extraction of different forms of silicon including silicon in amorphous silica and allophane in soils and sediments (Hashimoto and Jackson 1960; Beckwith and Reeve 1963, 1964; Schachtschabe and Heinemann 1967; Sauer et al. 2006). For computation of fertilizer recommendation, it makes sense to estimate the amount of silicon (in the soil) in forms that are soluble or readily available to plant and those in exchangeable form (Datnoff et al. 2001). Essentially, this is the monomeric silicic form in the soil solution and adsorbed on exchange sites. There is a long list of extraction procedures that were evaluated for capturing the fraction of silicon in the soil that is plant-available (Tubana and Heckman 2015). Numerous studies have been focused on evaluating the efficacy of 0.5 M acetic acid, sodium acetate-acetic acid, and 0.01 M calcium chloride for extracting plant-available silicon accompanied with outcomes such as the establishment of critical soil silicon levels. While the development of a standard extraction procedure for silicon remains a long-term standing goal in silicon research in agriculture, others continue the process of selecting extraction procedure that is more site-specific and suitable for the soils in a given region while some are working on simplifying the method and reducing the turn-around time of analysis (Crusciol et al. 2018). Thus, the research direction for gauging plant-available silicon in the soil is regarded as not unified.

Unlike in soil, the development of procedure for quantifying silicon in plant tissue had minor set-backs which were more focused on improvement and refinement to address issues on precision, time-consuming procedures, safety, and requirement for sophisticated instruments (Fox et al. 1969; Elliot and Snyder 1991; Ostatek-Boczynski and Haysom 2003). Silicon in plant tissue can be extracted using several methods. The choices include gravimetric, hydrofluoric acid solubilization, microwave-assisted digestion with nitric and hydrofluoric acids, and the most recent and widely used autoclave- or oven-induced digestion with a strong sodium hydroxide solution (Horwitz 1960; Yoshida et al. 1976; Novozamsky et al. 1984; Elliot and Snyder 1991; Feng et al. 1999; Kraska and Breitenbeck 2010). The autoclave-induced digestion method was developed by Elliot and Snyder (1991) and modified later by Kraska and Breitenbeck (2010) to address the vigorous foaming when hydrogen peroxide and sodium hydroxide are mixed. Excessive foaming can take out sample particles with it out of the tubes or foam with sample particles dries out on the upper wall of the tube preventing complete digestion, either case results in underestimation of silicon content in the plant samples. Kraska and Breitenbeck (2010) addressed this issue by pre-treating the ground plant tissue samples with five drops of octyl-alcohol prior to hydrogen peroxide addition. The modified procedure also includes the addition of 1 mL of 5 mM ammonium fluoride in the digestion process to complete the ionization of the polysilicic acid, which stabilizes the absorbance reading of the solution for up to six hours during the colorimetric procedure.

Silicon in soil extracts and plant digests can be quantified based on molybdenum blue colorimetry (MBC) and inductively coupled plasma-optical emission spectrometry (ICP-OES) analysis (Hallmark et al. 1982; Wang et al. 2004). In soil extracts, the most relevant form of silicon to quantify is the H_4SiO_4 acid. For this reason, the MBC gives a better estimate of plant available silicon in soil extracts as opposed to ICP-OES since H_4SiO_4 acid is the only form of silicic acid that is reactive to molybdate to form silicon-molybdate blue complex. The presence of polysilicic acid essentially has no effect on the formation of this blue complex and thus will not lead to an overestimation of plant-available silicon in the soil. The ICP-OES measures all the forms of silicon in soil extracts including those that are not plant-available. On the other hand, the total silicon content is quantified in plant digest, thus both MBC and ICP-OES are viable techniques.

Critical silicon levels are required for interpretation of soil and plant test results. Tubana and Heckman (2015) summarized several established critical silicon levels in soil and plant. For soil, the critical levels varied significantly with soil type, crops, and soil extraction procedures. Calibration and validation studies for some of these soil testing procedures with established critical silicon levels were conducted (Korndörfer et al. 2001; Xu et al. 2001; De Lima et al. 2003). The results came back suggesting that soil silicon interpretation for a given soil testing procedure was valid for certain soil types and crops. Recent studies continue to show the site- and crop-specific nature of interpreting soil silicon level for a given soil testing procedure (Babu et al. 2016; Crusciol et al. 2018; Paye et al. 2018).

The interpretation of plant silicon analysis is crop specific and requires to strictly adhere to sampling protocol employed during the establishment of critical silicon concentration in plant. Plant silicon content varied with plant parts, growth stage, and leaf age thus, the process of establishing plant silicon testing also includes the standardization of sampling protocol. Due to this, the progress of plant tissue testing to derive fertilizer recommendation is slow. There are only few published critical silicon levels in plant tissue, and they are primarily for rice and sugarcane (Nair and Aiyer 1968; Takijima et al. 1970; Lian 1976; De Datta 1981; Snyder et al. 1986; Dobermann and Fairhurst 2000; Korndörfer et al. 2001; Narayanaswamy and Prakash 2009; Paye et al. 2018). For this reason, it is not common to use plant silicon testing and interpretation to estimate fertilizer application rates. In addition, silicon is immobile in the soil and should plant silicon testing indicates the crop in question is deemed deficient of silicon, in-season correction is not likely going to be effective. However, plant silicon content can be used to estimate silicon removal rate. Under the Crop Nutrient Requirement approach or also known as "feed-the-crop" approach, fertilizer rates will be applied at rates equivalent (plus a little extra) to the anticipated amount of nutrient (silicon) that will be removed from the soil (Hochmuth and Hanlon 2010).

4.6 Silicon Application Rates

Once a soil or plant test interpretation indicates deficiency and therefore needs fertilizer the next step is to conduct calibration studies to determine the appropriate application rate to change the concentration of H_4SiO_4 in the soil solution that garners a significant yield response (increase). Calibration studies are essentially lacking, and the few established ones are mainly for rice and sugarcane. Sugarcane in Florida organic and mineral soils has an existing calibration of silicon soil test values and application rate using calcium silicate as source recommending up to 6 Mg ha^{-1} calcium silicate for soil testing below 5 g m^{-3} (McCray et al. 2018; McCray 2022). For rice, the recommended application rates are 1120 and 1500 kg silicon ha^{-1} in Florida organic soils testing medium and low levels of soil silicon, respectively, based on a 0.5 M acetic acid extraction procedure (Korndörfer et al. 2001).

With limited availability of complete silicon fertilization guidelines, i.e., testing procedures with established critical levels and interpretation, prediction for the need for silicon fertilizer and the right application rate remains a challenging aspect in crop production. Others came with alternative approaches to make silicon fertilizer recommendation. For example, soil pH, percent base saturation, and need for liming are used to derive silicon application rates using sources such as calcium silicate slag and steel slag (Heckman 2017). In such approach, it is common to find recommendation rates of the actual material for correcting soil pH rather than to suffice the silicon requirement of a crop. After all, over-application of silicon to soil is generally a lesser of a concern than over-liming. Raising soil pH outside the optimum range has an immediate impact on solubility of plant-essential nutrients, especially the trace mineral nutrients. Heckman (2017) reported that calcium silicate slag application rates range from 2 to 12 Mg ha^{-1} depending on initial soil pH and the target pH range suitable for the crop in question. De Souza et al. (2021) reported the application of calcium and magnesium silicate at rates ranging from 2 to 8 Mg ha^{-1} resulted in increased rice grain yield. Other studies reported the elemental silicon application rates that led to significant yield improvement. Grain yield was significantly increased following the application of 9 Mg ha^{-1} (~1000 kg silicon ha^{-1}) of steel slag in wheat receiving rather an excessive rate of nitrogen fertilizer (White et al. 2017). Similarly, the application of 600 kg silicon ha^{-1} and 150 kg nitrogen ha^{-1} increased rice grain yield (Jafari et al. 2013). Others reported the actual silicon rates resulting in optimal yield of rice such as 120 kg ha^{-1} (Singh et al. 2006), 200 kg ha^{-1} (Swe et al. 2021), and 300 kg ha^{-1} (Chaiwong et al. 2022) with a higher application rate at 330 kg ha^{-1} (700 kg SiO_2 ha^{-1}) for (rice) production in

Oxisols (Sinegar et al. 2021). A wider application range, i.e., 300 to 800 kg ha^{-1}, of elemental silicon was reported from studies with focus on disease suppression (Datnoff et al. 1991; Korndörfer et al. 2001; Rodrigues et al. 2015; Seebold et al. 2001). Considering all these (wide-ranged) rates are for rice alone, calibration studies is indeed needed to refine the soil silicon test interpretation and recommendation for each crop. Makabe et al. (2009) provided an estimate of rice removal rate at 500 kg silicon ha^{-1} whereas the estimate by Tubana et al. (2016) based on the shoot silicon content was marked at 329 kg ha^{-1}. For other high- and intermediate-accumulators like barley, maize, sugarcane, and wheat, the estimated silicon removal rates at 99, 129, 160, and 108 kg Si ha^{-1}, respectively, are good basis to make silicon fertilizer recommendations. The suggestion made that root silicon content should be included for estimating plant silicon removal rate may result in higher silicon application rates for these crops.

4.7 Conclusion

Clearly a vast amount of information has been built across the world about the role of silicon in improving plant's resilience to abiotic and biotic stresses and yet silicon fertilization is an accepted agronomic practice in only a few regions of the world. Reviews were made on the research progress on silicon in agriculture to identify the roadblocks for adoption of silicon fertilization to manage plant health. The reviews indicated that silicon research in soil is lacking to build the database required for the (a) development of standardized procedure for gauging plant available silicon in the soil and (b) establishment of critical silicon level together with the corresponding interpretations and recommendations. This chapter revealed that indeed the database is inadequate, not unified, and only focused to certain crops, mainly rice and sugarcane. Thus, a universal testing procedure that will work across soil types remains a long-term research endeavor on silicon in agriculture. In fact, the diversity of soils, crops, and systems around the world are a point of argument such that a site- or region-specific testing procedure would likely be more effective than a standardized one. The grand scheme for any soil and plant tissue testing for silicon is to get good estimates of both the amount plant-usable form of silicon in the soil and plant silicon requirement. Currently, recommendations are decided based on lime requirement, average optimal silicon application rates from many crop-specific response studies, and plant silicon removal rate. As most silicon sources have high liming potential, the application of the material is done for lime and not for silicon wherein the rates vary from 2 to 12 Mg ha^{-1} depending on initial pH and the target pH range. Based on numerous studies, the average optimal silicon application rate established for rice was 300 kg ha^{-1} but with a range that is quite wide (120 to 800 kg ha^{-1}). The recommended application rates based on rice shoot silicon removal rate were between 300 to 500 kg ha^{-1}. For other crops (barley, maize, and wheat), the application rate based on shoot silicon removal rate ranged only from 100 to 160 kg ha^{-1} but was likely higher if root silicon uptake was accounted for.

References

Allmaras RR, Laird DA, Douglas CL et al (1991) Long-term tillage, residue management and nitrogen fertilizer influences on soluble silica in Haploxerol. Am Soc Agron, Madison, p 323

Babu T, Tubana B, Paye W et al (2016) Establishing soil silicon test procedure and critical silicon level for rice in Louisiana soils. Comm Soil Sci Plant Anal 47:1578–1597. https://doi.org/10.1080/00103624.2016.1194996

Baylis AD, Gragopoulou C, Davidson KJ et al (1994) Effects of silicon on the toxicity of aluminum to soybean. Commun Soil Sci Plant Anal 25:537–546. https://doi.org/10.1080/00103629409369061

Beckwith RS, Reeve R (1963) Studies on soluble silica in soils. I. The sorption of silicic acid by soils and minerals. Aust J Soil Res 1:157–168

Beckwith RS, Reeve R (1964) Studies on soluble silica in soils. II. The release of monosilicic acid from soils. Aust J Soil Res 2:33–45

Brown TH, Mahler RL (1987) Effects of phosphorus and acidity on levels of silica extracted from a Palouse silt loam. Soil Sci Soc Am J 51:674–677. https://doi.org/10.2136/sssaj1987.03615995005100030022x

Carré-Missio V, Rodrigues FA, Schurt DA et al (2009) Inefficiency of silicon in leaf rust control on coffee grown in nutrient solution. Trop Plant Pathol 34:416–421. https://doi.org/10.1590/S1982-56762009000600009

Chaiwong N, Pusadee T, Jamjod S et al (2022) Silicon application promotes productivity, silicon accumulation and upregulates silicon transporter gene expression in rice. Plan Theory 11:989. https://doi.org/10.3390/plants11070989

Chen HM, Zheng CR, Tu C, Shen ZG (2000) Chemical methods and phytoremediation of soil contaminated with heavy metals. Chemosphere 41:229–234. https://doi.org/10.1016/S0045-6535(99)00415-4

Crusciol CAC, de Arruda DP, Fernandes AM et al (2018) Methods and extractants to evaluate silicon availability for sugarcane. Sci Rep 8:916. https://doi.org/10.1038/s41598-018-19240-1

Datnoff LE, Deren CW, Snyder GH (1997) Silicon fertilization for disease management of rice in Florida. Crop Prot 16:525–531. https://doi.org/10.1016/S0261-2194(97)00033-1

Datnoff LE, Snyder GH, Korndörfer GH (2001) Silicon in agriculture. Elsevier

Datnoff LE, Snyder GH, Raid RN et al (1991) Effect of calcium silicate on blast and brown spot intensities and yields of rice. Plant Dis 75:729–732. https://doi.org/10.1094/PD-75-0729

De Datta SK (1981) Principles and practices of rice production. Wiley, p 357

De Lima RL, Daroub SH, Rice WR et al (2003) Comparison of three soil test methods for estimating plant-available silicon. Commun Soil Sci Plant Anal 34:2059–2071. https://doi.org/10.1081/CSS-120024048

De Saussure NT (1804) Recherches chimiques sur la végétation. Chez la veuve Nyon Ed, Gauthiers-Villars, Paris, p 112

De Souza ACA, de Filippi MCC, Nascente AS et al (2021) Silicon rates and beneficial microorganism on blast suppression and productivity of upland rice. J Plant Sci Phytopathol 5:020–027. https://doi.org/10.29328/journal.jpsp.1001057

Desplanques V, Cary L, Mouret JC et al (2006) Silicon transfers in a rice field in Camargue (France). J Geochem Explor 88:190–193. https://doi.org/10.1016/j.gexplo.2005.08.036

Dietrich D, Hinke S, Baumann W et al (2003) Silica accumulation in *Triticum aestivum* L. and *Dactylis glomerata* L. Anal Bioanal Chem 376:399–404. https://doi.org/10.1007/s00216-003-1847-8

Dietzel M (2000) Dissolution of silicates and the stability of polysilicic acid. Geochim Cosmochim Acta 64:3275–3281. https://doi.org/10.1016/S0016-7037(00)00426-9

Dietzel M (2002) Interaction of polysilicic and monosilicic acid with mineral surfaces. In: Stober I, Bucher K (eds) Water–rock interaction. Kluwer, pp 207–235

Dobermann A, Fairhurst TH (2000) Rice: nutrient disorders and nutrient management. Potash and Phosphate Inst, Singapore and IRRI

Dos Santos Sarah MM, de Mello PR, de Souza Júnior JP et al (2021) Silicon supplied via foliar application and root to attenuate potassium deficiency in common bean plants. Sci Rep 11:19690. https://doi.org/10.1038/s41598-021-99194-z

Elawad SH, Green VE (1979) Silicon and the rice plant environment: a review of recent research. Riso 28:235–253

Elliot CL, Snyder GH (1991) Autoclave-induced digestion for the colorimetric determination of silicon in rice straw. J Agric Food Chem 39:1118–1119

Epstein E (1999) Silicon. Annu Rev Plant Physiol Plant Mol Biol 50:641–664. https://doi.org/10.1146/annurev.arplant.50.1.641

Feng X, Xu S, Wharmby A et al (1999) Microwave digestion of plant and grain standard reference materials in nitric and hydrofluoric acids for multi-elemental determination by inductively coupled mass spectrometry. J Anal Atos Spectrom 14:939–946

Fox RL, Silva JA, Plucknett DL et al (1969) Soluble and total silicon in sugarcane. Plant Soil 30:81–91

Foy CD (1992) Soil chemical factors limiting plant root growth. Adv Soil Sci 19:97–149

French-Monar R, Rodrigues FA, Korndöfer GH et al (2010) Silicon suppresses *Phytophthora blight* development on bell pepper. J Phytopathol 158:554–560. https://doi.org/10.1111/j.1439-0434.2009.01665.x

Gladkova KF (1982) The role of silicon in phosphate plant nutrition. Agrochemistry 2:133

Gu H-H, Li FP, Guan X et al (2011) Effects of fly ash on heavy metal uptake of rice growing on multi-metal contaminated acidic soil. Adv Mater Res 680:94–99. https://doi.org/10.4028/www.scientific.net/AMR.680.94

Hajihashemi S, Kazemi S (2022) The potential of foliar application of nano-chitosan-encapsulated nano-silicon donor in amelioration the adverse effect of salinity in the wheat plant. BMC Plant Biol 22:148. https://doi.org/10.1186/s12870-022-03531-x

Hallmark CT, Wilding LP, Smeck NE (1982) Methods of soil analysis, Part 2. Chemical and microbiological properties, pp 263–274

Hansen HCB, Raben-Lange B, Raulund-Rasmussen K et al (1994) Monosilicate adsorption by ferrihydrite andgoethite at pH 3–6. Soil Sci 158:40–46

Hashimoto J, Jackson ML (1960) Rapid dissolution of allophane and kaolinite-halloysite after dehydration. In: Proceedings of the 7th national conference on clays and clay minerals, Washington, DC, pp 102–113

Heckman J (2017) Silicon needs of soils and crops. Cooperative Extension Fact Sheet FS1278. New Jersey Agricultural Experiment

Henriet C, Draye X, Oppitz I et al (2006) Effects of distribution and uptake of silicon in banana (Musa spp.) under controlled conditions. Plant Soil 287:359–374. https://doi.org/10.1007/s11104-006-9085-4

Hochmuth G, Hanlon E (2010) Principles of sound fertilizer recommendations. University of Florida, IFAS Extension SL315

Hodson MJ, Sangster AG (1988) Observations on the distribution of mineral elements in the leaf of wheat (*Triticum aestivum* L.) with particular reference to silicon. Ann Bot 62:463–471. https://doi.org/10.1093/oxfordjournals.aob.a087681

Horwitz W (1960) Official methods of analysis (12th 516 edn). Association of Official Analytical 517 Chemists, p 73

Huang CH, Roberts PD, Datnoff LE (2011) Silicon suppresses *Fusarium crown* and root rot of tomato. J Phytopathol 159:546–554. https://doi.org/10.1111/j.1439-0434.2011.01803.x

Iler RK (1979) The chemistry of silica. Wiley, p 621

Jafari H, Dastan S, Nasin AR et al (2013) Nitrogen and silicon application facts on rice growth parameters at Alborz Mountain range. Elec J Biol 9:72–76

Jones LHP, Handreck KA (1965) Studies of silica in the oat plant. III. Uptake of silica from soils by plant. Plant Soil 23:79–96. https://doi.org/10.1007/BF01349120

Jones LHP, Handreck KA (1967) Silica in soils, plants, and animals. Adv Agron 19:107–149

Knight CTG, Kinrade SD (2001) A primer on the aqueous chemistry of silicon. In: Datnoff LE, Snyder GH, Korndörfer GH (eds) Silicon in agriculture, pp 57–84

Komisarov ID, Panfilova LA (1987) The method for production of slowly soluble fertilizer. USSR patent 1353767, Patent pending 15 May 84. Published in Bull. Inventions, no 4

Korndörfer GH, Snyder GH, Ulloa M et al (2001) Calibration of soil and plant silicon analysis for rice production. J Plant Nutr 24:1071–1084. https://doi.org/10.1081/PLN-100103804

Kovda VA (1985) Biogeochemistry of soil cover. Nauka Publication, Moscow, pp 159–179

Kraska B, Breitenbeck G (2010) Simple, robust method for quantifying silicon in plant tissue. Comm Soil Sci Plant Anal 41:2075–2085. https://doi.org/10.1080/00103624.2010.498537

Kurtz AC, Derry LA, Chadwick OA (1987) Accretion of Asian dust to Hawaiian soils: isotopic, elemental and mineral mass balances. Geochim Cosmochim Acta 65:1971–1983

Laane HM (2018) The effects of foliar sprays with different silicon compounds. Plan Theory 7:45. https://doi.org/10.3390/plants7020045

Lewin J, Reimann BEF (1969) Silicon and plant growth. Annu Rev Plant Physiol 20:289–304

Lian S (1976) Silica fertilization of rice. In: The fertility of paddy soils and fertilizer application for rice. Food and Fertilizer Technology Center, pp 197–221

Liang Y, Wong JW, Wei L (2005) Silicon-mediated enhancement of cadmium tolerance in maize (*Zea mays* L.) grown in cadmium contaminated soil. Chemosphere 58:475–483. https://doi.org/10.1016/j.chemosphere.2004.09.034

Liang Y, Sun W, Zhu Y G, Christie P (2007) Mechanisms of silicon-mediated alleviation of abiotic stresses in higher plants: a review. Environ Pollut 147:422–428

Lindsay WL (1979) Chemical equilibria in soils. Wiley Interscience

Ma JF, Goto S, Tamai K et al (2001) Role of root hairs and lateral roots in silicon uptake by rice. Plant Physiol 127:1773–1780. https://doi.org/10.1104/pp.010271

Ma JF, Nishimura K, Takahashi E (1989) Effect of silicon on the growth of rice plant at different growth stages. Soil Sci Plant Nutr 35:347–356. https://doi.org/10.1080/00380768.1989.10434768

Ma JF, Takahashi E (2002) Soil, fertilizer and plant silicon research in Japan. Elsevier Science

Ma JF, Tamai K, Yamaji N et al (2006) A silicon transporter in rice. Nature 440:688–691. https://doi.org/10.1038/nature04590

Ma JF, Yamaji N (2006) Silicon uptake and accumulation in higher plants. Trends Plant Sci 11:392–397. https://doi.org/10.1016/j.tplants.2006.06.007

Ma JF, Yamaji N, Mitani N et al (2007) An efflux transporter of silicon in rice. Nature 448:209–212. https://doi.org/10.1038/nature05964

Makabe S, Kakuda K, Sasaki Y et al (2009) Relationship between mineral composition or soil texture and available silicon in alluvial paddy soils on the Shounai Plain, Japan. Soil Sci Plant Nutr 55:300–308. https://doi.org/10.1111/j.1747-0765.2008.00352.x

Matichencov VV, Bocharnikova EA (2001) The relationship between silicon and soil physical and chemical properties. In: Datnoff LE, Snyder GH, Korndörfer GH (eds) Silicon in agriculture. Elsevier, pp 209–219

Matichenkov VV, Ammosova YM (1996) Effect of amorphous silica on soil properties of a sod-podzolic soil. Eurasian Soil Sci 28:87–99

McCray JM, Rice RW, Baucum LE (2018) Calcium silicate recommendations for sugarcane on Florida organic soils. SS-AGR-350. University of Florida

McCray JB (2022) Calcium silicate recommendations for sugarcane on Florida mineral soils. SS-AGR-467. University of Florida. https://doi.org/10.32473/edis-SC109-2022

McKeague JA, Cline MG (1963) Silica in soil solutions. I. The form and concentration of dissolved silica in aqueous extracts of some soils. Can J Soil Sci 43:70–82. https://doi.org/10.4141/cjss63-010

Meunier JD (2003) Le role des plantes dans le transfert du silicium a la surface des continents. Compt Rendus Geosci 335:1199–1206. https://doi.org/10.1016/j.crte.2003.07.007

Meunier JD, Guntzer F, Kirman S et al (2008) Terrestrial plant-Si and environmental changes. Mineral Mag 72:263–267. https://doi.org/10.1180/minmag.2008.072.1.263

Meyer JH, Keeping MG (2001) Past, present and future research of the role of silicon for sugarcane in southern Africa. In: Datnoff LE, Snyder GH, Korndörfer GH (eds) Silicon in agriculture. Elsevier, pp 257–275

Milnes A, Twidale C (1983) An overview of silicification in Cainozoic landscapes of arid central and southern Australia. Soil Res 21:387–410. https://doi.org/10.1071/SR9830387

Mitani N, Ma JF (2005) Uptake system of silicon in different plant species. J Exp Bot 56:1255–1261. https://doi.org/10.1093/jxb/eri121

Monger HC, Kelly EF (2002) Silica minerals. In: Soil mineralogy with environmental applications. Soil Science Society of America, pp 611–636

Motomura K, Fuji T, Suzuki M (2004) Silica deposition in relation to ageing of leaf tissues in Sasaveitchii (Carriére) Rehder (Poaceae: Bambusoideae). Ann Bot 93:235–248. https://doi.org/10.1093/aob/mch034

Nair PK, Aiyer RS (1968) Study of available silicon in rice soils of Keralastate (India). II. Silicon uptake by different varieties of rice in relation to available silicon contributed by soil and irrigation water. Agric Res J Kerala 6:88–99

Narayanaswamy C, Prakash NB (2009) Calibration and categorization of plant available silicon in rice soils of South India. J Plant Nutr 32:1237–1254. https://doi.org/10.1080/01904160903005970

Nash DJ, Ullyot JS (2007) In: Nash DJ, McLaren SJ (eds) Geochemical sediments and landscapes, pp 95–143

Novozamsky I, van Eck R, Houba VJG (1984) A rapid determination of silicon in plant material. Commun Soil Sci Plant Anal 15:205–211. https://doi.org/10.1080/00103628409367470

O'Reilly SE, Sims JT (1995) Phosphorus adsorption and desorption in a sandy soil amended with high rates of coal fly ash. Commun Soil Sci Plant Anal 26:2983. https://doi.org/10.1080/00103629509369502

Okuda A, Takahashi E (1962a) Effect of silicon supply on the injuries due to excessive amounts of Fe, Mn, Cu, As, Al, Co of barley and rice plant. Jpn J Soil Sci Plant Nutr 33:1–8

Okuda A, Takahashi E (1962b) Studies on the physiological role of silicon in crop plants (Part 6). Effect of silicon on iron uptake by rice plant and oxidation power of roots. J Sci Soil Manure Japan 33:59–64

Opfergelt S, Cardinal D, André L et al (2010) Variations of 30Si and Ge/Si with weathering and biogenic input in tropical basaltic ash soils under monoculture. Geochim Cosmochim Acta 74:225–240

Ostatek-Boczynski ZA, Haysom MB (2003) The determination of silicon in sugarcane tissue using a microwave-assisted acid and alkali digestion procedure. In: Currie LP, Hanley JA (eds) Tools for nutrient and pollution management (Occasional report No 17). Fertilizer and Lime Research Centre, Massey University, pp 419–424

Panov NP, Goncharova NA, Rodionova LP (1982) The role of amorphous silicic acid in solonetz soil processes. Vestnik Agric Sci 11:18

Pavlovic J, Kostic L, Bosnic P et al (2021) Interactions of silicon with essential and beneficial elements in plants. Front Plant Sci 12:697592. https://doi.org/10.3389/fpls.2021.697592

Paye W, Tubana B, Babu T et al (2018) Determination of critical soil silicon levels for rice production in Louisiana using different extraction procedures. Comm Soil Sci Plant Anal 47:2091–2102. https://doi.org/10.1080/00103624.2018.1495731

Raven JA (1983) The transport and function of silicon in plants. Biol Rev 58:179–207. https://doi.org/10.1111/j.1469-185X.1983.tb00385.x

Rodrigues FA, Dallagnol LJ, Duarte HSS et al (2015) Silicon control of foliar diseases in monocots and dicots. In: Rodrigues FA, Datnoff LE (eds) Silicon and plant diseases. Springer, Switzerland, pp 67–108. https://doi.org/10.1007/978-3-319-22930-0_4

Sauer D, Saccone L, Conley DJ et al (2006) Review of methodologies for extracting plant-available and amorphous Si from soils and aquatic sediments. Biogeochemistry 80:89–108. https://doi.org/10.1007/s10533-005-5879-3

Savant NK, Snyder GH, Datnoff LE (1997) Silicon management and sustainable rice production. Adv Agron 58:151–199. https://doi.org/10.1016/S0065-2113(08)60255-2

Schachtschabe LP, Heinemann CG (1967) Wasserlo sliche Kieselsa ure in Lo ßbo den. Z Pflanzenern Bodenk 118:22–35

Schulthess CP, Tokunaga Y (1996) Metal and pH effects on adsorption of poly(vinilalcohol) by silicon oxide. Soil Sci Soc Am J 60:92. https://doi.org/10.2136/sssaj1996.03615995006000010016x

Sebastian D (2012) Silica SLV results, presented at the Association of American Plant Food Control Officials Lab Services Committee Meeting, San Antonio, 2012. http://www.aapfco.org/meetings_aapfco.html. Accessed 4 Oct 2022

Sebastian D, Rodrigues H, Kinsey C et al (2013) A 5-day method for determination of soluble silicon concentration in nonliquid fertilizer materials using a sodium carbonate-ammonium nitrate extractant followed by visible spectroscopy with heteropoly blue analysis: single laboratory validation. J AOAC Int 96:251–259. https://doi.org/10.5740/jaoacint.12-243

Seebold KW, Kucharek TA, Datnoff LE et al (2001) The influence of silicon on components of resistance to blast in susceptible, partially resistance, and resistant cultivars of rice. Phytopathology 91:63–69. https://doi.org/10.1094/PHYTO.2001.91.1.63

Sinegar AF, Sipahutar IA, Anggria L et al (2021) IOP Conf Ser Earth Environ Sci 648:012202. https://doi.org/10.1088/1755-1315/648/1/012202

Singh B, Gilkes RJ, Butt CRM (1992) An electron optical investigation of aluminosilicate cements in silcretes. Clay Mineral 40:707–721. https://doi.org/10.1346/CCMN.1992.0400610

Singh KK, Singh K, Singh R et al (2006) Response of nitrogen and silicon levels on growth, yield and nutrient uptake of rice (*Oryza sativa* L.). Oryza 43:220

Singh KP, Sarkar MC (1992) Phosphorus availability in soil as affected by fertilizer phosphorus, sodium silicate and farmyard manure. J Indian Soc Soil Sci 40:762–767

Snyder GH, Jones DB, Gascho GJ (1986) Silicon fertilization of rice on Everglades Histosols. Soil Sci Soc Am J 50:1259–1263. https://doi.org/10.2136/sssaj1986.03615995005000050035x

Snyder GH, Matichenkov VV, Datnoff LE (2007) Silicon. In: Barker AV, Pilbeam DJ (eds) Handbook of plant nutrition. Taylor and Francis Group/CRC Press, Boca Raton, pp 551–565

Street-Perrott FA, Barker P (2008) Biogenic silica: a neglected component of the coupled global continental biogeochemical cycles of carbon and silicon. Earth Surf Proc Land 33:1436–1414

Swe MM, Mar SS, Naing TT et al (2021) Effect of silicon application on growth, yield and uptake of rice (*Oryza sativa* L.) in two different soils. Open Access Lib J 8:1–15. https://doi.org/10.4236/oalib.1107937

Takahashi E, Ma JF, Miyake Y (1990) The possibility of silicon as an essential element for higher plants. Comments Agric Food Chem 2:99–122

Takijima Y, Wijayaratna HMS, Seneviratne CJ (1970) Nutrient deficiency and physiological disease of lowland rice in Ceylon. III. Effect of silicate fertilizers and dolomite for increasing rice yields. Soil Sci Plant Nutr 16:11–16

Tebow JB, Houston LL, Dickson RW (2021) Silicon foliar spray and substrate drench effects on plant growth, morphology, and resistance to wilting with container-grown edible species. Horticulturae 7:263. https://doi.org/10.3390/horticulturae7090263

Tokunaga Y (1991) Potassium silicate: a slow release potassium fertilizer. Fertil Res 30:55

Tubana B, Babu T, Datnoff LE (2016) A review of silicon in soils and plants and its role in US agriculture: history and future perspective. Soil Sci 181:393–411. https://doi.org/10.1097/SS.0000000000000179

Tubana B, Heckman J (2015) Silicon in soils and plants. In: Rodrigues FA, Datnoff LE (eds) Silicon and plant diseases. Springer, pp 7–52

Verma KK, Song X-P, Zeng Y et al (2021) Foliar application of silicon boosts growth, photosynthetic leaf gas exchange, antioxidative response and resistance to limited water irrigation in sugarcane (*Saccharum officinarum* L.). Plant Physiol Biochem 166:582–592. https://doi.org/10.1016/j.plaphy.2021.06.032

Volker H, Nudling W, Adam K (1985) Verfahren zur herstellung eines mehrnahrstofffdungers. Fels-werke Peine-Salzgitter GmBH, Pat. 3538411.5 Germany
Volkova VV (1980) Silicate content in soil solutions and natural waters of the Russian plain. In: Kovda VA (ed) Pedological and biogeocenotic research of the Russian lowland Centre. Nauka Publication, Moscow
Wallace A (1993) Participation of silicon in cation-anion balance as possible mechanism for aluminum and iron tolerance in some gramineae. J Plant Nutr 16:547–553. https://doi.org/10.1080/01904169309364555
Wang JJ, Dodla SK, Henderson RE (2004) Soil silicon extractability with seven selected extractants in relation to colorimetric and ICP determination. Soil Sci 169:861–870. https://doi.org/10.1097/00010694-200412000-00005
White B, Tubana BS, Babu T et al (2017) Effect of silicate slag application on wheat grown under two nitrogen rates. Plan Theory 6:47. https://doi.org/10.3390/plants6040047
Xu G, Zhan X, Chunhua L et al (2001) Assessing methods of available silicon in calcareous soils. Commun Soil Sci Plant Anal 32:787–801. https://doi.org/10.1081/CSS-100103909
Yoshida S, Forno DA, Cock JH et al (1976) Laboratory manual for physiological studies of rice, 3rd edn. International Rice Research Institute, pp 17–22
Yoshida S, Ohnishi Y, Kitagishi K (1962) Chemical forms, mobility, and deposition of silicon in the rice plant. Soil Sci Plant Nutr 8:107–111. https://doi.org/10.1080/00380768.1962.10430992
Zellner W, Tubana B, Rodrigues FA et al (2021) Silicon's role in plant stress reduction and why this element is not used routinely for managing plant health. Plant Dis 105:2033–2049. https://doi.org/10.1094/PDIS-08-20-1797-FE

Volker U, Nidling W, Adamck K (1985) Verfahren zur herstellung eines mehrnahrstoffdunge… Peine Salzgitter GmbH. Pat … Germany

Volkova VV (1980) Silicate content in soil solutions and natural waters of the Russian plain. In: Kovda VA (ed) Problems of biogeochemistry … of the Russian … Nauka Publishing …

[illegible]

Chapter 5
Innovative Sources and Ways of Applying Silicon to Plants

Rilner Alves Flores and Maxuel Fellipe Nunes Xavier

5.1 Introduction

Silicon (Si) is an element that benefits plants by mitigating biotic and abiotic stresses while maintaining or increasing the productive potential of plants through biochemical, morphological, and anatomical modifications (Mengel and Kirkby 2001; Ma and Yamaji 2006; Pavlovic et al. 2021). Given that it is the second most abundant element in the earth's crust, second only to oxygen, which, when bound to Si, forms silicon dioxide (SiO_2), responsible for 50–70% of soil mass (Ma and Yamaji 2006), little importance has been given to the element for agricultural production over time.

However, the chemical form in which plants absorb silicon is monosilicic acid (H_4SiO_4) (Ma et al. 2001; Pavlovic et al. 2021), which reflects in the amount of Si available to plants, which is relatively low because about 95 to 99% of the Si available to plants, is in the weakly adsorbed form to the soil, and only 1 to 5% is present in soil solution (Matichenkov et al. 1997; Soares et al. 2016).

In tropical regions, most soils are highly weathered, formed by processes consisting of removing silica and exchangeable bases, such as calcium, magnesium, and potassium (Resende et al. 1995; Soares et al. 2016). Thus, most of them are soils with the clay fraction essentially composed of kaolinite and sesquioxides, with low molecular SiO_2:Al_2O_3 ratio (Ki ratio), sometimes lower than 0.5 (Embrapa 1982), with soils having low available Si contents. Compared to soils in temperate regions, Si contents can be five to ten times lower (McKeague and Cline 1963).

Therefore, the low availability of Si in the form of H_4SiO_4 in tropical region soils has increased the demand for its supply in fertilization programs, mainly due to all the benefits this element promotes for plants, especially in inducing resistance to biotic and abiotic stresses (Prado et al. 2019).

R. A. Flores (✉) · M. F. N. Xavier
School of Agronomy, Federal University of Goiás, Goiania, Brazil
e-mail: rilner@ufg.br

R. de Mello Prado (ed.), *Benefits of Silicon in the Nutrition of Plants*,
https://doi.org/10.1007/978-3-031-26673-7_5

In order to ensure optimum silicate nutrition for plants, it is important to consider the silicon sources and application forms for crops, favoring the element's efficient use in agriculture.

5.2 Sources and Ways of Supplying Si to Tropical Crops

There are several silicon sources for plants, such as the soil itself, plant residues (rice husks, bagasse, and straw left under the soil surface after harvest), steel mill slag, sodium metasilicates (cement), calcium metasilicates, thermophosphates, magnesium silicate (serpentinites), calcium silicates, potassium silicates, and nanosilicas (Soares et al. 2016; Souza Junior et al. 2022).

In this sense, new sources have emerged as alternative supply forms that can be used through the foliar application such as monosilicic acid stabilized with PEG400 and sodium-potassium silicate stabilized with sorbitol (Souza Junior 2018).

5.2.1 Silicon Sources for Soil Application or Fertigation in Tropical Regions

The soil supply of silicon demands large amounts because the sources have low solubility, requiring its incorporation into the soil. It also has low feasibility in no-till systems, i.e., without disturbing the soil for further planting. In the soil, silicates are applied as a silicon source and can form polymers, reducing plant uptake efficiency (Carvalho et al. 2009).

However, most are insufficient sources to readily meet the demand for silicon in modern agriculture, especially for short-cycle crops, due to the slow release to the soil from some limited solubility sources (Soares et al. 2016). According to Pereira et al. (2007), using residues such as sugarcane bagasse and rice husk, besides presenting low silicon release to the soil, has more noble uses, such as steam generation.

One of the main silicon sources applied to the soil comprises the byproducts of the steel and pig iron industry, which are slag from steel mills rich in Si that can meet the plants' demands for this element and have a low cost for agricultural use (Korndörfer et al. 2002). The chemical composition of these slags presents several oxides of Ca, Mg, Si, Fe, and Mn, whose concentrations depend on the chemical constitution of the raw material used in the product's manufacturing process (Prado et al. 2001). However, in order to use these slags in agriculture, they must have very low heavy metal contents to meet the current legislation and avoid soil, environmental, and human health pollution risks (Soares et al. 2016).

The silicates applied as slag act similarly to carbonates in the soil because they can increase the pH value, neutralizing the Al in the soil's exchangeable phase and other elements considered toxic to plants (Alcarde 1992; Prado et al. 2002).

According to Alcarde (1992), the reactions of silicate materials that occur in the soil are:

$$CaSiO_3 \rightarrow Ca^{+2} + SiO_3^{-2}$$

$$SiO_3^{-2} + H_2O_{(soil)} \rightarrow HSiO_3^{-} + OH^{-}$$

$$HSiO_3^{-} + H_2O_{(soil)} \rightarrow H_2SiO_3 + OH^{-}$$

$$H_2SiO_3 + H_2O_{(soil)} \rightarrow H_4SiO_4$$

Thus, slag applied to the soil promotes acidity correction due to the calcium silicate ($CaSiO_3$) action (Soares et al. 2016). Then, chemical reactions similar to limestone occur, such as an increase in pH value and Al and Mn precipitation from the soil, avoiding their toxicity to the plants (Prado et al. 2001; Vidal and Prado 2011).

Studies conducted by Preston et al. (2021) using slag in melon culture observed that besides the increase in fruit production, it was possible to observe a significant reduction in plant resistance to bacterial spot (disease progression curve – 10%, disease index – 14%, and disease incidence in fruit – 12%). Furthermore, they observed that applying 5.0 t ha^{-1} in the form of slag improved fruit quality, with an increase in fruit pulp thickness (8%) and soluble solids (7%) compared to the treatment not receiving the silicate.

In another study by Galindo et al. (2021), they present the positive effects of using calcium-magnesium silicate as a Si source for corn and wheat production in tropical environments. They observed that silicon use improved plant chlorophyll index, agronomic efficiency, and N uptake for corn and wheat, increasing grain yield. The authors further suggested the possibility of reducing nitrogen fertilization in corn from 180–185 to 100 kg N ha^{-1} and from 195–200 to 100 kg N ha^{-1} while maintaining grain yield.

For hydroponic crops, using these slags is not very feasible, mainly because of their low solubility in water (Sousa et al. 2010), and depending on the concentration applied, there may be no response to the application as a Si source due to possible polymerization forming colloidal particles of hydrated silica ($SiO_2.H_2O$) (Birchall 1995; Zanão Jr. et al. 2010).

Thus, more stable sources that avoid or delay this polymerization effect as much as possible are emerging, such as using sodium silicates or potassium silicates stabilized with sorbitol (Sarah 2022). Sorbitol is a polyalcohol that can promote greater stability in solution by slowing down the polymerization rate of silicon (Kubicki and Heaney 2003), with a reduction in pH value, in which monomeric Si species (silicic acid) can occur.

The effect of Si polymerization in these nutrient solutions affects its uptake efficiency by plants. For hydroponic crops, studies show its supply in concentrations ranging between 1 and 2 mmol L^{-1} of Si through the nutrient solution (Pei et al. 2010).

Studies conducted with potassium silicate application via fertigation in strawberries by Figueiredo et al. (2010) observed positive effects on sucrose contents of fruits after applying 94 mg L^{-1} of Si in the form of potassium silicate. At the same time, Teixeira et al. (2021) evaluated the effect of Si application (in the form of sodium-potassium silicate stabilized with sorbitol) via fertigation under water deficit in sugarcane and energy cane. The authors observed that the application of 2.5 mmol L^{-1} of Si via fertigation could reduce the effect of severe water deficiency (30% of water holding capacity), modulating the antioxidant response system, and favoring the physiological metabolism of sugarcane and energy cane. In another study conducted with sugarcane and energy cane, Oliveira et al. (2021) evaluated the effect of applying 2.0 mmol L^{-1} of Si via fertigation in the form of sodium-potassium silicate stabilized with sorbitol in mitigating manganese deficiency. In this study, the authors observed that Si could promote a reduction in cellular degradation by reactive oxygen species (ROS) by 21%, with an increase in phenol content by 18% and carotenoids by 64%, and modulate SOD activity, improving physiological responses and plant growth.

In a similar study, but with the crops of *Panicum maximum* (Massai and BRS Zuri), Rocha et al. (2022) also observed that the application of silicon via fertigation with sodium-potassium silicate stabilized with sorbitol alleviated water deficiency in addition to promoting growth increases in the two cultivars evaluated. Furthermore, it stabilized the stoichiometric homeostasis of C:N and C:Si, which is responsible for increasing the capacity for converting accumulated C in the plant into dry matter.

In this regard, Lata-Tenesaca et al. (2021) observed that Si can modify the C:N:P stoichiometry and increase nutrient use efficiency and productivity of quinoa when applied up to 3.0 mmol L^{-1} of Si in the form of sodium-potassium silicate stabilized with sorbitol via fertigation. In this study, they observed that Si application improved the use efficiency of C, N, and P, since the plant produced more biomass per unit of N and P absorbed and reduced the C:N and C:P ratios. Sales et al. (2021) evaluated the effect of silicon via fertigation using sodium silicate under macronutrient deficiency on quinoa (*Chenopodium quinoa* Willd.) production. At the end of this study, the authors observed that Si could mitigate the effects of stress caused by Ca deficiency with increased synthesis and protection of plant chlorophyll. Furthermore, for the N and Ca deficient treatments, the presence of Si reduced cell wall damage by decreasing electrolyte leakage, with gains in the dry biomass yield of quinoa plants.

Silva et al. (2021a) evidenced the beneficial effects of Si using sodium-potassium silicate stabilized with sorbitol via fertigation on cabbage. In this study, the authors evaluated the effect of calcium suppression and Si addition on the physiological apparatus of the crop and the yield and post-harvest quality. At the end of this study, they observed that Si application alleviated the deleterious effects of Ca deficiency, promoting increases in vitamin C levels, decreased electrolyte leakage, and a smaller reduction in the firmness index of cabbage leaves, thus ensuring the crop's longer post-harvest storage time.

Another Si source used in fertigation is sodium silicate stabilized with sorbitol. Araújo et al. (2022) observed that using this source at a 2.0 mmol L^{-1} of Si concentration alleviated the nutritional stress of N, P, and Ca deficiency in forage species (*Urochloa brizantha* cv. Marandu and *Megathyrsus maximum* cv. Massai). Furthermore, Si improved antioxidant activity, net photosynthesis rates, and the use efficiency of these nutrients (N, P, and Ca) by forages.

Based on the latest research results related to Si use via fertigation, new sources have emerged, such as potassium silicate, sodium silicate, and sodium-potassium silicate. However, its technical feasibility has mainly occurred when associated with sorbitol, as previously mentioned, a compound that can promote stability in the solution and delay the Si polymerization.

5.2.2 Silicon Sources for Foliar Application in Tropical Regions

New sources for foliar application of Si to plants have emerged, especially those that are water-soluble or highly water dispersible.

Nanosilica consists of silicon oxide nanoparticles (SiO_2) in colloidal dispersion, with dimensions between 1 and 10 nanometers (Ball 2002; Roco 2003). This product shows positive effects related to the hydrophilic properties of the nanosilica, which allows better plant uptake (Suriyaphabha et al. 2012). However, its use requires knowledge, especially regarding the factors affecting its solubility and stability in the spray solution. This compound's solubility increases linearly with increasing temperature and pH up to 7.5, decreasing with higher values (Souza Junior et al. 2022). In addition, the presence of other ions in the solution can negatively affect the stability of the nanosilica, such as Mg^{+2}, Al^{+3}, Fe^{+3}, Fe^{+2}, and Ca^{+2} (Weng 1995). However, compared to other silicate sources used to supply Si, nanosilica stands out as a source with low polymerization. The lower polymerization rate of nanosilica is evidenced in different crops, as shown in Table 5.1.

The use of monosilicic acid stabilized with PEG400 shows a hydrolysis reaction when in contact with an aqueous solution, as shown below:

Table 5.1 Increase in silicon content after foliar application in the form of nanoparticles

Crop	References
Beta vulgaris	Khan and Siddiqui (2020)
Citrullua lanatus	Kang et al. (2021)
Fragaria vesca	Avestan et al. (2021)
Glycine max	Felisberto et al. (2020)
Oryza sativa	Felisberto et al. (2020)
Polianthes tuberosa	Karimian et al. (2020)
Saccharum officinarum	Santos et al. (2020)
Zea mays	Aqaei et al. (2020)

Adapted from Souza Junior et al. (2022)

$$Na_4SiO_4 + H_2O \rightarrow 2NaOH + NaSiO_3$$

Silva (2011) also finds that sodium disilicate (Na_2SiO_3) has a lower solubility in water and is dissociated as follows:

$$Na_2SiO_3 + H_2O \leftrightarrow NaHSiO_3 + NaOH$$

Furthermore, the concentration of OH^- ions increases considerably as the sodium silicate is diluted, as shown in the following equation:

$$NaHSiO_2 + OH^- \leftrightarrow Si(OH)_4 + Na$$

It is a pH-dependent reaction, where at pH 7, there is a predominance of [$Si(OH)_4$] formation, which is the form readily available for uptake by plants (Souza Junior 2018). At a pH greater than 7, polynuclear silicate species are predominant (Marinakis and Shergold 1985), while at a pH lower than 7, polymeric silicon species are formed (Yang et al. 2008). Souza Junior (2018) states that foliar application of silicon sources containing sodium silicate in syrup, with a pH value of 7, would have available Si in the form of $Si(OH)_4$, which may increase the beneficial element's agronomic feasibility.

Andrade et al. (2022) evaluated the effect of applying sodium-potassium silicate stabilized with sorbitol on common bean production. The authors observed there were increases in the crop's grain yield and the weight of 1000 grains up to a dose of 1.0 g L^{-1} of Si, and this dose promoted the highest economic return, US$ 494.28, with its application at higher doses not being an advantage. Another study conducted with the same silicon source in sunflower obtained increases in biomass production of 37% over the control treatment when applying 2.52 g L^{-1} of Si. The authors also state that at this dose, the production cost increases by 15% compared to the control and, if sunflower oil production increases in the same proportion, there may be a 22% increase in profit to producers with the application of 2.52 g L^{-1} of Si (Peixoto et al. 2022). Silicon also showed a significant response for rice crops grown under dryland conditions in tropical environments. In this regard, Flores et al. (2022) found that the application of Si in the form of sodium-potassium silicate could promote a 9% increase in rice grain productivity when applying 252 g of Si per ha with a differential yield over the control treatment (no Si added) of US$129.5 per hectare.

Another silicon source used for foliar spraying is potassium copper silicate. Thus, Bueno et al. (2022) evaluated this source's effect on common bean production as a function of nitrogen management when grown in a tropical environment. The authors observed gains in the relative chlorophyll index with Si application, regardless of nitrogen management. Furthermore, in all Si doses evaluated, there were increases in grain productivity. However, the most efficient dose was 2 g L^{-1} of Si, which promoted differential gains of US$283 per hectare associated with the application of 60 kg ha^{-1} of N at 20 and 40 days after planting the crop. Moraes et al. (2020) also conducted a study applying potassium and copper silicate stabilized with sorbitol for tomato crop production destined for industrial processing. In this study, the authors evaluated the effect of applying doses of silicon in association with two water potentials in the soil. At the end of the study, the authors observed that Si application could improve gas exchange in tomato crops when subjected to water stress (treatment with soil water potential of −60 KPa) at doses not exceeding 2.8 g L^{-1} of Si.

The beneficial effects of foliar application of silicon are evidenced by Tripathi et al. (2021). They observed an increase in the number and size of nodules present in soybean roots, with an average gain in productive yield of 20% after two harvests. In this study, they provided silicon through the soil associated with foliar spraying of sodium metasilicate (Na_2SiO_3) at a concentration of 2 mM at V5. Flores et al. (2021) evaluated the effect of sodium-potassium silicate application on the millet crop. The authors observed that the application of 2.52 g L^{-1} of Si promoted a return of US$45 per hectare, with gains of 76% in photosynthetic rate and 15% in biomass production compared to the control treatment.

Souza Junior et al. (2021) evaluated the effect of different silicon sources for foliar spraying on cotton. In this study, the authors observed that all sources (nanosilica, monosilicic acid stabilized with PEG400, potassium silicate without stabilizer, and sodium-potassium silicate stabilized with sorbitol) promoted gains in dry biomass, boll, and seed production in the cotton crop. Moreover, the application of silicon in the form of sodium-potassium silicate stabilized with sorbitol showed better performance at a concentration of 0.8 g L^{-1} of Si compared to the other sources evaluated.

A viable foliar spraying alternative with silicon is its possible association with other chemicals to increase the application and operation efficiency during a given crop cycle. In this regard, Silva et al. (2021b) evaluated the effect of silicon association (12 mmol L^{-1} of Si) to foliar spray syrup with boron in cauliflower crops. In the end, the authors observed increased crop dry matter productivity, with improvements in the photosynthetic apparatus, reducing oxidative stress by reducing MDA and H_2O_2 production, and increasing phenolic compounds and ascorbic acid. Furthermore, Si addition improved leaf tissue firmness, increased photosynthetic pigment production, PSII quantum efficiency, and plant growth.

5.3 Final Considerations

The use of silicon in tropical crops has increased, mainly because of all its benefits in mitigating biotic and abiotic stresses. However, the major difficulties for its large-scale diffusion are related to the following factors: calibration of doses to the different crop requirements and creation of stable sources that present less polymerization.

As research advances in this area, it is believed that soon, silicon, even if it is not considered a nutrient, will be included in fertilization programs aimed at greater production efficiencies, mitigating deleterious stresses to the crop production process.

References

Alcarde JC (1992) Soil acidity correctors: characteristics and technical interpretations, 2nd edn. ANDA, São Paulo

Andrade AF, Bueno AM, Carvalho AS, Lima ML, Flores RA, Abdala KO, Prado RM, Souza Junior JP (2022) Innovative soluble silicon leaf source increase gas exchange, grain yield and economic viability in common bean. SILICON 14:3739–3747. https://doi.org/10.1007/s12633-021-01142-z

Aqaei P, Weisany W, Diyanat M, Razmi J, Struik PC (2020) Response of maize (*Zea mays* L.) to potassium nano-silica application under drought stress. J Plant Nutr 43:1205–1216. https://doi.org/10.1080/01904167.2020.1727508

Araújo WBS, Teixeira GCM, Prado RM, Rocha MAS (2022) Silicon mitigates nutritional stress of nitrogen, phosphorus, and calcium deficiency in two forages plants. Sci Rep 12:6611. https://doi.org/10.1038/s41598-022-10615-z

Avestan S, Ghasemnezhad M, Esfahani M, Barker AV (2021) Effects of nanosilicon dioxide on leaf anatomy, chlorophyll fluorescence, and mineral element composition of strawberry under salinity stress. J Plant Nutri 44:3005–3019. https://doi.org/10.1080/01904167.2021.1936036

Ball P (2002) Natural strategies for the molecular engineer. Nanotechnology 13:15–28

Birchall JD (1995) The essentiality of silicon in biology. Chem Soc Rev 24:351–357

Bueno AM, Flores RA, Ferreira EPB, Andrade AF, Lima FRS, Souza Junior JP, Abdala KO, Mesquita M, Prado RM (2022) Effects of foliar silicon application, seed inoculation and splitting of N fertilization on yield, physiological quality, and economic viability of the common bean. SILICON 14:4169–4181. https://doi.org/10.1007/s12633-021-01208-y

Carvalho MP, Zanão Júnior LA, Grossi JA, Barbosa JG (2009) Silicon improves ornamental pot sunflower production and quality. Ciênc Rural 39:2304–2399. https://doi.org/10.1590/S0103-84782009005000194

Embrapa. Brazilian Agricultural Research Corporation (1982) Survey of medium intensity soil recognition and evaluation of agricultural suitability of lands in the Triângulo Mineiro. National Soil Survey and Conservation Service, Rio de Janeiro

Felisberto G, Prado RM, Oliveira RLL, Carvalho PA (2020) Are nanosilica, potassium silicate and new soluble sources of silicon effective for silicon foliar application to soybean and rice plants? SILICON 13:3217–3228. https://doi.org/10.1007/s12633-020-00668-y

Figueiredo FC, Botrel PP, Teixeira CP, Petrazzini LL, Locarno M, Carvalho JG (2010) Leaf spraying and fertirrigation with silicon on the physicochemical attributes of quality and coloration indices of strawberry. Ciênc Agrotec 34:1306–1311. https://doi.org/10.1590/S1413-70542010000500032

Flores RA, Pessoa-de-Souza MA, Andrade AF, Bueno AM, Abdala KO, Souza Junior JP, Prado RM, Santos GG, Mesquita M (2022) Does foliar application of silicon under natural water stress conditions increase rice yield in subtropical dry regions? SILICON 14:3591–3600. https://doi.org/10.1007/s12633-021-01109-0

Flores RA, Pessoa-de-Souza MA, Bueno AM, Andrade AF, Souza Junior JP, Abdala KO, Prado RM, Santos GG, Mesquita M (2021) Does foliar silicon application enhance the biomass yield of millet silage, and does it provide significant economic gains? Res. Soc Dev 10:e41610414232. https://doi.org/10.33448/rsd-v10i4.14232

Galindo FS, Pagliari PH, Rodrigues WL, Fernandes GC, Boleta EHM, Santini JMK, Jalal A, Buzetti S, Teixeira Filho MCM (2021) Silicon amendment enhances agronomic efficiency of nitrogen fertilization in maize and wheat crops under tropical conditions. Plan Theory 10:1329. https://doi.org/10.3390/plants10071329

Kang H, Elmer W, Shen Y, Zuverza-Mena N, Ma C, Botella P, White JC, Haynes CL (2021) Silica nanoparticle dissolution rate controls the suppression of Fusarium Wilt of watermelon (*Citrullus lanatus*). Environ Sci Technol 55:13513–13522. https://doi.org/10.1021/acs.est.0c07126

Karimian N, Nazari F, Samadi S (2020) Morphological and biochemical properties, leaf nutrient content, and vase life of tuberose (*Polianthes tuberosa* L.) affected by root or foliar applications

of silicon (Si) and silicon nanoparticles (SiNPs). J Plant Growth Regul 40:2221–2235. https://doi.org/10.1007/s00344-020-10272-4
Khan MR, Siddiqui ZA (2020) Use of silicon dioxide nanoparticles for the management of *Meloidogyne incognita*, *Pectobacterium betavasculorum* and *Rhizoctonia solani* disease complex of beetroot (*Beta vulgaris* L.). Sci Hortic 265:109211. https://doi.org/10.1016/j.scienta.2020.109211
Korndörfer GH, Pereira HS, Camargo MS (2002) Calcium and magnesium silicate in agriculture. Federal University of Uberlandia
Kubicki JD, Heaney PJ (2003) Molecular orbital modeling of aqueous organosilicon complexes: implications for silica biomineralization. Geochim Cosmochim Acta 67:4113–4121. https://doi.org/10.1016/S0016-7037(03)00093-0
Lata-Tenesaca LF, Prado RM, Piccolo MC, Silva DL, Silva JLF (2021) Silicon modifies C:N:P stoichiometry, and increases nutrient use efficiency and productivity of quinoa. Sci Rep 11:9893. https://doi.org/10.1038/s41598-021-89416-9
Ma JF, Miyake Y, Takahashi E (2001) Silicon as a beneficial element for crop plants. In: Datnoff LE, Snyder GH, Korndörfer GH (eds) Silicon in agriculture. Elsevier, pp 17–39
Ma JF, Yamaji N (2006) Silicon uptake and accumulation in higher plants. Trends Plant Sci 11:392–397. https://doi.org/10.1016/j.tplants.2006.06.007
Marinakis KI, Shergold HL (1985) Influence of sodium silicate addition on the adsorption of oleic acid by fluorite, calcite and barite. Int J Miner Process 14:177–193. https://doi.org/10.1016/0301-7516(85)90002-X
Matichenkov V, Ammosova Y, Bocharnikova E (1997) The method for determination of plant available silica in soil. Agrochemistry 1:76–84
McKeague JA, Cline MG (1963) Silica in soil solution. II The adsorption of monossilic acidy by soil by other substances. Can J Soil Sci 43:83–95
Mengel K, Kirkby EA (2001) Principles of plant nutrition. Kluwer Academic Publishers, Dordrecht
Moraes DHM, Mesquita M, Bueno AM, Flores RA, Oliveira HFE, Lima FRS, Prado RM, Battisti R (2020) Combined effects of induced water deficit and foliar application of silicon on the gas exchange of tomatoes for processing. Agronomy 10:1715. https://doi.org/10.3390/agronomy10111715
Oliveira KS, Prado RM, Checchio MV, Gratão PL (2021) Silicon via nutrient solution modulates deficient and sufficient manganese sugar and energy cane antioxidant systems. Sci Rep 11:16900. https://doi.org/10.1038/s41598-021-96427-z
Pavlovic J, Kostic L, Bosnic P, Kirkby EA, Nikolic M (2021) Interactions of silicon with essential and beneficial elements in plants. Front Plant Sci 12:697592. https://doi.org/10.3389/fpls.2021.697592
Pei ZF, Ming DF, Liu D, Wan GL, Geng XX, Gong HJ, Zhou WJ (2010) Silicon improves the tolerance to water-deficit stress induced by polyethylene glycol in wheat (*Triticum aestivum* L.) seedlings. J Plant Growth Regul 29:106–115. https://doi.org/10.1007/s00344-009-9120-9
Peixoto MM, Flores RA, Couto CA, Pacheco HDN, Prado RM, Souza Junior JP, Castro-Netto JA, Graciano-Ribeiro D (2022) Silicon application increases biomass yield in sunflower by improving the photosynthesizing leaf area. SILICON 14:275–280. https://doi.org/10.1007/s12633-020-00818-2
Pereira HS, Barbosa NC, Carneiro MAC, Korndörfer GH (2007) Evaluation of sources and extractors of silicon in the soil. Pesq Agropec Bras 42:239–247. https://doi.org/10.1590/S0100-204X2007000200013
Prado RM, Coutinho ELM, Roque CG, Villar MLP (2002) Evaluation of slag and calcareous rocks as corrective of the acidity of the ground in the culture of lettuce. Pesq Agropec Bras 37:539–546. https://doi.org/10.1590/S0100-204X2002000400016
Prado RM, Felisberto G, Flores RA (2019) Technical feasibility of foliar application of silicon in crops. In: Flores RM, Cunha PP, Marchão RL, Moraes MF (eds) Nutrition and fertilization of large crops in the Cerrado region, pp 97–138

Prado RM, Fernandes MP, Natale W (2001) Use of steel slag in Brazil: studies on sugarcane culture. FUNEP, Jaboticabal

Preston HAF, Nunes GHS, Preston W, Souza EB, Mariano RLR, Datnoff LE, Nascimento CWA (2021) Slag-based silicon fertilizer improves the resistance to bacterial fruit blotch and fruit quality of melon grown under field conditions. Crop Prot 147:105460

Resende M, Curi N, Resende SB, Corrêa GF (1995) Pedology: basis for distinguishing environments. NEPUT

Rocha JR, Prado RM, Piccolo MC (2022) New outcomes on how silicon enables the cultivation of *Panicum maximum* in soil with water restriction. Sci Rep 12:1897. https://doi.org/10.1038/s41598-022-05927-z

Roco MC (2003) Broader societal issues of nanotechnology. J Nanopart Res 5:181–189. https://doi.org/10.1023/A:1025548512438

Sales AC, Campos CNS, Souza Junior JP, Silva DL, Oliveira KS, Prado RM, Teodoro LP, Teodoro PE (2021) Silicon mitigates nutritional stress in quinoa (*Chenopodium quinoa* Willd.). Sci Rep 11:14665. https://doi.org/10.1038/s41598-021-94287-1

Santos LCN, Teixeira GCM, Prado RM, Rocha AMS, Pinto RCS (2020) Response of pre-sprouted sugarcane seedlings to foliar spraying of potassium silicate, sodium and potassium silicate, nanosilica and monosilicic acid. Sugar Tech 22:773–781. https://doi.org/10.1007/s12355-020-00833-y

Sarah MMS (2022) Silicon via fertigation alleviates the stresses caused by water deficit in sugarcane plants grown in different soils. Thesis, São Paulo State University

Silva DL, Prado RM, Lata-Tenesaca LF, Silva JLF, Mattiuz B-H (2021a) Silicon attenuates calcium deficiency by increasing ascorbic acid content, growth and quality of cabbage leaves. Sci Rep 11:1770. https://doi.org/10.1038/s41598-020-80934-6

Silva JLF, Prado RM, Souza Junior JP, Lata-Tenesaca LF, Silva DL, Pinsetta Junior JS (2021b) Feasibility of silicon addition to boron foliar spraying in cauliflowers. J Soil Sci Plant Nutr 21:2448–2455. https://doi.org/10.1007/s42729-021-00536-7

Silva JPP (2011) Mechanism of action of sodium silicate as a flotation depressant. Dissertation, Federal University of Pernambuco

Soares AAVL, Prado RM, Camargo MS (2016) Importance of silicon in plant nutrition in the Cerrado region. In: Flores RA, Cunha PP (eds) Soil management practices for adequate plant nutrition in the Cerrado, pp 447–480

Sousa JV, Rodrigues CR, Luz JMQ, Sousa VBF, de Carvalho PC, Rodrigues TM, de Brito CH (2010) Foliar application of the potassium silicate in corn: photosynthesis, growth and yield. Biosci J 26:502–513

Souza Junior JP (2018) Silicon as a mitigator of boron deficiency and toxicity in cotton grown in nutrient solution. Dissertation, São Paulo State University

Souza Junior JP, Prado RM, Campos CNS, Teixeira GCM (2022) Nanosilica-mediated plant growth and environmental stress tolerance in plants: mechanisms of action. In: Etesami H, El-Ramady H, Pessarakli M, Al Saeedi AH, Fujita M, Hossain MA (eds) Silicon and nano-silicon in environmental stress management and crop quality improvement. Elsevier Academic Press, pp 325–337

Souza Junior JP, Prado RM, Soares MB, Silva JLF, Guedes VHF, Sarah MMS, Cazetta JO (2021) Effect of different foliar silicon sources on cotton plants. SILICON 21:95–103. https://doi.org/10.1007/s42729-020-00345-4

Suriyaphabha R, Karunakaran G, Yuvakkumar R, Rajendran V, Kannan N (2012) Silica nanoparticles for increased silica availability in maize (*Zea mays* L.) seeds under hydroponic conditions. Curr Nanosci 8:902–908. https://doi.org/10.2174/157341312803989033

Teixeira GCM, Prado RM, Rocha AMS, Sousa Junior GS, Gratão PL (2021) Beneficial effect of silicon applied through fertigation attenuates damage caused by water deficit in sugarcane. J Plant Growth Regul. https://doi.org/10.1007/s00344-021-10510-3

Tripathi P, Na C-I, Kim Y (2021) Effect of silicon fertilizer treatment on nodule formation and yield in soybean (*Glycine max* L.). Eur J Agron 122:126172

Vidal AA, Prado RM (2011) Application of steel slag, limestone and urea in Oxisol cultivated with rice. Pesq Agropec Trop 41:264–272. https://doi.org/10.5216/pat.v41i2.13012

Weng PF (1995) Silica scale inhibition and colloidal silica dispersion for reverse osmosis systems. Desalination 103:59–67

Yang X, Roonasi P, Holmgren A (2008) A study of sodium silicate in aqueous solution and sorbed by synthetic magnetite using in situ ATR-FTIR spectroscopy. J Colloid Interface Sci 328:41–47. https://doi.org/10.1016/j.jcis.2008.08.061

Zanão LA Jr, Fontes RLF, Ávila VT (2010) Silicon application to increase rice resistance to brown spot. Pesq Agropec Bras 44:203–206. https://doi.org/10.1590/S0100-204X2009000200013

Chapter 6
Silicon Mitigates the Effects of Nitrogen Deficiency in Plants

Cid Naudi Silva Campos and Bianca Cavalcante da Silva

6.1 Introduction

Unsuitable agricultural practices and population growth have led to increased environmental impacts, which could threaten the world's food production (Etesami and Jeong 2018; Mu and Chen 2021). To achieve higher yields, it is necessary to use large amounts of fertilizers, which will result in increased production costs and less sustainability. On the other hand, the lack of fertilization or the supply of inadequate quantities, i.e., not meeting the nutritional requirements of agricultural crops may culminate in nutritional disorder by deficiency. This fact is very common, especially for macronutrients, such as nitrogen (N), since plants have a high demand for this element.

Nitrogen is one of the most widely distributed elements in nature. It is present in the atmosphere, the lithosphere, and the hydrosphere. The soil accounts for only a minute fraction of lithospheric N, and of this soil N, only a very small proportion is directly available to plants. Nitrogen is a very mobile element circulating between the atmosphere, the soil and living organisms. Many factors and processes are involved in this N-tumover, some of which are physico-chemical and others biological (Mengel et al. 2001). N plays a crucial role in the biochemical and physiological processes of plants, as it is a component of many molecules such as amino acids, proteins, nucleotides, nucleic acids, enzymes, and pigments, directly favoring crop growth and yield (Prado 2021; Xie et al. 2022).

C. N. S. Campos (✉)
Federal University of Mato Grosso do Sul, Chapadão do Sul, Brazil
e-mail: cid.campos@ufms.br

B. C. da Silva
São Paulo State University, Jaboticabal, Brazil
e-mail: bianca.cavalcante@unesp.br

R. de Mello Prado (ed.), *Benefits of Silicon in the Nutrition of Plants*,
https://doi.org/10.1007/978-3-031-26673-7_6

Thus, plants grown under N deficiency suffer damage to various physiological processes, particularly photosynthesis, which is severely affected. Besides the lower levels of pigments, N deficiency reduces the production and activity of the enzyme ribulose-1,5-bisphosphate carboxylase-oxygenase (RuBisCO), which is essential for carboxylation reactions in the biochemical photosynthesis stage. Furthermore, higher leaf senescence is observed under N deficiency, affecting the radiation interception (Shangguan et al. 2000; Sigua et al. 2018; Prado 2021; Kang et al. 2022). Furthermore, such deficiency causes the stomata closure, as there is an increase in abscisic acid (ABA) biosynthesis and production of hydrogen peroxide (H_2O_2), affecting gas exchange and hence carbohydrate production (Tewari et al. 2007; Mu et al. 2018a; Mu and Chen 2021).

Silicon (Si) is considered a beneficial element for plants (Li et al. 2020; Prado 2021; Barreto et al. 2022), acting as a mitigator of biotic and abiotic stresses. This element promotes the reduction of oxidative stress, decreasing H_2O_2 and malondialdehyde production (Souza Júnior et al. 2022), increases nutrient-use and accumulation efficiencies, and promotes higher shoot and dry mass (Araújo et al. 2022; Barreto et al. 2022), thus mitigating the stress caused by nutrient deficiency (Santos Sarah et al. 2022). Si plays an important role in nitrogen (N) nutrition, that is, absorption, assimilation, and remobilization (Pavlovic et al. 2021).

Si supply promotes increased N assimilation and chlorophyll production (Gou et al. 2020). Using Si fertilization in plants grown under N deficiency conditions leads to lower production of reactive oxygen species (ROS) and promotes increased N uptake, nitrate transporter gene expression, chlorophyll content, and photosynthetic activity (Haddad et al. 2018). It is worth mentioning that the Si application promotes increased production of phenolic compounds, higher quantum efficiency of photosystem II, higher N-use efficiency, and lower electrolyte leakage in plants grown under N deficiency (Souza Júnior et al. 2022). Furthermore, when the plant is submitted to this kind of stress, Si increases the lignin content and plays the role of performing carbon allocation, reducing the stoichiometric C:Si ratio by replacing organic C compounds by Si in photosynthetic tissues, which has a lower energy expenditure (Deus et al. 2020).

In light of the above, this chapter aims to highlight the effects of applying Si on plants grown under nutritional stress caused by N deficiency.

6.2 Biochemical and Physiological Effects of N Deficiency in Plants

The world population is expected to reach 9.2 billion people by 2050, which increases the demand for agricultural production due to the consumption of meat, dairy products, vegetables, and biofuels (Mu and Chen 2021). Among the inputs that contribute most to the increase in crop yields are fertilizers, especially nitrogen-based ones (Jiao et al. 2016). This is because N is one of the most demanding elements for most plants and is essential for their growth and development since it is a constituent of amino acids, proteins, enzymes, coenzymes, chlorophylls, nucleotides, and nucleic acids (Marschner 2011; Prado 2021).

N deficiency reduces photosynthesis and leaf area, promotes early plant senescence, and decreases crop yield (Chen et al. 2015; Mu et al. 2016), which in turn is highly dependent on the production and accumulation of biomass generated by photosynthesis (Mu et al. 2018a, b; Mu and Chen 2021).

One of the physiological processes most affected by N deficiency is photosynthesis since the lack of N causes a decrease in the protein and chlorophyll contents and changes the chlorophyll a/b ratio. (Lawlor et al. 1989; Shangguan et al. 2000). The low production and activity of RuBisCO under N deficiency conditions are other reasons for reduced photosynthesis (Kang et al. 2022). RuBisCO is a fundamental enzyme for fixing CO_2 in the Calvin cycle, and the activity of this enzyme is highly correlated with leaf N (Cheng and Fuchigami 2000; Shangguan et al. 2000).

When carrying out a study with adequate and low N supply, Sigua et al. (2018) observed reduced leaf development and leaf duration, i.e., increased senescence compared to the adequate N treatment. Under N deficiency, plants increase abscisic acid (ABA) biosynthesis (Mu et al. 2018b) and hydrogen peroxide production (H_2O_2), biochemical events that lead to the stomata closing (Tewari et al. 2007; Mu and Chen 2021), and thus impairing gas exchange.

Besides these biochemical events detrimental to physiological processes involving gas exchange, N-deficient plants show reduced photon capture, electron transport rate, ATP synthase enzyme activity, and carboxylation rate (Mu and Chen 2021), decreasing photosynthetic rates and consequently carbohydrate production.

6.3 Beneficial Effect of Si on Plants Under Nutrient Deficiency Stress

When undergoing abiotic stresses, plants suffer several alterations in their metabolism, culminating in biochemical and physiological disorders, such as a decreased catalytic activity of enzymes, destruction of membrane integrity, problems in uptake and assimilation of nutrients, impaired osmotic regulation, reduced water-use efficiency, which lead to losses in photosynthetic efficiency and yield. (Moradtalab et al. 2018; Ahanger et al. 2019; Khan et al. 2015, 2021; Kumar et al. 2022; Rocha et al. 2022; Teixeira et al. 2022).

As a strategy to mitigate abiotic stresses, plants drive physiological and molecular mechanisms to adapt to the changing environment. Once plants perceive external actions that cause metabolic instability, they activate a series of defense mechanisms (Khan et al. 2021). Among the external stimuli that activate these defense mechanisms to abiotic stress conditions, Si can act to modulate stress responses.

The benefits of Si on the growth and development of plants grown under stress conditions have been widely reported in the literature (Deus et al. 2020). Several studies have shown that Si alleviates the damaging effects of abiotic stress caused by nutritional disorders, such as deficiencies of potassium (K) (Santos Sarah et al. 2022), K and sulfur (S) (Barreto et al. 2022), phosphorus (P) and calcium (Ca) (Araújo et al. 2022), boron (Souza Júnior et al. 2022), and N (Deus et al. 2020; Sales et al. 2021; Araújo et al. 2022), as detailed in Table 6.1.

In a review paper, Ali et al. (2020) compiled the positive effects of Si on nutrient-deficient plants, evidencing that Si application promotes improved biomass production due to beneficial actions on gas exchange, regulation of nutrient uptake, and reduced oxidative stress. In this sense, using Si may be promising for mitigating the impact of nutrient deficiency in plants grown in tropical soils. Among the main nutritional disorders found in these growing regions, nitrogen deficiency stands out because of its high demand by plants and significant contribution to the formation of macromolecules (proteins and nucleic acids), pigments, phytohormones, and secondary compounds (Ali et al. 2020; Prado 2021).

6.4 Beneficial Action of Si in Tropical Plants Under N Deficiency: How Can Si Mitigate the Effects of N Deficiency?

Several studies have reported the beneficial effects of Si on the biochemistry and physiology of nutritionally stressed plants. Although the distinct action mechanisms of Si on plant metabolism still need to be better understood, we know that using Si can mitigate the effects of nutrient deficiency stress. Allied to this, attention is drawn to the effects of Si on plants under N deficiency stress, as shown in Fig. 6.1, which highlights some biochemical, physiological, anatomical, and morphological mechanisms of plants grown under N deficiency conditions in the presence (+Si) and absence of Si (−Si).

Although the effects of Si on mitigating N deficiency are still little reported in the scientific literature (Haddad et al. 2018), some research has evidenced this benefit of Si. The damage caused by N deficiency can be seen in (Fig. 6.2, Sales et al. 2021) in plants of *Chenopodium quinoa* Willd., it can be seen that the typical visual symptoms start with chlorosis in the oldest leaves and a reduction in plant growth and development.

In this research, the authors verified a lower electrolyte leakage with the Si supply (Fig. 6.3a, Sales et al. 2021), that is, the Si reduced the deleterious stress effect caused by N deficiency, reducing the damage to the plant cell wall and leading to a higher leaf area (Fig. 6.3b, Sales et al. 2021) and shoot and root dry mass of the plants (Fig. 6.3c, Sales et al. 2021). Thus, the Si supply attenuated the visual effects of N deficiency (−N−Si treatment), as it was reflected in increased shoot and root dry mass production compared to N-deficient plants that did not receive Si (−N−Si treatment).

Table 6.1 Biochemical and morpho-physiological effects of applying Si to mitigate abiotic stresses caused by nutrient deficiency in various plant species grown in tropical regions

Application way	Si concentration	Abiotic stress	Plant species	Biochemical and morpho-physiological attributes	References
Nutrient solution	2.0 mmol L^{-1}	K and S deficiency stress	*Ocimum basilicum* L.	Improving K-use efficiency, increasing shoot and root dry mass, increasing S accumulation, modulating the expression of sulfate transporters, attenuating damage caused by K and S deficiency	Barreto et al. (2022)
Sprayed on the leaves	1.00 g L^{-1}	B deficiency stress	*Gossypium hirsutum* L.	Reducing oxidative damage by increasing both proline and glycine-betaine, reducing the production of H_2O_2 and malondialdehyde	Souza Júnior et al. (2022)
Nutrient solution	Nutrient solution (2 mmol L^{-1} Si) or sprayed on the leaves (5.3 mmol L^{-1} Si)	K deficiency stress	*Zea mays* L.	Si improved the nutritional status and growth of the plants	Santos Sarah et al. (2022)
Sprayed on the leaves	3.45 g L^{-1} Si	N deficiency stress	*Oryza sativa*	Stoichiometry change in C with Si and increases of lignin synthesis	Deus et al. (2020)
Nutrient solution	1.5 mmol L^{-1}	N deficiency stress	*Chenopodium quinoa* Willd.	Si promoted decreased electrolyte leakage and reflected in a higher dry mass production	Sales et al. (2021)
Nutrient solution	2 mmol L^{-1}	N, P, and K deficiency stress	*U. brizantha* and *M. maximum*	Si decreased the cellular electrolyte leakage and increased the content of phenolic compounds, the green color index, the quantum efficiency of photosystem II, the use efficiencies of N, P, and Ca and the production of shoot dry mass	Araújo et al. (2022)

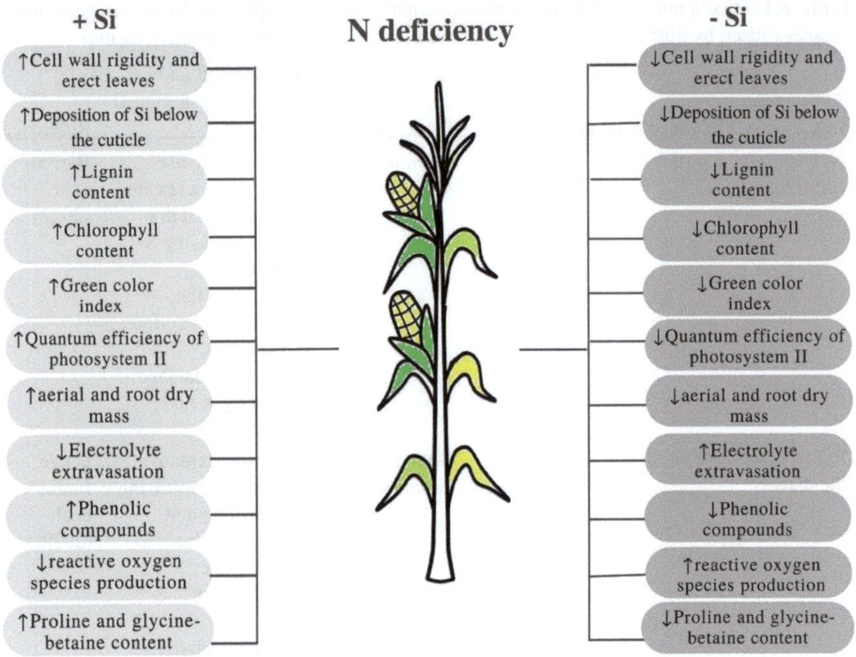

Fig. 6.1 Schematic representation identifying the biochemical, physiological, anatomical, and morphological effects of the presence (+Si) and absence of Si (−Si) in N-deficient plant

Fig. 6.2 N deficiency symptoms in *Chenopodium quinoa* Willd. plants grown in nutrient solution with low N content under absence (−N−Si) and presence of Si (−N+Si) and treatment with adequate N content in the absence (+N−Si) and presence (+N+Si) of Si. (Extracted and adapted from Sales et al. (2021))

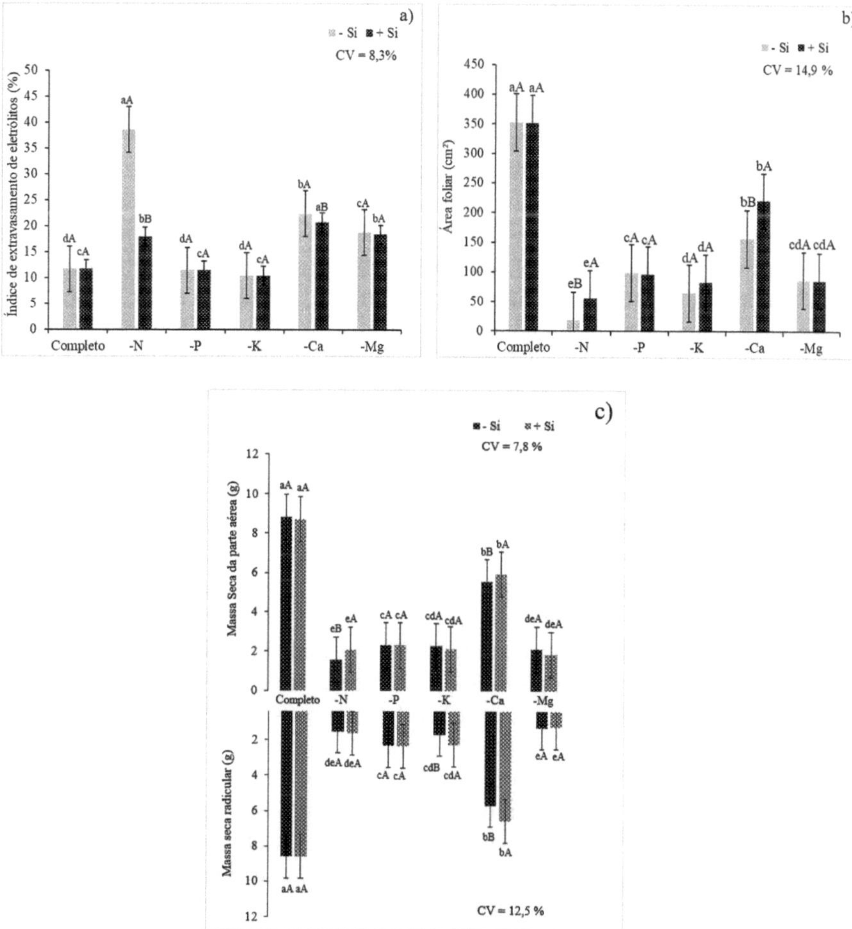

Fig. 6.3 Electrolyte leakage index and biomass production of quinoa plants submitted to Hoagland solution (Hoagland and Arnon 1950) under omission of macronutrients in the absence (−Si) and presence of silicon (+Si), (**a**) Electrolyte leakage index, (**b**) leaf area, (**c**) shoot and root dry mass. Equal lowercase letters demonstrate a similarity between the omitted and complete treatments by Tukey's test at 5% probability. Equal uppercase letters show a similarity between treatments (−Si and +Si) by Tukey's test at 5% probability. (Extracted and adapted from Sales et al. (2021))

In studies with N-deficient *Oryza sativa* plants, leaf application of Si altered carbon allocation, thus changing the stoichiometric C:Si ratio, besides increasing lignin concentration, decreasing electrolyte leakage, and reflecting directly on biomass production, Si accumulation, and grain yield of rice (Deus et al. 2020). One explanation for the increased biomass and yield of N-deficient plants that received Si application may be related to the effect on the nutritional state of these plants because Si promotes improved nutrient uptake, such as N, P, and K (Eneji et al. 2008).

The effects of Si on N-deficient plants can be observed in the study by Araújo et al. (2022), where the Si improved the performance of two forage species (*U. brizantha* and *M. maximum*). Through the application of the beneficial element, the effects of N deficiency were attenuated by increasing antioxidant activity and photosynthesis with a consequent increase in plant dry matter production. According to the authors, the benefits of Si in mitigating N deficiency in forage plants demonstrate the importance of this element for crop sustainability since most of these species are grown in tropical soils with poor fertility and few investments in fertilizers by the farmers. Another important point addressed in this research was the effect of Si on increasing the nutritional N efficiency, improving the production of these forages.

Another effect of Si in mitigating N deficiency stress would be in inducing a lower senescence rate of N deficient leaves. Such effect of Si was noticed in the study carried out by Haddad et al. (2018) evaluating *Brassica napus* plants grown under N deficiency. The pigmentation of *Brassica napus* leaves was totally different under Si application (+Si) compared to the absence of Si (−Si). In Fig. 6.4a, Haddad et al. (2018) shows the phenotypic evolution of the mature leaves in the −Si−N treatment, where yellowing, i.e., lower chlorophyll production, began on the sixth day and increased rapidly until the ninth day, when the leaf became entirely yellow. In the +Si−N treatment, the yellowing of mature leaves only began between the 9th and 12th day after N omission.

Given these findings, Haddad et al. (2018) concluded that in N-deficient leaves of *Brassica napus* with Si supply (+Si−N), there was delayed leaf senescence compared to plants grown without Si under N deficiency (−Si−N). These authors also pointed out that the levels of cytokinin, especially the active form (isopentenyladenine - iP), were higher in mature leaves of *Brassica napus* with Si application (+Si−N) compared to leaves without Si application (−Si−N) (Fig. 6.4b, Haddad et al. 2018). Moreover, another biochemical aspect evaluated by Haddad et al. (2018) was the oxidative stress of plants grown in N deficiency, with and without the Si supply. The authors quantified the contents of some oxidative stress markers (H_2O_2 and MDA) and found that they were significantly lower in mature Brassica napus leaves when the beneficial element (+Si) was applied, thus mitigating the oxidative damage caused by N deficiency (Fig. 6.4c, Haddad et al. 2018).

On the other hand, it should be noted that Si application reduces the damaging effects caused by nutrient deficiency only if it is not too severe (Buchelt et al. 2020; Barreto et al. 2022). The explanation for the no effect of Si in severe N deficiency conditions would be that the beneficial element will never substitute the functions that N plays in plant metabolism. The Si only attenuates the damaging effects caused by a deficiency in moderate cases, while if the deficiency is very severe, the beneficial effect of Si will be insignificant. This fact is evidenced in the study carried out by Buchelt et al. (2020), where severe N deficiency impaired dry matter production, nutrient accumulation, and nutrient use efficiency, limiting the growth of *Panicum maximum* and *Brachiaria* plants. However, no beneficial effect of Si application was observed to mitigate the damage caused by N deficiency (Fig. 6.5a–d, Buchelt et al. 2020), which leads to the assumption that when the plant is under

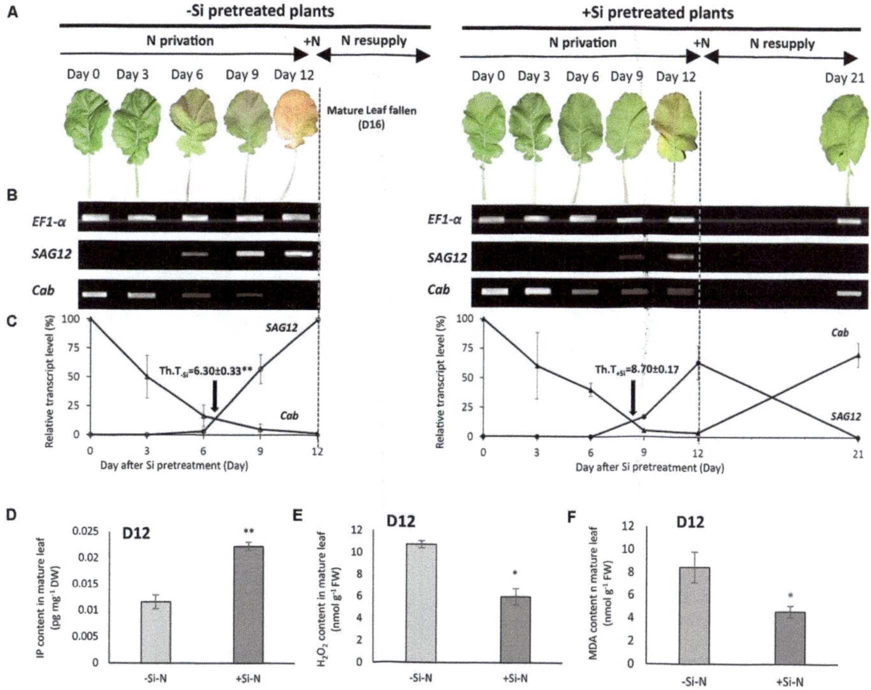

Fig. 6.4 Images, changes in *SAG12/Cab* molecular senescence indicators and active cytokinin, hydrogen peroxide, and malondialdehyde contents in mature leaf from N-deprived *Brassica napus* pretreated or not with Si. (**a**) Images of mature leaf from plants pretreated without (–Si) or with Si (+Si) cultivated during 12 days without N (–N; from D0 to Day 12) and resupply with N (+N; from Day 12). (**b**) *BnaSAG12* and *Cab* gene expressions monitored by RT-PCR in mature leaves harvested after 0, 3, 6, 9, and 12 days of N starvation and 9 days after N resupply (day 21). *EF1-α* was used as a cDNA synthesis and amplification control. (**c**) Quantification of *BnaSAG12* and *Cab* transcripts using ImageLabTM software after normalization with the *EF1-α* gene. Arrows indicate the theoretical time (Th.T.) of the onset of senescence of mature leaf determined as previously described by Gombert et al. (2006). (**d**) Active cytokinin (N_6-Isopentenyladenine: IP) content in mature leaf at 12 from plants pretreated without (–Si) or with (–Si) Si and cultivated without N (–N) during 12 days (From D0 to D12). (**e**) Hydrogen peroxide (H_2O_2) content in mature leaf at 12 from plants pretreated without (–Si) or with (–Si) Si and cultivated without N (–N) during 12 days (From D0 to D12). (**f**) Malondialdehyde (MDA) content in mature leaf at 12 from plants pretreated without (–Si) or with (–Si) Si and cultivated without N (–N) during 12 days (From D0 to D12). Data are means ± SE (for $n = 3$). * and ** indicate significant differences between –Si and +Si plants with $p < 0.01$ and $p < 0.05$, respectively. (Extracted from the paper by Haddad et al. (2018))

severe N deficiency conditions, Si will not be able to mitigate the damaging effects caused by the nutritional stress.

Likewise, Si did not attenuate the damage caused by severe N deficiency in *Ocimum basilicum* L. plants, as reported by Barreto et al. (2022). The authors concluded that, under severe deficiency of this macronutrient, there was no Si effect on improving N accumulation, N-use efficiency, and dry mass production of the plants (Fig. 6.6a, b, Barreto et al. 2022). Furthermore, we can observe in Fig. 6.6 that

Fig. 6.5 BRS Zuri forage (*Panicum maximum*) grown in complete nutrient solution with Si (+Si) and omission of N in the presence of Si (**a**), complete nutrient solution without Si (−Si) and N omission (−N) in the absence of Si (**b**), and forage BRS RB331 Ipyporã (*Brachiaria ruziziensis* × *Brachiaria brizantha*) grown in complete nutrient solution with Si and N omission in the presence of Si (**c**) and complete nutrient solution without Si and N omission in the absence of Si (**d**). (Extracted from the paper by Buchelt et al. (2020))

under N deficiency, regardless of the presence or absence of Si, there was chlorosis in the old leaves, in addition to plant growth being affected.

6.5 Concluding Remarks

From the above, we can conclude that Si mitigates stresses caused by nutritional disorders in plants, including nitrogen deficiency in different processes such as absorption, assimilation, and redistribution.

Therefore, applying Si is a strategy to minimize the damaging effects of N deficiency in agricultural crops, which in most cases is diagnosed only by chemical tissue analysis. Anyway, it is evident that Si mitigates the damaging effects of moderate nutrient stress, reducing its action in severe N deficiency conditions (Buchelt et al. 2020; Barreto et al. 2022).

Fig. 6.6 Basil plants grown in complete nutrient solution in the absence and presence of silicon (**a**) and with omission of N −Si and +Si (**b**). (Extracted from the paper by Barreto et al. (2022))

The benefits of Si on N metabolism draw attention by increasing the duration of the active leaf area, favoring the photosynthetic rate and sustainability of the crops and this needs to be more valued and discussed in research.

The optimal N metabolism in plants is dependent on C:N:P homeostasis in plant tissues, but in N deficiency, this homeostasis is disturbed and Si can reverse it, favoring plant growth, but research is lacking for a better understanding of the mechanisms involved.

The synergy of Si and amino acids can potentiate the response of N-deficient plants to the application of Si, decreasing the polymerization in solution and at the same time having a biostimulating effect of the amino acids, but there is a lack of research to prove this synergy.

In addition, we emphasize the need for further scientific investigations to understand better the nutritional, biochemical, and physiological effects of Si on plants with N deficiency. The mechanisms and processes by which Si acts and allows the mitigation of nutritional stress are still poorly understood, a fact also reported in the review by Pavlovic et al. (2021).

In order to elucidate the gaps in understanding the role of Si and its interaction with nutritional disorders, as already evidenced by Ali et al. (2020), future studies need to address the effects of Si on soil chemistry and differences between plant species regarding Si accumulation. There is also a clear need for further investigations at the molecular level, involving metabolomic, genomic, and proteomic approaches, for a better understanding of the metabolic changes caused by Si in plants grown under abiotic stress, such as the nutritional disorder caused by N deficiency.

References

Ahanger MA, Qin C, Maodong Q, Dong XX, Ahmad P, Abdallah EE, Zhang L (2019) Spermine application alleviates salinity induced growth and photosynthetic inhibition in Solanum lycopersicum by modulating osmolyte and secondary metabolite accumulation and differentially regulating antioxidant metabolism. Plant Physiol Biochem 144:1–13. https://doi.org/10.1016/j.plaphy.2019.09.021

Ali N, Réthoré E, Yvin J-C, Hosseini AS (2020) The regulatory role of silicon in mitigating plant nutritional stresses. Plan Theory 9:9121779. https://doi.org/10.3390/plants9121779

Araújo WBS, Teixeira GCM, De Mello PR et al (2022) Silicon mitigates nutritional stress of nitrogen, phosphorus, and calcium deficiency in two forages plants. Sci Rep 12:6611. https://doi.org/10.1038/s41598-022-10615-z

Barreto RF, Maier BR, Prado RM, Morais TCB, Felisberto G (2022) Silicon attenuates potassium and sulfur deficiency by increasing nutrient use efficiency in basil plants. Sci Hortic 291:110616. https://doi.org/10.1016/j.scienta.2021.110616

Buchelt AC, Teixeira GCM, Oliveira KS et al (2020) Silicon contribution via nutrient solution in forage plants to mitigate nitrogen, potassium, calcium, magnesium, and sulfur deficiency. Soil Sci Plant Nutr 20:1532–1548. https://doi.org/10.1007/s42729-020-00245-7

Cheng LL, Fuchigami LH (2000) Rubisco activation state decreases with increasing nitrogen content in apple leaves. J Exp Robô 51:1687–1694. https://doi.org/10.1093/jexbot/51.351.1687

Chen K, Kumudini SV, Tollenaar M, Vyn TJ (2015) Changes in plant biomass and nitrogen partition between heading and maturity in younger versus older maize hybrids. Field Crops Res 183:315–328

Deus ACF et al (2020) Role of silicon and salicylic acid in mitigating nitrogen deficiency stress in rice plants. SILICON 12:997–1005. https://doi.org/10.1007/s12633-019-00195-5

Eneji AE, Inanaga S, Satoru M, Li J, Hattori T, An P, Tsuji W (2008) Growth and nutrient use in four grasses under drought stress as mediated by silicon fertilizers. J Plant Nutr 31:355–365. https://doi.org/10.1080/01904160801894913

Etesami H, Jeong BR (2018) Silicon (Si): review and future perspectives on the mechanisms of action in relieving biotic and abiotic stresses in plants. Ecotoxicol Environ Saf 147:881–896. https://doi.org/10.1016/j.ecoenv.2017.09.063

Gombert J, Etienne P, Ourry A, Le Dily F (2006) The expression patterns of SAG12/Cab genes reveal the spatial and temporal progression of leaf senescence in Brassica napus L. with sensitivity to the environment. J Exp Bot 57:1949–1956. https://doi.org/10.1093/jxb/erj142

Gou T, Yang L, Hu W, Chen X, Zhu Y, Guo J, Gong H (2020) Silicon improves the growth of cucumber under excess nitrate stress by enhancing nitrogen assimilation and chlorophyll synthesis. Plant Physiol Bioch 152:53–61. https://doi.org/10.1016/j.plaphy.2020.04.031

Haddad C, Arkun M, Jamois F, Schwarzenberg A, Yvin JC, Etienne P, Laîné P (2018) Silicon promotes growth of *Brassica napus* L. and delays leaf senescence induced by nitrogen starvation. Front Plant Sci 9:516. https://doi.org/10.3389/fpls.2018.00516

Hoagland DR, Arnon DI (1950) The water-culture method for growing plants without soil. Circular Calif Agric Exper Station 347:42. https://www.cabdirect.org/cabdirect/abstract/19500302257

Jiao X, Lyu Y, Wu X, Li H, Cheng L, Zhang C, Yuan L, Jiang R, Jiang B, Rengel Z (2016) Grain production versus environmental and resource costs: to increase the sustainability of nutrient use in China. J Exp Robô 67:101–10.4935

Kang J, Chu Y, Ma G, Zhang Y, Zhang X, Wang M, Lu H, Wang L, Kang G, Ma D, Xie Y, Wang C (2022) Physiological mechanisms underlying reduced photosynthesis in wheat leaves grown in the field under conditions of nitrogen and water deficiency. Crop J 10(06):010. https://doi.org/10.1016/j.cj.2022.06.010

Khan MIR, Fatma M, Per TS, Anjum NA, Khan NA (2015) Salicylic acid-induced abiotic stress tolerance and underlying mechanisms in plant. Front Plant Sci 6:462. https://doi.org/10.3389/fpls.2015.00462

Khan MIR, Ashfaque F, Chhillar H, Irfan M, Khan NA (2021) The intricacy of silicon, plant growth regulators and other signaling molecules for abiotic stress tolerance: an entrancing crosstalk between stress alleviators. Plant Physiol Bioch 162:36–47. https://doi.org/10.1016/j.plaphy.2021.02.024

Kumar RR, Rai GK, Kota S et al (2022) Fascinating dynamics of silicon in alleviation of heat stress induced oxidative damage in plants. Plant Growth Regul 98:879. https://doi.org/10.1007/s10725-022-00879-w

Lawlor DW, Konturi M, Young AT (1989) Photosynthesis by flag leaves of wheat in relation to protein, ribulose bisphosphate carboxylase activity and nitrogen supply. J Exp Robô 40:43–52. https://doi.org/10.1093/jxb/40.1.43

Li Y, Heckman J, Wyenandt A, Mattson N, Durner E, Ambos AJ (2020) Potential benefits of silicon nutrition to hydroponically grown sweet basil. Hortic Sci 55:5. https://doi.org/10.21273/HORTSCI15320-20

Marschner H (2011) Mineral nutrition of Marschner's higher plants. Academic Press, Amsterdam

Mengel K, Kirkby EA, Kosegarten H, Appel T (2001) Nitrogen. In: Mengel K, Kirkby EA, Kosegarten H, Appel T (eds) Principles of plant nutrition. Springer. https://doi.org/10.1007/978-94-010-1009-2_7

Moradtalab N, Weinmann M, Walker F et al (2018) Si improves chilling tolerance during early growth of maize by effects on micronutrient homeostasis and hormonal balances. Front Plant Sci 9:420. https://doi.org/10.3389/fpls.2018.00420

Mu X, Chen Q, Chen F, Yuan L, Mi G (2016) Intraleaf allocation of nitrogen in adaptation to low nitrogen supply in corn during the grain filling phase. Front Plant Sci 7:699. https://doi.org/10.3389/fpls.2016.00699

Mu X, Chen Q, Wu X, Chen F, Yuan L, Mi G (2018a) Synthesis of gibberellins is involved in reducing cell flow and elemental growth rate in corn leaves under low nitrogen supply. Environ Exp Robot 150:198–208. https://doi.org/10.1016/j.envexbot.2018.03.012

Mu X, Chen Q, Chen F, Yuan L, Mi G (2018b) Dynamic remobilization of foliar nitrogen components in relation to photosynthetic rate during grain filling in maize. Plant Physiol Bioch 29:874–882. https://doi.org/10.1016/j.plaphy.2018.05.020

Mu X, Chen Y (2021) The physiological response of photosynthesis to nitrogen deficiency. Plant Physiol Biochem 158:76–82. https://doi.org/10.1016/j.plaphy.2020.11.019

Pavlovic J, Kostic L, Bosnic P, Kirkby EA, Nikolic M (2021) Interactions of silicon with essential and beneficial elements in plants. Front Plant Sci 12:697592. https://doi.org/10.3389/fpls.2021.697592

Prado RM (2021) Mineral nutrition of tropical plants. Springer Nature, Cham

Rocha JR, De Mello PR, De Cássia Piccolo M (2022) New outcomes on how silicon enables the cultivation of *Panicum maximum* in soil with water restriction. Sci Rep 12:1897. https://doi.org/10.1038/s41598-022-05927-z

Sales AC, Campos CNS, De Souza Junior JP et al (2021) O silício atenua o estresse nutricional em quinoa (*Chenopodium quinoa* Willd.). Sci Rep 11:14665. https://doi.org/10.1038/s41598-021-94287-1

Santos Sarah MM, Mello Prado R, Teixeira GCM et al (2022) Silicon supplied via roots or leaves relieves potassium deficiency in maize plants. SILICON 14:773–782. https://doi.org/10.1007/s12633-020-00908-1

Shangguan Z, Shao M, Dyckmans J (2000) Effects of nitrogen nutrition and water deficit on net photosynthetic rate and chlorophyll fluorescence in winter wheat. J Plant Physiol 156:46–51. https://doi.org/10.1016/S0176-1617(00)80271-0

Sigua GC, Stone KC, Bauer PJ, Szogi AA (2018) Biomass and nitrogen use efficiency of grain sorghum with nitrogen and supplemental irrigation. Agron J 110:1119–1127. https://doi.org/10.2134/agronj2017.09.0533

Souza Júnior JP, Prado RM, Campos CNS et al (2022) Addition of silicon to boron foliar spray in cotton plants modulates the antioxidative system attenuating boron deficiency and toxicity. BMC Plant Biol 22:338. https://doi.org/10.1186/s12870-022-03721-7

Teixeira GCM, Prado RM, Souza Rocha A, Piccolo MC (2022) Silicon as a sustainable option to increase biomass with less water by inducing carbon:nitrogen:phosphorus stoichiometric homeostasis in sugarcane and energy cane. Front Plant Sci 13:826512. https://doi.org/10.3389/fpls.2022.826512

Tewari RK, Kumar P, Sharma PN (2007) Oxidative stress and antioxidant responses in young leaves of mulberry plants grown under nitrogen, phosphorus or potassium deficiency. J Integr Biol plant 49:313–322. https://doi.org/10.1111/j.1744-7909.2007.00358.x

Xie K, Ren Y, Chen A, Yang C, Zheng Q, Chen J, Wang D, Li Y, Hu SX (2022) Plant nitrogen nutrition: the roles of arbuscular mycorrhizal fungi. J Plant Physiol 269:153591. https://doi.org/10.1016/j.jplph.2021.153591

Chapter 7
Silicon Alleviating Potassium and Phosphorus Deficiency in Plants

Gustavo Caione

7.1 Introduction

Adequate plant nutrition requires availability and absorption of chemical elements in the growth medium. However, nutrient availability fluctuates in time and space, and plants typically face nutrient-poor environments (Prado 2021). Factors such as rising fertilizer prices, stagnant crop prices, and growing concerns about environmental degradation with nutrient loss highlight the importance of increasing nutrient use efficiency so that plants have a higher conversion of each unit of nutrient absorbed into dry biomass (Sarkar et al. 2021).

Researchers have increasingly sought to increase fertilization efficiency. In this context, studies indicate the role of silicon (Si) in plant nutrition to mitigate the effects of nutritional deficiencies, especially potassium (K) and phosphorus (P) deficiency.

Silicon is a beneficial chemical element commonly occurring in plant tissues. It has several functions, including physical regulation of transpiration. Moreover, it forms a barrier of mechanical resistance to the entry of fungi and bacteria into the plant, also hindering the attack of sap sucking and herbivorous insects (Epstein 2001; Melo et al. 2003). Silicon accumulation in the cell wall in the form of amorphous silica makes this element a source of mechanical protection. In the leaf epidermis Si joins cellulose and may be present in stomata, trichomes, and vessel elements (Ma and Yamaji 2006).

In addition to the physical effects, it is important to emphasize the role of Si in biochemical and physiological processes (Miao et al. 2010). Silicon application indirectly correlates with the increase in chlorophyll levels (Jafarei et al. 2015) and photosynthetic rates (Zuccarini 2008; Chen et al. 2016b), the reduction of

G. Caione (✉)
State University of Mato Grosso, Alta Floresta, Brazil
e-mail: gcaione@unemat.br

R. de Mello Prado (ed.), *Benefits of Silicon in the Nutrition of Plants*,
https://doi.org/10.1007/978-3-031-26673-7_7

transpiration rates (Gao et al. 2006), and the increase in water content and nutrient uptake by plants (Mali and Aery 2008).

Research has shown that Si mitigates the symptoms of nutritional deficiencies in crops such as soybean (Miao et al. 2010), sorghum (Chen et al. 2016a, b), barley (Hosseini et al. 2017), pasture (Buchelt et al. 2020), bean (Sarah et al. 2021), barley (Akhter et al. 2022), basil (Barreto et al. 2022a), and corn (Sarah et al. 2022). We aim in this chapter to highlight the results of these studies that indicate the effects of Si in plants grown under K and P deficiency.

7.2 Silicon in the Plant

Depending on the ability of a plant to accumulate Si in its tissues, it can be classified as accumulator, nonaccumulator, and intermediate. This classification considers Si content in the plant leaf. Thus, accumulating plants, which actively absorb Si, have a Si content greater than 10 g kg^{-1}. In turn, intermediate plants, which passively absorb Si, maintain Si levels between 5 and 10 g kg^{-1}. Finally, plants that do not accumulate Si have levels below 5 g kg^{-1} (Ma and Takahashi 2002). The practical importance of this information is that Si benefits are expected to occur more significantly in Si-accumulating plants (Coskun et al. 2019).

Notwithstanding, some plants may present significant variation in Si content in plant tissues as a function of the availability of the element in the medium. Basil, for example, has been shown to be a nonaccumulating or intermediate plant depending on Si content in the nutrient solution (Li et al. 2020). Furthermore, in the presence of Si, basil plants showed an increase in the absorption of some nutrients (Li et al. 2020).

Silicon often increases nutrient use and efficiency (Ali et al. 2020) and mitigates visual symptoms of deficiencies (Buchelt et al. 2020). The symptoms of nutrient deficiency are similar, since nutrients have the same function. However, regardless of plant species, there may be specific responses between species caused by gene expression, which interferes with the distribution of elements and/or the sensitivity of metabolic systems (Prado 2021).

Indeed, as far as we are aware, Si appears to be the only known element that effectively alleviates both biotic (pathogens and pests) and abiotic (e.g., drought, salinity, heavy metals, UV irradiation, nutrient imbalance) stresses in many plant species (Römheld and Kirkby 2010).

7.3 The Role of Silicon in Potassium-Deficient Plants

Potassium activates more than 60 enzymes and directly participates in protein synthesis in plants (Barker and Pilbeam 2015; Prado 2021). It acts in the translocation of carbohydrates from the shoots to the roots and stimulates osmotic regulation, thus reducing water loss through transpiration, balancing the opening and closing of stomata, and increasing cell turgor and expansion (Prado 2021). Potassium deficiency in plants increases reactive oxygen species that degrade chlorophyll, decreasing photosynthesis and increasing transpiration due to inefficiency in water use (Prado 2021; Chen et al. 2016a, b). Thus, the adequate nutritional status of K is necessary to increase the stress tolerance of cultivated plants, in addition to the involvement of K in stress signaling (Römheld and Kirkby 2010).

Nutritional imbalance of K causes osmotic imbalance due to low water absorption, affecting photosynthetic rates and decreasing water use efficiency (Römheld and Kirkby 2010; Barker and Pilbeam 2015; Prado 2021). Furthermore, severe conditions of K deficiency lead to chlorophyll degradation due to increased levels of putrescine. This results in chlorosis and necrosis on the edge of older leaves (Barker and Pilbeam 2015).

Research indicates that Si can mitigate K-deficiency stress by increasing the levels of chlorophyll and antioxidant compounds (carotenoids). This favors photosynthesis (Miao et al. 2010; Chen et al. 2016a, b), regulates stomatal conductance (Sarah et al. 2022), and decreases transpiration, thus increasing water use efficiency (Chen et al. 2016b) and dry matter production (Miao et al. 2010; Chen et al. 2016b).

Furthermore, a study with sorghum plants under water deficit indicates that Si supply in the nutrient solution decreased transpiration rate (Chen et al. 2016b). However, this may indicate that Si can increase water use efficiency in the plant since this physiological aspect is compromised in K-deficient plants.

Some authors state that Si supply via the nutrient solution alleviates the symptoms of K deficiency in soybean (Miao et al. 2010), sorghum (Chen et al. 2016a, b), and barley (Hosseini et al. 2017).

In a more recent research on the bean crop, Sarah et al. (2021) assessed the effect of foliar and root application of Si in the nutrient solution in plants with or without K deficiency. Foliar spraying of Si proved to be agronomically viable for common bean, especially the solution containing stabilized sodium and potassium silicate at a concentration of 5.4 mmol L^{-1}.

Moreover, Si supplied via nutrient solution or foliar application mitigated K-deficiency stress in bean plants (Fig. 7.1). Plants grown in K-deficient nutrient solution showed chlorosis followed by necrosis on the edges of older leaves. These symptoms were alleviated with Si supply, especially via roots.

In forages BRS Zuri and BRS RB331 Ipyporã, silicon application attenuated the effects of potassium deficiency (Fig. 7.2) due to higher nutrient use efficiency (Buchelt et al. 2020). In that study, the authors also verified that Si supply in the solution (fertigation) can be a new strategy in forage grasses. These plants accumulate Si, thus attenuating a nutritional deficiency, which benefits forage production.

In corn, Si applied via roots or leaves mitigated K deficiency stress (Fig. 7.3). The element improved nutritional status and plant growth. The authors found that these benefits were due to Si increasing K use efficiency, water use efficiency, and total chlorophyll and carotenoid content (Sarah et al. 2022).

In common bean, Si supplied via nutrient solution or foliar application also mitigated K deficiency stress in common bean (Fig. 7.4) (Sarah et al. 2021). This

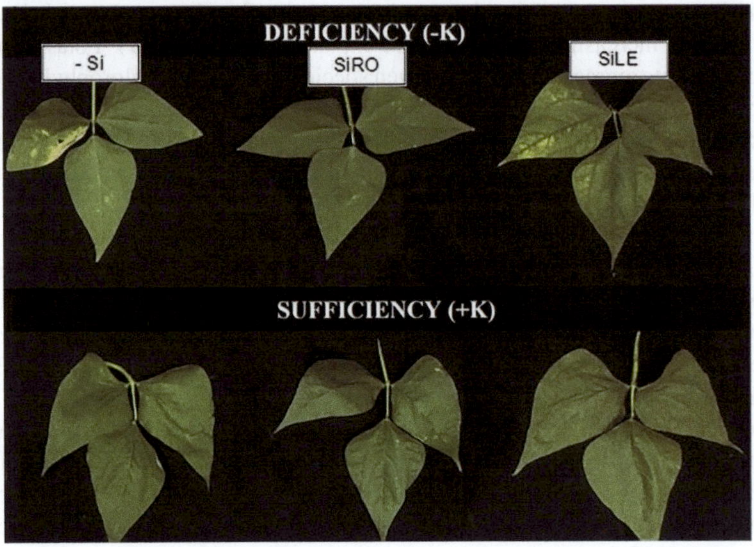

Fig. 7.1 Photo illustrating leaves of bean plants grown in nutrient solution with K deficiency (−K) and with K sufficiency (+K) and which received Si supplied via nutrient solution (SiRO), foliar route (SiLE) and control without Si (−Si)

Fig. 7.2 Photos of forage grown in pots with complete nutrient solution (all nutrients) plus Si (complete+Si) and nutrient solution with omission of K and inclusion of Si (−K+Si) ((**a**) cultivar Zuri; (**c**) cultivar Ipyporã); complete nutrient solution with omission of Si (Complete-Si) and nutrient solution with omission of K and Si (−K−Si) ((**b**) Zuri cultivar; (**d**) Ipyporã cultivar) illustrating the effect of Si in attenuating deficiency symptoms from K

Fig. 7.3 Photos illustrating leaves of K-deficient maize cultivated with Si omission (−Si) and Si supplied via root (SiRO) and foliar (SiLE) (left photo) and with K sufficiency (+K) cultivated with omission of Si and supply of Si via SiRO and SiLE (right photo)

occurred mainly by its effect in increasing nutritional, physiological, and crop growth variables.

Some authors have found similar results in basil (Fig. 7.5) (Barreto et al. 2022a). These authors suggest the inclusion of Si supply for basil, especially in production environments with K and S deficiency. Silicon supply would favor efficiency in the use of these elements for dry matter production, favoring the sustainability of crops that present nutritional limitations.

In barley grown under saline stress, Si supply increased plant growth, synthesis of photosynthetic pigments, leaf water content, and Si and K^+ absorption (Akhter et al. 2022). The dose of 200 ppm of Si was the one that increased plant growth (biomass and length) and photosynthetic pigments while maintaining ionic homeostasis (Na^+ and K^+ balance).

It is worth mentioning that this information on the relationship between Si and K deficiency in plants is still incipient and is limited to a few crops. The perspective is that further studies are carried out in Si accumulating and nonaccumulating crops to deepen the understanding of the responses of plants with different levels of K deficiency. A key response in this context is the increase in water use efficiency associated with mitigation of oxidative stress. A future strategy is to minimize the use of K from the use of Si in plants to preserve a nonrenewable natural resource without decreasing yield. It is also worth noting that Na can reduce K deficiency in plants (Barreto et al. 2022b), and the use of sodium silicate may be interesting.

Moreover, with climate change, field stresses such as water deficit and temperature increase can be more frequent. If combined with low-fertilized soils (especially soils with low K), this should exacerbate the decrease in yield at different levels depending on the species. Therefore, studies assessing the potential of Si to mitigate

Fig. 7.4 Photos of bean plants with K (−K) deficiency grown with Si omission (−Si) and Si supplied via root (SiRO) and foliar (SiLE) and with K sufficiency (+K) grown with omission of Si and supply of Si via SiRO and SiLE

multistress, including K deficiency, are important because of the high possibility of crops having more than one stress in the future.

7.4 The Role of Silicon in Phosphorus-Deficient Plants

Phosphorus deficiency after nitrogen is a key yield-limiting factor in tropical crops. This is because the soils of these crops are highly acid, with high P adsorption in colloids, reducing its availability to plants (Prado 2021). The deficiency causes many disturbances in plant metabolism, as P plays an important role in the energy metabolism of plants and in the formation of ATP and ADP (Lin et al. 2009). The lack of this element compromises the formation of new cells, reducing plant elongation (Sun et al. 2016). It further affects the process of photosynthesis and the activity of ribulose 1,5-bisphosphate carboxylase (Rubisco) (Santos et al. 2015). Phosphorus

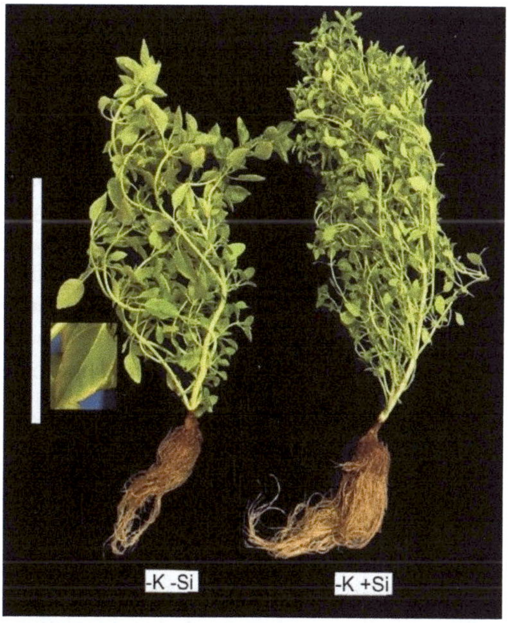

Fig. 7.5 Photo illustrating basil plants grown in nutrient solution with K deficiency and without Si (−K−Si) and with Si (−K+Si)

deficiency increases the production of reactive oxygen species that degrade the cell membrane due to lipid oxidation (Fu et al. 2014; Xu et al. 2007) and decreases the synthesis of plant defense compounds such as phenolic compounds (Ma et al. 2018). Sales et al. (2021) studied these effects on quinoa plants. The authors state that these disturbances in plant P metabolism lead to deficiency symptoms that begin with chlorosis and progress to necrosis in the lower leaves. Moreover, the upper leaves turn pale green and the plant develops slower. In turn, phosphorus deficiency in string beans causes purplish tones only in the older leaves of the lower third of the plant (Osório et al. 2020). The lack of this element initially stimulates root growth, but over time, this strategy fails and root growth is limited. Initially, plants under P limitation generally exhibit greater root branching, greater root length/weight ratio (i.e., specific root length), and greater investment in phosphatases to degrade organic P compounds (Venterink and Güsewell 2010; White et al. 2013). In addition, the roots are likely to present a hormonal imbalance, with a reduced concentration of auxin and an increased concentration of cytokinins and abscisic acid decreasing lateral roots (Vysotskaya et al. 2016). Osório et al. (2020) observed this fact in bean plants from a decrease in root dry matter production, probably due to the lower branching capacity of the root System. Part of these plant P deficiency disorders may decrease with Si supply (Pavlovic et al. 2021). Silicon can increase the production of phenolic compounds, improving the plant defense mechanism (Vega et al. 2019). The antioxidant mechanism stands out in this context (Kim et al. 2017), which prevents the degradation of organic compounds. This favors an increase in chlorophyll content (Cao et al. 2015) and, consequently, in the photosynthetic rate (Berger et al. 2007). In P-deficient sorghum plants, Si application via the nutrient solution increased P accumulation and its use efficiency in sorghum plants. The treatment reduced electrolyte leakage and increased relative chlorophyll, leaf content, phenolic compounds, and the quantum efficiency of photosystem II (Fv/Fm), thus increasing plant dry matter (Silva and Prado 2021). The effect of Si in increasing P uptake in P-deficient wheat plants has to do with an increased expression of genes related to P transporters (Kostic et al. 2017; Pontigo et al. 2015). Further reports show that this beneficial element may favor P uptake even in plants without macronutrient deficiency (Mehrabanjoubani et al. 2015). However, studies show better results of Si in P-deficient than in P-sufficient plants (Silva and Prado 2021). Silicon supply mitigates phosphorus deficiency, which favors the antioxidant system of plants by increasing phenolic compounds. The plasma membrane is thus preserved, with a lower rate of electrolyte leakage. In forage plants *U. brizantha* cv. Marandu and *M. maximum* cv Massai, for instance, this supply has increased the leaf green color index as well as P use efficiency (Araújo et al. 2022). These results are important because most tropical forage plants occupy vast areas worldwide, growing in soils with low P content. Therefore, this improvement in P use efficiency promoted by Si will have a relevant sustainable practical implication in these regions.

P-deficient quinoa plants grown in a nutrient solution that received Si had a decrease in fluorescence energy loss in relation to plants that did not receive Si. This reveals a likely beneficial effect of Si on the photosynthetic system of quinoa plants (Sales et al. 2021). Scarcity of energy in plant metabolism compromises biosynthesis processes. One example is the decrease in the formation of phospholipids, which are essential components of the cell membrane and many other plant metabolic pathways, thereby limiting plant growth. One study induced P deficiency in wheat plants grown in nutrient solution without P and supplemented with silicon, which increased phosphate levels after 30 days of cultivation (Cuyas et al. 2022). The authors found that the phosphate was recycled from phosphorus-containing metabolites in shoots (sugar phosphates, phospholipids, and nucleotides) due to the effect of silicon in promoting phosphate recycling. Alfalfa plants with P deficiency that received Si in association with a microorganism (P-solubilizing bacterium) had attenuation of oxidative damage induced by the stress of low P availability. The treatment decreased the accumulation of reactive oxygen species, leakage of electrolytes, and lipid peroxidation. Moreover, it increased leaf chlorophyll content. This has to do with the ability of the microorganism to modulate enzymatic and nonenzymatic antioxidant molecules, including superoxide dismutase activity, total polyphenols, flavonoids, and proline content (El Moukhtari et al. 2022). In *Holcus lanatus* grown under nutrient deficiency, the plants absorbed more of the nonessential element silicon. Along with it, these plants had decreasing levels of the essential element carbon. In other words, the plants assimilated less "high cost C" and absorbed more "low cost Si" under nutrient limitation, especially under P deficiency (Minden et al. 2021). Knowledge of the benefits of Si in improving the plant defense system against nutritional stress in association with greater C use efficiency can improve plant growth in P-deficient environments. A nutrient solution study assessed two wheat cultivars with contrasting sensitivity to P deficiency (Purple, sensitive;

Fritz, semi-tolerant) that were exposed to P (0, 0.01 or 0.1 mM) and Si (0 or 2 mM). In plants of the sensitive cultivar with P deficiency, Si application increased sucrose levels, starch breakdown, and shoot length, in addition to improving plant dry matter. In the semi-tolerant cultivar Fritz, exposure to Si reduced P concentration and increased shoot length and P use efficiency (PUE) under P scarcity. Interestingly, Si application altered the cell wall composition due to higher mesophyll conductance and net CO_2 assimilation in Fritz plants grown under P stress. In plants under phosphorus deficiency, silicon nutrition favors photosynthesis and carbohydrate levels in a genotype-dependent manner. Furthermore, these results suggest that Si plays an important role in maintaining high photosynthetic rates in P-deficient wheat plants (Ulloa et al. 2021). In tomato plants, silicon supply can increase the tolerance of tomato seedlings to phosphorus deficiency, improving root growth and leaf water status as well as the plant antioxidant capacity and the regulatory effect. These results are useful to better understand the synergistic effects of Si and P. This knowledge provides a theoretical basis to clarify the physiological mechanism of Si-mediated relief in P-deficient tomato seedlings (Liang et al. 2020).

In a review, Hu et al. (2021) reinforced that the beneficial effect of Si in attenuating P deficiency is due to different factors. These include improvement of soil P availability, since Si decreases P adsorption by occupying adsorption sites. Moreover, silicate is strongly basic and its application may thus increase soil pH value, including neutralizing toxic Al if any in the soil. The authors stated that silicon increases organic acids in the rhizosphere, contributing to the absorption of P as well as to the positive regulation of genes responsible for the expression of P transporters. Finally, their study emphasized that Si could indirectly improve nutrient use efficiency through a decrease in Fe and Mn uptake that could complex with P in the plant, further aggravating internal P availability. Silicon application in moderately acidic soils can be an effective method for reducing P deficiency stress and toxicity in oats, the latter by decreasing excess P uptake (Bakır and Horuz 2021). More studies under field conditions are important to improve knowledge on Si application in solution via fertigation in different soils and crops to use P more efficiently in agriculture (Silva and Prado 2021). Further research is also needed to assess variation in C assimilation and photosynthesis rates in response to light and nutrient conditions in relation to Si absorption (Minden et al. 2021).

References

Bakır R, Horuz A (2021) Role of silicon in the phosphorus nutrition and growth of the oats (*Avena sativa* L.) at different phosphorus levels. J Anim Plant Sci 31:1610–1621. https://doi.org/10.36899/JAPS.2021.6.0365

Akhter MS, Noreen S, Saleen N, Saeed M, Ahmad S, Khan TM, Saeed M, Mahmood S (2022) Silicon can alleviate toxic effect of NaCl stress by improving K^+ and Si uptake, photosynthetic efficiency with reduced Na+ toxicity in barley (*Hordeum vulgare* L.). Silicon 14:4991–5000. https://doi.org/10.1007/s12633-021-01270-6

Ali N, Réthoré E, Yvin JC, Hosseini SA (2020) The regulatory role of silicon in mitigating plant nutritional stresses. Plan Theory 9:1779. https://doi.org/10.3390/plants9121779

Araújo WBS, Teixeira GCM, Prado RM, Rocha AMS (2022) Silicon mitigates nutritional stress of nitrogen, phosphorus, and calcium deficiency in two forages plants. Sci Rep 12:6611. https://doi.org/10.1038/s41598-022-10615-z

Barker AV, Pilbeam DJ (2015) Handbook of plant nutrition. CRC Press

Barreto RF, Maier BR, Prado RM, Marais TCB, Felisberto G (2022a) Silicon attenuates potassium and sulfur deficiency by increasing nutrient use efficiency in basil plants. Sci Hortic 291:110616. https://doi.org/10.1016/j.scienta.2021.110616

Barreto RF, Prado RM, Bodelão NC et al (2022b) Na improves the growth of K-deficient but not K-sufficient kale. Food Chem 370:131017. https://doi.org/10.1016/j.foodchem.2021.131017

Berger S, Sinha AK, Roitsch T (2007) Plant physiology meets phytopathology: plant primary metabolism and plant-pathogen interactions. J Exp Bot 58:4019–4026. https://doi.org/10.1093/jxb/erm298

Buchelt AC, Teixeira GCM, Oliveira KS, Rocha AMS, Prado RM, Caione G (2020) Silicon contribution via nutrient solution in forage plants to mitigate nitrogen, potassium, calcium, magnesium, and sulfur deficiency. J Soil Sci Plant Nutr 20:1532–1548. https://doi.org/10.1007/s42729-020-00245-7

Cao B, Ma Q, Zhao Q, Wang L, Xu K (2015) Effects of silicon on absorbed light allocation, antioxidante enzymes and ultrastructure of chloroplasts in tomato leaves under simulated drought stress. Sci Hortic 194:53–62. https://doi.org/10.1016/j.scienta.2015.07.037

Coskun D, Deshmukh R, Sonah H, Menzies JG, Reynolds O, Ma JF, Kronzucker HJ, Bélanger RR (2019) The controversies of silicon's role in plant biology. New Phytol 221:67–85. https://doi.org/10.1111/nph.15343

Chen D, Cao B, Qi L, Yin L, Wang S, Deng X (2016a) Silicon-moderated K-deficiency-induced leaf chlorosis by decreasing putrescine accumulation in sorghum. Ann Bot 118:305–315. https://doi.org/10.1093/aob/mcw111

Chen D, Cao B, Wang S, Liu P, Deng X, Yin L, Shang S (2016b) Silicon moderated the K deficiency by improving the plant-water status in sorghum. Sci Rep 6:1–14. https://doi.org/10.1038/srep22882

Cuyas L, Jing L, Pluchon S, Arkoun M (2022) Effect of Si on P-containing compounds in Pi-sufficient and Pi-deprived wheat. J Soil Sci Plant Nutr 22:1873–1884. https://doi.org/10.1007/s42729-022-00778-z

El Moukhtari A, Lamsaadi N, Farssi O, Oubenali A, El Bzar I, Alami QL, Triqui ZEA, Lazali M, Farissi M (2022) Silicon- and phosphate-solubilizing *pseudomonas alkylphenolica* PF9 alleviate low phosphorus availability stress in alfalfa (*Medicago sativa* L.). Front Agron 4:823396. https://doi.org/10.3389/fagro.2022.823396

Epstein E (2001) Silicon in plants: facts vs concepts. In: Datnoff LE, Snyder GH, Korndörfer GH (eds) Silicon in agriculture, vol 8. Elsevier Science, pp 1–15

Fu YQ, Yang XJ, Shen H (2014) The physiological mechanism of enhanced oxidizing capacity of rice (*Oryza sativa* L.) roots induced by phosphorus deficiency. Acta Phys Plantarum 36:179–190. https://doi.org/10.1007/s11738-013-1398-3

Gao X, Zou C, Wang L (2006) Silicon decreases transpiration rate and conductance from stomata of maize plants. J Plant Nutr 29:1637–1647. https://doi.org/10.1080/01904160600851494

Hosseini SA, Maillard A, Hajirezaei MR, Ali N, Schwarzenberg A, Jamois F, Yvin JC (2017) Induction of barley silicon transporter HvLsi1 and HvLsi2, increased silicon concentration in the shoot and regulated starch and ABA homeostasis under osmotic stress and concomitant potassium deficiency. Front Plant Sci 8:1359. https://doi.org/10.3389/fpls.2017.01359

Hu AY, Xu SN, Qin DN, Li W, Zhao XQ (2021) Role of silicon in mediating phosphorus imbalance in plants. Plan Theory 10:51. https://doi.org/10.3390/plants10010051

Jafarei Y, Tabrizi EFM, Bybordi A (2015) Effect of different stages and times of silicon foliar spray on yield and yield components of bean. Cumhuriyet Sci J 36:81–92. https://doi.org/10.17776/CSJ.13430

Liang Y, Shi Y, Zhao X, Bai L, Hou L, Zhang Y (2020) Effects of silicon on the growth and physiological properties of tomato seedlings under low phosphorus condition. J Zhejiang Univ – Agric Life Sci 46:151–160. https://doi.org/10.3785/j.issn.1008-9209.2019.07.301

Lin Z, Chen LS, Chen RB, Zhang FZ, Jiang HX, Tang N (2009) CO_2 assimilation, ribulose-1,5-bisphosphate carboxylase/oxygenase, carbohydrates and photosynthetic electron transport probed by the JIP-test, of tea leaves in response to phosphorus supply. BMC Plant Biol 9:43. https://doi.org/10.1186/1471-2229-9-43

Kim YH, Khan AL, Waqas M, Lee IJ (2017) Silicon regulates antioxidant activities of crop plants under abiotic-induced oxidative stress: a review. Front Plant Sci 8:510. https://doi.org/10.3389/fpls.2017.00510

Kostic L, Nikolic N, Bosnic D, Samardzic J, Nikolic M (2017) Silicon increases phosphorus (P) uptake by wheat under low P acid soil conditions. Plant Soil 419:447–455. https://doi.org/10.1007/s11104-017-3364-0

Ma D, Zhang J, Hou J, Li Y, Huang X, Wang C, Lu H, Zhu Y, Guo T (2018) Evaluation of yield, processing quality, and nutritional quality in different-colored wheat grains under nitrogen and phosphorus fertilizer application. Crop Sci 58:402–415. https://doi.org/10.2135/cropsci2017.03.0152

Ma JF, Yamaji N (2006) Silicon uptake and accumulation in higher plants. Trends Plant Sci 11:392–397. https://doi.org/10.1016/j.tplants.2006.06.007

Mali M, Aery NC (2008) Influence of silicon on growth, relative water contents and uptake of silicon, calcium and potassium in wheat grown in nutrient solution. J Plant Nutr 31:1867–1876. https://doi.org/10.1080/01904160802402666

Mehrabanjoubani P, Abdolzadeh A, Sadeghipour HR, Aghdasi M (2015) Silicon affects transcellular and apoplastic uptake of some nutrients in plants. Pedosphere 25:192–201. https://doi.org/10.1016/S1002-0160(15)60004-2

Melo SPD, Korndörfer GH, Korndörfer CM, Lana RMQ, Santana DGD (2003) Silicon accumulation and water deficit tolerance in Brachiaria grasses. Sci Agric 60:755–759. https://doi.org/10.1590/S0103-90162003000400022

Miao BH, Han XG, Zhang WH (2010) The ameliorative effect of silicon on soybean seedlings grown in potassium-deficient medium. Ann Bot 105:967–973. https://doi.org/10.1093/aob/mcq063

Minden V, Schaller J, Old Venterink H (2021) Plants increase silicon content as a response to nitrogen or phosphorus limitation: a case study with *Holcus lanatus*. Plant Soil 462:95–108. https://doi.org/10.1007/s11104-020-04667-1

Li Y, Heckman J, Wyenandt A, Mattson N, Durner E, Both AJ (2020) Potential benefits of silicon nutrition to hydroponically grown sweet basil. Hortic Sci 55:1799–1803. https://doi.org/10.21273/HORTSCI15320-20

Ma JF, Takahashi E (2002) Chapter 5 silicon-accumulating plants in the plant kingdom. In: Soil, fertilizer, and plant silicon research in Japan, pp 63–71p. https://doi.org/10.1016/b978-044451166-9/50005-1

Osório CRWS, Teixeira GCM, Barreto RF, Campos CNS, Leal AJF, Teodoro PE (2020) Macronutrient deficiency in snap bean considering physiological, nutritional, and growth aspects. PLoS One 15:4512. https://doi.org/10.1371/journal.pone.0234512

Pavlovic J, Kostic L, Bosnic P, Kirkby EA, Nikolic M (2021) Interactions of silicon with essential and beneficial elements in plants. Front Plant Sci 12:697592. https://doi.org/10.3389/fpls.2021.697592

Prado RM (2021) Introduction to plant nutrition. In: Mineral nutrition of tropical plants. Springer International Publishing, Cham, pp 1–38. https://doi.org/10.1007/978-3-030-71262-4_1

Pontigo S, Ribera A, Gianfreda L, Mora ML, Nikolic M, Cartes P (2015) Silicon in vascular plants: uptake, transport and its influence on mineral stress under acidic conditions. Planta 242:23–37. https://doi.org/10.1007/s00425-015-2333-1

Römheld V, Kirkby EA (2010) Research on potassium in agriculture: needs and prospects. Plant Soil 335:155–180. https://doi.org/10.1007/s11104-010-0520-1

Sales AC, Campos CNS, de Souza Junior JP, da Silva DL, Oliveira KS, Prado RM, Teodoro LPR, Teodoro PE (2021) Silicon mitigates nutritional stress in quinoa (*Chenopodium quinoa* Willd.). Sci Rep 11:14665. https://doi.org/10.1038/s41598-021-94287-1
Santos KR, Pereira MP, Ferreira ACG, de Rodrigues LCA, de Castro EM, Corrêa FF, Pereira FJ (2015) *Typha domingensis* Pers. growth responses to leaf anatomy and photosynthesis as influenced by phosphorus. Aquat Bot 122:47–53. https://doi.org/10.1016/j.aquabot.2015.01.007
Sarah MMS, Prado RM, Teixeira GCM, Souza Júnior JP, Medeiros RLS, Barreto RF (2022) Silicon supplied via roots or leaves relieves potassium deficiency in maize plants. Silicon 14:773–782. https://doi.org/10.1007/s12633-020-00908-1
Sarah MM, Prado RM, Teixeira GCM, Souza Júnior JP, Medeiros RLS, Barreto RF (2021) Silicon supplied via roots or leaves relieves potassium deficiency in maize plants. Silicon 13:1–10. https://doi.org/10.1007/s12633-020-00908-1
Sarkar D, Rakshit A, Al-Turki AI, Sayyed RZ, Datta R (2021) Connecting biopriming approach with integrated nutrient management for improved nutrient use efficiency in crop species. Agriculture 11:372. https://doi.org/10.3390/agriculture11040372
Silva JLF, Prado RM (2021) Elucidating the action mechanisms of silicon in the mitigation of phosphorus deficiency and enhancement of its response in sorghum plants. J Plant Nutr 44:2572–2582. https://doi.org/10.1080/01904167.2021.1918155
Sun L, Tian J, Zhang H, Liao H (2016) Phytohormone regulation of root growth triggered by P deficiency or Al toxicity. J Exp Bot 67:3655–3664. https://doi.org/10.1093/jxb/erw188
Ulloa M, Nunes-Nesi A, Fonseca-Pereira P, Poblete-Grant P, Reyes-Díaz M, Cartes P (2021) The effect of silicon supply on photosynthesis and carbohydrate metabolism in two wheat (*Triticum aestivum* L.) cultivars contrasting in response to phosphorus nutrition. Plant Phys Biochem 169:236–248. https://doi.org/10.1016/j.plaphy.2021.11.022
Vega I, Nikolic M, Pontigo S, Godoy K, Mora MLL, Cartes P (2019) Silicon improves the production of high antioxidant or structural phenolic compounds in barley cultivars under aluminum stress. Agronomy 9:388–402. https://doi.org/10.3390/agronomy9070388
Venterink HO, Güsewell S (2010) Competitive interactions between two meadow grasses under nitrogen and phosphorus limitation. Funct Ecol 24:877–886. https://doi.org/10.1111/j.1365-2435.2010.01692.x
Vysotskaya LB, Trekozova AW, Kudoyarova GR (2016) Effect of phosphorus starvation on hormone content and growth of barley plants. Acta Physiol Plant 38:108. https://doi.org/10.1007/s11738-016-2127-5
White PJ, George TS, Dupuy LX, Karley AJ, Valentine TA, Wiesel L, Wishart J (2013) Root traits for infertile soils. Front Plant Sci 4:0193. https://doi.org/10.3389/fpls.2013.00193
Xu HX, Weng XY, Yang Y (2007) Effect of phosphorus deficiency on the photosynthetic characteristics of rice plants. Rus J Plant Phys 54:741–748. https://doi.org/10.1134/S1021443707060040
Zuccarini P (2008) Effects of silicon on photosynthesis, water relations and nutrient uptake of *Phaseolus vulgaris* under NaCl stress. Biol Plantarum 52:157–160. https://doi.org/10.1007/s10535-008-0034-3

Chapter 8
Silicon Mitigates the Effects of Calcium, Magnesium, and Sulfur in Plants

Dalila Lopes da Silva and Renato de Mello Prado

8.1 The Relationship Calcium and Silicon

8.1.1 General Aspects

Essential elements for plant growth are defined on the assumption that they are directly involved in the metabolism and cannot be replaced by other elements, and their absence may cause plants to not complete their life cycle (Marschner 2012). Among the plant nutrients, we have Ca, which plays essential roles for the growth and development of plants, whether in its main function of structuring and stabilizing the cell wall and membranes or in processes related to signaling in the face of some stress (Thor 2019). Calcium occurs as a relatively large, divalent ion which readily enters the apoplast and is bound in exchangeable form in cell walls and on the exterior surface of the plasmalemma. It occurs in only very low concentrations in the cytoplasm and chloroplasts and appears to have a limited role as an enzymatic cofactor (Kirkby and Pilbeam 1984).

Like Ca, silicon (Si) also acts in the defense against stresses, whether of biotic or abiotic origin, by increasing the production of antioxidants, which help combat oxidative stress and reduce cell leakage in plant tissue membranes (Kim et al. 2014). In addition, Ca and Si are absorbed via mass flow in the xylem, show greater accumulation in older leaves, and show low redistribution in the phloem (Brackhage et al. 2013).

D. L. da Silva (✉)
University of São Paulo, Center of Nuclear Energy in Agriculture, Piracicaba, Brazil
e-mail: dalila.ls@usp.br

R. de Mello Prado
São Paulo State University, Jaboticabal, São Paulo, Brazil
e-mail: rm.prado@unesp.br

R. de Mello Prado (ed.), *Benefits of Silicon in the Nutrition of Plants*,
https://doi.org/10.1007/978-3-031-26673-7_8

Although Si is not an essential element for plants like Ca, it can affect cell wall composition, especially in accumulating species (>10 g kg^{-1} dry matter), and the accumulation of other essential nutrients for plant development. The addition of Si in small concentrations can increase the concentration of Ca in the roots and shoots, as well as its accumulation, while also partially restoring membrane integrity and helping to maintain plant growth, but only under conditions of moderate to adequate Ca supply (Mali and Aery 2008).

Given the similarities between Ca and Si, it is important to study these elements together, but little is known about the effect of Si from a Ca point of view. In this chapter, we provide an updated overview of the relationship between Ca and Si in plants, considering the biochemical, physiological, and structural aspects regarding the importance of these elements in agricultural production and human health.

Calcium is an essential macronutrient for plants and constitutes about 5% of the dry weight in the shoot (Marschner 1995; White and Broadley 2003). It fulfills a dual function in the plant, being necessary in several structural roles for cell wall and membrane stability but also serving as an intracellular messenger in the cytosol, coordinating responses in many developmental and physiological processes, including the plant response to biotic stress and an counter-cation action for organic and inorganic anions in the vacuole (Marschner 1995). The main function of Ca in plants is to provide structural integrity to cell membranes, controlling their permeability, in addition to acting on cell signaling (Prado 2021).

Si is considered a beneficial element for plants, as it can stimulate their growth, but it is not essential (Ma and Yamaji 2006). Among the main benefits of Si for plants, especially those that are under stress, whether biotic or abiotic, are: decrease of cellular oxidation and of damage caused by water stress (Mali and Aery 2008; Kim et al. 2014; Sarto et al. 2019; Buchelt et al. 2020; Teixeira et al. 2020b), improvements in leaf architecture, increased photosynthetic efficiency, greater rigidity to leaf tissues due to silica deposition, greater accumulation of Ca in conditions of macronutrient deprivation, and in some cases potentiate Ca absorption in plants grown without any nutrient restriction (Silva et al. 2021a, b).

Calcium is absorbed by plants in its ion form, Ca^{2+}, from the soil solution via mass flow through the roots and delivered to the shoot passively, using the transpiration current, with transport carried out in a unidirectional way via xylem, and redistribution via phloem being very rare (Prado 2021). It has a higher proportion in the apoplast (30–40% total Ca of the plant) and a lower one in the cytoplasm, which is controlled due to the fact that most of the Ca in the vacuole is in the form of calcium oxalate, which generates an ionic cation balance that maintains a low concentration (Prado 2021).

In general, Ca deficiency in plants is not common; however, it is important to note that to fulfill its structural role, calcium concentrations in tissues must be sufficient (White and Broadley 2003). Since the redistribution of Ca via phloem in the plant is almost null, the supply of this nutrient in the plant is highly dependent on transpiration and transport via xylem. Therefore, Ca deficiency in plants occurs mainly in the youngest parts of the plant, where less transpiration occurs (White and Broadley 2003; Thor 2019).

Si is taken up by plants in the form of silicic acid [$Si(OH)_4$], transported by the transpiration stream and polymerized as water is lost by evaporation from the aerial parts of the plant in the form of amorphous silica [SiO_2 -nH_2O] (Ma et al. 2001), formed by the polymerization of high concentration silicic acid (>2 mmol L^{-1}) (Epstein 1994).

The absorption of Si by plants occurs in an active and passive way, in which each mechanism will depend on plant species, external concentration of Si in the solution, and activity of Si transporters (Lsi1 and Lsi2) in the root cell membranes, which decrease its activity according to the capacity of accumulation by plants (Ma and Yamaji 2006). Regarding the accumulation of Si in their tissues, plants can be considered Si accumulators (>10 g kg^{-1}), intermediate (5–10 g kg^{-1}), and non-Si accumulators (<5 g kg^{-1}) (Ma et al. 2001). After absorption and translocation to different plant organs, Si is accumulated or deposited as phytoliths, while water is lost through transpiration (Jian et al. 2006; Piperno 2006; Nawaz et al. 2019).

The effect of Si supply on Ca absorption is still controversial in the literature. Some studies show that plants enriched with Si tend to decrease Ca absorption at doses higher than 1.66 mmol L^{-1} (Ma and Takahashi 1993), whereas recent works have demonstrated a synergistic effect, already observed in research with cabbage and rocket produced in a hydroponic system, which, when receiving Si (2.5 mmol L^{-1} of stabilized potassium silicate with sorbitol) as a supplement, had Ca levels in the shoot increase significantly (Silva et al. 2021a, b).

This synergistic effect on Ca absorption mediated by the supply of Si, in theory, can be explained by the similarities that Ca and Si present, allowing the suggestion that Si can replace carbon (Schaller et al. 2012) in some of its functions, such as structural support (McNaughton et al. 1985), due to the ability of Si to form bonds in the carboxylic and phenolic radicals with the four atoms at once instead of forming long chains (polymers). However, the chains formed by Si, unlike those formed by carbon, are not so stable, although Si can bind to oxygen. In addition to that, amorphous silica compounds formed, known as phytoliths, can be deposited anywhere in the plant, within or between the cells, or even as part of the cell wall, but they cannot be reallocated after deposition (Inanaga et al. 1995; Cooke and Leishman 2011; Minden et al. 2021). Allied to these facts, the absorption of Si is energetically less expensive than that of carbon, being 1:10 to 1:20 by weight, respectively (O'Reagain and Mentis 1989; Raven 1983). This synergistic effect of the supply of Si in increasing absorption and accumulation of Ca in plants can be seen in Fig. 8.1, where the addition of Si promoted an increase in Ca absorption in conditions of adequate supply of the nutrient, while also promoting an increase in biomass of the aerial part.

8.1.2 Sources of Calcium and Silicon

The main source of calcium in agriculture is agricultural limestone ($CaCO_3$), originated from carbonate rocks, which, depending on the Ca and magnesium (Mg) oxide contents, can assume more calcic or magnesium characteristics, being

Fig. 8.1 Rocket (**a**) and cabbage (**b**) cultivated in nutritious solution with sufficiency and deficiency of Ca in the presence (+Si) and absence (−Si) of Si in the culture solution. Accumulation of Ca (**c**) and Si (**d**) in the shoot and dry mass of the shoot of rocket plants (**e**). Accumulation of Ca (**f**) and Si (**g**) in the shoot and dry mass of the shoot of cabbage plants (**h**). (Adapted from Silva et al. (2021a, b))

classified as calcitic, magnesian, or dolomitic limestone. The latter is the most used in agriculture, mainly for soil correction, with $CaCO_3$ contents greater than 50% and $MgCO_3$ contents between 40% and 43%. Agricultural gypsum, hydrated lime and quicklime are also very rich sources of Ca widely used in agriculture, as are steel

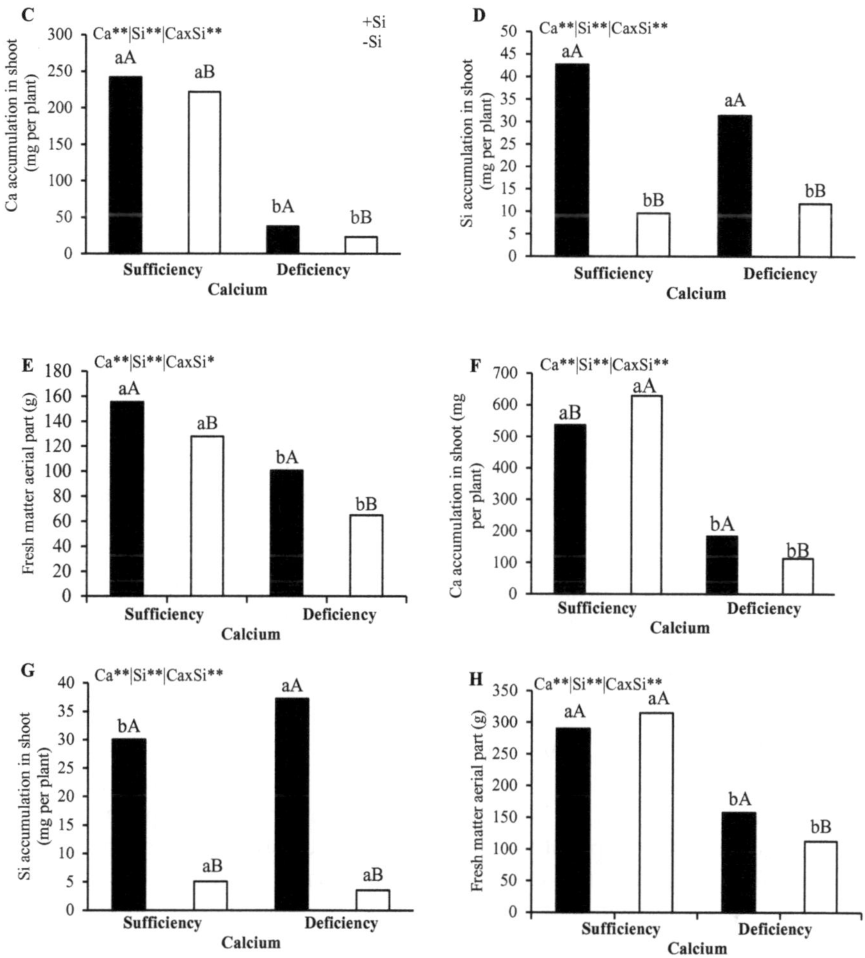

Fig. 8.1 (continued)

residues, which can be an alternative to correct soil acidity, since not only do they have Si in their composition but they also have Ca and Mg (Ma and Takahashi 1993).

The main silicate sources that provide Ca and Si are calcium silicate (Ca_2SiO_4) and calcium and magnesium silicate ($MgSiCO_3$) which, in addition to correcting soil acidity, promote increased absorption of other plant nutrients, photosynthetic efficiency and gas exchange, and biomass production, while also giving plants greater tissue resistance against pathogens and pests and mitigating the effects of heavy metal toxicity (Bokhtiar et al. 2012; Elisa et al. 2016; Lu et al. 2017; Sarto et al. 2019). However, when it comes to choosing the source of Si, whether it is accompanied by Ca or not, we must take into account the following criteria: Si content in the source, solubility, reactivity, and cost-effectiveness.

Considering the effects the Ca_2SiO_4 and $MgSiCO_3$, despite having numerous advantages for plant nutrition for providing Ca, Mg, and Si, which is a mitigating agent with numerous benefits, they have low solubility when compared to sources such as potassium silicate, for example, in addition to the fact that alkaline sources tend to increase the pH of the soil, decreasing the solubility and absorption of plant nutrients, which can lead to losses for the plants. This is similar to what happens when using monosilicic acid, since depending on pH variations and the concentration of Si in free state, this source also presents high rates of polymerization (Annenkov et al. 2017; Mantovani et al. 2020), transforming it into if in a form that plants cannot absorb.

In addition to the sources of Si that provide Ca as well, we can find several Si sources: potassium silicate, sodium and potassium silicate, sodium silicate, magnesium silicate, thermophosphates, and steel residues. Among these, potassium silicate (K_2SiO_3) is the most used, due to its high solubility and positive responses regarding growth, plus increased photosynthetic efficiency and mitigation of abiotic stresses such as water deficit, nutritional deficiency, salinity, and heavy metal toxicity (Buchelt et al. 2020; Hurtado et al. 2020; Teixeira et al. 2020a).

The different sources of Si have their specificities regarding the risk of polymerization and/or solubility, which directly affect the availability of the element to the plant. Recent studies have highlighted the effects of additives to Si sources in order to increase the availability and stability of the Si solution for plants, reducing the risk of polymerization and increasing the efficiency of the beneficial element in plant nutrition with the use of stabilizers in its composition, such as sorbitol (Silva et al. 2021b; Souza et al. 2019; Tenesaca et al. 2021). However, we still find a big difficulty, which is the difficulty of implementing the use of this element in the producer's day-to-day, given its incompatibility with other ions that polymerize the spray solution, since the pH of the solution gets extremely alkaline upon receiving Si. Therefore, research in this segment is still necessary.

8.1.3 Physiological and Biochemical Benefits of Silicon in Mitigating Nutritional Calcium Deficiency

Calcium deficiency enhances the oxidative stress in the plant, increasing the levels of cellular leakage, due to membrane instability under Ca limitation. In situations like this, supplying silicon to improve the absorption of Ca can be advantageous for plant performance, as the use of Si requires less energy during the formation of cellular compounds (Raven 1983) and helps to reduce foliar transpiration, due to silica deposition and to an increase in components such as lignin, cellulose, and hemicellulose in the epidermal walls, thus limiting water losses to the atmosphere, which can result in a more expansive post-harvest life of vegetables (Galati et al. 2015).

This effect provided by Si, if observed from the point of view of plants with nutritional Ca deficiency, can present several benefits, especially considering that these plants present structural damage in membranes, since the main function of Ca in the plant is structuring and cell wall formation, so its lack can lead to increased water loss through tissue transpiration (White and Broadley 2003); therefore, supplementing Si may slow down the rate of water loss.

Regarding crops of vegetables, where the visual aspect is of great importance for commercialization, the damage caused by Ca deficiency can reduce consumer interest in the final product. This effect was verified in cabbage and rocket crops grown in a hydroponic system that received Si in the culture solution to fight the negative effects of nutritional Ca deficiency, increasing the production of non-enzymatic antioxidants (Silva et al. 2021a, b) and directly reflecting the visual aspect of the plants, as can be seen in Fig. 8.1a, b, showing plants with less leaf damage and greater biomass.

The effect of Si supply on visual quality was also observed in tomato fruits grown in Ca sufficiency and deficiency with and without Si spraying (Alonso et al. 2020). Si supply on the shoot, in addition to incrementing the content of photosynthetic pigments, promoted an increase in the number of fruits per plant, which had greater weight and a healthier visual appearance in the Ca deficiency condition, as shown in Fig. 8.2.

The decrease in transpiration and water loss from leaf tissues was also observed in wheat, but it is important to note that this effect has greater impacts when plants are supplemented with lower concentrations of Si (25–200 ppm), since these not only promote less water loss but also hasten the recovery of tissue turgidity because there is an increase in the relative water content in addition to higher levels of Ca in the leaf tissues. This way, greater cell wall stability is guaranteed, with the Si supply partially restoring the integrity of membranes, as observed in hydroponic wheat grown with some Ca restriction (Mali and Aery 2008).

The effect of Si in increasing enzymatic and non-enzymatic antioxidant activity, which helps to combat reactive oxygen species (ROS), is already known (Shen et al. 2010; Kim et al. 2017; Wu et al. 2017; Vega et al. 2019; Souza et al. 2019). If we consider this effect in the increase of non-enzymatic antioxidant compounds such as ascorbate (vitamin C) and carotenoids (vitamin A) (Souza et al. 2019; Silva et al. 2021a), both essential for human health, the effect of Si stands out not only as a stress reliever but also as a potential agent for use in agronomic biofortification.

The supply of Si is also beneficial for increasing photosynthetic pigments and gas exchange efficiency (Harizanova and Koleva-Valkova 2019; Maghsoudi et al. 2016; Teixeira et al. 2020b) in addition to promoting better leaf architecture, which benefits in the capture of light (Ma and Yamaji 2006).

When searching for works that showcase the relationship of Ca and Si and their effects on biochemical and physiological parameters, we found few results; however, those found highlight that there is a synergism which biochemically benefits the plant, demonstrating that more research should be carried out to elucidate how this dynamic is established.

The supply of Si has shown benefits for root growth and increased nutrient uptake, (Mali and Aery 2008; Alonso et al. 2020), and since the roots are the main organs for absorbing water and nutrients for plants, research on this topic is of great

Fig. 8.2 Tomato plants grown under calcium sufficiency (**a**) and deficiency (**b**), tomato plants with Ca deficiency and foliar supply of sodium silicate (**c**), and tomato plants with Ca deficiency and foliar supply of Ca and Si (**d**). The plants were sprayed at intervals of 3 or 4 days, totaling 19 applications, starting 32 days after transplanting

importance. Recent studies in cabbage, rocket, and tomato plants show that the supply of potassium silicate stabilized with sorbitol at a concentration of 2.5 mmol L^{-1} and sodium silicate 7.1 mmol L^{-1} via foliar sprays promotes an increase in biomass of roots in plants with Ca deficiency, in addition to increasing the absorption of the macronutrient in deficient plants and potentiating when sufficient (Alonso et al. 2020; Silva et al. 2021a, b). This increase in root growth and biomass can be attributed to the fact that Si favors some morphological characteristics of plants, such as cell elongation (Hossain et al. 2002); however, this issue still needs more detailed research for a better understanding of the theme.

8.1.4 Calcium and Silicon Essential for Human Health

Calcium is as essential for man as it is for plants, being the third most abundant element in the human body, vital for bone health, and also predominant in teeth, in addition to playing a fundamental role in muscle contraction, blood clotting, etc. When the human body is deficient in this mineral, diseases such as osteoporosis, rickets, and others become common (Sharma et al. 2017). To avoid Ca deficiency, a daily intake of 800–1300 mg is necessary (National Institutes of Health 2022); hence, a diet rich in Ca becomes essential. To achieve this goal, agronomic biofortification has been widely used to increase the content of certain minerals in the edible parts of plant foods.

As mentioned, although Ca deficiency in plants is rare due to the abundance of this element in soils, its absorption occurs in the meristematic growth regions, such as root tips, which makes its absorption zone restricted, not to mention the competition with other cationic elements. Therefore, the enrichment of the soil solution with calcium is important so that its absorption occurs in the appropriate proportions. Based on this and on the relevance of this element for human health, the need of increasing Ca levels in plant foods, especially in raw ones, is highlighted, in order to ensure a fulfilling diet.

Recent studies with Si in vegetables showed that its enrichment can potentiate the absorption, including of Ca, by plants under conditions of adequate supply of the nutrient, also reducing the deleterious effects in conditions of deficiency, plus being an alternative to increase the levels of both Ca and Si in the edible parts of the vegetables (Silva et al. 2021a, b).

In addition to playing important roles mainly in the mitigation of stress in plants, Si is one of the most important trace elements for human health, and contrary to what is observed in plants, for humans Si is essential and is involved in numerous biochemical reactions, also being important for bone health (Martin 2013; D'Imperio et al. 2016; Prescha et al. 2019). Si acts in the prevention and treatment of osteoporosis (Price et al. 2013), and it is present in all body tissues, but with higher concentrations in bones and connective tissues, including skin, hair, arteries and nails (Jugdaohsingh 2007). It increases the anti-inflammatory immune response in the synthesis of lymphocytes as well (Nielsen 2008, 2014), acting as a non-enzymatic antioxidant, but it cannot be synthesized by man (Ma et al. 2021).

Considering the importance of this element for man, the deficiency of some minerals turns into hidden hunger, which represents a big problem in public health, especially in developing countries. In this scenario, several strategies have been adopted to overcome this situation, including agronomic biofortification of plants (increased concentration of nutrients in edible organs of plants through fertilization via soil or leaves). This is a simple and viable alternative because of its ease and low cost in increasing the contents of a given nutrient in the edible portions of plants and other compounds specific to human health (Buturi et al. 2021). In addition to improving the nutritional status of the plant, this technique represents a viable method to sustainably provide nutrients or elements to the population, promoting greater nutritional balance and helping to overcome mineral deficiencies in the body.

Several studies have shown an increase in the concentration of Si in plants (D'Imperio et al. 2016; Montesano et al. 2016; Souza et al. 2019); however, these studies did not aim to evaluate the combined effect of Ca and Si in plant biofortification, which is an opportunity for future research, especially when it comes to leafy and tuberous vegetables, where the final product is consumed raw or pre-cooked.

Research involving the use of Si without biofortification purposes has brought to attention that the increase in the absorption of the element occurs, therefore increasing its concentration in the edible parts, as observed in tomato fruits, in which the foliar spraying of Si (0.2 g L mmol L^{-1}) was an efficient strategy for supplementation in a plant considered not to accumulate this element, in addition to attenuating stress caused by Ca deficiency (Alonso et al. 2020). Recent studies also indicate that it is possible to produce cabbage and rocket biofortified with Si in soilless cultivation, using 2.5 mmol L^{-1} of Si, in addition to enhancing the production and post-harvest quality of these vegetables (Silva et al. 2021a, b). The authors also point out that Si supplementation increases the content of ascorbate (vitamin C) in Swiss chard and cabbage plants as well.

8.2 The Relationship Between Magnesium and Silicon

The magnesium concentration of most soils is generally around 0.5 g kg^{-1} for sandy soils and 5 g kg^{-1} for clayey soils (Mengel et al. 2001). Mg deficiency is common in tropical lands, even more so with inappropriate management that occurs in acidic ones, where corrector materials with very low Mg levels are applied for many years, combined with the use of exceedingly high doses of potassium (K) that can inhibit Mg absorption.

The massive accumulation of carbohydrates and the weakened fixation of CO_2 in leaves with Mg deficiency during photosynthesis cause an excessive reduction in the photosynthetic electron transport chain, which enhance the production of highly reactive O_2 species (ROS) (Cakmak and Kirkby 2008). The authors suggested that maintaining a state of high nutritional value of Mg in plants is therefore essential to avoid the generation of ROS, which occurs at the expense of phloem sugars export inhibition and impairment of CO_2 fixation, particularly under conditions of high luminosity.

According to the literature available, studies on the interactions of Si with Mg are limited, and it is noted that the effect of Si seems to depend on plant species and environmental circumstances (Pavlovic et al. 2021). Greger et al. (2018) reported that the addition of Si increased the absorption and accumulation of Mg in the shoots of several plant species cultivated in culture in solution with optimal supply of nutrients. Other studies also described Si-mediated Mg accumulation in plants exposed to stress, i.e., under excess ammonium in broccoli and cauliflower (Barreto et al. 2017), or under drought in eight sunflower cultivars (Gunes et al. 2009).

However, this effect of Si in increasing the accumulation of Mg needs to be better studied and understood, especially at the level of Mg transporters.

At the moment, research involving Si benefits on Mg mitigation is very restricted. The supply of Si favored the growth of Mg deficient corn plants due to an increase in chlorophyll levels and by regulating sugar metabolism and hormonal balance (Hosseini et al. 2019). In a study with two foraging plants, Buchelt et al. (2020) reported that Si attenuated Mg deficiency in plants, improving dry mass production due to a better efficiency of Mg use, and not to its absorption.

Another study in quinoa plants found that despite adding Si to a nutritious solution for plants with Mg deficiency, the supplementation was not enough to influence plant development (Sales et al. 2021). The authors added that despite this, in the plants deficient in Mg, the supply of Si in the solution promoted larger accumulation of Si in the aerial part, in addition to greater accumulation of N and K. The largest K accumulation under Mg deficiency with Si supply was possibly due to increased H^+-ATPase activity.

The plant's lack of response to the application of Si was probably due to the beneficial element not increasing chlorophyll content and photosynthesis efficiency, because it did not decrease oxidative stress nor leakage of electrolytes from the cell. The Mg deficit led to a decrease close to 80% in shoot dry mass and reached 85% in the decrease of the plant root system, regardless of Si supply. It is possible that the benefit of Si in Mg deficiency mitigation has not been effective because the macronutrient deficiency was much too severe.

8.3 The Relationship Between Sulfur and Silicon

Under long-term sulfur deficiency, the ability of plants to regulate S homeostasis by increasing lateral root growth and to induce S transport/assimilation is overloaded. The photosynthetic activity of plants gradually decreases due to a reduction in chlorophyll content, protein abundance, photosystem I enzymatic activity, and the Rubisco protein (Lunde et al. 2008), and may shorten leaf useful life span (Hawkesford and De Kok 2006). There are indications that the sulfate and molybdate compete for the same carrier and transport locations at absorption, and that sulfate deficiency leads to excess absorption of Mo, but not enough to cause toxicity in the plant (Alhendawi et al. 2005). Sulfur deficiency decreases the efficiency of N use in eggplants, due to the participation of both nutrients in protein synthesis for being constituents of essential amino acids, and of K use as well (Teixeira et al. 2021).

Research on the interactions of Si with S is preliminary (Pavlovic et al. 2021). The deficiency of S allied to Si supplementation resulted in delay in leaf senescence under combined stress of S deficiency and osmotic stress, increasing chlorophyll levels, suppressing the expression of the barley senescence gene HvS40, and creating an even larger relative water and aerial part biomass content. In roots under combined stress, Si induced the expression of the sulfur transporter HvST1-1, which resulted in a larger absorption of SO_4^{2-} and NO_3^-, favoring an increase in the

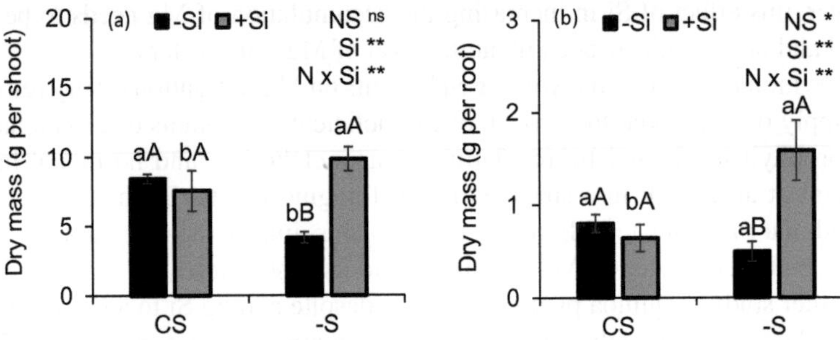

Fig. 8.3 Effect of silicon supply on the attenuation of sulfur deficiency in shoot (**a**) and root (**b**) dry mass production of basil grown in complete nutrient solution (SC) and with omission of S (−S). Capital letters indicate differences with and without Si and lowercase letters indicate differences between nutrient solutions

concentration of glutamine and proline and consequently a larger root biomass (Maillard et al. 2018).

The first study assessing only the deficiency of S and its relationship with Si was carried out in two foraging plant species (Buchelt et al. 2020). The authors observed that the absorption and accumulation of S was not affected by Si supply, and consequently there was no effect on dry mass production of *Panicum maximum* and *Brachiaria ruziziensis* × *Brachiaria brizantha*. However, another study suggested that the use of Si effectively decreased the stress level of S deficiency supported by smaller accumulation of stress phytohormones, and therefore, maintaining plant growth and development (Réthoré et al. 2020).

A recent study evaluated the effect of Si supply on the absorption of S in *Brassica napus* L., a culture of great agronomic interest with high sulfur requirements and low Si accumulation. The authors observed that, although the offering of Si did not lead to changes in sulfur status, a significant effect on gene expression responsible for encoding sulfate transporters was observed (Laîné et al. 2022).

In vegetables such as basil, the use of Si in nutritious solution attenuated damages caused by S deficiency, increasing the efficiency of use of this nutrient and consequently dry mass production in the shoot and root (Fig. 8.3).

8.4 Conclusions and Future Perspectives

Studies in the literature have shown use of silicon is viable, given its benefits in incrementing the production of biomass in view of the increase in defense mechanisms, whether enzymatic or non-enzymatic; accretion of photosynthetic pigments; and enhancing the quality of vegetables and their potential for use in biofortification. However, when it comes to the joint effect of the supply of silicon and calcium, either via root or via foliar sprays, research is still scarce. Nevertheless, existing research demonstrates a synergistic effect between the elements regarding the increase in Ca absorption, in the production of shoot and root biomass as well as of vitamin C, and in post-harvest conservation, in addition to other benefits, requiring further studies to detail the effects of the benefits of Si supply under Ca optics.

Research on the interactions of Si with Mg and S is still incipient, and studies at the level of transporters are necessary to better understand the effects of Si on Mg and S. In addition to knowing the effect of the beneficial element on enzyme activation and other defense mechanisms regarding the generation of reactive oxygen species, especially under stress conditions.

References

Alhendawi RA, Kirkby EA, Pilbeam DJ (2005) Evidence that sulfur deficiency enhances molybdenum transport in xylem sap of tomato plants. J Plant Nutr 28:1347–1353. https://doi.org/10.1081/PLN-200067449

Alonso TAS, Barreto RF, Prado RM et al (2020) Silicon spraying alleviates calcium deficiency in tomato plants, but Ca-EDTA is toxic. J Plant Nutr Soil Sci 183:659–664. https://doi.org/10.1002/jpln.202000055

Annenkov VV, Danilovtseva EN, Pal'shin VA et al (2017) Silicic acid condensation under the influence of water-soluble polymers: from biology to new materials. RSC Adv 7:20995–21027. https://doi.org/10.1039/c7ra01310h

Barreto FR, Schiavon Júnior AA, Maggio MA et al (2017) Silicon alleviates ammonium toxicity in cauliflower and in broccoli. Sci Hortic 225:743–750. https://doi.org/10.1016/j.scienta.2017.08.014

Bokhtiar SM, Huang HR, Li YR (2012) Response of sugarcane to calcium silicate on yield, gas exchange characteristics, leaf nutrient concentrations, and soil properties in two different soils. Commun Soil Sci Plant Anal 43:1363–1381. https://doi.org/10.1080/00103624.2012.670516

Brackhage C, Schaller J, Bäucker E et al (2013) Silicon availability affects the stoichiometry and content of calcium and micronutrients in the leaves of common reed. Silicon 5:199–204. https://doi.org/10.1007/s12633-013-9145-3

Buchelt AC, Teixeira GCM, Oliveira KS et al (2020) Silicon contribution via nutrient solution in forage plants to mitigate nitrogen, potassium, calcium, magnesium, and sulfur deficiency. J Soil Sci Plant Nutr 20:1532–1548. https://doi.org/10.1007/s42729-020-00245-7

Buturi CV, Mauro RP, Fogliano V et al (2021) Mineral biofortification of vegetables as a tool to improve human diet. Foods 10:223. https://doi.org/10.3390/foods10020223

Cakmak I, Kirkby EA (2008) Role of magnesium in carbon partitioning and alleviating photooxidative damage. Physiol Plant 113:692–704. https://doi.org/10.1111/j.1399-3054.2007.01042.x

Cooke J, Leishman MR (2011) Is plant ecology more siliceous than we realise? Trends Plant Sci 16:61–68. https://doi.org/10.1016/j.tplants.2010.10.003

D'Imperio M, Renna M, Cardinali A et al (2016) Silicon biofortification of leafy vegetables and its bioaccessibility in the edible parts. J Sci Food Agric 96:751–756. https://doi.org/10.1002/jsfa.7142

Elisa AA, Ninomiya S, Shamshuddin J et al (2016) Alleviating aluminum toxicity in an acid sulfate soil from Peninsular Malaysia by calcium silicate application. Solid Earth 7:367–374. https://doi.org/10.5194/se-7-367-2016

Epstein E (1994) The anomaly of silicon in plant biology. Proc Natl Acad Sci U S A 91:11–17. https://doi.org/10.1073/pnas.91.1.11

Galati VC, Guimarães JER, Marques KM et al (2015) Silicon, in hydroponics, in postharvest American 'Lucy Brown' lettuce minimally processed. Cienc Rural 45:1932–1938. https://doi.org/10.1590/0103-8478cr20140334

Greger M, Landberg T, Vaculík M (2018) Silicon influences soil availability and accumulation of mineral nutrients in various plant species. Plan Theory 7:41. https://doi.org/10.3390/plants7020041

Gunes A, Kadioglu YK, Pilbeam DJ et al (2009) Influence of silicon on sunflower cultivars under drought stress II: essential and non essential element uptake determined by polarized energy dispersive x-ray fluorescence. Commun Soil Sci Plant Anal 39:1904–1927. https://doi.org/10.1080/00103620802134719

Harizanova A, Koleva-Valkova L (2019) Effect of silicon on photosynthetic rate and the chlorophyll fluorescence parameters at hydroponically grown cucumber plants under salinity stress. J Cent Eur Agric 20:953–960. https://doi.org/10.5513/JCEA01/20.3.2312

Hawkesford MJ, De Kok LJ (2006) Managing sulphur metabolism in plants. Plant Cell Environ 29:382–395. https://doi.org/10.1111/j.1365-3040.2005.01470.x

Hossain MT, Mori R, Soga K et al (2002) Growth promotion and an increase in cell wall extensibility by silicon in rice and some other Poaceae seedlings. J Plant Res 115:23–27. https://doi.org/10.1007/s102650200004

Hosseini SA, Naseri RS, Ali N et al (2019) The ameliorative effect of silicon on maize plants grown in Mg-deficient conditions. Int J Mol Sci 20:969. https://doi.org/10.3390/ijms20040969

Hurtado AC, Chiconato DA, Prado RM et al (2020) Silicon alleviates sodium toxicity in sorghum and sunflower plants by enhancing ionic homeostasis in roots and shoots and increasing dry matter accumulation. Silicon 1–12. https://doi.org/10.1007/s12633-020-00449-7

Inanaga S, Okasaka A, Tanaka S (1995) Does silicon exist in association with organic compounds in rice plant? Soil Sci Plant Nutr 41:111–117. https://doi.org/10.1080/00380768.1995.10419564J

Jian FM, Tamai K, Yamaji N et al (2006) A silicon transporter in rice. Nature 440:688–691. https://doi.org/10.1038/nature04590

Jugdaohsingh R (2007) Silicon and bone health. J Nutr Health Aging 11:99–110. https://www.ncbi.nlm.nih.gov/pmc/articles/PMC2658806/

Kim YH, Khan AL, Waqas M et al (2014) Silicon application to rice root zone influenced the phytohormonal and antioxidant responses under salinity stress. J Plant Growth Regul 33:137–149. https://doi.org/10.1007/s00344-013-9356-2

Kim YH, Khan AL, Waqas M et al (2017) Silicon regulates antioxidant activities of crop plants under abiotic-induced oxidative stress: a review. Front Plant Sci 8:1–7. https://doi.org/10.3389/fpls.2017.00510

Kirkby EA, Pilbeam DJ (1984) Calcium as a plant nutrient. Plant Cell Environ 7:397–405. https://doi.org/10.1111/j.1365-3040.1984.tb01429.x

Laîné P, Coquerel R, Arkoun M et al (2022) Assessing the effect of silicon supply on root sulfur uptake in S-fed and S-deprived *Brassica napus* L. Plants 11:1606. https://doi.org/10.3390/plants11121606

Lu H, Li Z, Wu J et al (2017) Influences of calcium silicate on chemical forms and subcellular distribution of cadmium in *Amaranthus hypochondriacus* L. Sci Rep 7:40583. https://doi.org/10.1038/srep40583

Lunde C, Zygadlo A, Simonsen HT et al (2008) Sulfur starvation in rice: the effect on photosynthesis, carbohydrate metabolism, and oxidative stress protective pathways. Physiol Plant 134:508–521. https://doi.org/10.1111/j.1399-3054.2008.01159.x

Ma JF, Takahashi E (1993) Interaction between calcium and silicon in water-cultured rice plants. Plant Soil 148:107–113. https://doi.org/10.1007/BF02185390

Ma JF, Yamaji N (2006) Silicon uptake and accumulation in higher plants. Trends Plant Sci 11:392–397. https://doi.org/10.1016/j.tplants.2006.06.007

Ma JF, Miyake Y, Takahashi E (2001) Silicon as a beneficial element for crop plants. In: Silicon in agriculture. Elsevier, Stud Plant Sci, pp 17–39. https://doi.org/10.1016/S0928-3420(01)80006-9

Ma F, Luo J, Li X et al (2021) A "switch-on" fluorescence assay based on silicon quantum dots for determination of ascorbic acid. Spectrochim Acta A Mol Biomol Spectrosc 249:119343. https://doi.org/10.1016/j.saa.2020.119343

Maghsoudi K, Emam Y, Pessarakli M (2016) Effect of silicon on photosynthetic gas exchange, photosynthetic pigments, cell membrane stability and relative water content of different wheat cultivars under drought stress conditions. J Plant Nutr 39:1001–1015. https://doi.org/10.1080/01904167.2015.1109108

Maillard A, Ali N, Schwarzenberg A et al (2018) Silicon transcriptionally regulates sulfur and ABA metabolism and delays leaf senescence in barley under combined sulfur deficiency and osmotic stress. Environ Exp Bot 155:394–410. https://doi.org/10.1016/j.envexpbot.2018.07.026

Mali M, Aery NC (2008) Influence of silicon on growth, relative water contents and uptake of silicon, calcium and potassium in wheat grown in nutrient solution. J Plant Nutr 31:1867–1876. https://doi.org/10.1080/01904160802402666

Mantovani C, Pivetta KFL, Prado RM et al (2020) Silicon toxicity induced by different concentrations and sources added to in vitro culture of epiphytic orchids. Sci Hortic 265:109272. https://doi.org/10.1016/j.scienta.2020.109272

Marschner H (1995) Mineral nutrition of higher plants, 3rd edn. Academic Press, London

Marschner P (2012) Mineral nutrition of higher plants. Academic Press, London

Martin KR (2013) Silicon: the health benefits of a metalloid. In: Interrelations between essential metal ions and human diseases. Metal ions in life sciences. Springer, pp 451–473. https://doi.org/10.1007/978-94-007-7500-8

McNaughton SJ, Tarrants JL, McNaughton MM et al (1985) Silica as a defense against herbivory and a growth promotor in African grasses auth. Ecol Soc Am 66:528–535. https://doi.org/10.2307/1940401

Mengel K, Kirkby EA, Kosegarten H et al (2001) Magnesium. In: Mengel K, Kirkby EA, Kosegarten H, Appel T (eds) Principles of plant nutrition. Springer. https://doi.org/10.1007/978-94-010-1009-2_12

Minden V, Schaller J, Olde Venterink H (2021) Plants increase silicon content as a response to nitrogen or phosphorus limitation: a case study with *Holcus lanatus*. Plant Soil 462:95–108. https://doi.org/10.1007/s11104-020-04667-1

Montesano FF, D'Imperio M, Parente A et al (2016) Green bean biofortification for Si through soilless cultivation: Plant response and Si bioaccessibility in pods. Sci Rep 6:1–9 https://doi.org/10.1038/srep31662

National Institutes of Health (2022) Calcium fact sheet for consumers. The ISSN Register. https://ods.od.nih.gov/. Accessed 12 July 2022

Nawaz MA, Zakharenko AM, Zemchenko IV et al (2019) Phytolith formation in plants: from soil to cell. Plants 8:1–38. https://doi.org/10.3390/plants8080249

Nielsen FH (2008) A novel silicon complex is as effective as sodium metasilicate in enhancing the collagen-induced inflammatory response of silicon-deprived rats. J Trace Elem Med Biol 22:39–49. https://doi.org/10.1016/j.jtemb.2007.11.004

Nielsen FH (2014) Update on the possible nutritional importance of silicon. J Trace Elem Med Biol 28:379–382. https://doi.org/10.1016/j.jtemb.2014.06.024

O'Reagain PJ, Mentis MT (1989) Leaf silicification in grasses – a review. J Grassl Soc South Afr 6:37–43. https://doi.org/10.1080/02566702.1989.9648158

Pavlovic J, Kostic L, Bosnic P et al (2021) Interactions of silicon with essential and beneficial elements in plants. Front Plant Sci 12:697592. https://doi.org/10.3389/fpls.2021.697592

Piperno DR (2006) Phytoliths: a comprehensive guide for archaeologists and paleoecologists. Altamira Press, Lanham, p 304

Prado RM (2021) Mineral nutrition of tropical plants. Springer. https://doi.org/10.1007/978-3-030-71262-4

Prescha A, Zabłocka-Słowinska K, Grajeta H (2019) Dietary silicon and its impact on plasma silicon. Nutrients 11:980. https://doi.org/10.3390/nu11050980

Price CT, Koval KJ, Langford JR (2013) Silicon: a review of its potential role in the prevention and treatment of postmenopausal osteoporosis. Int J Endocrinol 2013:316783. https://doi.org/10.1155/2013/316783

Raven JA (1983) The transport and function of silicon in plants. Biol Rev 58:179–207. https://doi.org/10.1111/j.1469-185x.1983.tb00385.x

Réthoré E, Ali N, Yvin JC et al (2020) Silicon regulates source to sink metabolic homeostasis and promotes growth of rice plants under sulfur deficiency. Int J Mol Sci 21:3677. https://doi.org/10.3390/ijms21103677

Sales AC, Campos CNS, Souza Junior JP et al (2021) Silicon mitigates nutritional stress in quinoa (*Chenopodium quinoa* Willd.). Sci Rep 11:14665. https://doi.org/10.1038/s41598-021-94287-1

Sarto MVM, Carmo LM, Rampim L et al (2019) Effects of calcium and magnesium silicate on the absorption of silicon and nutrients in wheat. Semina: Cienc Agrar 40:67–80. https://doi.org/10.5433/1679-0359.2019v40n1p67

Schaller J, Brackhage C, Gessner MO et al (2012) Silicon supply modifies C:N:P stoichiometry and growth of *Phragmites australis*. Plant Biol 14:392–396. https://doi.org/10.1111/j.1438-8677.2011.00537.x

Sharma D, Jamra G, Singh UM et al (2017) Calcium biofortification: three pronged molecular approaches for dissecting complex trait of calcium nutrition in finger millet (*Eleusine coracana*) for devising strategies of enrichment of food crops. Front Plant Sci 7:1–18. https://doi.org/10.3389/fpls.2016.02028

Shen X, Zhou Y, Duan L et al (2010) Silicon effects on photosynthesis and antioxidant parameters of soybean seedlings under drought and ultraviolet-B radiation. J Plant Physiol 167:1248–1252. https://doi.org/10.1016/j.jplph.2010.04.011

Silva DL, Prado RM, Tenesaca LFL et al (2021a) Silicon attenuates calcium deficiency by increasing ascorbic acid content, growth and quality of cabbage leaves. Sci Rep 11:1770. https://doi.org/10.1038/s41598-020-80934-6

Silva DL, Prado RM, Tenesaca LFL et al (2021b) Silicon attenuates calcium deficiency in rocket plants by increasing the production of non-enzymatic antioxidants compounds. Sci Hortic 285:110169. https://doi.org/10.1016/j.scienta.2021.110169

Souza JZ, Prado RM, Silva SLO et al (2019) Silicon leaf fertilization promotes biofortification and increases dry matter, ascorbate content, and decreases post-harvest leaf water loss of chard and kale. Commun Soil Sci Plant Anal 50:164–172. https://doi.org/10.1080/00103624.2018.1556288

Teixeira GCM, Prado RM, Oliveira KS et al (2020a) Silicon increases leaf chlorophyll content and iron nutritional efficiency and reduces iron deficiency in sorghum plant. J Soil Sci Plant Nutr 20:1311–1320. https://doi.org/10.1007/s42729-020-00214-0

Teixeira GCM, Prado RM, Rocha AMS et al (2020b) Silicon in pre-sprouted sugarcane seedlings mitigates the effects of water deficit after transplanting. J Soil Sci Plant Nutr 20:849–859. https://doi.org/10.1007/s42729-019-00170-4

Teixeira GCM, Prado RM, Oliveira KS, Buchelt AC, Rocha AMS, Santos MS (2021) Nutritional deficiency in scarlet eggplant limits its growth by modifying the absorption and use efficiency of macronutrients. PLoS One 16:0252866. https://doi.org/10.1371/journal.pone.0252866

Tenesaca LFL, Prado RDM, Piccolo MDC et al (2021) Silicon modifies C:N:P stoichiometry, and increases nutrient use efficiency and productivity of quinoa. Sci Rep:1–9. https://doi.org/10.1038/s41598-021-89416-9

Thor K (2019) Calcium—nutrient and messenger. Front Plant Sci 10:7. https://doi.org/10.3389/fpls.2019.00440

Vega I, Nikolic M, Pontigo S et al (2019) Silicon improves the production of high antioxidant or structural phenolic compounds in barley cultivars under aluminum stress. Agronomy 8:15. https://doi.org/10.3390/agronomy9070388

White PJ, Broadley MR (2003) Calcium in plants. Ann Bot 92:487–511. https://doi.org/10.1093/aob/mcg164

Wu Z, Liu S, Zhao J et al (2017) Comparative responses to silicon and selenium in relation to antioxidant enzyme system and the glutathione-ascorbate cycle in flowering Chinese cabbage (*Brassica campestris* L. ssp. chinensis var. utilis) under cadmium stress. Environ Exp Bot 133:1–11. https://doi.org/10.1016/j.envexpbot.2016.09.005

Chapter 9
Silicon Mitigates the Effects of Zinc and Manganese Deficiency in Plants

Kamilla Silva Oliveira, Guilherme Felisberto, and Renato de Mello Prado

9.1 Zinc Deficiency in Tropical Plants

Zinc (Zn) is an essential element for plant growth and development (Marschner 2012). It is the second most abundant transition metal in living organisms, showing contents between 15 and 20 mg kg^{-1} of leaf dry matter in crop plants (Broadley et al. 2007).

Only a small portion of the Zn in the soil is available to plants. Zinc deficiency is one of the most limiting factors in agricultural production worldwide, especially in tropical plants (Alloway 2009; Broadley et al. 2007). Low Zn availability occurs in soils that are naturally poor in this element, alkaline (pH > 7) or that have received excessive liming, low redox potential, frequently flooded, high organic matter (complexation), high carbonate and bicarbonate contents, high iron and manganese oxide contents, and high phosphorus availability (Kabata-Pendias 2010; Zeng et al. 2021).

Soil fertilization, foliar application, and genetic improvement are tools for minimizing the effects of low Zn availability to plants. However, in some cases, it may be technically and economically unfeasible. Some species are naturally more efficient at absorbing and accumulating Zn than others (De Mello Prado 2021). It is also true for varieties of the same species. Tropical plants such as sorghum, corn, rice, and beans are highly susceptible to Zn deficiency (Khan et al. 2021).

K. S. Oliveira (✉) · R. de Mello Prado
São Paulo State University, Jaboticabal, São Paulo, Brazil

G. Felisberto
Coordination of Integral Technical Support, São Paulo, Brazil

R. de Mello Prado (ed.), *Benefits of Silicon in the Nutrition of Plants*,
https://doi.org/10.1007/978-3-031-26673-7_9

According to De Mello Prado (2021), Zn has no defined structural function. However, it is an activator of several enzymes, although it can also be part of some metalloenzymes' constitution (Castillo-González et al. 2018).

Zinc is a cofactor of over 300 proteins and the only micronutrient present in all enzyme classes (oxidoreductases, transferases, hydrolases, lyases, isomerases, and ligases) (Gupta et al. 2016; Zlobin 2021). Thanks to such a characteristic, it plays a key role in numerous physiological and biochemical processes as a regulator in gene expression, biosynthesis of photosynthetic pigments, photosynthesis, respiration, hormone regulation, signaling, and stress tolerance (Marschner 2012; Castillo-González et al. 2018; Zeng et al. 2021). Moreover, it is a constituent of superoxide dismutase (Cu-Zn SOD), one of the most important enzymes in the antioxidant defense mechanism of plants (Fig. 9.1) (Cakmak et al. 1997; Marschner 2012).

Zinc deficiency triggers an imbalance of reactive oxygen species (ROS), such as "singlet" oxygen ($^1O_2$), superoxide radical ($O_2^{\bullet-}$), hydrogen peroxide (H_2O_2), and hydroxyl radical ($OH^{\bullet}$), which leads to the so-called oxidative stress (Sharma et al. 2004). Reactive oxygen species degrade nucleic acids, proteins, and membrane lipids, compromising cell structure and function, increasing the leakage of electrolytes from the cell, and inhibiting plant development (Bailey and Mittler 2006). Therefore, Zn acts as a protector and a stabilizing component of biomembranes against ROS by

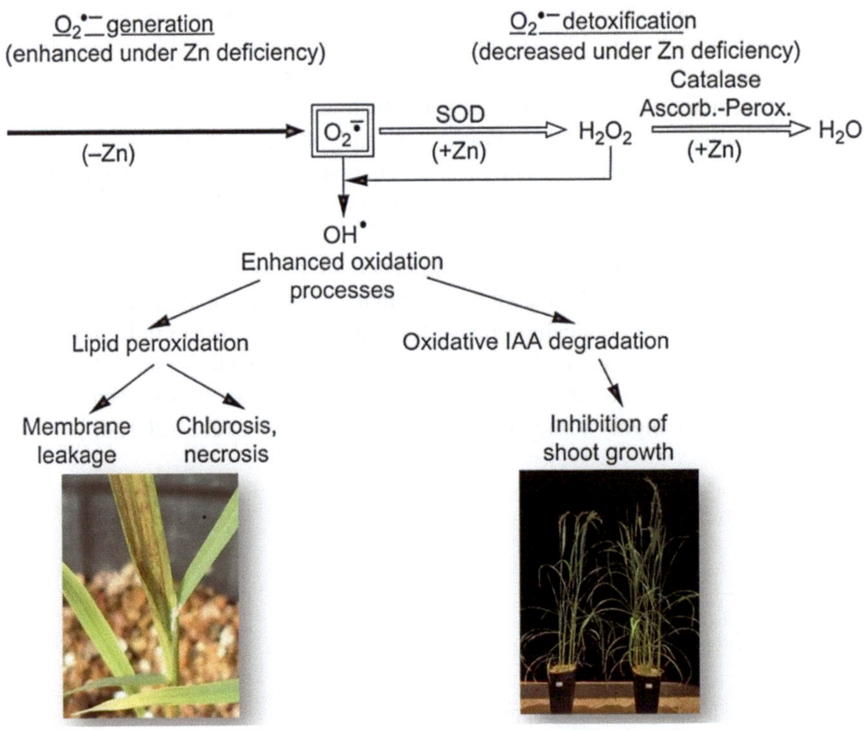

Fig. 9.1 Zn deficiency effects on oxidative stress in plants. (Adapted from Marschner (2012))

controlling the generation and detoxification (Alloway 2009). Thus, oxidative stress in plants with Zn deficiency tends to be more severe because, besides triggering the ROS imbalance, the Cu-Zn SOD exhibits a decreased activity when Zn is absent.

Another important Zn role is to act as a cofactor of plant growth hormones such as auxin, thus regulating plant growth (De Mello Prado 2021). It also has a special role in reproduction, as pollens have a high concentration of Zn (Pandey et al. 2006). Khan et al. (2021) found other important Zn roles in plants.

Zinc deficiency symptoms appear in young leaves as interveinal chlorosis and/or leaf margins, brown spots, slight curvature, size reduction, and lanceolate appearance in leaves and leaflets. Furthermore, there is a reduction in height and dry mass, epinasty, and shortening of the internodes (Marschner 2012; De Mello Prado 2021).

Zinc deficiency causes plant growth and development disturbances due to its role in enzyme activity, carbohydrate and protein metabolism, and phytohormones (Fig. 9.1) (Cakmak 2000; Marschner 2012). Notably, plants show reduced development because it affects auxin synthesis, leaf, flower, fruit, and seed malformation due to reduced carbohydrate synthesis and DNA damage (Khan et al. 2021).

Normally, Zn deficiency in different species results in the shortening of internodes, and the leaves become small. However, there are yellowish (or white) bands between the rib and the edges of leaves, with subsequent purple to brown spots followed by necrosis (De Mello Prado 2021).

9.2 Silicon Mitigates the Effects of Zinc Deficiency in Tropical Plants

Silicon (Si) is commonly found in plant tissues, even though it is not considered a nutrient. The beneficial effects of Si have been reported in many crops and are most prominent in plants subjected to stress (Ma and Yamaji 2006). Recently, some authors have dedicated themselves to studying the mitigating effect of Si to stress in the plant caused by micronutrient deficiency (Hernandez-Apaolaza 2014). However, the effect of silicon in mitigating micronutrient deficiencies is still poorly studied (Pavlovic et al. 2021).

The first studies reporting the effect of Si as a mitigator of micronutrient deficiency, such as Fe, were Bityutskii et al. (2010), Gonzalo et al. (2013), and Pavlovic et al. (2013). Bityutskii et al. (2014) were the first researchers to observe that Si could mitigate the effects of Zn deficiency in plants.

In this review, we detail the proposed mechanisms that indicate the role of Si as a mitigator and attenuator of Zn deficiency. It begins with the plants' acquisition of the micronutrient and the physiological and biochemical benefits that ultimately lead to increased production and productivity of the species compared to the damage they face when exposed to Zn deficiency.

9.2.1 Silicon Influences Zinc Uptake and Accumulation

Bityutskii et al. (2014), when working with Zn-deficient cucumber plants and the supply of Si in the nutrient solution (1.5 mM), observed that the beneficial element minimized the effects of Zn deficiency. However, it did not increase the micronutrient concentration in the different plant compartments. The authors found increased concentrations of metal-mobilizing compounds, such as citrate and fumarate in the roots and fumarate and oxalate in the leaves. They proposed that oxalate acts as a chelate of Zn due to its high affinity, thus increasing the nutrient use efficiency.

Mehrabanjoubani et al. (2015) observed in Zn-deficient rice plants that Si via nutrient solution (1.5 mM) increased the Zn concentration in the aerial part to the extent that it also increased the aerial part/root Zn ratio. The same authors reported an 85% increase in the Zn concentration of rice seeds when plants received 1 μg L^{-1} of Zn. Silicon increased Zn uptake and translocation from roots to aerial part, bark, and seeds.

Again in rice, Felisberto (2018) observed higher Zn accumulation and use efficiency in the aerial part in Zn-deficient and Zn-sufficient plants with Si application via nutrient solution (2 mM). In soybean, the same author observed a similar effect. Silicon application promoted the same Zn accumulation in Zn-deficient and Zn-sufficient plants, while it was higher than in Zn-deficient plants without Si application. According to Pascual et al. (2016), Si doubled the Zn content of the aerial part by increasing Zn remobilization from the root apoplast of soybean plants.

Therefore, it is possible to say that Si can promote a better Zn distribution in the plant, decreasing the nutritional deficiency effects (Bityutskii et al. 2014; Mehrabanjoubani et al. 2015; Pascual et al. 2016). This higher root uptake and better Zn distribution in Zn-deficient plants may be related to the fact that Si affects the development of Casparian strips in Si-accumulating and non-accumulating plants (Fleck et al. 2015), facilitating solute transport.

9.2.2 Silicon Acts on Oxidative Metabolism and Reduces Zinc Deficiency Symptoms

Zinc deficiency symptoms are very characteristic, with a decrease in size and, above all, chlorosis followed by necrotic spots on the plants' youngest leaves. The reduction in size and change in shape is related to damage to auxin, carbohydrate, and protein synthesis. On the other hand, chlorosis and necrosis effects are related to the impacts of the oxidative stress generated.

When provided to plants, Si maintains chlorophyll at adequate levels in stressed plants since chloroplasts are sites that generate ROS, which damage this important organelle. Silicon seems to increase the expression of genes that regulate the synthesis and activity of enzymes important in the antioxidant metabolism and system,

Fig. 9.2 Effect of Si supply on the development of visual symptoms of Zn deficiency in cucumber plants subjected to the micronutrient deficiency. (Taken from Bityutskii et al. (2014))

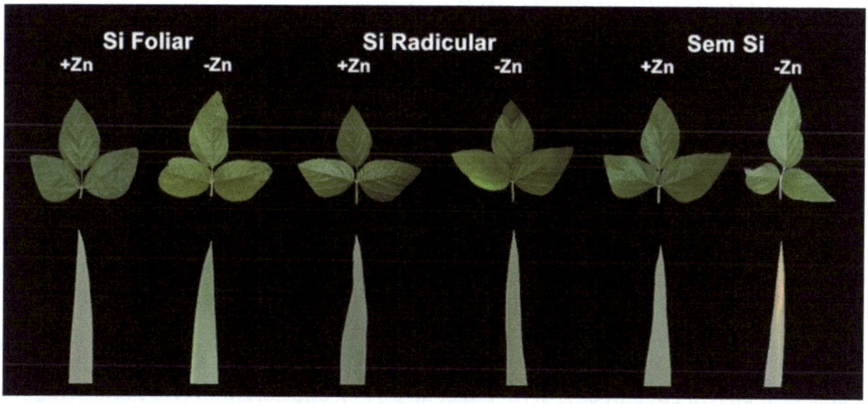

Fig. 9.3 Soybean (top) and rice (bottom) leaves as a function of foliar (Foliar Si), root (Root Si), and no Si application (No Si), and zinc presence (+Zn) and absence (−Zn). (Taken from Felisberto (2018))

responsible for fighting ROS by decreasing oxidative stress damage (Brunings et al. 2009; Cao et al. 2015; Ghareeb et al. 2011; Li et al. 2012).

In this regard, Si prevented the necrotic spots on the leaves caused by Zn deficiency in cucumber plants (Fig. 9.2) (Bityutskii et al. 2014), although it showed a reduction in plant size. According to the authors, Si may have indirectly promoted increased antioxidant defense in plants.

Felisberto (2018) found that Si increased SOD activity and consequently decreased H_2O_2 contents and electrolyte leakage, which enabled increased leaf green color index, regardless of whether Zn-deficient or Zn-sufficient. In soybean, the same author observed that malondialdehyde (MDA) contents, a byproduct of oxidative stress, were reduced with foliar Si application but not with root Si application. It resulted in Si-treated plants showing a higher green color index and less visual deficiency effect (Fig. 9.3), corroborating what Pascual et al. (2016) found in Zn-deficient soybean plants.

In sorghum, a species highly susceptible to Zn deficiency, De Farias Guedes et al. (2020) observed that foliar Si application reduced Zn deficiency effects. There was reduced cellular electrolyte leakage and increased chlorophyll, carotenoid, and total protein contents in the leaves due to the Si's capability to reduce ROS damage and induce the plant's antioxidant defense system.

There is still a need for further studies to define the role of Si in mitigating Zn deficiency, especially in oxidative stress mitigation, hormone synthesis, and photosynthetic pigment maintenance in the various tropical crops, which are as important as a food source for humans and animals of zootechnical interest.

9.2.3 Silicon Improves Physiological Responses and Increases Production in Zn-Deficient Plants

Although still incipient, studies evaluating the effects of Si application on Zn-deficient tropical plants present conclusive results and refer to the beneficial element's capability to increase the plant defense system and to the Si's capability to promote greater Zn mobility in the plant (Bityutskii et al. 2014; Mehrabanjoubani et al. 2015; Pascual et al. 2016; Felisberto 2018; De Farias Guedes et al. 2020).

Therefore, this Si effect on Zn-deficient plants is due to molecular, biochemical, and physiological changes, which increase agronomic characteristics. In this context, there has been an increase in root dry mass (Pascual et al. 2016) and the dry mass of the aerial part of soybean plants (Felisberto 2018). There was also an increase in the dry mass of the aerial part in rice (Mehrabanjoubani et al. 2015; Felisberto 2018), cucumber (Bityutskii et al. 2014), and sorghum (De Farias Guedes et al. 2020). Mehrabanjoubani et al. (2015) observed in Zn-deficient rice plants that Si via nutrient solution (1.5 mM) increased the Zn concentration in the aerial part to the extent that it also increased the aerial part/root Zn ratio.

9.3 Manganese Deficiency in Tropical Plants

Manganese (Mn) is an abundant micronutrient in tropical soils with an originally acidic reaction. However, its deficiency has been identified with relative frequency in crops due to inadequate management of replenishment of nutrients extracted by crops and acidity correction with excessive or poorly incorporated lime (De Oliveira et al. 2019). In addition, soils with low Mn parent materials (Broadley et al. 2012), high organic matter contents (Alejandro et al. 2020), and sandy (Uthman et al. 2020) have recurring Mn deficiency problems. They are found in many tropical soils around the world. These specific conditions affect soil pH, redox reactions, and the ligands that interact with Mn affecting its availability (Norvell 1988).

Plants absorb manganese in the bivalent form (Mn^{2+}). However, in soil solution, it is also found as Mn^{3+}, which is unstable, and Mn^{4+}, which forms insoluble oxides and precipitates (Schmidt and Husted 2019). In soils with alkaline pH, the Mn^{2+} is oxidized and converted to insoluble Mn oxides (Schmidt et al. 2016). In soils with high organic matter contents, micronutrient adsorption occurs with the negatively charged particles forming complexes (Millaleo et al. 2010).

The roles of Mn in plant metabolism mainly involve enzyme composition and activation, acting in processes such as photosynthesis, respiration, ionic uptake, hormone control, protein synthesis, and disease resistance (De Mello Prado 2021). One of Mn's most prominent functions is in the chloroplasts, where the photosynthetic process occurs. Manganese makes up the water evolution complex in photosystem II (PSII) in chloroplasts. The Mn_4CaO_5 provides the energy required for the water molecule breakdown, releasing electrons that will be transported through photosystems II and I (PSI), and makes up PSII stabilizing proteins (De Bang et al. 2015; Schmidt et al. 2016).

In addition, Mn makes up SOD (Mn-SOD), which detoxifies $O_2^{\bullet-}$ radicals in a reaction with the formation of H_2O_2 and water (Millaleo et al. 2010; Broadley et al. 2012; Sharma et al. 2012). It participates in the activation of several enzymes, some involved in the synthesis route of phenols and lignin (Burnell 1988), isoprenoids (Wilkinson and Ohki 1988), lipids, and proteins (Broadley et al. 2012; Ness and Woolhouse 1980; Basu et al. 2015). Manganese is also present in the enzyme oxalate oxidase (Requena and Bornemann 1999), acting in the defense against pathogens by destroying fungal toxins and assisting in lignification by producing the H_2O_2 used in synthesis reactions (Darvill et al. 1980). Thus, Mn deficiency causes severe disorders in plant metabolism, affecting the growth and productivity of crops.

Most species have critical foliar Mn concentrations between 10 and 20 mg kg^{-1} (Broadley et al. 2012). For some tropical crops such as corn (*Zea mays*) and soybean (*Glycine max* L.), foliar Mn contents lower than 20 mg kg^{-1} are considered below the appropriate for these species' development. While for wheat (*Triticum aestivum* L.), cotton (*Gossypium hirsutum* L.), and sugarcane (*Saccharum officinarum* L.), the limit leaf content is 25 mg kg^{-1} of Mn in the collected leaves, according to Raij et al. (1996). Therefore, contents below these concentrations in the leaves induce a plant deficiency, triggering several physiological and biochemical disorders.

Manganese deficiency affects the electron transfer process, resulting in energy loss between the photosystems, reducing photochemical efficiency. In barley, Mn deficiency has been reported to cause chlorophyll *a* fluorescence and reductions in PSII D1 protein (Schmidt et al. 2013) and PSII quantum efficiency in barley (Chen et al. 2019) and soybean (Kwano et al. 2016).

Given the loss of electrons in chloroplasts, the formation of reactive oxygen species (ROS) occurs (Sharma et al. 2012), including singlet oxygen, which is formed by transferring excess energy from chlorophylls in PSII to molecular oxygen formed through the breakdown of the water molecule (Dogra and Kim 2020). Carotenoids are pigments that help dissipate the high energy received in the PSII, prevent damage to chlorophylls, and reduce energy losses (Sharma et al. 2012). However, Mn deficiency reduces this pigment synthesis, as demonstrated in sugarcane and energy cane type I (*Saccharum spontaneum* L.) (Oliveira et al. 2021) and type II (Oliveira et al. 2022), and amplifies the formation of singlet oxygen, which mainly targets the PSII component protein D1 (Krieger-Liszkay et al. 2008).

Changes in SOD activity upon Mn deficiency have been reported, both positive in type I energy cane (Oliveira et al. 2021) and negative in sorghum (De Oliveira et al. 2019), sugarcane (Oliveira et al. 2021), and type II energy cane (Oliveira et al. 2022). These changes are attributed to stress levels, species-intrinsic changes, and changes in other isoforms of the SOD enzyme. Overall, both actions lead to accumulations of H_2O_2 and $O_2^{\bullet-}$, causing oxidative stress. In Mn-deficient corn and sorghum plants, there was an accumulation of H_2O_2 in the leaves, causing lipid peroxidation, which is expressed through the content of MDA, formed as the final byproduct of the ROS action in the peroxidation of polyunsaturated acids (Morales and Munné-Bosch 2019), reflecting the damage to plant cells. Similarly, Tahura and Kabir (2021) reported an accumulation of H_2O_2 and $O_2^{\bullet-}$ in Mn-deficient wheat plants.

Photosynthesis, which is dependent on light absorption by chlorophylls and electron transport, tends to be affected by Mn deficiency. In soybean without Mn supply, there have been reductions in CO_2 photosynthesis flux (Moreira et al. 2018) and in net photosynthetic rate in sorghum plants (De Oliveira et al. 2019). Furthermore, the protein content of soybeans (De Oliveira Rocha et al. 2022) and energy cane plants (Oliveira et al. 2021) reduced upon Mn deficiency. Possibly, this damage is associated with this micronutrient's action as an activator of the RNA polymerase enzyme and glycolytic enzymes (Ness and Woolhouse 1980; Strasser et al. 2007), but these studies have not evaluated them.

Plants under Mn deficiency show reduced chlorophyll contents due to the damage to chloroplasts and chlorophyll synthesis (De Bang et al. 2015; Schmidt et al. 2016). In Mn-deficient barley plants, there were drastic reductions in the SPAD index, which reflect an indirect measure of chlorophyll (Tahura and Kabir 2021). The same occurred in corn and sorghum plants (Oliveira et al. 2020) and sugarcane and type I energy cane (Oliveira 2022). In barley, reductions in chlorophyll contents were as high as 34% in Mn-deficient plants (Schmidt et al. 2013). It reflects in the visual symptom of Mn deficiency in most plant species, which is characterized by chlorosis (yellowing) of young leaves, progressing between the veins (the veins turn green under a yellow background (De Mello Prado 2021).

As a consequence of all these factors, the productivity of tropical crops has suffered losses in Mn deficiency, as is the case of soybean (Kwano et al. 2016; Alt et al. 2018; De Oliveira Rocha et al. 2022), sorghum (De Oliveira et al. 2019; Oliveira et al. 2020), corn (Oliveira et al. 2020), *Saccharum* spp. (Oliveira et al. 2021, 2022), cotton (Dordas 2009), wheat (Barman et al. 2017; Tahura and Kabir 2021), rice (Snyder et al. 1990), and barley (Schmidt et al. 2013).

Therefore, Mn nutritional disorder affects numerous metabolic processes in several tropical plant species, expressing the damage at physiological and biochemical levels, reflecting on crop production.

9.4 Silicon Mitigates the Effects of Manganese Deficiency in Tropical Plants

Studies with Si in plants under micronutrient deficiency, especially Mn, are recent and scarce. Manganese and Silicon interactions were, for many years, the subject of investigations only in situations of micronutrient toxicity (Pavlovic et al. 2021). However, recent reports point to various mitigation mechanisms induced by using Si in plants. Below, we describe the mitigating and alleviating mechanisms of Si in Mn deficiency, ranging from the acquisition of the micronutrient by the plants to physiological and biochemical benefits that ultimately lead to the species' increased production and productivity (Fig. 9.4).

9.4.1 Silicon Influences Manganese Uptake and Accumulation

Manganese uptake by plants under stress conditions from low soil Mn concentration is an important mechanism for plant tolerance to deficiency (Alejandro et al. 2020). There were already indications of Si influencing Mn uptake, as indicated by Gunes et al. (2008) in sunflower (*Helianthus annuus* L.) plants under water stress and by Moradtalab et al. (2018) in corn plants. In the latter, the authors found that corn plants with limited Mn uptake, when subjected to heat stress, increased the Mn contents in the leaves due to the presence of Si. However, studies evaluating the dynamics of Si in plants under specific Mn deficiency with effects on the uptake of the micronutrient are recent.

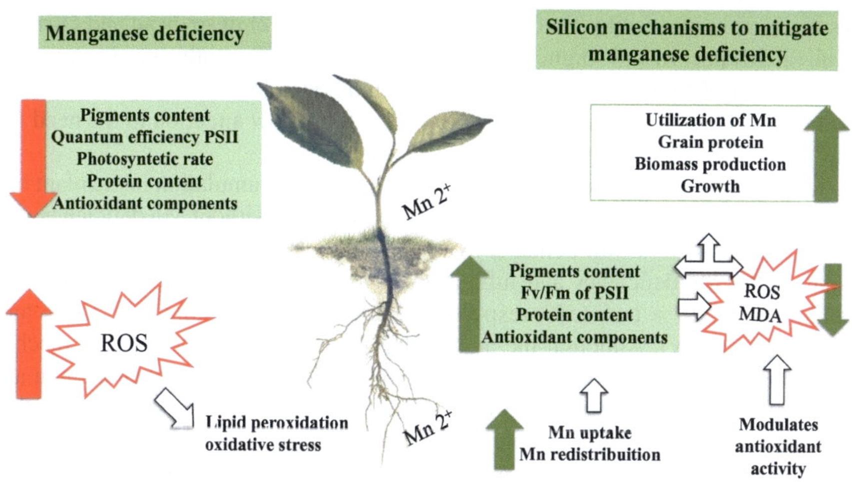

Fig. 9.4 Scheme for elucidating the different mechanisms involved in mitigating and alleviating Mn deficiency by Si in tropical plants. Red arrows indicate the damaging effects of Mn deficiency. Green arrows indicate the benefits of Si on the metabolism of deficient plants

Research has shown that in the presence of Si, several plant species grown under low Mn conditions tend to increase the nutrient accumulation in the plant with both leaf and root Si supply (Oliveira et al. 2020, 2021; De Oliveira Rocha et al. 2022). It is worth noting that the mechanisms involved in both forms of Si application in increasing Mn accumulation may differ. These results are presented given the proven absorption of the beneficial element by the plants.

Following this theory, Oliveira et al. (2020) observed that when Si was applied via foliar application (0.476 g L^{-1}) with enrichment of the aerial part of Mn-deficient corn and sorghum plants, it induced greater Mn accumulation in the aerial part of sorghum plants and in new leaves of corn plants that emerged after the applications. A similar result was recently demonstrated in Mn-deficient soybean plants by De Oliveira Rocha et al. (2022). In these cases, the most accepted theory for the increased accumulation of Mn in the aerial part and young leaves and explored by the authors is that in the presence of Si, Mn can be remobilized, a fact already highlighted by other studies involving other micronutrients (Bityutskii et al. 2014; Pavlovic et al. 2016; Pascual et al. 2016). In these studies, Si favors this aspect by increasing the concentration of metal-mobilizing compounds. This response can occur with Mn bound to the cell wall or precipitated in the apoplast (Zia-ur-Rehman et al. 2016), favoring the nutrient's translocation and redistribution.

The increased Mn accumulation in Si-deficient plants has also been highlighted in recent studies of deficiency correction in plants through foliar sprayings with the two elements combined. In a recent study carried out with sugarcane and type I energy cane plants with foliar applications of Mn and Si, it indicated that when applied in association, the presence of Si amplifies the accumulation of Mn in sub-optimal and optimal concentrations by 74% and 140%, respectively, in sugarcane plants, and in optimal concentration in type I energy cane (56%), amplifying the response of foliar correction of Mn in deficient plants (Oliveira 2022). Similar results were evidenced by De Oliveira Rocha et al. (2022) in soybean and Oliveira et al. (2020) in corn and sorghum plants. To some extent, the authors indicate a possible benefit of Si on cell signaling, which could have induced the action of transporters involved in Mn redistribution. However, they draw attention to the need for further studies to help understand these answers.

In root Si supply, the benefit of Si in increasing Mn accumulation in the plant has been demonstrated in type I energy cane plants with and without Mn deficiency and in sugar cane without Mn deficiency (Oliveira et al. 2021). In type II energy cane plants under Mn deficiency and sufficiency (Oliveira et al. 2022), Si increased Mn accumulation (97% and 15%, respectively) and Mn uptake efficiency (53% and 18%, respectively). Although the authors have not evaluated the internal mechanisms involved in these responses, theories that best explain this phenomenon are suggested, such as: (1) increased expression of genes encoding specific proteins involved in Mn acquisition through roots (Socha and Guerinot 2014; Nikolic et al. 2019; Alejandro et al. 2020); (2) induction of the activity of H+-ATPases as a generator of the electrochemical gradient for nutrient uptake (Liang et al. 2006); (3) induction in transporters that unload the nutrient into phloem vessels (Pavlovic et al.

2013); and (4) increased translocation of the nutrient to the aerial part (Teixeira et al. 2020).

Further studies at the proteomic level are still lacking to concretely elucidate the direct effect of Si on the action of Mn transporters in Mn deficiency. However, it is known that Si has already positively activated Mn transporters in plants in other studies (Bao et al. 2021). Coskun et al. (2018), in a broad review on Si and its action mechanisms, raised an interesting question about whether the effects of Si on element accumulation in plants occur only under stress conditions. It suggested that the benefits of Si are of an indirect order on genes involved in transcription or transporter activity since no effect was observed when plants were under normal conditions.

These indications reflect the results of studies with several elements and nutrients. However, specifically for Mn, we can mention Bityutskii et al. (2014), when they observed in Mn-deficient cucumber plants that Si supply did not affect Mn accumulation despite partial improvement in deficiency symptoms, the same as in sorghum plants studied by De Oliveira et al. (2019). Nevertheless, increased Mn in the aerial part of unstressed plants has been reported by Greger et al. (2018) in corn and wheat on different soils and recently in sugarcane and energy cane type I (Oliveira et al. 2021) and type II (Oliveira et al. 2022).

The results are in line with this discussion and may suggest that Si has some direct action on the uptake mechanism for Mn that has yet to be unraveled. The time of exposure to the beneficial element by the roots and the species studied may be responsible for the recent results.

9.4.2 Silicon Acts on Oxidative Metabolism and Reduces Manganese Deficiency Symptoms

Manganese deficiency affects oxidative metabolism, and in the results of research with Mn, this important plant defense metabolism pathway has been favored by the Si action. Oxidative stress occurs due to the imbalance between production and elimination of oxygen radicals and their derivatives amid stress imposed on plants since production during the life cycle is normal and metabolically necessary (Hasanuzzaman et al. 2020). In a review by Kim et al. (2017) on the action of Si on a wide range of abiotic stresses, except nutritional deficiency, it was elucidated that Si supply has reduced ROS production and influenced antioxidant enzyme activity, especially those involved in eliminating H_2O_2 and reducing MDA. However, these mechanisms are still poorly elucidated in micronutrient deficiency.

Recent studies involving Mn deficiency have also pointed out the large role of Si in this metabolic pathway. They suggest that one of the most important mitigation pathways of Si deficiency is the reduction of oxidative stress, resulting from reduced production or accumulation of ROS. Mainly H_2O_2, such as seen in sorghum plants with Si supply both via root and leaf (De Oliveira et al. 2019) and which are also perceived through reduced leakage of cell contents in corn, sorghum, and soybean plants (Oliveira et al. 2020; De Oliveira Rocha et al. 2022), reflecting lower lipid peroxidation. This mechanism was evidenced in studies with *Saccharum* spp. deficient in Mn and Si applications via root (Oliveira et al. 2021, 2022) and foliar (Oliveira 2022).

The reduction in oxidative stress seems to reflect modulations in the activity of antioxidant enzymes in these plants and the concentration of non-enzymatic compounds responsible for maintaining homeostasis between the production and detoxification of ROS. De Oliveira et al. (2019) investigated Si supply via foliar and root in the nutrient solution of sorghum plants lacking Mn. They proposed that the mitigation of the deficiency occurred by the positive regulation of the activity of the enzyme superoxide dismutase (SOD, EC 1. 15.1.1.1) by Si, which induced lower contents of H_2O_2 and MDA without altering Mn nutrition. This fact was also previously observed by Bityutskii et al. (2014) in cucumbers. Recently, it was also shown in sugarcane and type I energy cane plants that Si applied via foliar sprayings to Mn deficient plants provided lower MDA levels, indicating less lipid peroxidation and oxidative stress (Oliveira 2022).

On the other hand, Oliveira et al. (2021) and De Oliveira Rocha et al. (2022) indicated that reductions in oxidative stress in type I energy cane, type II energy cane, and Mn-deficient soybean plants, respectively, were associated with increased Mn accumulation when in the presence of Si. These authors suggest that increasing Mn accumulation increased the contents of total phenols and carotenoids, which are non-enzymatic antioxidant agents, resulting in less oxidative damage due to detoxification of ROS from metabolism and lower production. In addition to the increase in phenol and carotenoid contents, SOD and guaiacol peroxidase (GPOX, EC 1.11.1.7) enzyme activities were positively regulated in type II energy cane plants, reducing H_2O_2 and MDA contents in Mn-deficient plants but receiving Si via nutrient solution (Oliveira et al. 2022). The justifications are based on the peculiarity of Mn regarding its roles in metabolism and the interaction with Si. Besides being part of the prosthetic group of the SOD enzyme (Millaleo et al. 2010; Broadley et al. 2012; Sharma et al. 2012), Mn activates enzymes of the phenol synthesis route such as phenylalanine ammonia lyase (PAL) (Burnell 1988) and phytoene synthase, important for carotenoid synthesis (Wilkinson and Ohki 1988). Thus, an increase in Mn content maintained the synthesis of these antioxidants to some extent.

It is worth noting that under conditions where no micronutrient increase by Si was reported but benefited the defense system, other aspects of Si action are proposed. The induction of antioxidant system response in the presence of Si in several stresses has been further investigated in the metabolic pathways at transcriptional levels. These studies have proven that Si induces the transcription of genes involved in the synthesis and gene transcription of molecules with enzymatic and non-enzymatic antioxidant functions (Manivannan and Ahn 2017). Studies reveal the induction of the activity of enzymes involved in the phenylpropanoid pathway, such as peroxidases, polyphenol oxidases, and the PAL enzyme (Rodrigues et al. 2004). Therefore, they indicate a direct action of Si in reducing the formation of ROS, detoxification by the defense system, and maintenance of plant membrane integrity.

Oxidative stress is responsible for numerous physiological disorders in plant metabolism under Mn deficiency. One of them is chlorophyll degradation resulting in the interveinal chlorosis symptom characteristic of Mn deficiency (Broadley et al. 2012) observed in several plant species, cucurbits (Bityutskii et al. 2014), grasses (Oliveira et al. 2020), and legumes (De Oliveira Rocha et al. 2022). Therefore, when reducing plant oxidative stress in plants, Si has alleviated interveinal Mn chlorosis, as reported in sorghum plants (de Oliveira 2017, Fig. 9.5), corn plants (Oliveira et al. 2020, Fig. 9.6a, b), and cucumber plants (*Cucumis sativus* L.) (Bityutskii et al. 2014, Fig. 9.6c), causing a reduction of symptoms with exogenous Si supply. In soybean plants (De Oliveira Rocha et al. 2022, Fig. 9.7), reductions in chlorosis and necrosis at leaf edges were also seen in plants that received no Mn (Fig. 9.7a) or insufficient Mn (0.29 g L^{-1} of Mn) (Fig. 9.7b).

Therefore, the benefits of Si on the biochemistry of plants under Mn deficiency do not seem to be restricted to Si accumulating species or the application through the root, not depending on the micronutrient accumulation.

Fig. 9.5 Leaves of sorghum plants grown without manganese and silicon (**a**), with manganese and no silicon (**b**), without manganese with foliar silicon (**c**), and without manganese with root silicon (**d**). (Taken from de Oliveira (2017))

Fig. 9.6 Interveinal chlorosis in Mn-deficient corn plants that did not receive foliar Si (**a**) and that received Si (0.476 g L^{-1}) (**b**). Interveinal chlorosis from Mn deficiency in cucumber plants grown in a nutrient solution without Si (−Si) and with Si (+Si) (**c**). (Taken from Oliveira et al. (2020) and Bityutskii et al. (2014), respectively)

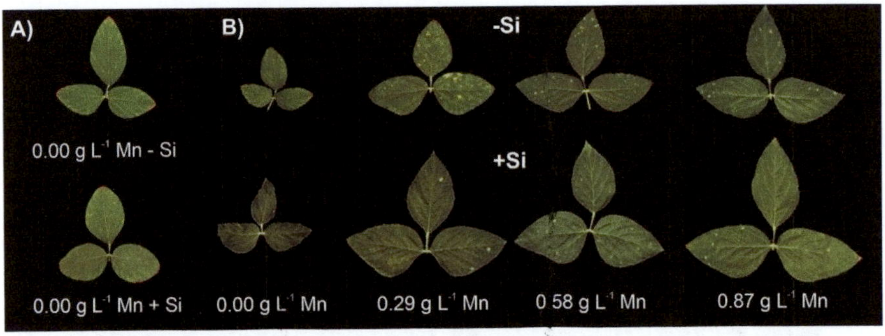

Fig. 9.7 Soybean leaves with interveinal chlorosis and edge necrosis caused by Mn deficiency (**a**) and that received foliar applications of varying concentrations of Mn with Si (0.476 g L^{-1}) and without Si (**b**). (Taken from De Oliveira Rocha et al. (2022))

9.4.3 Silicon Improves Physiological Responses and Increases Production in Mn-Deficient Plants

When it comes to the Mn micronutrient, the mitigating mechanisms of Si in the damage caused by Mn deficiency in crops also reach the physiology, which, added or not with other actions, increases the production, productivity, and product quality of tropical crops. In this regard, studies have highlighted that in Mn deficiency, when Si is applied, plants have improved net photosynthesis (De Oliveira et al. 2019), photochemical efficiency, pigment contents (mainly chlorophyll), dry mass production (De Oliveira et al. 2019; Oliveira et al. 2020, 2021; De Oliveira Rocha et al. 2022), and grain quality (De Oliveira Rocha et al. 2022).

It is worth noting that the improvement in the growth of species in response to Mn deficiency by Si has often been attributed to the two main mechanisms cited above: (1) reduction of oxidative stress and, therefore, reduced degradation of plant membranes; and/or (2) increased accumulation of Mn, thus boosting physiological and biochemical mechanisms (Fig. 9.4), which may occur individually or simultaneously, as verified by Oliveira et al. (2021) in type I and type II energy cane by Oliveira et al. (2022).

Silicon supply in Mn-deficient sorghum plants promoted net photosynthesis and dry mass production equal to plants without the deficiency stress (De Oliveira et al. 2019) without affecting micronutrient accumulation. In this study, the authors suggest that Si reduces oxidative damage, the degradation of molecules that make up pigments and proteins, and also reduces damage to chlorophyll and electron transport in PSII, thus enhancing the photosynthetic response. Chlorophyll is essential for photosynthesis, capturing, distributing, and transforming light. Therefore, its maintenance by the application of Si contributes to increasing photosynthesis. In addition, Si, in general, has kept more PSII reaction centers open, favoring greater energy use of the electron transport chain and increasing photochemical efficiency (Rastogi et al. 2021).

These facts are further consolidated by Oliveira et al. (2020), Oliveira et al. (2021, 2022), and De Oliveira Rocha et al. (2022) when quantifying the electron transport efficiency (Fv/Fm) and pigment content in Mn-deficient corn, sorghum, type I energy cane, type II energy cane, sugar cane, and soybean plants, indicating these Si benefits regarding plants without Si. In these previously cited studies, it is complemented that the Si effect in mitigating Mn deficiency is also due to the increment in Mn accumulation, which was responsible for the observed physiological improvements and associated antioxidant responses that resulted in greater plant dry mass and leaf area.

Given the fact that Mn is mainly involved in photosynthesis and antioxidant defense processes (Alejandro et al. 2020), improvements in these two important points occur with Si supply for plant growth maintenance. De Oliveira et al. (2019) suggested that the increased Mn use efficiency by Mn-deficient sorghum plants in the presence of Si is due to higher photoassimilate conversion by increased photosynthesis and antioxidant defense. Similarly, Si also increased Mn use efficiency in deficient energy cane plants (Oliveira et al. 2021, 2022). Increased plant nutrient efficiency, in addition to physiological benefits in deficiency situations, is of interest when seeking to reduce environmental degradation and fertilizer losses by extending the conversion of nutrients into the harvested product (Sarkar et al. 2021).

Nutrient use efficiency (NUE) is based on two basic principles: (1) the efficiency of nutrient uptake and acquisition by the roots, and (2) the conversion efficiency of nutrients absorbed into biomass and is referred to as utilization efficiency (Nieves-Cordones et al. 2020). Based on the methodology followed by the studies involving the efficient use of Mn and the role of Si, they have emphasized that Si increases plant efficiency in Mn use. Given that the utilization efficiency is associated with the remobilization of the nutrient in the whole plant during its development (Baligar and Fageria 2015), it reinforces the indication that Si plays a role in the remobilization of Mn within plants reflected in productivity increase.

In addition to these facts, in the attenuation of Mn deficiency by Si in soybean, a tropical grain producing crop, Si favoring Mn accumulation also provided higher protein content in the grains of plants that did not receive Mn application, or that received insufficient concentrations of the micronutrient (De Oliveira Rocha et al. 2022). The authors related that Si may have mobilized the micronutrient favoring its redistribution and, consequently, physiological aspects.

Oliveira et al. (2021) observed similar results in Mn-deficient energy cane plants but with root Si supply. Thus, Si not only favored Mn accumulation, which triggered greater protein synthesis, but the increased protein content also reflected the decreased oxidative stress, which preserved the plants' protein structure.

Given the many mechanisms of Si involved in mitigating Mn deficiency, plant dry mass and growth have been favored by this nutritional disorder, either with root or foliar supply of the beneficial element, indicating that using Si under Mn-limiting conditions can increase the productive performance of tropical plant crops.

The future perspectives are to study more species involving the interaction between Si and Mn and to conduct further studies on the underlying proteomic level that Si promotes in fundamental physiological processes in Mn-deficient plants.

References

Alejandro S, Höller S, Meier B et al (2020) Manganese in plants: from acquisition to subcellular allocation. Front Plant Sci 11:300. https://doi.org/10.3389/fpls.2020.00300

Alloway B (2009) Soil factors associated with zinc deficiency in crops and humans. Environ Geochem Health 31:537–548. https://doi.org/10.1007/s10653-009-9255-4

Alt D, Ng SJ, Grusenmeyer J et al (2018) Seed yield and quality of transgenic high-oleic and conventional soybean as influenced by foliar manganese application. Crop Sci 58:874–879. https://doi.org/10.2135/cropsci2017.06.0391

Bailey JS, Mittler R (2006) The role of reactive oxygen species in plant cells. Plant Physiol 141:311. https://doi.org/10.1104/pp.104.900191

Baligar VC, Fageria NK (2015) Nutrient use efficiency in plants: an overview. In: Rakshit A, Singh HB, Sen A (eds) Nutrient use efficiency: from basics to advances. Springer, New Delhi, pp 1–14

Bao Q, Bao W, Li Y et al (2021) Silicon combined with foliar melatonin for reducing the absorption and translocation of Cd and As by *Oryza sativa* L. in two contaminated soils. J Environ Manag 287:112343. https://doi.org/10.1016/j.jenvman.2021.112343

Barman A, Pandey RA, Singh B et al (2017) Manganese deficiency in wheat genotypes: physiological responses and manganese deficiency tolerance index. J Plant Nutr 40:2691–2708. https://doi.org/10.1080/01904167.2017.1381717

Basu D, Tian L, Wang W et al (2015) A small multigene hydroxyproline-O-galactosyltransferase family functions in arabinogalactan-protein glycosylation, growth and development in Arabidopsis. BMC Plant Biol 15:295. https://doi.org/10.1186/s12870-015-0670-7

Bityutskii N, Yakkonen K, Zlotina M (2010) Vliyanie kremniya na proyavlenie khlorosa rasteniy v usloviyakh defizita zhelesa i marganza. Agrokhimiia 2:e51

Bityutskii N, Pavlovic J, Yakkonen K et al (2014) Contrasting effect of silicon on iron, zinc and manganese status and accumulation of metal-mobilizing compounds in micronutrient deficient cucumber. Plant Physiol Biochem 74:205–211. https://doi.org/10.1016/j.plaphy.2013.11.015.9

Broadley MR, White PJ, Hammond JP et al (2007) Zinc in plants. New Phytol 173:677–702. https://doi.org/10.1111/j.1469-8137.2007.01996.x

Broadley M, Brown P, Cakmak I et al (2012) Function of nutrients: micronutrients. In: Marschner P (ed) Mineral nutrition of higher plants. Elsevier, Oxford, pp 191–249

Brunings AM, Datnoff LE, Ma JF et al (2009) Differential gene expression of rice in response to silicon and rice blast fungus *Magnaporthe oryzae*. Ann Appl Biol 155:161–170. https://doi.org/10.1111/j.1744-7348.2009.00347.x

Burnell JN (1988) The biochemistry of manganese in plants. In: Graham RD, Hannam RJ, Uren NC (eds) Manganese in soils and plants. Springer, pp 125–137. https://doi.org/10.1007/978-94-009-2817-6_10

Cakmak I (2000) Possible role of zinc in protecting plant cells from damage by reactive oxygen species. New Phytol 146:185–205. https://doi.org/10.1046/j.1469-8137.2000.00630.x

Cakmak I, Öztürk L, Eker S et al (1997) Concentration of zinc and activity of copper/zinc-superoxide dismutase in leaves of rye and wheat cultivars differing in sensitivity to zinc deficiency. J Plant Physiol 151:91–95. https://doi.org/10.1016/S0176-1617(97)80042-9

Cao B, Ma Q, Zhao Q et al (2015) Effects of silicon on absorbed light allocation, antioxidant enzymes and ultrastructure of chloroplasts in tomato leaves under simulated drought stress. Sci Hortic 194:53–62. https://doi.org/10.1016/j.scienta.2015.07.037

Castillo-González J, Ojeda-Barrios D, Hernández-Rodríguez A et al (2018) Zinc metalloenzymes in plants. Interciencia 43:242–248. https://www.redalyc.org/articulo.oa?id=33957240004

Chen A, Husted S, Salt DE et al (2019) The intensity of manganese deficiency strongly affects root endodermal suberization and ion homeostasis. Plant Physiol 181:729–742. https://doi.org/10.1104/pp.19.00507

Coskun D, Deshmukh R, Sonah H et al (2018) The controversies of silicon's role in plant biology. New Phytol 221:67–85. https://doi.org/10.1111/nph.15343

Darvill JE, McNeil M, Darvill AG et al (1980) Structure of plant cell walls: XI. Glucuronoarabinoxylan, a second hemicellulose in the primary cell walls of suspension-cultured sycamore cells. Plant Physiol 66:1135–1139

De Bang TC, Petersen J, Pedas PR et al (2015) A laser ablation ICP-MS based method for multiplexed immunoblot analysis: applications to manganese-dependent protein dynamics of photosystem II in barley (*Hordeum vulgare* L.). Plant J 83:555–565. https://doi.org/10.1111/tpj.12906

De Farias Guedes VH, Mello Prado R, Frazão JJ et al (2020) Foliar-applied silicon in sorghum (*Sorghum bicolor* L.) alleviate zinc deficiency. Silicon 14:281–287. https://doi.org/10.1007/s12633-020-00825-3

De Mello Prado R (ed) (2021) Mineral nutrition of tropical plants. Springer. https://doi.org/10.1007/978-3-030-71262-4_12

de Oliveira RLL (2017) Application of silicon in the physiology, production and mitigation of stress caused by manganese deficiency in grain sorghum plants. Dissertation, São Paulo State University. http://hdl.handle.net/11449/151146

De Oliveira RLL, Mello Prado M, Felisberto G et al (2019) Silicon mitigates manganese deficiency stress by regulating the physiology and activity of antioxidant enzymes in sorghum plants. J Soil Sci Plant Nutr 19:524–534. https://doi.org/10.1007/s42729-019-00051-w

De Oliveira Rocha IL, Mello Prado R, Oliveira KS et al (2022) Foliar spraying of Mn with addition of Si increases phenolic compound, photosynthetic efficiency, productivity and the protein content of the soybean crop. J Soil Sci Plant Nutr. https://doi.org/10.1007/s42729-022-00780-5

Dogra V, Kim C (2020) Singlet oxygen metabolism: from genesis to signaling. Front Plant Sci 10:1640. https://doi.org/10.3389/fpls.2019.01640

Dordas C (2009) Foliar application of manganese increases seed yield and improves seed quality of cotton grown on calcareous soils. J Plant Nutr 32:160–176. https://doi.org/10.1080/01904160802609013

Felisberto G (2018) Silicon in mitigating zinc deficiency stress in rice and soybean plants. Ph.D. Thesis, São Paulo University State. http://hdl.handle.net/11449/157511

Fleck AT, Schulze S, Hinrichs M et al (2015) Silicon promotes exodermal casparian band formation in Si-accumulating and Si-excluding species by forming phenol complexes. PLoS One 10:e0138555. https://doi.org/10.1371/journal.pone.0138555

Ghareeb H, Bozsó Z, Ott GP et al (2011) Transcriptome of silicon-induced resistance against *Ralstonia solanacearum* in the silicon non-accumulator tomato implicates priming effect. Physiol Mol Plant Pathol 75:83–89. https://doi.org/10.1016/j.pmpp.2010.11.004

Gonzalo MJ, Lucena JJ, Hernandez-Apaolaza L (2013) Effect of silicon addition on soybean (*Glycine max*) and cucumber (*Cucumis sativus*) plants grown under iron deficiency. Plant Physiol Biochem 70:455–461. https://doi.org/10.1016/j.plaphy.2013.06.007

Greger M, Landberg T, Vaculík M (2018) Silicon influences soil availability and accumulation of mineral nutrients in various plant species. Plan Theory 7:41. https://doi.org/10.3390/plants7020041

Gunes A, Kadioglu YK, Pilbeam DJ et al (2008) Influence of silicon on sunflower cultivars under drought stress, ii: essential and nonessential element uptake determined by polarized energy dispersive x-ray fluorescence. Commun Soil Sci Plant Anal 39:1904–1927. https://doi.org/10.1080/00103620802134719

Gupta N, Ram H, Kumar B (2016) Mechanism of zinc absorption in plants: uptake, transport, translocation and accumulation. Rev Environ Sci Biotechnol 15:89–109. https://doi.org/10.1007/s11157-016-9390-1

Hasanuzzaman M, Bhuyan M, Zulfiqar F et al (2020) Reactive oxygen species and antioxidant defense in plants under abiotic stress: revisiting the crucial role of a universal defense regulator. Antioxidants 9:681. https://doi.org/10.3390/antiox9080681

Hernandez-Apaolaza L (2014) Can silicon partially alleviate micronutrient deficiency in plants? A review. Planta 240:447–458. https://doi.org/10.1007/s00425-014-2119-x

Kabata-Pendias A (2010) Trace elements in soils and plants. CRC Press/Taylor & Francis Group, Boca Raton. https://doi.org/10.1201/9781420039900

Khan ST, Malik A, Alwarthan A et al (2021) The enormity of the zinc deficiency problem and available solutions; an overview. Arab J Chem 15:103668. https://doi.org/10.1016/j.arabjc.2021.103668

Kim Y-H, Khan AL, Waqas M, Lee I-J (2017) Silicon regulates antioxidant activities of crop plants under abiotic-induced oxidative stress: a review. Front Plant Sci 8:510. https://doi.org/10.3389/fpls.2017.00510

Krieger-Liszkay A, Fufezan C, Trebst A (2008) Singlet oxygen production in photosystem II and related protection mechanism. Photosynth Res 98:551–564. https://doi.org/10.1007/s11120-008-9349-3

Kwano BH, Moreira A, Moraes LAC et al (2016) Magnesium-manganese interaction in soybean cultivars with different nutritional requirements. J Plant Nutr 40:372–381. https://doi.org/10.1080/01904167.2016.1240198

Li P, Song A, Li Z et al (2012) Silicon ameliorates manganese toxicity by regulating manganese transport and antioxidant reactions in rice (*Oryza sativa* L.). Plant Soil 354:407–419. https://doi.org/10.1007/s11104-011-1076-4

Liang Y, Zhang W, Chen Q, et al (2006) Effect of exogenous silicon (Si) on H+-ATPase activity, phospholipids and fluidity of plasma membrane in leaves of salt-stressed barley (Hordeum vulgare L.). Environ Exp Bot 57:212–219. https://doi.org/10.1016/j.envexpbot.2005.05.012

Ma JF, Yamaji N (2006) Silicon uptake and accumulation in higher plants. Trends Plant Sci 11:392–397. https://doi.org/10.1016/j.tplants.2006.06.007

Manivannan A, Ahn YK (2017) Silicon regulates potential genes involved in major physiological processes in plants to combat stress. Front Plant Sci 8:1346. https://doi.org/10.3389/fpls.2017.01346

Marschner H (2012) Mineral nutrition of higher plants. Elsevier, Amsterdam

Mehrabanjoubani P, Abdolzadeh A, Sadeghipour H et al (2015) Impacts of silicon nutrition on growth and nutrient status of rice plants grown under varying zinc regimes. Theor Exp Plant Physiol 27:19–29. https://doi.org/10.1007/s40626-014-0028-9

Millaleo M, Reyes-Diaz M, Ivanov AG et al (2010) Manganese as essential and toxic element for plants transport, accumulation and resistance mechanisms. J Soil Sci Plant Nutr 10:470–481. https://doi.org/10.4067/S0718-95162010000200008

Moradtalab N, Weinmann M, Walker F et al (2018) Silicon improves chilling tolerance during early growth of maize by effects on micronutrient homeostasis and hormonal balances. Front Plant Sci 9:420. https://doi.org/10.3389/fpls.2018.00420

Morales M, Munné-Bosch S (2019) Malondialdehyde: facts and artifacts. Plant Physiol 180:1246–1250. https://doi.org/10.1104/pp.19.00405

Moreira A, Moraes LAC, Aquino GS (2018) Iron and manganese effect on soil chemical properties, yield components, and nutritional status of soybean. Commun Soil Sci Plant Anal 49:1844–1854. https://doi.org/10.1080/00103624.2018.1475564

Ness PJ, Woolhouse HW (1980) RNA synthesis in Phaseolus chloroplasts: I. Ribonucleic acid synthesis in chloroplast preparations from *Phaseolus vulgaris* L. Leaves and solubilization of the RNA polymerase. J Exp Bot 31:223–233

Nieves-Cordones M, Rubio F, Santa-María GE (2020) Editorial: nutrient use-efficiency in plants: an integrative approach. Front Plant Sci 11:623976. https://doi.org/10.3389/fpls.2020.623976

Nikolic DB, Nesic S, Bosnic D et al (2019) Silicon alleviates iron deficiency in barley by enhancing expression of strategy ii genes and metal redistribution. Front Plant Sci 10:416. https://doi.org/10.3389/fpls.2019.00416

Norvell WA (1988) Inorganic reactions of manganese in soils. In: Graham RD, Hannam RJ, Uren NC (eds) Manganese in soils and plants. Springer, pp 37–58. https://doi.org/10.1007/978-94-009-2817-6_4

Oliveira KS (2022) Responses of sugarcane and energy cane to the application of manganese with silicon via root and leaf. Ph.D. Thesis, São Paulo State University. http://hdl.handle.net/11449/235126

Oliveira KS, Mello Prado R, Guedes VH (2020) Leaf spraying of manganese with silicon addition is agronomically viable for corn and sorghum plants. J Soil Sci Plant Nutr 20:872–880. https://doi.org/10.1007/s42729-020-00173-6

Oliveira KS, Mello Prado R, Checchio MV et al (2021) Silicon via nutrient solution modulates deficient and sufficient manganese sugar and energy cane antioxidant systems. Sci Rep 11:16900. https://doi.org/10.1038/s41598-021-96427-z

Oliveira KS, Mello Prado R, Checchio MV et al (2022) Interaction of silicon and manganese in nutritional and physiological aspects of energy cane with high fiber content. BMC Plant Biol 22:334. https://doi.org/10.1186/s12870-022-03766-8

Pandey N, Pathak GC, Sharma CP (2006) Zinc is critically required for pollen function and fertilisation in lentil. J Trace Elem Med Biol 20:89–96. https://doi.org/10.1016/j.jtemb.2005.09.006

Pascual MB, Echevarria V, Gonzalo MJ et al (2016) Silicon addition to soybean (*Glycine max* L) plants alleviates zinc deficiency. Plant Physiol Biochem 108:132–138. https://doi.org/10.1016/j.plaphy.2016.07.008

Pavlovic J, Samardzic J, Maksimović V et al (2013) Silicon alleviates iron deficiency in cucumber by promoting mobilization of iron in the root apoplast. New Phytol 199:1096–1107. https://doi.org/10.1111/nph.12213

Pavlovic J, Samardzic J, Kostic L et al (2016) Silicon enhances leaf remobilization of iron in cucumber under limited iron conditions. Ann Bot 118:271–280. https://doi.org/10.1093/aob/mcw105

Pavlovic J, Kostic L, Bosnic P et al (2021) Interactions of silicon with essential and beneficial elements in plants. Front Plant Sci 12:1224. https://doi.org/10.3389/fpls.2021.697592

Raij BV, Cantarella H, Quaggio JA et al (1996) Fertilization and liming recommendations for the State of São Paulo. Instituto Agronômico & Fundação IAC, Campinas

Rastogi A, Yadav S, Hussain S et al (2021) Does silicon really matter for the photosynthetic machinery in plant? Plant Physiol Biochem 169:40–48. https://doi.org/10.1016/j.plaphy.2021.11.004

Requena L, Bornemann S (1999) Barley (*Hordeum vulgare*) oxalate oxidase is a manganese-containing enzyme. Biochem J 343:185–190. https://doi.org/10.1042/bj3430185

Rodrigues FA, McNally DJ, Datnoff LE et al (2004) Silicon enhances the accumulation of diterpenoid phytoalexins in rice: a potential mechanism for blast resistance. Phytopathology 94:177–183. https://doi.org/10.1094/PHYTO.2004.94.2.177
Sarkar A, Rakshit AI, Al-Turki RZ et al (2021) Connecting bio-priming approach with integrated nutrient management for improved nutrient use efficiency in crop species. Agriculture 11:372. https://doi.org/10.3390/agriculture11040372
Schmidt SB, Husted S (2019) The biochemical properties of manganese in plants. Plants 8:381. https://doi.org/10.3390/plants8100381
Schmidt SB, Pedas P, Laursen KH et al (2013) Latent manganese deficiency in barley can be diagnosed and remediated on the basis of chlorophyll a fluorescence measurements. Plant Soil 372:417–429. https://doi.org/10.1007/s11104-013-1702-4
Schmidt SB, Jensen PE, Husted S (2016) Manganese deficiency in plants: the impact on photosystem II. Trends Plant Sci 21:622–632. https://doi.org/10.1016/j.tplants.2016.03.001
Sharma PN, Kumar P, Tewari RK (2004) Early signs of oxidative stress in wheat plants subjected to zinc deficiency. J Plant Nutr 27:451–463. https://doi.org/10.1081/PLN-120028873
Sharma P, Jha AB, Dubey RS et al (2012) Reactive oxygen species, oxidative damage, and antioxidative defense mechanism in plants under stressful conditions. J Bot 2012:217037. https://doi.org/10.1155/2012/217037
Snyder GH, Jones DB, Coale FJ (1990) Occurrence and correction of manganese deficiency in histosol-grown rice. Soil Sci Soc Am J 54:1634. https://doi.org/10.2136/sssaj1990.03615995005
Socha AL, Guerinot ML (2014) Mn-euvering manganese: the role of transporter gene family members in manganese uptake and mobilization in plants. Front Plant Sci 5:106. https://doi.org/10.3389/fpls.2014.00106
Strasser R, Bondili JS, Vavra U et al (2007) A unique β1,3-galactosyltransferase is indispensable for the biosynthesis of N-glycans containing Lewis a structures in *Arabidopsis thaliana*. Plant Cell 19:2278–2292. https://doi.org/10.1105/tpc.107.052985
Tahura S, Kabir AH (2021) Physiological responses and genome-wide characterization of TaNRAMP1 gene in Mn-deficient wheat. Plant Physiol Biochem 162:280–290. https://doi.org/10.1016/j.plaphy.2021.02.038
Teixeira GCM, Mello Prado R, Oliveira KS et al (2020) Silicon increases leaf chlorophyll content and iron nutritional efficiency and reduces iron deficiency in sorghum plants. J Soil Sci Plant Nutr 20:1311–1320. https://doi.org/10.1007/s42729-020-00214-0
Uthman QO, Kadyampakeni DM, Nkedi-Kizza P (2020) Manganese adsorption, availability, and uptake in citrus under microsprinkler irrigation. Agrosyst Geosci Environ 3:20061. https://doi.org/10.1002/agg2.20061
Wilkinson RE, Ohki K (1988) Influence of manganese deficiency and toxicity on isoprenoid syntheses. Plant Physiol 87:841–846. https://doi.org/10.1104/pp.87.4.841
Zeng H, Wu H, Yan F et al (2021) Molecular regulation of zinc deficiency responses in plants. J Plant Physiol 261:153419. https://doi.org/10.1016/j.jplph.2021.153419
Zia-ur-Rehman M, Akmal F et al (2016) Role of silicon under nutrient deficiency: advances and future prospects. In: Tripathi DK, Singh VP, Ahmad P, Chauhan DK, Prasad SM (eds.) Silicon in Plant Advances and Future Prospects. Boca Raton: FL:CRC Press, pp. 29–46. https://doi.org/10.1201/9781315369310-3
Zlobin IE (2021) Current understanding of plant zinc homeostasis regulation mechanisms. Plant Physiol Biochem 162:327–335. https://doi.org/10.1016/j.plaphy.2021.03.003

Chapter 10
Silicon Mitigates the Effects of Boron Deficiency and Toxicity in Plants

Davie Kadyampakeni and Jonas Pereira de Souza Júnior

10.1 Introduction

Boron (B) content in tropical soils is generally low, limiting agricultural production (Shorrocks 1997; Kirkby et al. 2001; Furlani et al. 2003; Brdar-Jokanović 2020; Atique-ur-Rehman et al. 2020), because of the low content of organic matter, which is the main source of the element in tropical soils. In addition, the micronutrient has high mobility in the soil and may be in the form of boric acid, easily lost by leaching (Tanaka and Fujiwara 2008; Uthman et al. 2020). In order to solve this low availability of the micronutrient in tropical soils, producers employ the foliar application of B, which is a common practice due to its high efficiency in increasing the B supply to plants (Atique-ur-Rehman et al. 2018). However, the difference between the appropriate and toxic concentrations of this micronutrient is tenuous, and it is possible to verify the phytotoxicity of this element in several cultures after the supply of high B concentrations (Atique-ur-Rehman et al. 2018; Hua et al. 2020; Wahid et al. 2020).

In this scenario of nutritional disorder due to B deficiency or toxicity, the application of the beneficial element silicon (Si) has shown promise in plants grown in tropical climates (de Souza Junior et al. 2019; Hua et al. 2020; Pavlovic et al. 2021; de Souza Júnior et al. 2022b). The benefits of Si can alleviate the deleterious effects of nutritional disorders by B. They are associated with the chemical characteristics of the beneficial element in the soil and roots and its structural, biochemical, and physiological functions in the shoots, reproductive organs, fruits, and fibers.

D. Kadyampakeni
Citrus Research and Education Center, University of Florida, Lake Alfred, FL, USA

J. P. de Souza Júnior (✉)
School of Agricultural and Veterinary Sciences, São Paulo State University, Sao Paulo, Brazil

R. de Mello Prado (ed.), *Benefits of Silicon in the Nutrition of Plants*,
https://doi.org/10.1007/978-3-031-26673-7_10

In soil solution and roots, in case of B deficiency, Si helps maintain the development of the root system and water relations, increasing biomass production and alleviating stress (Kasajima et al. 2010; Wimmer and Eichert 2013; González-Fontes et al. 2016). In cases of B toxicity, Si can form complexes with B, limiting its absorption (Nozawa et al. 2018). In addition, Si can increase the transcription of B efflux genes, contributing to plant detoxification (Akcay and Erkan 2016).

In shots, B deficiency is related to impairment in nucleic acid biosynthesis (Martin et al. 2013) and the metabolism of carbohydrates, proteins, indole-acetic acids, and phenols (Pilbeam and Kirkby 1983; Kirkby et al. 2001) that lead to reduced shoot development. In this case, Si can alleviate the deleterious effects by increasing the production of these compounds (Dragišić Maksimović et al. 2007; Schaller et al. 2012; Manivannan et al. 2016; Zhu et al. 2016; Souri et al. 2020; Tripathi et al. 2021; Sharma et al. 2022).

At the molecular level, B is important for cell wall stability, and when it is deficient, cell wall integrity decreases, impairing the process of cell division and expansion (Tanaka and Fujiwara 2008). Furthermore, silicon can contribute to B in its structural function, alleviating the deleterious effects of B deficiency (Sheng and Chen 2020), resulting in the development of plants without deficiency symptoms. Another important effect of the nutritional B disorder (deficiency or toxicity) is the increased production of reactive oxygen species (ROS) (Dordas and Brown 2005; Camacho-Cristóbal et al. 2008, 2011; Oliveira et al. 2020). The literature widely reports the Si role in the enzymatic antioxidant system in leaves of plants cultivated under abiotic stresses, increasing enzymatic (Liang et al. 2007; Khandekar and Leisner 2011; Moldes et al. 2013) and non-enzymatic defense mechanisms (Hua et al. 2020; de Souza Júnior et al. 2022a).

Plants require much B during the reproductive phase (Mengel and Kirkby 2012; de Mello Prado 2021). In this case, Si can contribute to the formation of pectins present in the structure of the cell wall of pollen grains, favoring the physical resistance of the cell wall (Kido et al. 2015), the flower formation, and the production and viability of pollen grains (Polster et al. 1978).

In this chapter, we seek to summarize the main action mechanisms of the B and Si interaction in the development of tropical plants; the synergistic and combined effects on the soil solution and the root system; and the impacts on biomass production, shoot development, and the development of reproductive organs of plants cultivated in tropical climates under B nutritional disorder.

10.2 Boron and Silicon Interaction in the Development of Tropical Crops

10.2.1 Effect on Soil Solution and Root System Development

Boron deficiency is related to lower development of the root system, with consequent damage to water relations and a reduction in transpiration rates and the absorption of water and other nutrients (Kasajima et al. 2010; Wimmer and Eichert

2013; González-Fontes et al. 2016). In this scenario, the supply of Si favors the development of the root system of plants cultivated under B deficiency (de Souza Junior et al. 2019). In addition, the beneficial effect of Si is known to increase water relations, transpiration rates, and water and nutrient absorption (Chen et al. 2018; Zargar et al. 2019; Atta et al. 2020; Wang et al. 2021). Consequently, it can act in mitigating the deleterious effect of B deficiency in the root system.

In a situation of B excess stress, there are indications that Si forms a complex with B in the soil solution or inside the roots, reducing its availability and phytotoxic potential (Nozawa et al. 2018). Orthosilicic acid $Si(OH)_4$ is the simplest soluble form of Si in soil solution. It is a weakly acidic molecule with silicon tetrahedrally coordinated to four hydroxyl groups (Schaller et al. 2021). There are several mechanisms proposed to explain the condensation reaction of silicon in solution, the simplest being the condensation of two silicic acid molecules, forming siloxane (Si-O-Si) with the release of a water molecule (Eq. 10.1) (Perry 2003).

$$\mathrm{Si(OH)_4 + Si(OH)_4 \leftrightarrow (HO)_3\,Si{-}O{-}Si(OH)_3 + H_2O} \tag{10.1}$$

In general, silicic acid molecules condense, producing a high number of Si-O-Si bonds, favoring the formation of cyclic oligomers with rings containing predominantly 3–6 Si atoms (Perry 2003).

Boron predominates in aqueous solutions in the form of boric acid (H_3BO_3). However, depending on the pH, it is in the form of $H_2BO_3^-$ (Wolska and Bryjak 2013). Moreover, boric acid forms a strong interaction with hydroxyl groups in the solution (Fig. 10.1) (Gadhave et al. 2018).

In laboratory studies, considering only B and Si in solution, at the concentration of 1.15 and 1.5 g L^{-1}, respectively, the formation of precipitates was observed depending on the pH of the solution (Fig. 10.2).

Thus, there may be the formation of B-Si complexes in the soil solution and inside the plant roots, leading to lower absorption of B when in case of toxicity, a fact observed in several species (Gunes et al. 2007, 2008; Inal et al. 2009; Soylemezoglu et al. 2009; Kaya et al. 2011; de Souza Junior et al. 2019). In addition, studies in *Hordeum vulgare* L. indicate that Si can increase the transcription of genes encoding the BOR_2 efflux transport, thus causing B detoxification in the apoplast (Akcay and Erkan 2016).

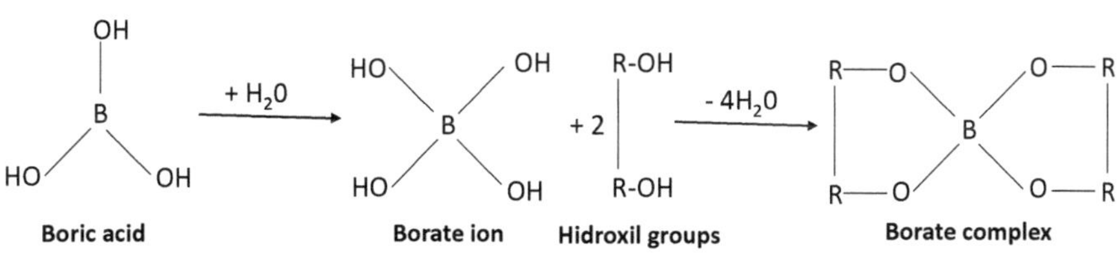

Fig. 10.1 Cross-linking mechanism of hydroxyl groups with boric acid

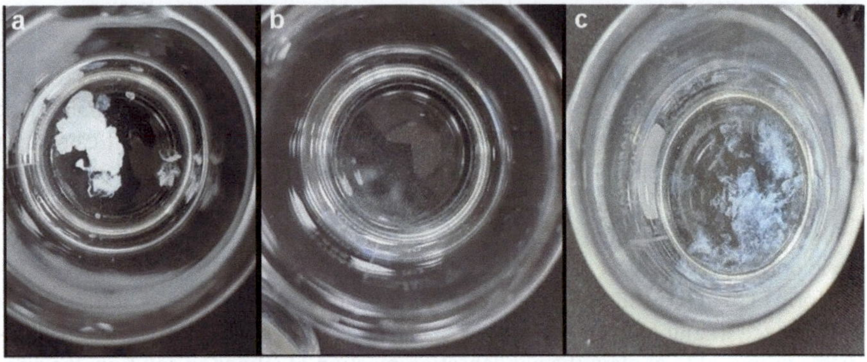

Fig. 10.2 Formation of precipitates in solution B + Si (1.5 g L^{-1} and 1.15 g L^{-1} of B and Si, respectively) at 25 °C at different pH: (**a**) pH = 8.0 ± 0.2; (**b**) pH = 9.2 ± 0.2; (**c**) pH = 10.0 ± 0.2. (Photo recorded immediately after preparing the syrup (Author's data))

Fig. 10.3 Visual appearance of cotton plants (phenological stage F7) as a function of different B concentrations in the nutrient solution (Author's data)

10.2.2 Effect on Shoot Growth and Biomass Production

Boron nutritional disorder, either due to deficiency or excess, is associated with a reduction in shoot development (Fig. 10.3).

The importance of B in plant production is mainly due to its biological role in the metabolism of nucleic acids (Martin et al. 2013), acting in the synthesis of uracil (Birnbaum et al. 1977), and the metabolism of carbohydrates, proteins, indole-acetic acids, and phenols (Pilbeam and Kirkby 1983). All these functions are essential for the good development of plants and, in case of B deficiency, result in a reduction in biomass production. Although there are no reports of whether Si can assist B in the synthesis of uracil, the beneficial effect of the element in increasing the production of carbohydrates, proteins, indole-acetic acids, and phenols is known in several cultures (Dragišić Maksimović et al. 2007; Schaller et al. 2012; Manivannan et al. 2016; Zhu et al. 2016; Souri et al. 2020; Tripathi et al. 2021; Sharma et al. 2022). Furthermore, it may contribute to alleviating the deleterious effects of B deficiency.

At the molecular level, studies demonstrate the importance of B in the binding of pectin in the cell wall, which is essential for leaf development (Tanaka and Fujiwara 2008). Pectins are important cellular components of plants formed by homogalacturonans and rhamnogalacturonans I and II (RG-I and RG-II) (Kobayashi et al. 1996). The synthesis of RG-II is cross-linked by a 1:2 borate-diol diester, which forms the dimeric RG-II, and in further cross-linking with borate, RG-II is synthesized by a stable three-dimensional pectic network (Tanaka and Fujiwara 2008). When in deficiency of B, the biosynthesis of RG-II is impaired. Consequently, the integrity of the cell wall decreases, impairing the process of cell division and expansion, leading to the deformation of new leaves with downward curvatures (Fig. 10.4).

Silicon can contribute to B in its structural function (Sheng and Chen 2020) because monosilicic acid can form cross-linked complexes with cell wall components, such as lignin, due to the presence of hydroxyl groups (Williams 2007). Moreover, the chemical properties of monosilicic acid and boric acid are similar in solutions with physiological pH, where H_3BO_3 is converted to tetrahydroxyborate [$B(OH)_4^-$] (Fig. 10.5a) (Sheng and Chen 2020). In a study on Si complexes in the plant cell wall, Sheng and Chen (2020) infer that the possible binding pattern of Si-pectin complexes may be similar to the formation of borate-diol esters (Fig. 10.5b); hemicelluloses show cis-diol structure for interaction with monosilicic acid (Fig. 10.5c); lignin is structurally similar to pyrocatechol, which can also form Si-lignin complexes (Fig. 10.5d); and there may be differences between the binding patterns of cell wall complexes with Si and cell wall complexes with B, with, for example, the formation of hexa-coordinated Si complexes in plants (Fig. 10.5e), while B forms tri-coordinated complexes (Sheng and Chen 2020).

Due to the structural function of Si similar to B, the supply of the beneficial element can alleviate the deleterious effects of B deficiency, resulting in normal leaf development (Fig. 10.6) (Silva 2020).

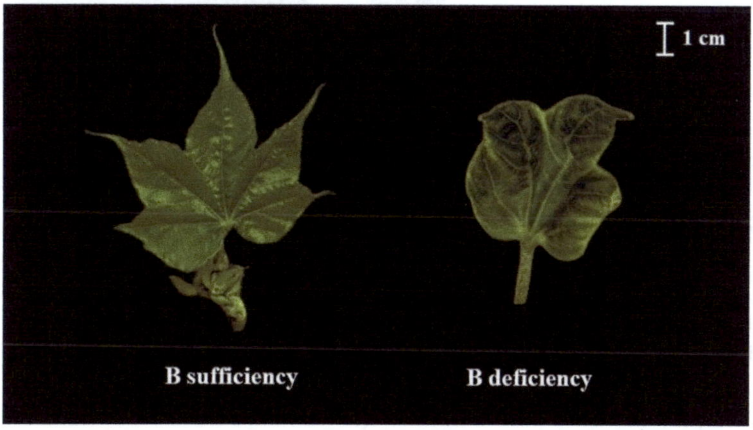

Fig. 10.4 New leaves of cotton plants with normal development (**a**) and deformed (**b**) due to B deficiency (Author's data)

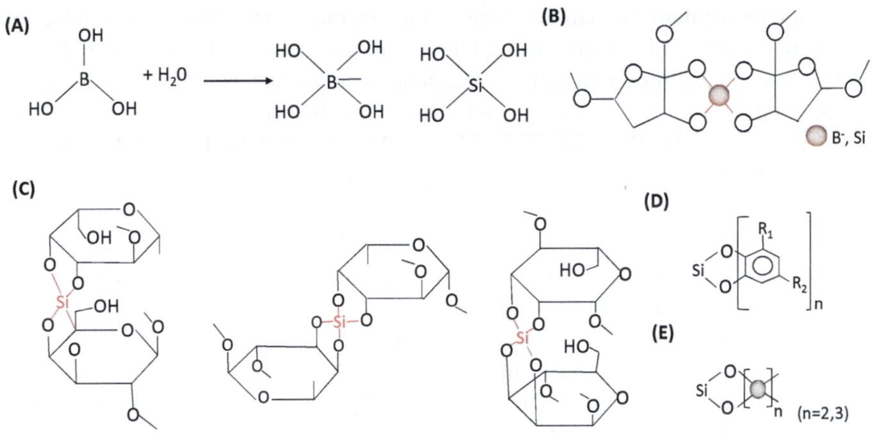

Fig. 10.5 Possible linkage patterns of Si-cell wall complexes. (**a**) Boric acid is converted to borate [B(OH)4–] in solutions at physiological pH (≈ 7.0 ± 0.2). Structural similarity implies the comparability of chemical properties of Si and B. (**b**) The reactions of 1,3-linked apiosyl residues in rhamnogalacturonan II (RG-II, pectin) with boric acid and monosilicic acid. (**c**) Hydroxyl complexation between H_4SiO_4 and cis-diols in hemicelluloses. (**d**) Possible cross-linking of Si and phenolic units in lignins. (**e**) Poly-coordinated Si in plants. In contrast to B, Si can form up to six coordinate bonds (Sheng and Chen 2020)

Fig. 10.6 Newly developed leaves of boron (−B) deficient cauliflower plants without Si (−Si) and with foliar application of Si (+Si) (Silva 2020)

Another important effect of the nutritional disorder of B, whether due to deficiency or toxicity, is the increase in the production of reactive oxygen species (ROS), which cause oxidative damage in cells, reducing the development of cultures (Dordas and Brown 2005; Camacho-Cristóbal et al. 2008, 2011; Oliveira et al. 2020). During the reduction of O_2 to H_2O in the photosynthetic process, energy can be transferred to O_2, forming singlet oxygen ($^1O_2$), which has high reactivity compared to O_2. In addition, one, two, or three electrons can also be transferred to O_2 and form ROS, superoxide ($O_2^{\bullet}$), hydrogen peroxide (H_2O_2), and hydroxyl radical ($^{\bullet}OH$), respectively, which can cause damage due to lipid peroxidation (Gratão et al. 2005). The main mechanisms of ROS elimination from plants include enzymes such as superoxide dismutase (SOD), catalase (CAT), ascorbate peroxidase (APX), and glutathione peroxidase (GPX). Superoxide dismutase (SOD) acts on the dismutase of $O_2^{\bullet}$ into H_2O_2, while CAT, APX, and GPX detoxify H_2O_2, transforming it into H_2O (Gratão et al. 2005; Hua et al. 2020).

An adequate supply of B alleviates oxidative stress, stimulating the first mechanisms of stress responses and favoring the activation of enzymes such as APX and GPX (Aydin et al. 2019; Hua et al. 2020). However, boron deficiency reduces the antioxidant system by decreasing the metabolic pathways for the biosynthesis of these enzymes, leading to oxidative damage (Cakmak and Römheld 1997; Brown et al. 2002).

The literature widely reports the Si role in the enzymatic antioxidant system in leaves of plants cultivated under abiotic stresses (Kim et al. 2017; Ali et al. 2020; Mostofa et al. 2021). The Si supply in plants grown under stress reduces oxidative damage by decreasing the H_2O_2 content, lipid peroxidation, and electrolyte leakage (Moldes et al. 2013). By extension, the beneficial element has an important role in the modulation of antioxidant enzymes, with an increase in the transcription of genes involved in the defense response of plants in several species (Liang et al. 2007; Khandekar and Leisner 2011; Moldes et al. 2013).

The literature widely explores and discusses the role of B and Si in the enzymatic antioxidant system in leaves. Despite this, the non-enzymatic antioxidant system must also be considered. In a review on B toxicity, Hua et al. (2020) indicated that the non-enzymatic antioxidant process is meritorious for regulating ROS neutralization in plant cells. Among the antioxidant enzymatic compounds, the authors highlight proline and glycine-betaine (GB). Proline is one of the most common osmolytes in plants related to tolerance to various abiotic stresses (Gill and Tuteja 2010) and neutralization of $^1O_2$ and $^{\bullet}OH$. Furthermore, it alleviates damage from lipid peroxidation (Das and Roychoudhury 2014). Similar to proline, GB has an osmotic and antioxidant function related to its contribution to maintaining cellular structural and functional integrity, possibly due to the interaction between GB and hydrophilic and hydrophobic domains of proteins and membranous complexes in cells (Sun et al. 2020).

The adequate supply of B also contributes to maintaining the integrity of the cell membrane, reducing electrolyte leakage (Ahmed et al. 2014). The supply of Si to plants also increases the membrane's physical integrity, reducing the extravasation of electrolytes caused by lipid peroxidation. After absorption, the beneficial element is deposited in the form of opal in the extracellular spaces and on the cell wall, which may increase the resistance of cells (Zargar et al. 2019), and, consequently, reduce the deleterious effect of B.

10.2.3 *Effect on the Development of Reproductive Organs*

During the reproductive stage, plants have a higher nutritional requirement of B compared to the vegetative stage (de Mello Prado 2021). It is indicated by the high concentration of the micronutrient in the pistil and stamen compared to the root, leaf, and stem (Dell and Huang 1997; Rerkasem et al. 2020).

Boron plays a fundamental role in the formation of pollen grains (O'Neill et al. 2004; Voxeur and Fry 2014; Fang et al. 2019) that occurs in the anthers in a process dependent on the differentiation and interaction of microsporocytes and the "carpet" formed by somatic cells (Clément et al. 1998). Microsporocytes give rise to microspores, while "carpet" cells are important for the maturation of microspores, providing materials for the construction of the cell wall of the pollen grain, composed mainly of pectin, callose, and sporopollenin. (Clément et al. 1998). As pollen develops, pectin is secreted from the "carpet" cells into the anther space, where it is used to build the pollen grain walls (Golaszweska and Bednarska 1999; Aouali et al. 2001). In this process, B forms borate complexes with several polyhydroxy compounds, including pectin and rhamnogalacturonan-II (O'Neill et al. 2004; Voxeur and Fry 2014; Fang et al. 2019), and acts by regulating the porosity and rigidity of the immature wall of the developing pollen grains (Willats et al. 2001; Iwai et al. 2006). As a result of B deficiency, pectin formed with rhamnogalacturonan-II, without borate complex, can be released from the walls of the apex of pollen tubes, but without formation of a rigid polysaccharide network, inducing malformation of pollen grains (Nyomora et al. 2000; Iwai et al. 2006).

In a classic study on the relationship between B and Si in the floral development of *Lilium longiflorum* (Thunb), Polster et al. (1978) inferred that Si has a synergistic action with B, stimulating the production and germination of pollen grains. Studies on *Gerbera jamesonii* (Savvas et al. 2002; Kamenidou et al. 2010), *Rosa damascena* (Farahani et al. 2020), *Solanum lycopersicum* (Miyake and Takahashi 1978), and *Triticum* spp. (Maghsoudi et al. 2015) showed the importance of Si for reproductive development, increasing the viability and germination of pollen grains and the growth of the pollen tube. In *Gossypium hirsutum*, the supply of B and Si, isolated or associated, in plants cultivated under B deficiency, favored the development of the flower's reproductive organs by reducing oxidative stress and increasing the production of phenol and proline (Fig. 10.7).

The supply of Si via foliar favors floral growth, increasing the diameter and height of flowers of *Gerbera jamesonii* (Fig. 10.8) and *Zinnia elegans* (Kamenidou et al. 2010).

After being absorbed by the plant, the element can bind to organic compounds (Neu et al. 2017; Klotzbücher et al. 2018), especially components of the cell wall, such as hemicellulose (He et al. 2015). This Si effect can generate a "low-cost energy" strategy, saving up to 80% of metabolic energy (Raven 1983). In order to assimilate 1 mole of SiO_2, the need for 1 mole of ATP is admitted (Werner 1977), while to assimilate 1 mole of C, the need for 5 moles of ATP and 2 moles of NADPH is assumed (Raven and Beardall 1981). Therefore, the beneficial element can favor floral growth by increasing the efficiency of boron in this organ and favoring plant metabolism, especially in processes linked to the cell wall.

In reproductive structures, Si can contribute to B in its structural function, as this beneficial element is highlighted as one of the elementary components present in the structure of the pollen grain wall (Crang and May 1974). During Si metabolism, complexes with pectins, calluses, tannins, and starches can occur (Boylston et al. 1990), which are compounds used to form the wall of pollen grains (Golaszweska

Fig. 10.7 Reproductive structures of cotton flowers grown under normal growing conditions (control) and B deficiency without methods of alleviating the deficiency (water) and with the application of silicon (Si) and boron (B) isolated or associated (B + Si). H_2O_2, hydrogen peroxide; MDA, malondialdehyde

Fig. 10.8 Gerbera plants supplemented with sodium silicate foliar sprays at 50 mg L^{-1} Si (right) showed an increased flower diameter and peduncle height compared to Si non-supplemented controls (left) (Kamenidou et al. 2010)

and Bednarska 1999; Aouali et al. 2001). In addition, the formation of mixed bonds of Si with (1;3,1;4)-β-D-glucan can occur, increasing the physical resistance of the cell wall of the pollen grains (Kido et al. 2015), which, in the case of B deficiency, have their tissue rigidity compromised (Nyomora et al. 2000; Iwai et al. 2006).

Increased oxidative stress, which can occur due to B deficiency or toxicity, also impairs pollen grain development (Fu et al. 2011; Jia et al. 2017; Ren et al. 2019). It occurs when there is an imbalance between the production of ROS and the antioxidant defense system present in the plant, causing oxidative damage, which is common in plants grown under biotic and abiotic stress (Foyer and Noctor 2000). Studies in plants grown under stress (water deficit) claim that the increase in the sterility of pollen grains is associated with increased oxidative stress (Fu et al. 2011). Ren et al. (2019) and Jia et al. (2017) reported similar results when studying the deterioration of pollen grains of several ornamental species after cryogenic storage. Cryogenic storage acts as abiotic stress, increasing the production of ROS that denature proteins, decompose nucleic acids, and destroy the integrity of the cell membrane of the cells comprising them, spoiling pollen grains. Conversely, the increase in the antioxidant defense system decreases the ROS and favors pollen grains' production and viability (Fu et al. 2011; Jia et al. 2017; Ren et al. 2019).

The foliar spraying of B, applied at the beginning of the reproductive stage of the plants, is efficient in providing the micronutrient to the cultures, increasing the production and viability of pollen grains, as well as the maintenance of flowering, as observed in *Malus domestica* (Nyomora et al. 2000) and *Olea europaea* (Perica et al. 2001). The authors attribute the beneficial effect of foliar application of B to the important role of the micronutrient in the pollen development and growth of the pollen tube.

The pollen tube is formed by an outer layer of pectocellulose, formed mainly by pectins, secreted by the Golgi complex (Hasegawa et al. 1998), and an inner layer of callose (Cai et al. 2011), with no pectins (Hao et al. 2013). The apex wall of pollen tubes consists of pectic substances (Mollet et al. 2002), and rhamnogalacturonan-II is also present in growing pollen tubes (Tanaka et al. 2013). Pollen tube growth is a process dependent on actin cytoskeletons (Rounds and Bezanilla 2013; Guan et al. 2013; Gao et al. 2016) that are located along the pollen tube as longitudinal actin cords, in the subapical region as short and dense filaments, and at the apex of the pollen tube, as highly dynamic actin filaments (Cheung and Wu 2008). The B role in pollen tube growth is associated with the structure of the cytoskeletons, increasing actin proteins and contributing to the maintenance of the arrangement of actin filaments in the subapical region of pollen tubes (Fang et al. 2019).

Boron deficiency inhibits pollen grain germination and delays pollen tube growth (Nyomora et al. 2000; Fang et al. 2019; Sharafi and Raina 2020), forming calluses at the ends of the pollen tube. Thus, it leads to stiffening, with a consequent decrease in pollen tube growth, as observed in *Lilium longiflorum* (Li et al. 1996) and *Picea meveri* (Wang et al. 2003). Furthermore, B deficiency promotes changes in pollen tube morphology (Dickinson 1978; Polster et al. 1978; Fang et al. 2019), with the formation of tubes with a smaller diameter and smaller growth (Fig. 10.9b), contributing to a decrease in flower fertilization, when compared to pollen tubes with no B deficiency (Fig. 10.9a).

The growth of the pollen tube is a fundamental process for fertilization. For flower fertilization, the pollen tube grows and penetrates, from the style, transmitting tissue into the ovule present in the ovary (Sharafi and Raina 2020). Silicon can

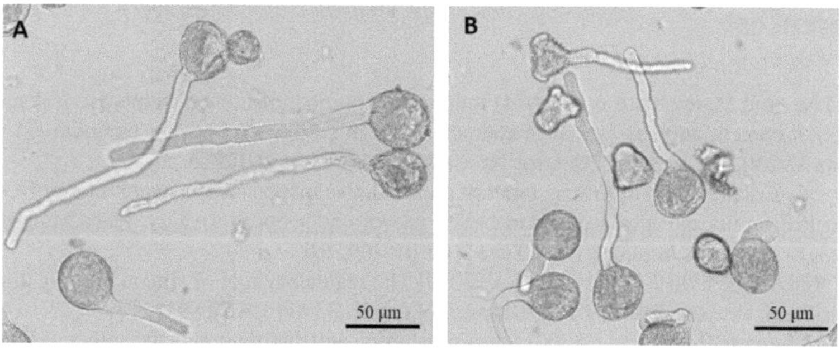

Fig. 10.9 (**a**) Control pollen tubes with a slender diameter and straight shape. (**b**) Pollen tubes under B deficiency (Fang et al. 2019)

also work in synergy with B to promote pollen tube growth. The literature rarely reports the direct role of Si in pollen tube growth. However, there are indications that, during the development of the pollen tube cell wall, Si is trapped in the pores of the matrix of this wall in a self-condensation process (Leroux et al. 2013). In this process, Si reacts with hydroxyl groups on the cell wall and increases its plasticity, facilitating cell expansion (Boylston et al. 1990) as observed during the elongation of leaf cells in rice (Hossain et al. 2002, 2007) and cotton fibers (Boylston et al. 1990), which also could occur in pollen tubes.

10.3 Final Considerations

Boron nutritional disorders have been reported in several crop-production regions in tropical climates. Silicon application has shown promise in mitigating these stressor effects. New evaluation techniques have allowed us to observe the Si effect in the soil solution and inside the roots, forming complexes and limiting the absorption of excess elements, as in the case of B. In the aerial part, more studies need to be carried out considering the effect of nutritional disorder on oxidative stress and the antioxidant mechanisms affected by B deficiency and toxicity. Silicon is a stress alleviator because it activates enzymatic and non-enzymatic antioxidant defense mechanisms. Recently, studies aimed to verify the use of Si in its structural function in maintaining the stoichiometric relationship with other structural nutrients (carbon, nitrogen, and phosphorus, mainly). However, the Si role in maintaining the stoichiometric relationship in plants grown under B nutritional disorder remains unknown. In flowers, studies on the relationship between B and Si in the development of reproductive organs, pollen grain formation, and pollen tube growth in B-deficient plants are well advanced. However, it is interesting to start studying the joint application of B + Si and reinforce the stoichiometric functions of Si in plants.

References

Ahmed N, Abid M, Rashid A et al (2014) Influence of boron nutrition on membrane leakage, chlorophyll content and gas exchange characteristics in cotton (*Gossypium hirsutum* L.). J Plant Nutr 37:2302–2315. https://doi.org/10.1080/01904167.2014.912325

Akcay UC, Erkan IE (2016) Silicon induced antioxidative responses and expression of BOR2 and two PIP family aquaporin genes in barley grown under boron toxicity. Plant Mol Biol Rep 34:318–326. https://doi.org/10.1007/s11105-015-0923-5

Ali N, Réthoré E, Yvin J-C, Hosseini SA (2020) The regulatory role of silicon in mitigating plant nutritional stresses. Plants 9:1779. https://doi.org/10.3390/PLANTS9121779

Aouali N, Laporte P, Clément C (2001) Pectin secretion and distribution in the anther during pollen development in Lilium. Planta 213:71–79. https://doi.org/10.1007/s004250000469

Atique-ur-Rehman FM, Rashid A et al (2018) Boron nutrition of rice in different production systems: a review. Agron Sustain Dev 38:25. https://doi.org/10.1007/s13593-018-0504-8

Atique-ur-Rehman QR, Hussain A et al (2020) Soil applied boron (B) improves growth, yield and fiber quality traits of cotton grown on calcareous saline soil. PLoS One 15:e0231805. https://doi.org/10.1371/journal.pone.0231805

Atta AA, Morgan KT, Hamido SA et al (2020) Water and soil nutrient dynamics of huanglongbing-affected citrus trees as impacted by ground-applied nutrients. Agronomy 10:1485. https://doi.org/10.3390/agronomy10101485

Aydin M, Tombuloglu G, Sakcali MS et al (2019) Boron alleviates drought stress by enhancing gene expression and antioxidant enzyme activity. J Soil Sci Plant Nutr 19:545–555. https://doi.org/10.1007/s42729-019-00053-8

Birnbaum EH, Dugger WM, Beasley BCA (1977) Interaction of boron with components of nucleic acid metabolism in cotton ovules cultured in vitro. Plant Physiol 59:1034–1038. https://doi.org/10.1104/pp.59.6.1034

Boylston EK, Hebert JJ, Hensarling TP et al (1990) Role of silicon in developing cotton fibers. J Plant Nutr 13:131–148. https://doi.org/10.1080/01904169009364063

Brdar-Jokanović M (2020) Boron toxicity and deficiency in agricultural plants. Int J Mol Sci 21:1424. https://doi.org/10.3390/ijms21041424

Brown PH, Bellaloui N, Wimmer MA et al (2002) Boron in plant biology. Plant Biol 4:205–223. https://doi.org/10.1055/s-2002-25740

Cai G, Faleri C, Casino CDC et al (2011) Distribution of callose synthase, cellulose synthase, and sucrose synthase in tobacco pollen tube is controlled in dissimilar ways by actin filaments and microtubules. Plant Physiol 155:1169–1190. https://doi.org/10.1104/pp.110.171371

Cakmak I, Römheld V (1997) Boron deficiency-induced impairments of cellular functions in plants. Plant Soil 193:71–83. https://doi.org/10.1023/A:1004259808322

Camacho-Cristóbal JJ, Rexach J, González-Fontes A (2008) Boron in plants: deficiency and toxicity. J Integr Plant Biol 50:1247–1255. https://doi.org/10.1111/j.1744-7909.2008.00742.x

Camacho-Cristóbal JJ, Rexach J, Herrera-Rodríguez MB et al (2011) Boron deficiency and transcript level changes. Plant Sci 181:85–89. https://doi.org/10.1016/j.plantsci.2011.05.001

Chen D, Wang S, Yin L, Deng X (2018) How does silicon mediate plant water uptake and loss under water deficiency? Front Plant Sci 9:281. https://doi.org/10.3389/fpls.2018.00281

Cheung AY, Wu H (2008) Structural and signaling networks for the polar cell growth machinery in pollen tubes. Annu Rev Plant Biol 59:547–572. https://doi.org/10.1146/annurev.arplant.59.032607.092921

Clément C, Laporte P, Audran JC (1998) The loculus content and tapetum during pollen development in Lilium. Sex Plant Reprod 11:94–106. https://doi.org/10.1007/s004970050125

Crang RE, May G (1974) Evidence for silicon as a prevalent elemental component in pollen wall structure. Can J Bot 52:2171–2174. https://doi.org/10.1139/b74-280

Das K, Roychoudhury A (2014) Reactive oxygen species (ROS) and response of antioxidants as ROS-scavengers during environmental stress in plants. Front Environ Sci 2:53

de Mello Prado R (2021) Mineral nutrition of tropical plants. Springer Nature, Switzerland.

de Souza Junior JP, de Mello Prado R, dos Santos Sarah MM, Felisberto G (2019) Silicon mitigates boron deficiency and toxicity in cotton cultivated in nutrient solution. J Plant Nutr Soil Sci 182:805–814. https://doi.org/10.1002/jpln.201800398

de Souza Júnior JP, Prado R d M, Campos CNS et al (2022a) Addition of silicon to boron foliar spray in cotton plants modulates the antioxidative system attenuating boron deficiency and toxicity. BMC Plant Biol 22:338. https://doi.org/10.1186/s12870-022-03721-7

de Souza Júnior JP, de Mello Prado R, Campos CNS et al (2022b) Silicon foliar spraying in the reproductive stage of cotton plays an equivalent role to boron in increasing yield, and combined boron-silicon application, without polymerization, increases fiber quality. Ind Crop Prod 182:114888. https://doi.org/10.1016/j.indcrop.2022.114888

Dell B, Huang L (1997) Physiological response of plants to low boron. Plant Soil 193:103–120. https://doi.org/10.1023/A:1004264009230

Dickinson DB (1978) Influence of borate and pentaerythritol concentrations on germination and tube growth of *Lilium longiflorum* pollen. J Am Soc Hortic Sci 103:413–416. https://doi.org/10.21273/JASHS.103.3.413

Dordas C, Brown PH (2005) Boron deficiency affects cell viability, phenolic leakage and oxidative burst in rose cell cultures. Plant Soil 268:293–301. https://doi.org/10.1007/s11104-004-0309-1

Dragišić Maksimović J, Bogdanović J, Maksimović V, Nikolic M (2007) Silicon modulates the metabolism and utilization of phenolic compounds in cucumber (*Cucumis sativus* L.) grown at excess manganese. J Plant Nutr Soil Sci 170:739–744. https://doi.org/10.1002/jpln.200700101

Fang KF, Du BS, Zhang Q et al (2019) Boron deficiency alters cytosolic Ca^{2+} concentration and affects the cell wall components of pollen tubes in *Malus domestica*. Plant Biol 21:343–351. https://doi.org/10.1111/plb.12941

Farahani H, Sajedi NA, Madani H et al (2020) Effect of foliar-applied silicon on flower yield and essential oil composition of damask rose (*Rosa damascena* Miller) under water deficit stress. Silicon:1–10. https://doi.org/10.1007/s12633-020-00762-1

Foyer CH, Noctor G (2000) Oxygen processing in photosynthesis: regulation and signalling. New Phytol 146:359–388. https://doi.org/10.1046/j.1469-8137.2000.00667.x

Fu G, Song J, Xiong J et al (2011) Changes of oxidative stress and soluble sugar in anthers involve in rice pollen abortion under drought stress. Agric Sci China 10:1016–1025. https://doi.org/10.1016/S1671-2927(11)60089-8

Furlani ÂMC, Carvalho CP, de Freitas JG, Verdial MF (2003) Wheat cultivar tolerance to boron deficiency and toxicity in nutrient solution. Sci Agric 60:359–370. https://doi.org/10.1590/S0103-90162003000200022

Gadhave RV, Kasbe PS, Mahanwar PA, Gadekar PT (2018) To study the effect of boric acid modification on starch–polyvinyl alcohol blend wood adhesive. J Indian Acad Wood Sci 15:190–198. https://doi.org/10.1007/s13196-018-0225-2

Gao QF, Gu LL, Wang HQ et al (2016) Cyclic nucleotide-gated channel 18 is an essential Ca2+ channel in pollen tube tips for pollen tube guidance to ovules in Arabidopsis. Proc Natl Acad Sci U S A 113:3096–3101. https://doi.org/10.1073/pnas.1524629113

Gill SS, Tuteja N (2010) Reactive oxygen species and antioxidant machinery in abiotic stress tolerance in crop plants. Plant Physiol Biochem 48:909–930. https://doi.org/10.1016/j.plaphy.2010.08.016

Golaszweska B, Bednarska E (1999) Immunocytochemical localization of pectins in the maturing anther of *Allium cepa* L. Folia Histochem Cytobiol 37:199

González-Fontes A, Herrera-Rodríguez MB, Martín-Rejano EM et al (2016) Root responses to boron deficiency mediated by ethylene. Front Plant Sci 6:1103. https://doi.org/10.3389/fpls.2015.01103

Gratão PL, Polle A, Lea PJ, Azevedo RA (2005) Making the life of heavy metal-stressed plants a little easier. Funct Plant Biol 32:481–494. https://doi.org/10.1071/FP05016

Guan Y, Guo J, Li H, Yang Z (2013) Signaling in pollen tube growth: crosstalk, feedback, and missing links. Mol Plant 6:1053–1064. https://doi.org/10.1093/mp/sst070

Gunes A, Inal A, Bagci EG et al (2007) Silicon increases boron tolerance and reduces oxidative damage of wheat grown in soil with excess boron. Biol Plant 51:571–574. https://doi.org/10.1007/s10535-007-0125-6

Gunes A, Pilbeam DJ, Inal A, Coban S (2008) Influence of silicon on sunflower cultivars under drought stress, I: growth, antioxidant mechanisms, and lipid peroxidation. Commun Soil Sci Plant Anal 39:1885–1903. https://doi.org/10.1080/00103620802134651

Hao H, Chen T, Fan L et al (2013) 2, 6-dichlorobenzonitrile causes multiple effects on pollen tube growth beyond altering cellulose synthesis in *Pinus bungeana* Zucc. PLoS One 8:e76660. https://doi.org/10.1371/journal.pone.0076660

Hasegawa Y, Nakamura S, Kakizoe S et al (1998) Immunocytochemical and chemical analyses of Golgi vesicles isolated from the germinated pollen of *Camellia japonica*. J Plant Res 111:421–429. https://doi.org/10.1007/BF02507807

He C, Ma J, Wang L (2015) A hemicellulose-bound form of silicon with potential to improve the mechanical properties and regeneration of the cell wall of rice. New Phytol 206:1051–1062. https://doi.org/10.2307/newphytologist.206.3.1051

Hossain MT, Mori R, Soga K et al (2002) Growth promotion and an increase in cell wall extensibility by silicon in rice and some other Poaceae seedlings. J Plant Res 115:23–27. https://doi.org/10.1007/s102650200004

Hossain MT, Soga K, Wakabayashi K et al (2007) Modification of chemical properties of cell walls by silicon and its role in regulation of the cell wall extensibility in oat leaves. J Plant Physiol 164:385–393. https://doi.org/10.1016/j.jplph.2006.02.003

Hua T, Zhang R, Sun H, Liu C (2020) Alleviation of boron toxicity in plants: mechanisms and approaches. Crit Rev Environ Sci Technol. https://doi.org/10.1080/10643389.2020.1807451

Inal A, Pilbeam DJ, Gunes A (2009) Silicon increases tolerance to boron toxicity and reduces oxidative damage in barley. J Plant Nutr 32:112–128. https://doi.org/10.1080/01904160802533767

Iwai H, Hokura A, Oishi M et al (2006) The gene responsible for borate cross-linking of pectin Rhamnogalacturonan-II is required for plant reproductive tissue development and fertilization. Proc Natl Acad Sci U S A 103:16592–16597. https://doi.org/10.1073/pnas.0605141103

Jia MX, Shi Y, Di W et al (2017) ROS-induced oxidative stress is closely related to pollen deterioration following cryopreservation. In Vitro Cell Dev Biol Plant 53:433–439. https://doi.org/10.1007/s11627-017-9844-3

Kamenidou S, Cavins TJ, Marek S (2010) Silicon supplements affect floricultural quality traits and elemental nutrient concentrations of greenhouse produced gerbera. Sci Hortic 123:390–394. https://doi.org/10.1016/j.scienta.2009.09.008

Kasajima I, Ide Y, Yokota Hirai M, Fujiwara T (2010) WRKY6 is involved in the response to boron deficiency in *Arabidopsis thaliana*. Physiol Plant 139:80–92. https://doi.org/10.1111/j.1399-3054.2010.01349.x

Kaya C, Tuna AL, Guneri M, Ashraf M (2011) Mitigation effects of silicon on tomato plants bearing fruit grown at high boron levels. J Plant Nutr 34:1985–1994. https://doi.org/10.1080/01904167.2011.610485

Khandekar S, Leisner S (2011) Soluble silicon modulates expression of *Arabidopsis thaliana* genes involved in copper stress. J Plant Physiol 168:699–705. https://doi.org/10.1016/j.jplph.2010.09.009

Kido N, Yokoyama R, Yamamoto T et al (2015) The matrix polysaccharide (1;3,1;4)-2- d -glucan is involved in silicon-dependent strengthening of rice cell wall. Plant Cell Physiol 56:268–276. https://doi.org/10.1093/pcp/pcu162

Kim Y-H, Khan AL, Waqas M, Lee I-J (2017) Silicon regulates antioxidant activities of crop plants under abiotic-induced oxidative stress: a review. Front Plant Sci 8:510. https://doi.org/10.3389/fpls.2017.00510

Kirkby M, Kirkby EA, Kosegarten H, Appel T (2001) Boron. In: Mengel K, Kirkby EA (eds) Principles of plant nutrition. Springer, Dordrecht, pp 621–638

Klotzbücher T, Klotzbücher A, Kaiser K et al (2018) Variable silicon accumulation in plants affects terrestrial carbon cycling by controlling lignin synthesis. Glob Chang Biol 24:e183–e189. https://doi.org/10.1111/gcb.13845

Kobayashi M, Matoh T, Azuma J (1996) Two chains of rhamnogalacturonan II are cross-linked by borate-diol ester bonds in higher plant cell walls. Plant Physiol 110:1017–1020. https://doi.org/10.1104/pp.110.3.1017

Leroux O, Leroux F, Mastroberti AA et al (2013) Heterogeneity of silica and glycan-epitope distribution in epidermal idioblast cell walls in *Adiantum raddianum* laminae. Planta 237:1453–1464. https://doi.org/10.1007/s00425-013-1856-6

Li YQ, Zhang HQ, Pierson ES et al (1996) Enforced growth-rate fluctuation causes pectin ring formation in the cell wall of *Lilium longiflorum* pollen tubes. Planta 200:41–49. https://doi.org/10.1007/BF00196647

Liang Y, Sun W, Zhu YG, Christie P (2007) Mechanisms of silicon-mediated alleviation of abiotic stresses in higher plants: a review. Environ Pollut 147:422–428. https://doi.org/10.1016/j.envpol.2006.06.008

Maghsoudi K, Yahya E, Ashraf M (2015) Influence of foliar application of silicon on chlorophyll fluorescence, photosynthetic pigments, and growth in water-stressed wheat cultivars differing in drought tolerance. Turk J Botany 39:625–634. https://doi.org/10.3906/bot-1407-11

Manivannan A, Soundararajan P, Muneer S et al (2016) Silicon mitigates salinity stress by regulating the physiology, antioxidant enzyme activities, and protein expression in *Capsicum annuum* 'Bugwang'. Biomed Res Int 2016:1–14. https://doi.org/10.1155/2016/3076357

Martin AR, Vasseur J-J, Smietana M (2013) Boron and nucleic acid chemistries: merging the best of both worlds. Chem Soc Rev 42:5684. https://doi.org/10.1039/c3cs60038f

Mengel K, Kirkby EA (2012) Principles of plant nutrition. Springer Science and Business Media, Dordrecht, Netherlands

Miyake Y, Takahashi E (1978) Silicon deficiency of tomato plant. 24:175–189. https://doi.org/10.1080/00380768197810433094

Moldes CA, Lima Filho OF, Manuel Camiña J et al (2013) Assessment of the effect of silicon on antioxidant enzymes in cotton plants by multivariate analysis. J Agric Food Chem 61:11243–11249. https://doi.org/10.1021/jf4039088

Mollet J-C, Kim S, Jauh G-Y, Lord EM (2002) Arabinogalactan proteins, pollen tube growth, and the reversible effects of Yariv phenylglycoside. Protoplasma 219:89–98. https://doi.org/10.1007/s007090200009

Mostofa MG, Rahman MM, Ansary MMU et al (2021) Silicon in mitigation of abiotic stress-induced oxidative damage in plants. Crit Rev Biotechnol 41:918–934. https://doi.org/10.1080/07388551.2021.1892582

Neu S, Schaller J, Dudel EG (2017) Silicon availability modifies nutrient use efficiency and content, C:N:P stoichiometry, and productivity of winter wheat (*Triticum aestivum* L.). Sci Rep 7:1–8. https://doi.org/10.1038/srep40829

Nozawa S, Sato T, Otake T (2018) Effect of dissolved silica on immobilization of boron by magnesium oxide. Fortschr Mineral 8:76. https://doi.org/10.3390/min8020076

Nyomora AMS, Brown PH, Pinney K, Polito VS (2000) Foliar application of boron to almond trees affects pollen quality. J Am Soc Hortic Sci 125:265–270. https://doi.org/10.21273/JASHS.125.2.265

O'Neill MA, Ishii T, Albersheim P, Darvill AG (2004) Rhamnogalacturonan II: structure and function of a borate cross-linked cell wall pectic polysaccharide. Annu Rev Plant Biol 55:109–139. https://doi.org/10.1146/annurev.arplant.55.031903.141750

Oliveira KR, Souza Junior JP, Bennett SJ et al (2020) Exogenous silicon and salicylic acid applications improve tolerance to boron toxicity in field pea cultivars by intensifying antioxidant defence systems. Ecotoxicol Environ Saf 201:110778. https://doi.org/10.1016/j.ecoenv.2020.110778

Pavlovic J, Kostic L, Bosnic P et al (2021) Interactions of silicon with essential and beneficial elements in plants. Front Plant Sci 12:1224. https://doi.org/10.3389/FPLS.2021.697592/BIBTEX

Perica S, Brown PH, Connell JH et al (2001) Foliar boron application improves flower fertility and fruit set of olive. HortScience 36:714–716. https://doi.org/10.21273/hortsci.36.4.714

Perry CC (2003) Silicification: the processes by which organisms capture and mineralize silica. Rev Mineral Geochem 54:291–327. https://doi.org/10.2113/0540291

Pilbeam DJ, Kirkby EA (1983) The physiological role of boron in plants. J Plant Nutr 6:563–582. https://doi.org/10.1080/01904168309363126

Polster J, Schwenk M, Bengsch E (1978) The role of boron, silicon and nucleic bases on pollen tube growth of *Lilium longiflorum* (L.). Z Naturforsch – Sect C J Biosci 47:102–108. https://doi.org/10.1515/znc-1992-1-218

Raven JA (1983) The transport and funcion of silicon in plants. Biol Rev 58:179–207. https://doi.org/10.1111/j.1469-185x.1983.tb00385.x

Raven JA, Beardall J (1981) Respiration and photorespiration. In: Platt T (ed) Physiological bases of phytoplankton ecology. Canadian Bulletin of Fisheries and Aquatic Sciences, pp 55–82

Ren R, Jiang X, Di W et al (2019) HSP70 improves the viability of cryopreserved *Paeonia lactiflora* pollen by regulating oxidative stress and apoptosis-like programmed cell death events. Plant Cell Tissue Organ Cult 139:53–64. https://doi.org/10.1007/s11240-019-01661-z

Rerkasem B, Jamjod S, Pusadee T (2020) Productivity limiting impacts of boron deficiency, a review. Plant Soil 455:23–40

Rounds CM, Bezanilla M (2013) Growth mechanisms in tip-growing plant cells. Annu Rev Plant Biol 64:243–265. https://doi.org/10.1146/annurev-arplant-050312-120150

Savvas D, Manos G, Kotsiras A, Souvaliotis S (2002) Effects of silicon and nutrient-induced salinity on yield, flower quality and nutrient uptake of gerbera grown in a closed hycroponic system. J Appl Bot 76:153–158

Schaller J, Brackhage C, Dudel EG (2012) Silicon availability changes structural carbon ratio and phenol content of grasses. Environ Exp Bot 77:283–287. https://doi.org/10.1016/j.envexpbot.2011.12.009

Schaller J, Puppe D, Kaczorek D et al (2021) Silicon cycling in soils revisited. Plants 10:295. https://doi.org/10.3390/PLANTS10020295

Sharafi Y, Raina M (2020) Effect of boron on pollen attributes in different cultivars of *Malus domestica* L. Natl Acad Sci Lett:1–6. https://doi.org/10.1007/s40009-020-00986-0

Sharma A, Vishwakarma K, Singh NK et al (2022) Synergistic action of silicon nanoparticles and indole acetic acid in alleviation of chromium (CrVI) toxicity in *Oryza sativa* seedlings. J Biotechnol 343:71–82. https://doi.org/10.1016/j.jbiotec.2021.09.005

Sheng H, Chen S (2020) Plant silicon-cell wall complexes: identification, model of covalent bond formation and biofunction. Plant Physiol Biochem 155:13–19. https://doi.org/10.1016/j.plaphy.2020.07.020

Shorrocks VM (1997) The occurrence and correction of boron deficiency. Plant Soil 193:121–148. https://doi.org/10.1023/A:1004216126069

Silva JLF (2020) Foliar spray of boron and silicon is agronomically viable for cauliflower plant. São Paulo State University, Sao Paulo, Brazil.

Souri Z, Khanna K, Karimi N, Ahmad P (2020) Silicon and plants: current knowledge and future prospects. J Plant Growth Regul 40:906–925. https://doi.org/10.1007/S00344-020-10172-7

Soylemezoglu G, Demir K, Inal A, Gunes A (2009) Effect of silicon on antioxidant and stomatal response of two grapevine (*Vitis vinifera* L.) rootstocks grown in boron toxic, saline and boron toxic-saline soil. Sci Hortic 123:240–246. https://doi.org/10.1016/J.SCIENTA.2009.09.005

Sun H, Luo M, Zhou X et al (2020) Exogenous glycine betaine treatment alleviates low temperature-induced pericarp browning of 'Nanguo' pears by regulating antioxidant enzymes and proline metabolism. Food Chem 306:125626. https://doi.org/10.1016/j.foodchem.2019.125626

Tanaka M, Fujiwara T (2008) Physiological roles and transport mechanisms of boron: perspectives from plants. Eur J Phys 456:671–677. https://doi.org/10.1007/s00424-007-0370-8

Tanaka N, Uraguchi S, Saito A et al (2013) Roles of pollen-specific boron efflux transporter, OsBOR4, in the rice fertilization process. Plant Cell Physiol 54:2011–2019. https://doi.org/10.1093/pcp/pct136

Tripathi DK, Rai P, Guerriero G et al (2021) Silicon induces adventitious root formation in rice under arsenate stress with involvement of nitric oxide and indole-3-acetic acid. J Exp Bot 72:4457–4471. https://doi.org/10.1093/jxb/eraa488

Uthman QO, Kadyampakeni DM, Nkedi-Kizza P (2020) Boron availability and uptake in huanglongbing-affected citrus trees on a Florida entisol. J Plant Nutr 43:1248–1258. https://doi.org/10.1080/01904167.2020.1727510

Voxeur A, Fry SC (2014) Glycosylinositol phosphorylceramides from *Rosa* cell cultures are boron-bridged in the plasma membrane and form complexes with rhamnogalacturonan II. Plant J 79:139–149. https://doi.org/10.1111/tpj.12547

Wahid MA, Saleem M, Irshad S et al (2020) Foliar feeding of boron improves the productivity of cotton cultivars with enhanced boll retention percentage. J Plant Nutr 43:2411–2424. https://doi.org/10.1080/01904167.2020.1783300

Wang Q, Lu L, Wu X et al (2003) Boron influences pollen germination and pollen tube growth in *Picea meyeri*. Tree Physiol 23:345–351. https://doi.org/10.1093/treephys/23.5.345

Wang M, Wang R, Mur LAJ et al (2021) Functions of silicon in plant drought stress responses. Hortic Res 8:254. https://doi.org/10.1038/s41438-021-00681-1

Werner D (1977) Silicate metabolism. In: Werner D (ed) The biology of diatoms. Blackwells Scientific Publications, Oxford, pp 10–149

Willats WGT, Mccartney L, Mackie W, Knox JP (2001) Pectin: cell biology and prospects for functional analysis. Plant Mol Biol 47:9–27

Williams RJP (2007) Introduction to silicon chemistry and biochemistry Ciba Found Symposium Proceedings. pp 24–39

Wimmer MA, Eichert T (2013) Review: mechanisms for boron deficiency-mediated changes in plant water relations. Plant Sci 203–204:25–32. https://doi.org/10.1016/j.plantsci.2012.12.012

Wolska J, Bryjak M (2013) Methods for boron removal from aqueous solutions – a review. Desalination 310:18–24. https://doi.org/10.1016/j.desal.2012.08.003

Zargar SM, Mahajan R, Bhat JA et al (2019) Role of silicon in plant stress tolerance: opportunities to achieve a sustainable cropping system. 3 Biotech 9:1–16. https://doi.org/10.1007/s13205-019-1613-z

Zhu Y, Guo J, Feng R et al (2016) The regulatory role of silicon on carbohydrate metabolism in *Cucumis sativus* L. under salt stress. Plant Soil 406:231–249. https://doi.org/10.1007/s11104-016-2877-2

Chapter 11
Effect of Silicon in Mitigating Iron Deficiency

Luis Felipe Lata-Tenesaca and Diego Ricardo Villaseñor Ortiz

11.1 Introduction

The biological significance of iron (Fe) for plants cannot be disregarded, as the low availability of this nutrient affects plant nutrition and essential physiological processes, such as photosynthesis, mitochondrial respiration, carbon metabolism, nitrogen (N) assimilation, the synthesis of enzyme components, and regulation of reactive oxygen species production, among others.

Iron deficiency in soils with high bicarbonate contents and a high pH value, where Fe precipitates with phosphate or hydroxyl, is one of the major disturbances in plant nutrition (Alcántara et al. 2000). Alkaline soils occur in many regions of the world and reduce plant growth, besides causing severe losses in productivity and yield quality. Thus, more than 60% of the world's population is deficient in Fe (White and Broadley 2005), which causes various diseases in humans. This deficiency occurs in underdeveloped and developing countries and also in developed nations.

Under Fe-limiting conditions, plants have developed sophisticated mechanisms to maintain Fe homeostasis through adaptation strategies at the physiological, biochemical, morphological, and gene expression levels in order to increase Fe accumulation in their tissues (Kobayashi and Nishizawa 2014; Li and Lan 2017), including even light modulates Fe homeostasis in the plant (D'Amico-Damião et al. 2022; Gavassi et al. 2023).

The different strategies to correct Fe deficiency, such as the development and use of genotypes efficient in Fe uptake and the use of Fe chelating agents selected on the basis of their high stability and ability to provide Fe to plants, have been implemented both independently or integrated (Fernández and Ebert 2005; Lucena 2006; López-Millán et al. 2013). However, genetic approaches for obtaining Fe-efficient

L. F. Lata-Tenesaca (✉) · D. R. V. Ortiz
Universidad Técnica de Machala, Machala, Ecuador

R. de Mello Prado (ed.), *Benefits of Silicon in the Nutrition of Plants*,
https://doi.org/10.1007/978-3-031-26673-7_11

cultivars may present methodological limitations due to the complex regulation of the metabolic responses induced by Fe deficiency. Fertilization using chelate sources that are best suited for alkaline soils is costly and also environmentally risky, since common chelates (EDTA or EDDHA) degrade very slowly in the soil, while organic sources of chelates (without environmental burden) have no proven agronomic efficiency yet.

A strategy to mitigate iron deficiency in plants that is more economically viable and does not harm the environment would be to use silicon (Si) as a mitigating agent for multiple stresses. The use of Si as a sustainable strategy and emerging technique in agriculture can alleviate the oxidative stress caused by iron deficiency, also increasing crop growth and productivity. The mechanisms of action of Si that may enhance plant responses to iron deficiency tolerance include the regulation of essential physiological processes, such as Fe uptake and redistribution, as well as improvements in the photosynthetic process. In addition, Si can also improve biochemical processes by modulating reactive oxygen species and phenolic compounds and molecular processes by regulating the genes of Fe binding proteins.

Fe deficiency is one of the major challenges in important agricultural regions of the world. Therefore, this review aims to discuss the mechanisms involved in the Si-mediated mitigation of Fe deficiency, as well as future research perspectives.

11.2 Iron Uptake and the Benefits of Si

Iron is one of the most abundant elements in the earth's crust, but the uptake and availability of Fe to plants is frequently insufficient, particularly in calcareous soils with low Fe availability. This is due to the high pH value, which results in the non-absorption of poorly soluble forms of Fe [$Fe(OH)_3$] by the plants, causing nutritional disturbance and losses in crop yield and nutritional quality (Zuo and Zhang 2011).

In general, plants adopted two strategies related to the processes of Fe uptake by the roots. These plant adaptation strategies for Fe uptake involve physiological, biochemical, and morphological changes, in order to favor the efficiency of Fe uptake. When Fe is scarce, non-grass dicot and monocot plants use Strategy I, which is based on different mechanisms, such as an increase in root surface area, the acidification of the rhizosphere due to the release of protons by the roots (action of H+-ATPases), reduction of Fe(III) to Fe(II) (which occurs in the apoplast) and subsequent influx of Fe(II) into the cell by a high-affinity transporter (*IRT1*), and release of low molecular weight compounds such as phenols and carboxylates (Santi and Schmidt 2009; Kobayashi and Nishizawa 2014; Rajniak et al. 2018; Vert et al. 2021).

On the other hand, grass plants have developed Strategy II, which is based on chelation to acquire Fe^{3+} from the exudation of phytosiderophores (2′-deoxymugenic acid), forming soluble complexes that are taken up by the roots (Römheld and Marschner 1986). This occurs because the resulting Fe(III)-photosiderophore complex is taken up by the Yellow Stripe1-Like 15 (*OsYSL15*) transporter (Inoue et al.

2009; Lee et al. 2009). After uptake, a portion of Fe is sequestered in vacuoles by the Vacuolar Iron Transporters 1 and 2 (*OsVIT1* and *OsVIT2*) (Zhang et al. 2012), while the rest of the Fe via the symplastic pathway is carried in the xylem by unknown transporters until it reaches the leaves by transpiration flow. Even studies involving nutrient transporters (including Fe) are restricted to a few species (*Arabidopsis* and rice), and this second species may use both strategies of Fe uptake.

The multiple beneficial effects of Si on the physiological processes of plants to adapt to environmental stresses have been widely evidenced. Over the past few years, significant advances have been elucidated on the mechanisms by which Si nutrition influences Fe uptake responses in plants. These responses may depend on the plant species and its mechanism for Fe acquisition (Strategy I or II), as well as on the adequate or insufficient supply of Fe for root uptake. One of the first studies on the subject demonstrated that Si application increases the ability of the roots of *Valerianella locusta* (L.) to absorb Fe from the external environment (Gottardi et al. 2012), suggesting that this element plays the role of increasing the gene expression in Strategy I.

Initially, it has been suggested that the effect of Si depends on the uptake strategy, with little information available on Strategy II plants (Bityutskii et al. 2010). One study indicated that Si increased the expression of genes involved in Strategy II Fe acquisition, such as *HvNAS1* and *HvDMAS1*, which are involved in phytosiderophore biosynthesis in barley plants (Nikolic et al. 2019).

Taking into consideration that two-thirds of the root Fe are located in the apoplasm, the deposition and mobilization of Fe promoted by Si will impact the processes of absorption and transport of the element in the root into the xylem, consequently impacting its redistribution. Some studies analyzing rice grown in nutrient solution have indicated that Si stimulates the formation of Caspary bands in the exodermis, besides stimulating the biosynthesis of suberin and lignin. In addition, this barrier reduces the flux of Fe into the apoplastic space, hindering its transport to the xylem and later to the plant area (Carrasco-Gil et al. 2018; Becker et al. 2020). It is possible that this barrier is formed by the precipitation of negatively charged silica particles that allow the sequestration of positively charged elements such as Fe (Khan et al. 2016). The obstruction of Fe in the root apoplasm promoted by Si may be related to the need to ensure the homeostasis of the nutrient by avoiding that an eventual excess in the culture medium could cause toxicity in the plant. At the same time, a low concentration of Fe in the leaf could control the expression of genes linked to the efficient absorption of Fe by increasing the synthesis of chelates, for example. The Fe content in the leaf and phloem have also been attributed to long-range signaling to modulate the plant response against Fe deficiency in the root (Khan et al. 2018), consequently favoring its uptake. This compartmentalization of Fe in the plant promoted by Si induces a plant response to Fe deficiency by increasing the translocation of this nutrient from the root to the upper part, although it is not known if this is sufficient to meet the Fe demand of the leaves (Carrasco-Gil et al. 2018).

Molecular studies of the signaling pathways of Si and Fe in the regulation of the Si transporter in rice have further strengthened the interaction of these two elements in plants. Chaiwong et al. (2020) established that the expression of the Si transporter gene (Lsi1) in rice is regulated by the Si concentration in the root, being affected by Fe availability. Therefore, the regulation of Lsi1 would allow significant maintenance between the homeostasis of Si and Fe in rice plants.

The effect of Si on Fe uptake by plants may also be dependent on the pH value, as associations between Si and metallic elements can be found in minerals and living organisms. Under physiological pH conditions (pH 6–8), Fe becomes unavailable to the plant due to the formation of Fe oxide aggregates. Si may prevent this aggregation by forming a coating with silicic acid around soil particles (Perry and Keeling-Tucker 1998), which favors Fe availability for root uptake, but further research is necessary to gauge whether this effect occurs inside the cells.

The biosynthesis of metabolites that play an important role in facilitating Fe uptake, such as succinic, citric, and glutamic acids, was also favored by Si when cucumber plants were grown in the absence of Fe at higher pH values (pH 6 versus 5) (Bityutskii et al. 2018). Thus, Si increases the content of the chelated nutrient in Fe deficient plants, favoring uptake and preventing its reaction in the apoplasm. This occurs as Si induces the gene expression that regulates the synthesis of phenolic compounds such as gallic acid, which can chelate Fe and also contribute to the reduction of Fe(III) to Fe(II) in the apoplasm, making more Fe available to be transported to the plant shoot.

It is observed that the effects of Si on root uptake and radial transport of Fe are complex, involving many factors linked to the rhizosphere, root, and leaves, besides having been studied in only a few species, highlighting the need for further studies to indicate the main mechanism that favors the most the process of Fe acquisition by plants.

11.3 Iron Redistribution and the Benefits of Si

After being absorbed, Fe is transported to the plant shoot by the transpiration flow. This may occur by the action of the FRD3-like transporter, a protein 1 (*OsFRDL1*) and citrate efflux transporter that is responsible for the efficient translocation of Fe from the root to the shoot in rice, forming the Fe-citrate complex (Yokosho et al. 2009) and also nicotianamine complexes (Curie et al. 2009; Ivanov et al. 2012). Nicotianamine (a non-proteinogenic amino acid) is synthesized by the action of the enzyme nicotianamine synthase, which complexes to Fe, having high affinity. Members of the Yellow S tripe-Like (YSL) family and the ferric reductase defective 3 (FRD3) have been proposed as the transporters responsible for carrying and redistributing Fe(II)-nicotianamine and Fe(III)-citrate complexes, which provide xylem loading (Durrett et al. 2007) and also originate from the ferroportin (FPN) gene family (Guerinot and Morrissey 2009).

The Fe reaching the leaves can be redistributed to different plant organs. The redistribution of Fe takes place from the discharge of Fe into the phloem, being mainly carried out by Fe complexed nicotianamine and transported by the YSL proteins that are present in both the xylem and phloem (Pich and Scholz 1996; Conte and Walker 2011). YSL transporters are involved in the unloading of Fe from the xylem and participate in the regulation of Fe homeostasis in tissues and in the lateral movement of Fe within conducting vessels (Curie et al. 2001). Despite having distinct Fe uptake systems, it was indicated that plants that use both Strategies seem to share the same mechanisms that allow the translocation and redistribution of Fe after uptake (Muneer and Jeong 2015). The *OsFRDL1* transporter is also required to distribute Fe to the panicles, releasing citrate to solubilize apoplastic Fe deposited in parenchyma cells at the node (Yokosho et al. 2016).

Research to increase Fe redistribution in plants is limited since there are as stable mechanisms of transporter regulation to maintain nutrient balance at the cellular and tissue level in different plant organs, and various transporters are regulated at the transcriptional and/or protein level. In the case of rice, the *IRT* Fe transporter has a high activity in Fe-deficient plants, but if there is a high concentration of the micronutrient, this transporter will have a low activity precisely to avoid Fe toxicity (Dubeaux et al. 2018).

The plant response to Fe deficiency may occur due to the biosynthesis of organic acids (citrate, malate, and succinate), which tend to increase in the xylem (Conte and Walker 2011) and possibly tend to increase in the phloem. It has been suggested that Si supply may increase the utilization of apoplastic Fe and promote Fe mobilization via xylem due to the up-regulation of key genes (*ICD* for citric acid and *MDH* for malic acid) involved in the biosynthesis of Fe-mobilizing compounds in Fe-deficient cucumber roots (Pavlovic et al. 2013). The increase in the biosynthesis of Fe-mobilizing compounds such as citrate in roots and leaves and catechins in roots favored by Si seems to be of great importance for improving the redistribution and utilization of Fe in cucumber (Bityutskii et al. 2014; Nikolic et al. 2019).

In barley under Fe deficiency, the accumulation of this metal in young leaves was higher in plants that received Si application, which is explained by the fact that this element increased the expression of *HvYS1* and *HvDMAS1* genes that act in the remobilization of Fe from the root to the shoot organs (Nikolic et al. 2019). Si has been shown to improve the translocation and utilization efficiency of Fe in sorghum plants subjected to Fe deficiency (Teixeira et al. 2020). In soybean plants (inefficient in Fe uptake) grown under Fe deficiency, it has been shown that Si can assist in the formation of Fe pools in the root apoplast by favoring the transport of Fe to the shoot organs (Gonzalo et al. 2013). The same study demonstrated that there was no translocation of Fe to the leaves with the addition of Si in cucumber plants (efficient in Fe uptake). This indicates that the mediating role played by Si in Fe redistribution depends on the strategy for micronutrient uptake adopted by the plant and on the level of gene expression of the proteins involved in Fe mobilization.

The role played by Si in Fe redistribution was not only indicated in plants growing under low Fe availability, as Si has regulated Fe status in plants grown under Fe-sufficient conditions. As demonstrated in quinoa plants under adequate nutritional supply, different methods of Si application affected Fe accumulation in the plant. The root application of Si in nutrient solution in combination with foliar application had a similar effect to the root application of Si alone in increasing the

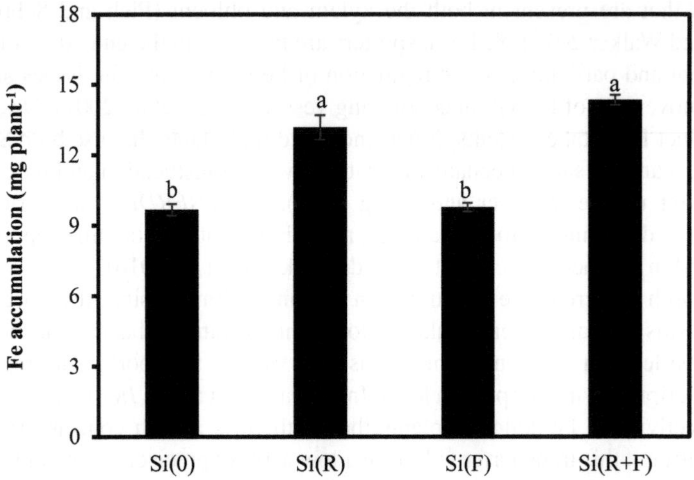

Fig. 11.1 Fe accumulation in quinoa plants according to different Si application methods; no Si [Si(0)]; foliar application of Si [Si(F)]; root application of Si [Si(R)]; combination of root and foliar application [Si(F+R)]

accumulation of Fe in the biomass of quinoa leaves (Lata-Tenesaca et al. 2023) (Fig. 11.1). This could be explained by the role played by Si in maintaining Fe homeostasis in plants through the formation of Fe-Si complexes and through the maintenance of the redox potential in the solution of the root apoplast and the xylem sap, which favors Fe transport through the xylem and its subsequent accumulation in the shoot tissues of the cucumber plant (Stevic et al. 2016). Therefore, Si could impact the bioavailability of Fe in plants at the rhizospheric and apoplasmic levels of root cells, which was verified by increased Fe concentrations in different plant tissues. However, further research is still needed to understand if these mechanisms of action of Si could be similar among the different strategies of Fe acquisition in plants.

11.4 Effect of Si on Oxidative Stress in Fe-Deficient Plants

It is well known that the photosynthetic metabolism depends on Fe nutrition and the deficiency of this micronutrient can result in oxidative damage induced by increased reactive oxygen species (ROS; e.g. H_2O_2, $OH^\bullet$, $1O_2$, $O_2^{\bullet -}$) (Molassiotis et al. 2006; Donnini et al. 2011). Photosystem II (PS-II) contains Fe proteins and loses its efficiency in Fe-deprived plants due to the low supply of photosynthetic electrons (Bertamini et al. 2002), thus decreasing the energy to optimize the photosynthetic process.

Normally, iron-deficiency-tolerant plants induce an increase in the activities of superoxide dismutase (SOD, Cu/Zn-SOD, and Mn-SOD) when subjected to deficiency cultivation by increasing the expression of the copper chaperone gene, as well as by accumulating the phenolic compound isoquercetin (Kallala et al. 2021).

Thus, iron is required by many antioxidant enzymes for being an important catalyst for electron transfer reactions. Potential targets include reactive oxygen species (ROS) such as H_2O_2 (hydrogen peroxide) and $OH^{\bullet}$ (hydroxyl radical), which are reduced to H_2O (antioxidant activity) or oxidized, such as $O_2^{\bullet -}$ (superoxide anion radical), with the concomitant reduction of Fe(III) to Fe(II) (Tewari et al. 2013).

Fe deficiency is clearly identified by interveinal chlorosis, which is a yellowing of the youngest leaves of a given plant due to delayed and insufficient chlorophyll production caused by increased oxidative stress, which causes root deformation and impairs plant development (Kabir et al. 2012; Barker and Stratton 2015).

Although several recent reports highlighted the role played by Si in the regulation of nutritional disorders caused by micronutrient deficiency in plants (Farias et al. 2022; Souza et al. 2022; Oliveira et al. 2022), the effect of Si in attenuating the oxidative stress caused by Fe deficiency has been little explored. In general, several results suggest that Si attenuates the deleterious effects caused by Fe deficiency by favoring Fe availability in the rhizosphere and in root apoplasm. In addition, Si contributes to Fe accumulation and redistribution in different plant tissues (Nikolic et al. 2019; Hernández-Apaolaza et al. 2020; Pavlovic et al. 2021).

Therefore, there are indications of the effect of Si in regulating these processes. Ksiaa et al. (2021) demonstrated that the full or partial application of Si application restored the effects caused by Fe deficiency in barley plants due to improved water status, gas exchange, increased photosynthetic pigments, and functioning of the photosynthetic apparatus. The beneficial role played by Si in increasing the chlorophyll content was also evident in sorghum plants under low Fe availability in the nutrient solution (Teixeira et al. 2020). One explanation for this is that Si favors Fe remobilization and accumulation in leaves (Nikolic et al. 2019), confirming the essentiality of Fe in chlorophyll biosynthesis. Furthermore, it has been recognized that Si plays a key role in the positive regulation of photosynthesis-related genes, the rate of electron transport, and its action as a protective layer in the chloroplast structure during stress induced by mineral deficiency (Manivannan and Ahn 2017; Rastogi et al. 2021). It has been observed that Si supply in Fe deficient soybean (Muneer and Jeong 2015) and cucumber (Gonzalo et al. 2013) plants can improve the composition of chloroplast protein complexes (such as RuBisCO), net photosynthesis rate, stomatal conductance, transpiration, and the dark green color index. The effect of Si in alleviating Fe deficiency symptoms can be seen visually in quinoa and sorghum plants, presenting a slight increase in the green color hue (Fig. 11.2).

Biochemical approaches revealed that Si supply can attenuate oxidative stress in Fe-deficient rice plants by increasing the activity of enzyme components such as catalase (CAT), polyphenol oxidase (PPO), peroxidase (POX), and guaicol peroxidase (GPX), consequently reducing membrane lipid peroxidation and H_2O_2 (Zadeh and Chalmardi 2013; Chalmardi et al. 2016). Sorghum grown under Fe deficiency and in the presence of Si in the nutrient solution showed a reduction of lipid peroxidation in leaves and roots and an increase in leaf chlorophyll content (Teixeira et al.

Fig. 11.2 Photos of leaves of quinoa (**a**) and sorghum (**b**) with moderate iron deficiency grown in pots with nutrient solution in the absence (−Si) and presence of 2 mM Si (+Si)

2020). As mentioned in the previous section, one of the mechanisms used by plants grown under Fe-limited conditions is the secretion of phenolic compounds, such as protocacechol acid and caffeic acid, to make apoplastic Fe available. Thus, Martín-Esquinas and Hernández-Apaolaza (2021) revealed that Si promotes the biosynthesis of phenolic compounds in rice plants as an antioxidant mechanism to decrease ROS produced by Fe deficiency. Overall, the mechanisms of action of Si modulate different physiological processes in plant gene expression by attenuating the damage caused by iron deficiency (Fig. 11.3).

11.5 Final Considerations and Future Perspectives

The role played by Si in plant growth under Fe deficiency conditions is recognized through regulations that occur at the physiological and biochemical level that directly affect the processes of Fe uptake and redistribution in plants and are mediated by changes in gene expression involved in these processes. The deleterious effects caused by Fe deficiency are attenuated by Si, which favors the development of a more robust antioxidant system to inactivate reactive oxygen species. Si can reduce the detrimental effects of Fe deficiency by improving the photosynthetic process and by increasing the action of transporters and the synthesis of organic chelating compounds involved in the mitigation of nutritional disorders, favoring Fe mobilization between the root and the shoot.

In light of this, we made some notes to guide future research on the role played by Si in mitigating iron deficiency in plants:

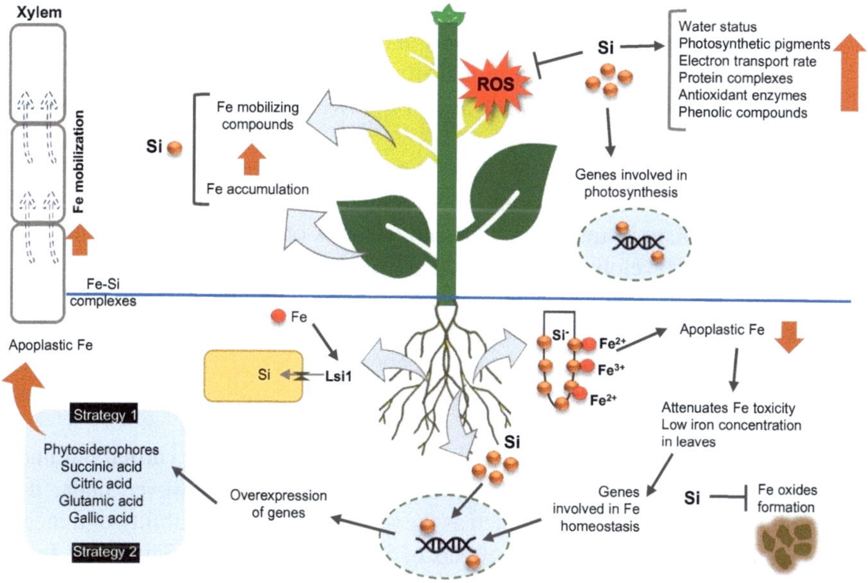

Fig. 11.3 Scheme illustrating the mechanisms of action of Si in mitigating Fe deficiency in plants

(i) Most studies involving the role played by Si in iron deficiency do not identify the level of this nutritional disorder in the plant and new research is needed to address this response. For example, the level of damage to the metabolism of a plant under Fe deficiency that suffered a decrease of 60% in dry mass production compared to a plant with sufficient Fe differs from the level of damage caused to the metabolism of a plant that suffered a decrease of 20% in dry mass production. Greater effects of Si are expected on moderate deficiency than on severe deficiency. In addition, moderate deficiency can be generated when supplementing the plant with Fe and then omitting the nutrient from the nutrient solution.

(ii) The potential for proteomic approaches must be further studied and better explored, as studying the interaction between stress-responsive proteins and Si could provide new insights into crop adaptation under moderate and severe Fe deficiency and in Fe-sufficient plants.

(iii) Understanding the mechanisms involved in Fe uptake, transport, and redistribution when modulated by Si is essential for the improvement of cultivars that are more tolerant to Fe deficiency in different species. Therefore, species-specific changes in the differential accumulation of Si and in Fe uptake strategies deserve further attention.

(iv) Studies with Si aiming at a better response from the plant with iron deficiency should potentiate its benefits in the absorption processes conciliated with the redistribution process and not only in one process. The redistribution of Fe in the plant is mediated mainly by transporters located in the upper node of the plant. Therefore, evaluations of the effects of Si on the transporters located in this node may promote greater gains in the internal cycling of Fe in the plant, consequently improving nutritional efficiency and productivity, as well as providing grains biofortified with Si and Fe, two elements essential for human nutrition.

(v) In addition, further studies on the effects of Si on the activity of enzymes associated with N assimilation are important, as iron deficiency causes losses in the metabolism of this macronutrient, which is one of the nutrients that most impacts the development and productivity of crops.

(vi) Auxin acts as a signaling molecule that detects Fe deficiency and enhances ferric reductase activity in the root of *Arabidopsis* (Chen et al. 2010), but little is known on the effect of Si on auxin deposition in the root and on the effect of other hormones related to improved iron use efficiency or increased tolerance to iron deficiency.

(vii) Studies indicating the relationship between Si and Fe were carried out under controlled conditions in hydroponic growing systems. However, under field conditions different factors such as the soil pH value and solubility of minerals could modify the interactive effects of Si on Fe availability. Therefore, future research could address the study of these effects, especially in alkaline soils that have limitations in Fe content. We indicate these studies to be initiated especially in irrigated systems where soluble Si (up to 10 kg ha^{-1} Si) can be employed to enhance the uptake of Si and the response of this beneficial element. In non-irrigated areas, low soluble and solid sources of Si should be avoided at high doses, as the silicate base increases the pH value of the soil, further aggravating the low Fe availability in the soil.

References

Alcántara E, Romera FJ, Cañete M, De la Guardia MD (2000) Effects of bicarbonate and iron supply on Fe(III) reducing capacity of roots and leaf chlorosis of the susceptible peach rootsrock "Nemaguard". J Plant Nutr 23:1607–1617. https://doi.org/10.1080/01904160009382127

Barker AV, Stratton M (2015) Iron. In: Barker A, Pilbeam D (eds) Handbook of plant nutrition, 2nd edn. CRC Press, Boca Raton, pp 399–426

Becker M, Ngo NS, Schenk MKA (2020) Silicon reduces the iron uptake in rice and induces iron homeostasis related genes. Sci Rep 10:2–11. https://doi.org/10.1038/s41598-020-61718-4

Bertamini M, Muthuchelian K, Nedunchezhian N (2002) Iron deficiency induced changes on the donor side of PS II in field grown grapevine (*Vitis vinifera* L. cv. Pinot noir) leaves. Plant Sci 162:599–605. https://doi.org/10.1016/S0168-9452(01)00604-5

Bityutskii N, Yakkonen K, Zlotina M (2010) Vliyanie kremniya na proyavlenie khlorosa rasteniy v usloviyakh defizita zhelesa i marganza. Agrokhimiia 2:45–51

Bityutskii N, Pavlovic J, Yakkonen K (2014) Contrasting effect of silicon on iron, zinc and manganese status and accumulation of metal-mobilizing compounds in micronutrient-deficient cucumber. Plant Physiol Biochem 74:205–211. https://doi.org/10.1016/j.plaphy.2013.11.015

Bityutskii NP, Yakkonen KL, Petrova AI (2018) Silicon ameliorates iron deficiency of cucumber in a pH-dependent manner. J Plant Physiol 231:364–373. https://doi.org/10.1016/j.jplph.2018.10.017

Carrasco-Gil S, Rodríguez-Menéndez S, Fernández B, Pereiro R, Fuente V, Hernandez-Apaloza L (2018) Silicon induced Fe deficiency affects Fe, Mn, Cu and Zn distribution in rice (*Oryza sativa* L.) growth in calcareous conditions. Plant Physiol Biochem 125:153–163. https://doi.org/10.1016/j.plaphy.2018.01.033

Chaiwong N, Bouain N, Prom-u-thai C, Rouached H (2020) Interplay between silicon and iron signaling pathways to regulate silicon transporter Lsi1 expression in rice. Front Plant Sci 11:1–9. https://doi.org/10.3389/fpls.2020.01065

Chalmardi Z, Abdolzadeh A, Sadeghipour HR (2016) Evaluation of the effects of silicon nutrition on alleviation of iron deficiency in rice plants (*Oriza sativa* L.) with emphasis on growth and antioxidant enzymes activity. Iran J Plant Biol 4:61–74

Chen WW, Yang JL, Qin C, Jin CW, Mo JH, Ye T, Zheng SJ (2010) Nitric oxide acts downstream of auxin to trigger root ferric-chelate reductase activity in response to iron deficiency in Arabidopsis. Plant Physiol 154:810–819. https://doi.org/10.1104/pp.110.161109

Conte SS, Walker EL (2011) Transporters contributing to iron trafficking in plants. Mol Plant 4:464–476. https://doi.org/10.1093/mp/ssr015

Curie C, Panaviene Z, Loulergue C, Dellaporta S, Briat JF, Walker EL (2001) Maize yellow stripe1 encodes a membrane protein directly involved in Fe(III) uptake. Nature 409:346–349. https://doi.org/10.1038/35053080

Curie C, Cassin G, Couch D, Divol F, Higuchi K, Jean ML, Misson J, Schikora A, Czernic P, Mari S (2009) Metal movement within the plant: contribution of nicotianamine and yellow stripe 1-like transporters. Ann Bot 103:1–11. https://doi.org/10.1093/aob/mcn207

D'Amico-Damião V, Barreto RF, Garcia LFO et al (2022) Cryptochrome 1a of tomato modulates nutritional deficiency responses. Sci Hortic 291:110577. https://doi.org/10.1016/j.scienta.2021.110577

Donnini S, Dell'Orto M, Zocchi G (2011) Oxidative stress responses and root lignification induced by Fe deficiency conditions in pear and quince genotypes. Tree Physiol 31:102–113. https://doi.org/10.1093/treephys/tpq105

Dubeaux G, Neveu J, Zelazny E, Vert G (2018) Metal sensing by the IRT1 transporter-receptor orchestrates its own degradation and plant metal nutrition. Mol Cell 69:953–964.e5. https://doi.org/10.1016/j.molcel.2018.02.009

Durrett TP, Gassmann W, Rogers EE (2007) The FRD3-mediated efflux of citrate into the root vasculature is necessary for efficient iron translocation. Plant Physiol 144:197–205. https://doi.org/10.1104/pp.107.097162

Farias GVH, Prado RM, Frazão JJ, Oliveira KS, Cazetta JO (2022) Foliar-applied silicon in sorghum (*Sorghum bicolor* L.) alleviate zinc deficiency. Silicon 14:281–287. https://doi.org/10.1007/s12633-020-00825-3

Fernández V, Ebert G (2005) Foliar iron fertilization: a critical review. J Plant Nutr 28:2113–2124. https://doi.org/10.1080/01904160500320954

Gavassi MA, Alves FRR, Monteiro CC (2023) Photomorphogenic tomato mutants high-pigment 1 and aurea responses to iron deficiency. Sci Hortic 307:111502. https://doi.org/10.1016/j.scienta.2022.111502

Gonzalo MJ, Lucena JJ, Hernández-Apaolaza L (2013) Effect of silicon addition on soybean (*Glycine max*) and cucumber (*Cucumis sativus*) plants grown under iron deficiency. Plant Physiol Biochem 70:455–461. https://doi.org/10.1016/j.plaphy.2013.06.007

Gottardi S, Iacuzzo F, Tomasi N et al (2012) Beneficial effects of silicon on hydroponically grown corn salad (*Valerianella locusta* L. Laterr) plants. Plant Physiol Biochem 56:14–23. https://doi.org/10.1016/j.plaphy.2012.04.002

Guerinot M, Morrissey J (2009) Iron uptake and transport in plants: the good, the bad, and the ionome. Chem Rev 109:4553–4567

Hernández-Apaolaza L, Escribano L, Zamarreño ÁM, García-Mina JM, Cano C, Carrasco-Gil S (2020) Root silicon addition induces Fe deficiency in cucumber plants, but facilitates their recovery after Fe resupply. A comparison with Si foliar sprays. Front Plant Sci 11:552. https://doi.org/10.3389/fpls.2020.580552

Inoue H, Kobayashi T, Nozoye T, Nakanishial H, Mori S, Nishizawa N (2009) Rice OsYSL15 is an iron-regulated iron (III)-deoxymugineic acid transporter expressed in the roots and is essential for iron uptake in early growth of the seedlings. J Biol Chem 284:3470–3479. https://doi.org/10.1074/jbc.M806042200

Ivanov R, Brumbarova T, Bauer P (2012) Fitting into the harsh reality: regulation of iron-deficiency responses in dicotyledonous plants. Mol Plant 5:27–42. https://doi.org/10.1093/mp/ssr065

Kabir AH, Paltridge NG, Roessner U, Stangoulis JCR (2012) Mechanisms associated with Fe-deficiency tolerance and signaling in shoots of *Pisum sativum*. Physiol Plant 147:381–395. https://doi.org/10.1111/j.1399-3054.2012.01682.x

Kallala N, M'sehli W, Hammi K, Abid G, Mhadhbi H (2021) Efficiency of antioxidant system in barrel medic (*Medicago truncatula*) genotypes and the orchestration of their key components under iron deficiency. Crop Pasture Sci 73:138–148. https://doi.org/10.1071/CP21041

Khan N, Seshadri B, Bolan N, Saint CP, Kirkham MB, Chowdhury S, Yamaguchi N, Lee DY, Kunhikrishnan A, Qi F, Karunanithi R, Qiu R, Zhu YG, Syu CH (2016) Root iron plaque on wetland plants as a dynamic pool of nutrients and contaminants. Adv Agron 138:1–96. https://doi.org/10.1016/bs.agron.2016.04.002

Khan MA, Castro-Guerrero NA, McInturf SA, Nguyen NT, Dame AN, Wang J, Bindbeutel RK, Joshi T, Jurisson SS, Nusinow DA (2018) Changes in iron availability in *Arabidopsis* are rapidly sensed in the leaf vasculature and impaired sensing leads to opposite transcriptional programs in leaves and roots. Plant Cell Environ 41:2263–2276. https://doi.org/10.1111/pce.13192

Kobayashi T, Nishizawa NK (2014) Iron sensors and signals in response to iron deficiency. Plant Sci 224:36–43. https://doi.org/10.1016/j.plantsci.2014.04.002

Ksiaa M, Farhat N, Rabhi M, Elkhouni A, Smaoui A, Debez A, Cabassa-Hourton C, Savouré A, Abdelly C, Zorrig W (2021) Silicon (Si) alleviates iron deficiency effects in sea barley (*Hordeum marinum*) by enhancing iron accumulation and photosystem activities. Silicon. https://doi.org/10.1007/s12633-021-01376-x

Lata-Tenesaca LF, Prado RM, Ajila-Celi GE, Silva D, Pinzetta Junior JS, Mattiuz B (2023) Iron biofortification in quinoa: effect of iron application methods on nutritional quality, anti-nutrient composition, and grain productivity. Food Chem 404:134573. https://doi.org/10.1016/j.foodchem.2022.134573

Lee S, Chiecko JC, Kim SA, Walker EL, Lee Y, Guerinot ML, An G (2009) Disruption of OsYSL15 leads to iron inefficiency in rice plants. Plant Physiol 150:786–800. https://doi.org/10.1104/pp.109.135418

Li W, Lan P (2017) The understanding of the plant iron deficiency responses in Strategy I plants and the role of ethylene in this process by omic approaches. Front Plant Sci 8:040. https://doi.org/10.3389/fpls.2017.00040

López-Millán AF, Grusak MA, Abadía A, Abadía J (2013) Iron deficiency in plants: an insight from proteomic approaches. Front Plant Sci 4:0254. https://doi.org/10.3389/fpls.2013.00254

Lucena JJ (2006) Synthetic iron chelates to correct iron deficiency in plants. In: Barton LL, Abadia J (eds) Iron nutrition in plants and rhizospheric microorganisms. Springer, Dordrecht, pp 103–128

Manivannan A, Ahn YK (2017) Silicon regulates potential genes involved in major physiological processes in plants to combat stress. Front Plant Sci 8:1346. https://doi.org/10.3389/fpls.2017.01346

Martín-Esquinas A, Hernández-Apaolaza L (2021) Rice responses to silicon addition at different Fe status and growth pH. Plant Physiol Biochem 163:296–307. https://doi.org/10.1016/j.plaphy.2021.04.012

Molassiotis A, Tanou G, Diamantidis G, Patakas A, Therios I (2006) Effects of 4-month Fe deficiency exposure on Fe reduction mechanism, photosynthetic gas exchange, chlorophyll fluorescence and antioxidant defense in two peach rootstocks differing in Fe deficiency tolerance. J Plant Physiol 163:176–185. https://doi.org/10.1016/j.jplph.2004.11.016

Muneer S, Jeong BR (2015) Silicon decreases Fe deficiency responses by improving photosynthesis and maintaining composition of thylakoid multiprotein complex proteins in soybean plants (*Glycine max* L.). J Plant Growth Regul 34:485–498. https://doi.org/10.1007/s00344-015-9484-y

Nikolic DB, Nesic S, Bosnic D, Kostic L, Nikolic M, Samardzic JT (2019) Silicon alleviates iron deficiency in barley by enhancing expression of Strategy II genes and metal redistribution. Front Plant Sci 10:0416. https://doi.org/10.3389/fpls.2019.00416

Oliveira KS, Prado RDM, Checchio MV, Gratão PL (2022) Interaction of silicon and manganese in nutritional and physiological aspects of energy cane with high fiber content. BMC Plant Biol:1–13. https://doi.org/10.1186/s12870-022-03766-8

Pavlovic J, Samardzic J, Maksimović V, Timotijevic G, Stevic N, Laursen KH, Hansen TH, Husted S, Schjoerring JK, Liang Y, Nikolic M (2013) Silicon alleviates iron deficiency in cucumber by promoting mobilization of iron in the root apoplast. New Phytol 198:1096–1107. https://doi.org/10.1111/nph.12213

Pavlovic J, Kostic L, Bosnic P, Kirkby EA, Nikolic M (2021) Interactions of silicon with essential and beneficial elements in plants. Front Plant Sci 12:697592. https://doi.org/10.3389/fpls.2021.697592

Perry CC, Keeling-Tucker T (1998) Aspects of the bioinorganic chemistry of silicon in conjunction with the biometals calcium, iron and aluminium. J Inorg Biochem 69:181–191. https://doi.org/10.1016/S0162-0134(97)10017-4

Pich A, Scholz G (1996) Translocation of copper and other micronutrients in tomato plants (*Lycopersicon esculentum* Mill.): nicotianamine-stimulated copper transport in the xylem. J Exp Bot 47:41–47. https://doi.org/10.1093/jxb/47.1.41

Rajniak J, Giehl RFH, Chang E, Murgia I, Wirén N, Sattely ES (2018) Biosynthesis of redox-active metabolites in response to iron deficiency in plants. Nat Chem Biol 14:442–450. https://doi.org/10.1038/s41589-018-0019-2

Rastogi A, Yadav S, Hussain S, Kataria S, Hajihashemi S, Kumari P, Yang X, Brestic M (2021) Does silicon really matter for the photosynthetic machinery in plant? Plant Physiol Biochem 169:40–48. https://doi.org/10.1016/j.plaphy.2021.11.004

Römheld V, Marschner H (1986) Evidence for a specific uptake system for iron phytosiderophores in roots of grasses. Plant Physiol 80:175–180. https://doi.org/10.1104/pp.80.1.175

Santi S, Schmidt W (2009) Dissecting iron deficiency-induced proton extrusion in Arabidopsis roots. New Phytol 183:1072–1084. https://doi.org/10.1111/j.1469-8137.2009.02908.x

Souza J, Prado RDM, Campos CNS, Junior GSS (2022) Addition of silicon to boron foliar spray in cotton plants modulates the antioxidative system attenuating boron deficiency and toxicity. BMC Plant Biol 22:1–13. https://doi.org/10.1186/s12870-022-03721-7

Stevic N, Korac J, Pavlovic J, Nikolic M (2016) Binding of transition metals to monosilicic acid in aqueous and xylem (*Cucumis sativus* L.) solutions: a low-T electron paramagnetic resonance study. Biometals 29:945–951. https://doi.org/10.1007/s10534-016-9966-9

Teixeira GCM, de Mello PR, Oliveira KS, D'Amico-Damião V, Sousa GS (2020) Silicon increases leaf chlorophyll content and iron nutritional efficiency and reduces iron deficiency in sorghum plants. J Soil Sci Plant Nutr 20:1311–1320. https://doi.org/10.1007/s42729-020-00214-0

Tewari RK, Hadacek F, Sassmann S, Lang I (2013) Iron deprivation-induced reactive oxygen species generation leads to non-autolytic PCD in *Brassica napus* leaves. Environ Exp Bot 91:74–83. https://doi.org/10.1016/j.envexpbot.2013.03.006

Vert G, Grotz N, Dédaldéchamp F, Gaymard F, Guerinot ML, Briat JF, Curie C (2021) IRT1, an arabidopsis transporter essential for iron uptake from the soil and for plant growth. Plant Cell 33:439–440. https://doi.org/10.1093/plcell/koaa033

White PJ, Broadley MR (2005) Biofortifying crops with essential mineral elements. Trends Plant Sci 10:586–593. https://doi.org/10.1016/j.tplants.2005.10.001

Yokosho K, Yamaji N, Ueno D, Mitani N, Ma JF (2009) OsFRDL1 is a citrate transporter required for efficient translocation of iron in rice. Plant Physiol 149:297–305. https://doi.org/10.1104/pp.108.128132

Yokosho K, Yamaji N, Ma JF (2016) OsFRDL1 expressed in nodes is required for distribution of iron to grains in rice. J Exp Bot 67:5485–5494. https://doi.org/10.1093/jxb/erw314

Zadeh A, Chalmardi Z (2013) Role of silicon in alleviation of iron deficiency and toxicity in hydroponically-grown rice (*Oryza sativa* L.) plants. J Sci Technol Greenh Cult 3:79–89

Zhang Y, Xu YH, Yi HY, Gong JM (2012) Vacuolar membrane transporters OsVIT1 and OsVIT2 modulate iron translocation between flag leaves and seeds in rice. Plant J 72:400–410. https://doi.org/10.1111/j.1365-313X.2012.05088.x

Zuo Y, Zhang F (2011) Soil and crop management strategies to prevent iron deficiency in crops. Plant Soil 339:83–95. https://doi.org/10.1007/s11104-010-0566-0

Chapter 12
Silicon Mitigates the Effects of Aluminium Toxicity

Martin J. Hodson

12.1 Introduction

Aluminium toxicity is a major problem for plant growth in acidic mineral soils (Marschner 1991). In some parts of the world, soils are naturally acidic, and the vegetation is adapted to these soils (Hodson 2012). A considerable percentage of the world's potentially arable soils are acidic, maybe as much as 40–50% (Kochian et al. 2015). In the 1970s and 1980s, soil acidification brought about by anthropogenic acid rain was a major problem in Europe and North America. Acid rain was largely caused by sulphur dioxide emissions from the burning of coal which had a high sulphur content. Fortunately, international legislation was brought in, and scrubbers were used to remove the sulphur dioxide from power station flu gasses (Grennfelt et al. 2020). The problem, if not solved, was brought under better control, at least in Europe and North America.

Whatever the cause of the acidity, in mineral soils where the pH drops below 5.0, aluminium (Al) toxicity can become a problem. The reason is that the trivalent Al^{3+} species becomes increasingly available (see below for more detail), and this is highly toxic. Aluminium toxicity in plants causes a number of key symptoms, including decreased root growth; decreased uptake of some nutrients (e.g. Ca and Mg); damage to the root apex; callose and lignin deposition in roots; and membrane damage. See Wei et al. (2021) for a recent review.

It has become increasingly evident that silicon (Si) is involved in the mitigation of a wide range of plant stresses. A number of other chapters in this book concern mitigation of other stresses. Here we will consider the mitigation of Al toxicity by Si, possibly one of the best characterised mitigation effects, and one that has been

M. J. Hodson (✉)
Department of Biological and Medical Sciences, Faculty of Health and Life Sciences, Oxford Brookes University, Headington, Oxford, UK
e-mail: mjhodson@brookes.ac.uk

R. de Mello Prado (ed.), *Benefits of Silicon in the Nutrition of Plants*,
https://doi.org/10.1007/978-3-031-26673-7_12

extensively studied for more than 30 years. I will aim to give a fairly comprehensive account, particularly concentrating on experiments conducted in hydroponic cultures, the co-deposition of Al and Si, and the mechanism of the mitigation effect. But first, we will give a brief historical perspective and spend a short time on Al and Si in the soil.

12.2 A Historical Perspective

As far as I am aware the first investigation of the mitigation of Al toxicity by Si ever carried out in hydroponic culture that was reported in Japanese by Okuda and Takahashi (1962), and was then summarized in English by Okuda and Takahashi (1965). They worked on lowland rice, used a full nutrient solution at pH 5.5, and could not observe mitigation of Al toxicity by Si, or any effect of Si on Al uptake. Perhaps not too surprisingly, given the negative results from this work, there then seemed to be a long gap before other workers looked at this topic again. In what I would regard as the first paper of the modern era, Galvez et al. (1987) investigated the mitigation of Al toxicity by Si with sorghum plants growing in a full nutrient solution at pH 4.0. They found that Si treatment could restore the root growth of one cultivar to 50% of control and another to 30%.

Sometimes in science it happens that workers in a very different field produce work that is relevant in a broader context. In the 1980s and 1990s, the chemists, Birchall and Exley, working primarily on Al/Si interactions in animal systems, gave plant scientists some of the tools to investigate interactions in plant systems. Birchall et al. (1989) found that Si could mitigate the toxic effects of Al on salmon. Chemical investigations by this group gave us some information on how Al and Si interact in solution (Exley and Birchall 1992, 1993). This work provided considerable impetus for studies on plant systems.

The early work of Bartoli (1985) had suggested that some phytoliths (solid amorphous silica bodies deposited in plant cells), particularly those originating from the conifers, could be high in Al. The final building block for studies of Al/Si interactions in plants involved microanalytical studies of Al/Si co-deposition. As far as I can tell, the first such investigation was by Godde et al. (1988) working on Norway spruce needles from a polluted forest in Germany. They found that Al and Si were co-localised in the stellar tissue of the needles. Soon after, Hodson and Wilkins (1991), working on the roots of Norway spruce, showed that Al and Si were co-localised in the cortical cell walls.

By the early 1990s, we knew that under some circumstances Si could mitigate Al toxicity for plants grown in hydroponic cultures, we knew that this could also happen in animal systems, we had some knowledge of Al/Si interactions in solution, and the first microanalytical results showing Al/Si co-deposition were coming in. This was when I started working on this problem. My colleague, David Evans, and I soon wrote a review of the topic (Hodson and Evans 1995), and for many years we worked on it together. In 2020 we updated our review (Hodson and Evans 2020), exactly 25 years later. For more details concerning the history of Al/Si interactions in plants see Hodson (2020).

12.3 A Brief Consideration of Silicon and Aluminium in Soils

In many mineral soils, aluminosilicates make up a large proportion of the solid mineral matter. This mineral matter is usually derived from bedrock. At 27.7%, Si is only lower than oxygen in the Earth's crust, with Al at 8.2%, the third most abundant element, and the most abundant metal (Exley 1998). Both Si and Al are sparingly soluble, and the amounts present in soluble form represent a very small fraction of the total mineral matter.

I will not spend very long on Si in soil as it is covered in great detail in several other chapters of this book. Most of the Si in soil is present in aluminosilicates, quartz, and other minerals that are not very soluble (Tubana et al. 2016). Monosilicic acid ($Si(OH)_4$), is the major soluble form with a maximum solubility of about 2.0 mM, but the range is between 0.1 and 0.6 mM in most soils. In neutral and acidic soils, pH does not affect Si speciation, but H_3SiO_4 may also be present in soils with a high pH. Biogenic solid amorphous silica (SiO_2. nH_2O), originating mostly from plants, is now recognised as an important source of soluble silica in soils (Haynes 2017).

In acid soils, the direct toxicity of H^+ ions is probably not the most important factor in limiting plant root growth in most cases (Foy 1992). Rather it is the effect of pH on the availability of nutrient elements (e.g. calcium, magnesium, phosphorus, molybdenum), and the increase in availability of toxic elements such as manganese and Al that is the bigger problem. The solubility of Al is highly affected by pH, as is its speciation. It seems that hydroxyaluminium species such as $Al(OH)_3$, $Al(OH)_2^+$, and $Al(OH)^{2+}$ are relatively non-toxic, and it is only Al^{3+} that is the toxic species under most circumstances. While the hydroxyaluminium species are dominant in the soil solutions of moderately acidic soils, Al^{3+} is mainly available below pH 4.5.

The interaction between the various Al species and Si in solution is thought to be an important factor in the mitigation of Al toxicity, and is covered in Sect. 12.7.1.

12.4 Silicon and Aluminium Uptake and Accumulation by Plants

Both Si and Al are not regarded as essential elements for plant growth. Most scientists see Si as a beneficial element, whilst Al is almost universally regarded as a toxic element. However, both elements are taken up by plants to varying extents. In this section, we will look at the uptake and accumulation of each element separately, and then consider how the two interact.

12.4.1 Silicon Uptake and Accumulation

Uptake of Si by plants is as $Si(OH)_4$ from the soil solution (Raven 1983; Epstein 1994). For many years, plant scientists debated about the mechanism of Si uptake, and whether it was passive or active, but it is only relatively recently, with the advent of molecular techniques that we have begun to have a clearer picture. Ma et al. (2006) reported on a Si transporter, Lsi1, from the rice root, and in the following year, Ma et al. (2007) discovered an efflux transporter Lsi2, also in the rice root, which seems to be an active transporter. Since then there have been a considerable number of investigations focusing on these two transporters, and rather less on others. Lsi1 is an aquaporin and a member of the Nodulin26-like intrinsic protein-III (NIP-III) subgroup of the Major Intrinsic Proteins. Its properties are now relatively well-known (see Ma and Yamaji 2015; Hodson and Evans 2020). The same cannot be said for Lsi2, which is still somewhat shrouded in mystery (Coskun et al. 2021). It is still not entirely clear how the presence or absence of these transporters leads to the considerable variation in plant Si concentrations that are observed in plants. We will look at the topic of differential Si accumulation by plants next.

It has been known for many years that plant species vary in the amount of Si that they take up. Striegel (1912) noted that in plants grown under the same conditions, the monocots (all grasses) had high Si concentrations and low calcium concentrations, while the reverse was observed in dicots. Over 60 years ago, Baker et al. (1961) wrote, "Plants differ in the total amount of silica which they absorb, and subsequently secrete as phytoliths. Quantitative determinations in the present study show that Wimmera ryegrass contains 23 times as much insoluble matter as lucerne. This difference, which is of the same order as the difference in the silica contents of the two plants, reflects the difference in their opal phytolith contents." Takahashi conducted a whole series of investigations from 1976 to 1981 to characterise Si accumulation in nearly 500 species (summarised in Ma and Takahashi 2002). They defined a Si accumulator species as one with a Si content above 1% and Si/Ca ratio higher than 1.0. Si excluders were those plants with Si content below 1% and Si/Ca ratio below 0.5. They also saw an intermediate category where one of the two criteria was not satisfied.

By the time Epstein (1994) wrote his classic review, he was able to state that Si in plants ranged from 0.1 to 10%, and that in some cases the concentration exceeded that for macronutrients. More recently, it has become evident that there are no clear demarcations between Si accumulators and excluders and between those plants that have active Si uptake (or exclusion) and those that show passive uptake. Rather it seems that there is a continuous spectrum of plant species with respect to Si concentration (Hodson et al. 2005).

12.4.2 Aluminium Uptake and Accumulation

Aluminium transporters in plants have also been characterized (Zhang et al. 2019), but they are much less studied than Si transporters. Transport of Al has been associated with ATP-binding cassette (ABC), aquaporin proteins, and natural resistance-associated macrophage protein (Nramp) families. Hodson and Evans (2020) summarized progress in this area, feeling that other transporters, such as those associated with malate transport that is linked to Al tolerance, had received more attention.

As with Si, it is now well-known that Al accumulation varies considerably across the plant kingdom. The early work of Chenery (1949, 1950a, b) considered Al concentrations in plant shoots. He defined an Al accumulator as a plant that had more than 1000 ppm (37.1 μmol g^{-1} dwt) in its shoot tissues. Very few monocots and herbaceous dicots were found to be Al accumulators, but this was certainly not the case with the arborescent dicots, some of which can accumulate very high concentrations. The most famous Al-accumulating plant is tea, which can accumulate up to 1137.5 μmol g^{-1} dwt in its old leaves (Matsumoto et al. 1976). A number of ferns are major Al accumulators, whilst some of the gymnosperms take up moderate amounts. Jansen et al. (2002) looked in considerable detail at the phylogenetic significance of Al hyper-accumulation in plants.

12.4.3 The Interaction Between Silicon and Aluminium Uptake and Accumulation

In the last 15 years, we have made great progress in our understanding of Si transporters in plants, and rather less in understanding Al transporters. But what happens to these transporters when plants are treated with toxic amounts of Al or both Al and Si? Pontigo et al. (2017) studied the effect of Si on Si and Al transport in ryegrass, grown in hydroponic culture, where plants were exposed to differing Al and Si concentrations. The presence of Si caused a decrease in Al uptake. When Al and Si were applied singly both Lsi1 and Lsi2 genes were down-regulated in roots. But up-regulation of both transporter genes occurred when Si was applied to Al-treated plants (this observation was repeated by Xiao et al. (2021) who were working on rice). When plants are stressed by Al toxicity it seems there is an increased Si requirement in the tissue. With only two publications, proof of the direct effects of Si on Al transporters (and vice versa) is inadequate at present, and it is uncertain whether these processes are involved in the amelioration of Al toxicity.

We have seen above that there is considerable variation in both Si and Al accumulation by plants, but how do these two factors interact with each other? Table 12.1 is reproduced from Hodson and Evans (1995), and some of the data included are now quite old, and have, to some extent, been superseded. We will consider some more recent data next, but Table 12.1 still gives a good overview.

Table 12.1 A comparison between silica and aluminium uptake into the shoots of plants from a variety of taxonomic groups (Hodson and Evans 1995)

Taxonomic group	Silica uptake (% SiO_2) into shoot	Al uptake into shoot
Wetland grasses	10–15 (a)	Limited – rice grown at 11.1 μmol l^{-1} Al in solution at pH 4.0 contained 11.8 μmol g^{-1} dwt Al in the shoots (e)
Dryland grasses	1–3 (a)	No plants accumulate more than 37.1 μmol g^{-1} dwt Al (f)
Herbaceous dicots	<1 (a)	Very few plants accumulate more than 37.1 μmol g^{-1} dwt Al (g)
Arborescent dicots	0.01–3.79 (b)	Considerable variation, but in some plants can be very high (e.g. *Symplocus spicata* at 2679.8 μmol g^{-1} dwt Al) (g)
Gymnosperms	0.09–1.4 (c)	Can be quite high (e.g. *Pinus* spp. at 37.1–74.2 μmol g^{-1} dwt Al) (h)
Ferns and allies	Can be high in some species (e.g. 12% in *Equisetum arvense*) (d)	Some species are heavy accumulators (e.g. *Danaea fendleri* at 1086 μmol g^{-1} dwt Al) (i)

Authors are: (a) Jones and Handreck (1967); (b) Geis (1973); (c) Klein and Geis (1978); (d) Chen and Lewin (1969); (e) Tanaka and Navasero (1966); (f) Chenery (1950a); (g) Chenery (1949); (h) Bollard (1983); and (i) Chenery (1950b)

Firstly, let us consider the Angiosperms. The separation of grasses by Jones and Handreck (1967) into wetland and dryland grasses has now given way to a more taxonomically precise division, and that includes all of the monocots. Hodson et al. (2005) collated a large data set, and from that we can determine that for Commelinid monocots (which include the Poales) the average percentage Si is 3.21 (where $n = 408$ measurements), while for non-Commelinid monocots this figure is 0.53 ($n = 41$). The Poales include some of the heaviest Si accumulators known, such as rice. In contrast, the monocots are nearly all low Al accumulators, with Jansen et al. (2002) reporting that only members of the Rapateaceae were high Al accumulators. Transport of Al into grass and cereal shoots is rarely measured, but Cocker et al. (1997) reported that in wheat plants treated with 100 μmol l^{-1} Al in their hydroponic solutions, shoot Al concentrations were below 5 μmol g^{-1} dwt Al.

As might be expected with this analysis, the percentage silica in all of the dicots (0.9, $n = 414$) came out much lower than that in the Commelinid monocots. Confirming earlier work on dicots, Jansen et al. (2002) state that, "Aluminum-hyperaccumulating plants are confined generally to woody, perennial species, although there are several examples of herbaceous taxa that hyperaccumulate aluminum in their leaves: for example, Coccocypselum (Rubiaceae), Genlisea and Utricularia (Lentibulariaceae)…."

Hodson and Sangster (1999) surveyed the data available on both Si and Al concentrations in conifer needles. For percentage silica, the data varied from a low of 0.04 in *Juniperus* to a high of 2.43 in *Picea*. In those conifers, like *Picea*, that retain their needles for several years, silica accumulates with time, and the older needles can reach quite high concentrations. In the case of Al, Hodson and Sangster (1999) reported that Al concentrations were moderate even in needles collected from polluted environments.

According to the data of Hodson et al. (2005), the Polypodiidae, commonly called leptosporangiate ferns, are moderate Si accumulators and average 2.42% ($n = 85$). Not surprisingly, the four *Equisetum* samples came out at 9.82% Si ($n = 4$). Schmitt et al. (2017) produced a comprehensive analysis of Al accumulation in fern species. Heavy Al accumulation was most common outside of the Polypodiales, and in tropical and sub-tropical ferns.

So, what patterns do we see in Si and Al accumulation and how are they related? The Commelinid monocots (which include the Poales, the grasses, and cereals) tend to be fairly heavy Si accumulators in their shoots, with some species, such as rice, being particularly high. On the other hand, within the Commelinid monocots, only members of the Rapateaceae that are found in tropical South America and West Africa appear to be Al accumulators. All other Commelinid monocots appear to exclude Al from the shoot. As we will see shortly, much of the work on Al/Si interactions in plants has concerned grasses and cereals, and this has meant that considerable attention has been focused on their root systems, and on the bulk solutions surrounding the roots.

For the non-Commelinid monocots and the dicots, relatively few species are heavy Si accumulators in their shoots. Most non-Commelinid monocots and herbaceous dicots accumulate low concentrations of Al in their shoots. Many of the greatest known Al accumulators (e.g. tea) are to be found in the woody dicots. In this group, much of the research focus has been on the mechanisms whereby these plants are able to accumulate toxic quantities of Al in their shoots.

The Gymnosperms tend to be low to moderate Si accumulators, although over time the needles of species like *Picea* (spruce) can reach quite high concentrations (Hodson and Sangster 1999). None of the gymnosperms appear to be heavy accumulators of Al, but moderate amounts do accumulate in the needles, and co-deposition of Al with Si has been a focus for researchers (see Sect. 12.6.2). The ferns and their allies can be heavy Si accumulators, with *Equisetum* species being particularly notable in this respect. There are also some ferns that are very heavy Al accumulators.

Hodson and Evans (1995) concluded that, "It would appear that heavy Al accumulation and heavy Si accumulation are mutually exclusive." The simultaneous transport of both Al and Si in the xylem without major co-precipitation would seem a likely reason why this is the case. The 1995 statement still appears to be largely correct, and in the vast majority of species they either accumulate neither of these elements or just one of them, and not both. However, Liu et al. (2019) recently produced a fascinating paper reporting that the fern *Dicranopteris linearis* was a heavy accumulator of both Si and Al (and rare earth elements). We will return to consider this case, and its implications, in Sect. 12.6.4.

12.5 The Amelioration of Aluminium Toxicity by Silicon in Experiments Carried Out in Hydroponic Cultures

Now we come to a core topic within this chapter – experiments to investigate Al/Si interactions that have been conducted in hydroponic cultures. Largely because of the difficulties of controlling the chemical reactions within soil systems, almost all investigations looking at mitigation of Al toxicity by Si have been in water culture (for the relatively few studies conducted in sand culture or soil systems see Sect. 12.8).

12.5.1 Plant Growth

As we saw in Sect. 12.2 it appears that Okuda and Takahashi (1962) were the first to carry out experiments in hydroponic culture to investigate Al/Si interactions. It took a little while for further work on this topic to happen, but after the work of Galvez et al. (1987) on sorghum we have seen many more papers that at least included some basic experiments on mitigation of Al toxicity by Si. In Table 12.2, I will attempt to summarise most of these publications. I hope that the table is reasonably comprehensive, but there are now so many papers on the topic, some of which have been published in relatively obscure places. I will certainly have missed some, particularly those published in languages other than English. Apologies to the authors concerned. Sadly, I found a few papers that were of such poor quality that I have not included them. I have decided to arrange the papers in chronological order starting with Okuda and Takahashi (1962), and working through to the most recent publications. At this point, we will only consider the effects on growth, and the discussion of the mechanism(s) behind the mitigation effect will be in Sect. 12.7.

Table 12.2 covers 33 papers, but in a few cases, I have split their results into two where cultivars or species investigated differed markedly in their responses. Of the papers considered 25 investigated grasses and cereals, five looked at herbaceous dicots, and one both sorghum and soybean (Kopittke et al. 2017). Only one paper considered a woody dicot Al hyperaccumulator (Watanabe et al. 1997), and there is just one paper on a conifer (Ryder et al. 2003). Some species have several papers devoted to them: rice has eight; maize has five; and barley, sorghum, and wheat three each. It is immediately obvious that we have a huge bias towards grasses and cereals, and that in future work we need to include more species from other groups. There have been calls for more Si studies on plants other than grasses (Katz 2014), but this has yet to happen.

Of the 33 papers in Table 12.2, a total of 15 used "full" (F) background solutions. That is, they used standard formulations that are used for hydroponic investigations, whether at full strength or diluted, and contained a complete suite of macronutrients and micronutrients. The problem with doing this is that Al reacts with some components of the media, particularly phosphorus, to produce insoluble precipitates, or at

Table 12.2 The effects of silicon on the amelioration of aluminium toxicity in hydroponic culture

Species	Culture media	Treatment time	[Al] tested	[Si] tested	Observations on growth	Researchers
Rice	F, pH 5.5	?	0–140 ppm $AlCl_3$	100 ppm SiO_2	Si had no ameliorative effect	Okuda and Takahashi (1962)
Sorghum	F, pH 4.0	16–20 d	296 μM	0–3600 μM	Restores root growth of one cv. to 50%; another cv. to 30%; shoot growth is not stimulated, but shoot/root ratio is restored	Galvez et al. (1987)
Cotton	0.5 mM CN, pH 4.5	4 d	0–100 μM	0–2800 μM	No effect	Li et al. (1989)
Teosinte	F, pH 4.0	7–21 d	0–120 μM	0–4 μM	Ameliorates inhibition of root and shoot length	Barceló et al. (1993)
Sorghum	0.5 mM CN, pH 4.5	8 d	0–100 μM	0–2800 μM	Si ameliorated the effects of Al on root growth and shoot/root ratio	Hodson and Sangster (1993)
Soybean	F, pH 4.0–7.0	3–4 d	0–1000 μM	0–2000 μM	Si ameliorates inhibition of root length	Baylis et al. (1994)
Barley	0.5 mM CN, pH 4.5	6 d	0–50 μM	0–2800 μM	Si ameliorates root growth more than shoot	Hammond et al. (1995)
Wheat (cv. Celtic)	0.5 mM CN, 31 μM KCl, pH 4.2 & 4.6	4 d	0–100 μM	0–2000 μM	Si ameliorates Al toxicity effects on root length in the pH 4.6 treatment, but not at pH 4.2	Cocker et al. (1997)
Maize	F, pH 4.4	24 h	0–50 μM (free Al = 0–17.3 μM)	72 h pre-treatment with 0 or 1000 μM	Si pre-treatment ameliorated Al toxicity effects on root length	Corrales et al. (1997)
Maize	100 μM $CaCl_2$, pH 4.3	20 h	0–20 μM	0–2000 μM	Si ameliorates Al toxicity effects on root elongation rate	Ma et al. (1997)
Melastoma malabathricum	F, pH 3.8	28 d	0–400 μM	0–650 μM	Si had very little effect on plant growth in Al treated plants	Watanabe et al. (1997)
Wheat (cv. Scout 66)	0.5 mM CN, 31 μM KCl, pH 4.6 or 5.0	4 d	0–1.5 μM	0–5 μM	Si ameliorates Al toxicity effects on root length in the pH 5.0 treatment, but not at pH 4.6	Cocker et al. (1998a)

(continued)

Table 12.2 (continued)

Species	Culture media	Treatment time	[Al] tested	[Si] tested	Observations on growth	Researchers
Wheat (cv. Atlas 66)	0.5 mM CN, 31 μM KCl, pH 4.2 or 4.6	4 d	0–100 μM	0–2000 μM	Si ameliorates Al toxicity effects on root length in the pH 4.6 treatment, but not at pH 4.2	Cocker et al. (1998a)
Rice (four varieties)	0.5 mM CN, pH 4.5	5 d	0–50 μM	0–2000 μM	Si ameliorates Al toxicity effects on root dry weight, with the effect being greater in some varieties than others	Rahman et al. (1998)
Rice	0.5 mM CN, pH 4.5	3 or 4 d	0, 200 or 800 μM	(1) 3.0 mM low molecular Si and 0.2 mM high molecular Si, OR (2) 3.0 mM low molecular Si and 21.7 mM high molecular Si	Si ameliorates Al toxicity effects on root length	Hara et al. (1999)
Maize (three cultivars)	F or 0.4 mM $CaCl_2$	48 h	0–50 μM	72 h pre-treatment with 1000 μM	Si ameliorates Al toxicity effects on root length, but only with $CaCl_2$ as background solution	Kidd et al. (2001)
Barley	F, pH 4.2	30 d	0–150 μM	0–1000 μM	Si ameliorates Al toxicity at a lower concentration of Al, but exacerbated it at a higher concentration of Al	Liang et al. (2001)
Norway Spruce	0.5 mM CN, pH 4.0, 4.25, 4.5, 4.75 or 5.0	7 d	0–100 μM	0–2000 μM	Si ameliorates Al toxicity effects on root length in the pH 4.75 or 5.0 treatments, but not at the lower pH values	Ryder et al. (2003)
Triticale and Rye	0.5 mM $CaSO_4$, pH 4.1	7 d	0–100 μM	0–1500 μM	Si ameliorates Al toxicity in triticale, but not in rye	Zsoldos et al. (2003a)

(continued)

Table 12.2 (continued)

Species	Culture media	Treatment time	[Al] tested	[Si] tested	Observations on growth	Researchers
Common and Durum Wheat	0.5 mM $CaSO_4$, pH 4.1	7 d	0–100 μM	0–1500 μM	Si ameliorates Al toxicity effects on root length in common wheat, but little in durum	Zsoldos et al. (2003b)
Maize	0.5 mM $CaCl_2$ and 8 μM H_3BO_3, pH 4.3	1 and 12 h	0–25 μM	0–1400 μM (both as a pre-treatment and throughout the experiment)	Si treatment but not Si pre-treatment ameliorated Al toxicity effects on root growth	Wang et al. (2004)
Rice	F, pH not stated	7 d	0–10 μM	0–50 μM	Si ameliorates Al toxicity effects on root and shoot fresh weight	Singh et al. (2011)
Peanut	F, pH 5.0	7 d	160 mg l^{-1}	80 mg l^{-1} sodium metasilicate	Si ameliorates Al toxicity effects on root growth	Shen et al. (2014)
Maize	F, pH 4.5	14 d	0–150 μM	0–2000 μM	Si has little ameliorative effect	Delavar et al. (2016)
Upland Rice	F. pH 4.0	19 d	1.4 mM	0–1700 μM	Si ameliorates Al toxicity effects on shoot growth	Freitas et al. (2017)
Sorghum	1 mM $CaCl_2$ and 5 μM H_3BO_3, pH 4.5	48 h	0–25 μM	0–2000 μM	Si ameliorates Al toxicity effects on root elongation rate	Kopittke et al. (2017)
Soybean	1 mM $CaCl_2$ and 5 μM H_3BO_3, pH 4.5 and 5.0	48 h	0–30 μM	0–2000 μM	Si ameliorates Al toxicity effects on root elongation rate at Al concentration up to 10 μM	Kopittke et al. (2017)
Ryegrass	F, pH 4.5	10 d	0–2.0 mM (free Al = 0–85 μM)	0–2000 μM	Si ameliorates Al toxicity effects on root yield	Pontigo et al. (2017)
Potato (cv. SMIJ319-7)	F(-P), pH 4.5	14 d	0–1.85 mM	0–1000 μM	Si has little ameliorative effect	Dorneles et al. (2019)
Potato (cv. SMIF212-3)	F(-P), pH 4.5	14 d	0–1.85 mM	0–1000 μM	Si ameliorates Al toxicity effects on root growth	Dorneles et al. (2019)
Barley	F, pH 4.5	21 d	0–200 μM	0–2000 μM	Si ameliorates Al toxicity effects on root dry weight (in two cultivars)	Vega et al. (2020)

(continued)

Table 12.2 (continued)

Species	Culture media	Treatment time	[Al] tested	[Si] tested	Observations on growth	Researchers
Sugarcane	F, pH 4.5	35 d	0–20 mg l^{-1}	0–2000 μM	Si ameliorates Al toxicity effects on shoot dry weight (in two cultivars)	Junior et al. (2021)
Rice	0.5 mM $CaCl_2$, pH 4.5	24 and 48 h	0–50 μM	0–1000 μM	Si ameliorates Al toxicity effects on root elongation	Xiao et al. (2021)
Rice	0.5 mM $CaCl_2$, pH 4.5	6 d treatment followed by 10 d in F	0–50 μM	0–1000 μM	Si ameliorates Al toxicity effects on root elongation and weight	Xiao et al. (2022)
Soybean	1 mM $CaCl_2$ and 5 μM H_3BO_3, pH 4.8	48 h	0–50 μM	0–2000 μM	Si ameliorates Al toxicity effects on root elongation	Li et al. (2022)
Rice	0.5 mM $CaCl_2$, pH 4.5	24 h	0–500 μM	0–1000 μM	Si reversed Al-induced decreases of the cell wall elongation	Jiang et al. (2022)

Key:
F Full nutrient as background solution (includes half strength and all solutions with all nutrients present), *F(-P)* Full nutrient minus phosphorus as background solution, *CN* Calcium nitrate as background solution

least the available Al is seriously reduced. There are several ways around this problem. The first is to leave phosphorus out of an otherwise "full" medium, and this was the approach adopted by Dorneles et al. (2019). The second is to use a full medium, but use a speciation program to determine the free available Al^{3+}. Corrales et al. (1997) used this methodology in their work. The third, and most widely used idea (17 papers in Table 12.2), is to use a much-simplified medium with a background solution of a calcium salt (nitrate, chloride, or sulphate). Some workers have also included boron salts in this formulation, concerned that it was needed for good root growth (Wang et al. 2004; Kopittke et al. 2017; Li et al. 2022). Of course, the difficulty with using such simple formulations is that plants might become deficient in certain nutrients and that this could affect their growth. It follows that the experiments using a calcium salt as the background solution have tended to be of relatively short duration (8 days is the longest).

A key factor in determining whether we see amelioration or not is solution pH. Often only very small changes in pH can have a very marked effect on amelioration. So, for example, Si was found to ameliorate Al toxicity effects on wheat root length at pH 4.6, but not at pH 4.2 (Cocker et al. 1997). Likewise, Ryder et al. (2003) found that Si ameliorated Al toxicity effects on root length of Norway spruce in pH 4.75 or 5.0 treatments, but not at the lower pH values. This may explain why the early work of Li et al. (1989) showed no amelioration effect; the authors grew cotton plants at pH 4.5 and this might be too low for amelioration.

One of the strange features of Table 12.2 is that there was only one paper (Singh et al. 2011) published in the ten years between 2004 and 2014. The reasons for this are obscure. One other notable fact is that only a few authors have published more than one paper in this area. This is regrettable as we need workers to gain a better understanding of the limitations of this kind of system. If they are continually moving on to other topics before they have really got to grips with all the factors involved, it means the work is unlikely to advance in a consistent way.

12.5.2 *Effects on Mineral Nutrition*

In addition to effects on growth, there is quite a lot of evidence that Si treatment can mitigate the effects of Al toxicity on mineral uptake.

Shoot and/or root Al was decreased by Si treatment in barley (Hammond et al. 1995; Vega et al. 2019), rice (Singh et al. 2011), ryegrass (Pontigo et al. 2017), and upland rice (Freitas et al. 2017).

Hammond et al. (1995) found that the calcium content of barley plants was decreased by Al treatment, but that adding Si mitigated this effect. Likewise, Delavar et al. (2016) found a similar result with calcium content of maize plants and also noted that magnesium showed this effect. Similarly, magnesium and zinc content of rice seedlings was decreased by Al treatment, but Si reduced this effect (Singh et al. 2011). Zsoldos et al. (2003b) showed that Al treatment increased potassium uptake in durum wheat, but that Si reduced this effect. Delavar et al. (2016) also reported this effect, and found a similar result with iron.

There are also some studies that have found that shoot and root Si is decreased by Al treatment: barley (Vega et al. 2019); ryegrass (Pontigo et al. 2017): and upland rice (Freitas et al. 2017).

In the most comprehensive study of the effects of Al toxicity on mineral nutrition in sugarcane (Junior et al. 2021), covering a wide range of macro- and micronutrients, Si was often found to ameliorate any decreases in mineral content caused by Al. There has only been one study looking at Si effects in an Al accumulator. In *Melastoma malabathricum* it was found that Si treatment had very little effect on shoot Al concentration or other mineral nutrients (Watanabe et al. 1997).

12.5.3 *Effects on Oxidative Damage*

There have been a number of reports that Si treatment can decrease the oxidative damage that is often caused by Al toxicity (Shen et al. 2014; Dorneles et al. 2019; Vega et al. 2019, 2020; Pontigo et al. 2017; Xiao et al. 2021). Si seems to act through increasing radical scavenging activity, and the production of phenolics and lignin.

There is no doubt that Si has real effects on decreasing oxidative damage under Al stress, and also on plant mineral nutrition. However, I suspect that all of these effects are secondary, and that the primary mechanisms for mitigation of Al toxicity by Si lie in those topics that will be discussed in Sect. 12.7.

12.6 Co-deposition of Silicon and Aluminium

As we saw in Sect. 12.2, it was in the 1980s that the first reports on co-deposition of Al and Si in plants started to appear. At first these reports were a curiosity as most phytoliths consist of fairly pure amorphous silica, but over time more and more cases were found, and we now have a fairly good idea of the types of situations this phenomenon is likely to occur. The majority of the reported cases have either been found in plant roots or in conifer needles. We will look at these first before briefly considering a few of the more unusual cases.

12.6.1 Co-deposition in Roots

Using X-ray microanalysis, Hodson and Wilkins (1991) were the first to observe Al/Si codeposition in roots, specifically the cortical cell walls of roots from Norway spruce grown in perlite culture. We saw in Sect. 12.4.3 that Al is largely excluded from the shoots in the grasses and cereals. Al/Si interactions in these species seem to be confined mostly or entirely to the roots. As far as I am aware there have only been three cases where Al/Si co-deposition has been observed in cereal roots, and most of the interactions involve soluble HAS complexes (See Sect. 12.7). Deposits containing Al and Si were located in the root epidermal cell walls in sorghum (Hodson and Sangster 1993) and wheat (Cocker et al. 1997) grown in hydroponic culture. More recently, Qian et al. (2016) also found such deposits in the roots of wheat plants where acidic soil had been amended with Si-rich biochar.

It is possible that co-deposition in roots may prevent some Al from penetrating further into the root cortex, and then to the shoot, and is therefore part of a detoxification mechanism, but there is not enough concrete evidence to be certain. We should also be aware that this phenomenon has mostly been observed in plants growing in hydroponic and perlite culture systems, and only the work of Qian et al. (2016) begins to suggest that it might be relevant in field situations.

12.6.2 Co-deposition in Conifer Needles

We have already noted that Godde et al. (1988) were the first to observe Al/Si co-deposition in conifer needles using X-ray microanalysis. At the time there was a lot of interest in Al toxicity that was caused by acid rain which acidified soils and made

Table 12.3 Summary of silicon and aluminium contents in tissues of needles of six conifers as determined by X-ray microanalysis

	Conifer species											
	White spruce		White pine		Eastern larch		Balsam fir		Douglas fir		Eastern Hemlock	
Needle age (yr.)	3		1		1		4		2		3	
Element	Si	Al	Si	Al	Si	Al	Si	Al	Si	Al	Si	Al
Epidermis	+	++	++	+	+++	+++	+	+++	+++	–	+	–
Hypodermis	++	–	+	+	++	+	–	–	++	–	++	–
Mesophyll	nd	nd	+++	+	++	+	–	–	++	–	–	–
Endodermis	++	+	++	+	–	–	–	–	–	–	–	–

White spruce, white pine, balsam fir, and Eastern hemlock are from Muskoka (pH 4.2 at 40 cm) Eastern larch, from Toronto (pH 6.7 at 40 cm), and Douglas fir from Breuil Forest, Morvan, France (pH 4.0 to 4.5). Site descriptions have been previously published (Hodson and Sangster 1998). Tissue contents were determined by x-ray microanalysis using previously described methodology. Relative amounts of elements are compared using the following key: – negligible; + present; ++ high; +++ very high; nd, not determined. Analyses were conducted at the middle of the needle (spruce) or the tip (other genera). Data shown are for white spruce [*Picea glauca* (Moench.) Voss] (Hodson and Sangster 1998); white pine [*Pinus strobus* L.] (Hodson and Sangster 2002), Eastern larch [*Larix laricina* (Du Roi) K. Koch.] (Sangster et al. 2001), balsam fir [*Abies balsamea* (L.) Mill.] (unpub.) Douglas fir [*Pseudotsuga menziesii* (Mirb.) Franco] (Sangster et al. 2007) and Eastern hemlock [*Tsuga canadensis* (L.) Carr.] (Sangster et al. 2009).

Al more available for uptake by trees. Working on Norway spruce needles from a polluted forest in Germany, they found that Al and Si were co-deposited in the transfusion tissue.

I was involved in an extensive survey of mineralisation in conifer needles, working with my late colleague, Allan Sangster from Toronto in Canada (Table 12.3). We spent over ten years gradually building up a picture of Al and Si distribution. By the end of our studies, it was clear that there were considerable differences between species with Al/Si co-deposition evident in white pine, Eastern hemlock, balsam fir, and the larch species, and less obvious in white spruce and Douglas fir. The epidermis and the transfusion tissue were major locations for Al/Si co-deposition. One prerequisite for Al/Si co-deposition to occur was that the plants needed to be growing in acid soils with readily available Al. Hodson and Sangster (2000), working on white pine, showed Al/Si co-deposition in the transfusion tissue near the tips of the needles, but this was markedly increased with low pH soils. The other major study of Al/Si co-deposition in needles was that of Carnelli et al. (2002), who included five conifers in their study of phytoliths isolated from plants growing in the Valaisan Swiss Alps. Three species (*Juniperus nana*, *Pinus cembra*, and *Pinus mugo*) all had phytoliths containing Al.

12.6.3 Co-deposition in the Leaves of Dicot Trees

There is relatively little work on the leaves of dicot trees. Carnelli et al. (2002) also included seven woody shrub species in their study of Al/Si co-deposition and Al was detected in phytoliths from leaves in all cases. Using microanalysis, Turnau et al. (2007) showed that both iron and Al were co-deposited with Si in the epidermis and glandular hairs of the leaves in *Erica andevalensis*.

There are a few Al accumulator species that potentially use Si in detoxification. Haruyama et al. (2019) found Al/Si co-deposition in the leaf epidermis of tea (*Camellia sinensis*) is a well-known Al accumulator plant. This is in contradiction to the work of Tolrà et al. (2011) who reported no co-deposition. It was possible that this difference was due to the ages of the leaves investigated and the conditions under which the plants grew. As we will see in Sect. 12.7.3, other Al accumulators seem to use soluble Si rather than co-deposition as a toxicity avoidance mechanism.

12.6.4 Co-deposition in Other Systems

As we saw in Sect. 12.4.3, Hodson and Evans (1995) thought that heavy accumulation of Al and Si was mutually exclusive. This is still mainly correct, but *Dicranopteris linearis*, a fern which is an Al hyperaccumulator in China, seems to be an exception (Liu et al. 2019). This species is an Al accumulator (up to 9660 mg kg^{-1}), a Si accumulator (up to 20,300 mg kg^{-1}), and a rare earth element (REEs) accumulator (up to 3830 mg kg^{-1}). The authors also showed Al/Si co-deposition and co-deposition with rare earth elements (REEs). I am not aware of any other species that is even close to *D. linearis* in its ability to accumulate both Al and Si. We would normally expect that high concentrations of both Al and Si in the xylem sap would cause precipitation of both elements, and this would prevent transport to the shoot. However, the following work by Liu et al. (2021) may have provided an explanation. They analysed the xylem sap of *D. linearis* and then entered the data into a speciation model. This suggested that almost 100% of REEs were present as $REEH_3SiO_4^{2+}$, while 24–45% of Al was in the sap as Al-citrate and only 1.7–16% of Al was $AlH_3SiO_4^{2+}$. Is it possible that the reason Al can be transported in the xylem which has a high Si concentration is that most of the Si is complexed with REEs? Are there other plant species like *D. linearis*? Only further research will determine this.

The only example of Al/Si co-deposition in the bryophytes that I am aware was discovered by Pressel et al. (2011) in the thalloid liverwort, *Mizutania riccardioides*. Using X-ray microanalysis, they showed that high concentrations of Al and Si were to be found in irregular blocks of dense material in the outer part of the papillae walls.

The mechanism(s) of silica deposition in plants are only just beginning to be worked out (Hodson 2016, 2018), and as yet we have little idea of the processes involved in Al/Si co-deposition, except maybe in roots. Exactly what role co-deposition has is difficult to determine. Many instances of co-deposition may be entirely fortuitous, but it is also possible that in some circumstances it represents a detoxification mechanism.

12.7 Possible Mechanisms for the Mitigation Effect

We have seen above that there are now many cases where Al/Si co-deposition has been reported in plants. But this is not the only possible way in which Al/Si interactions occur in plants. Before we look at internal mitigation mechanisms that do not seem to involve co-deposition, we will discuss the interactions between Al and Si in solution.

12.7.1 Solution Effects

In Sect. 12.3, we briefly considered Al and Si in soils. We saw that the speciation and availability of Al was affected by solution pH, with the toxic Al^{3+} species becoming more important at lower pH. Very early in work on Al/Si interactions in solution, it was recognised that the formation of HAS species was also influenced by pH (Exley and Birchall 1992, 1993). In this case, the formation of HAS is increased at higher pH. Workers have often used speciation models to determine what is happening in the solutions in which they are growing their plants. For example, the equilibrium speciation model (EQ3NR) was used by Ryder et al. (2003) to predict the speciation of Al and Si between pH 4.0 and 5.0. The Al^{3+} species was decreased by the addition of Si (1000 μmol l^{-1}) to Al (100 μmol l^{-1}). This decrease in Al^{3+} was less at pH 4.00, from 92.4 to 83.3%, whilst at pH 5.00 it was from 54.6 to 17.7%. So the formation of HAS species causes a decrease in Al^{3+}. This is undoubtedly one of the reasons for the mitigation of Al toxicity by Si that has often been observed in hydroponic solutions, as seen in Table 12.2. However, there is good reason to believe that this is not the only factor in the mitigation effect, and we will look at this topic next.

12.7.2 Mitigation in Root Systems

Now we come to a key area in this whole chapter and one that has been the focus of research for many years. As we saw in Sect. 12.5 and Table 12.2, there have been many experiments using hydroponic cultures to investigate Al/Si interactions. Most of these experiments were conducted on grasses and cereals, and as we observed in Sect. 12.4 these plants do not transport very much Al to their shoots. It follows that any mitigation effects caused by Si are therefore likely to be found in the roots or in the solutions that they are growing in. Ma et al. (1997) considered that the mitigation of Al toxicity by Si in maize plants was mainly or entirely due to solution effects and the formation of HAS species. However, other workers have suggested that although solution effects are important, they do not tell the whole story, and some type of within-plant (*in planta*) effects are also involved in mitigation phenomena.

One way of avoiding the complications of solution effects is to use pre-treatment of plants with Si before they are exposed to Al. So, under these conditions HAS formation in solution cannot happen as they are not present at the same time. Corrales et al. (1997), working on maize, used this approach, and demonstrated that pre-treatment of roots with 1000 μmol l^{-1} Si considerably increased the growth of plants that were later treated with Al (up to 100 μmol l^{-1}). The presence of Si decreased the uptake of Al into the roots, and the authors suggested that this was the cause of the better growth in the plants pre-treated with Si. The other approach that has been used is to work with ultrasensitive plants where Al toxicity is caused by very low concentrations, and where solution effects could be ruled out. This idea was used by Cocker et al. (1998a) who worked on the Al ultrasensitive wheat cultivar, Scout 66. Here 1.5 μmol l^{-1} Al, a very low concentration, decreased root growth of Scout 66, but a mitigation effect was detected with only 5 μmol l^{-1} Si. The Al concentration used is the lowest in Table 12.2, and HAS formation has not been observed at such low Al and Si concentrations. So, this is another case where *in planta* effects are almost certainly involved as solution interactions are ruled out.

Two papers, Cocker et al. (1998a) and Wang et al. (2004) have been foundational in our understanding of Si mitigation of Al toxicity in hydroponic culture systems. The Atlas 66 cultivar of wheat, is highly Al tolerant and made an interesting comparison with the highly susceptible Scout 66 for Cocker et al. (1998a). Atlas 66 had already been the focus of much attention and it was known that Al tolerance in this cultivar involved malate exudation from the roots. Using this cultivar Al toxicity (100 μmol l^{-1}) was mitigated by 2000 μmol l^{-1} Si at pH 4.6. Not surprisingly, 100 μmol l^{-1} Al caused malate exudation from the roots. But interestingly, when plants were grown with both 100 μmol l^{-1} Al and 2000 μmol l^{-1} Si, malate exudation was very similar. This suggested that although the Al toxicity response of malate exudation was triggered in the presence of Al, Si did not reduce the response. Cocker et al. (1998a) went on the try adding citrate, a well-known chelator of Al, to the solutions of plants being stressed with Al. Malate exudation was decreased at 5–40 μmol l^{-1} citrate, and completely prevented at 100 μmol l^{-1}. So a chelating agent (citrate) prevented the malate response, but the formation of HAS in bulk solution did not. This implies that HAS formation in the bulk solutions was not the reason for the mitigation of Al toxicity, which must have taken place *in planta*.

Cocker et al. (1998b) then assembled a hypothesis taking all of the available data into account. It was proposed that HAS formed in the root cell walls were the main mechanism behind Al detoxification, and mitigation by Si. High apoplastic pH, organic substances such as malate that were exuded into the apoplast, and local concentrations of Al and Si on or within the cell wall were thought to be responsible for the mitigation effect. Wang et al. (2004) conducted a very detailed investigation of the mitigation effect, working with an Al-sensitive maize cultivar, Lixis, and using 25 μmol l^{-1} Al with or without 1400 μmol l^{-1} Si, and much shorter exposure times. Significant mitigation was observed by Si, but monomeric Al concentrations were not reduced, and there were no interactions in the bulk solutions. Again, this suggested an *in planta* phenomenon. In 1 cm root apices, more than 85% of Al was sequestered within the cell wall. Silicon treatment had no effect on Al contents of the apoplastic sap, the symplastic sap, and the cell wall of root apices. Exudation of organic acids and phenols induced by Al was not affected by Si treatment. Increased accumulation of Si in the cell wall fraction appeared to be the major effect of Al treatment. The authors considered that this decreased the mobility of Al in the apoplast. So, the work of Wang et al. (2004) largely confirmed the hypothesis set out by Cocker et al. (1998b). Then, as we saw in Sect. 12.5.1, there was a considerable hiatus in work on Al/Si interactions on plants grown in hydroponic solution in the ten years between 2004 and 2014. It is only in the last few years that this work has restarted.

Kopittke et al. (2017) used highly sensitive synchrotron-based low-energy X-ray fluorescence (LEXRF) to determine Al and Si distribution in the roots of sorghum plants treated with Al and Si. They found that Al and Si were colocated in the mucigel and apoplast of the roots, and again they suggested that the formation of HAS and aluminosilicates was responsible for the Si mitigation effect.

Up until 2017, the hypothesis of Cocker et al. (1998b) seemed to be widely accepted, and the formation of HAS in the apoplast of plant roots was seen as the most likely mechanism for mitigation of Al toxicity by Si. However, the recent investigations into mitigation in rice by Xiao, Liang, and their colleagues have cast some doubt on this idea, and we will complete this sub-section by considering this work. Xiao et al. (2021) found that Si decreased the toxic effects of Al on cell elongation and oxidative injury and reduced the Al content of root cell walls by 41%. This reduction was brought about by Si inhibiting pectin methylesterase (PME) activity and its gene expression. This increased pectin methyl-esterification, and reduced hemicellulose content. Hemicellulose is the main Al binding site in cell walls under Al stress. This work was continued by Xiao et al. (2022), who showed that one of the effects of decreased uptake and transport of Al, caused by Si treatment, was to reduce Al-induced suberisation of roots. Finally, Xiao and Liang (2022) looked at the effects of Al and Si treatments on root border cells (RBCs) in rice. They found that Si decreased Al toxicity to RBCs by the formation of Al/Si complexes on the RBCs. This increased RBCs viability and promoted mucilage secretion. The combination of Si, RBCs, and their mucilage then prevented Al from moving into the root tip and thereby decreased toxicity.

Xiao and colleagues have conducted a very meticulous set of experiments that convincingly show that Si treatment decreases Al transport into rice roots. These results appear to be in conflict with those obtained by Cocker et al. (1998a) who worked on wheat, Wang et al. (2004) who investigated maize, and Kopittke et al. (2017) who studied sorghum. They also do not fit with the hypothesis proposed by Cocker et al. (1998b), which suggested that the formation of HAS in the apoplast of roots was the main mechanism. Why the difference? I think it is possible that it is because rice is unlike any of the other species in the way that it deals with Si. It is a much higher Si accumulator than the other species (Hodson et al. 2005). Moreover, rice differs very considerably from other species in the arrangement and distribution of Si transporters in the roots (Mitani-Ueno and Ma 2021). If rice is so different from other species in these features, should we necessarily expect that Si will be used in the same way in the mitigation of Al toxicity? However, Jiang et al. (2022), working on the root-apex transition zone of rice, found that Al and Si had similar distribution patterns in the cell walls, and suggested HAS formation in the apoplast was a significant factor in mitigation of Al toxicity. To complicate matters even more, the recent work on Li et al. (2022), on soybean, has suggested that mitigation of Al toxicity by Si involved increased pectin, hemicellulose, and cellulose concentrations. At least for hemicellulose, this is the opposite of what Xiao et al. (2021) found, working on rice. Clearly, more work is required in this area to resolve the differences between these various ideas.

12.7.3 Mitigation in Shoot Systems

We saw in Sect. 12.6 that Al and Si can be co-deposited in a whole variety of plant species and organs including the shoot. But there also appear to be some Al accumulator species where Al/Si co-deposition does not occur, but Si is implicated in decreasing the availability of Al in shoot systems. The majority of Al accumulators use organic ligands such as citrate or oxalate to form complexes with Al (Watanabe 2022), but there are at least two species where Si seems to be used for detoxification. Britez et al. (2002) studied the aluminium accumulator, *Faramea marginata* (Rubiaceae), a species that the authors described as "no more than an intermediate or weak Si accumulator"; it does not form phytoliths. Britez et al. showed that Al and Si form complexes, probably including HAS in the shoot tissues of this species. *Rudgea viburnoides*, another Al accumulator in the Rubiaceae, was investigated by Malta et al. (2016) using X-ray microanalysis, and they found that Al and Si co-localised in the roots, stems, and leaves, predominantly in the epidermis.

12.7.4 Mitigation in Tissue Culture Systems

There have also been a few publications that have investigated the effects of Si on Al toxicity in tissue culture systems. Here we would expect that none of the complications that arise from studying complex organs or tissues will be present, and that we should be able to see what happens to individual cells. As far as I can determine, Rahman et al. (1999) were the first workers to investigate Al/Si interactions in suspension-cultured cells. Working on rice they found that Si only slightly decreased Al toxicity effects in a sensitive variety of rice and had no effect in a tolerant variety.

Quintal-Tun et al. (2007) showed that the toxic effects of Al on the phospholipid signal transduction pathway were ameliorated by Si in coffee suspension cultures. Al/Si interactions in Norway spruce suspension cultures were investigated by Prabagar et al. (2011). Silicon reduced Al toxicity, and the effect was greater at pH 5.0 than at pH 4.2. Aluminium treatment caused considerable ultrastructural damage, and this was mitigated by Si. The fluorescent dye Morin was used to assess Al localisation, and Si treatment reduced fluorescence. It appeared that Si decreased free Al in the cell walls of the suspension culture, and the authors suggested that HAS formation in the walls was responsible for decreased cell damage in Si treatments. Writing in 2022, it appears that there have been no further publications using this approach.

12.8 Mitigation in Plants Grown in Soil

Compared with the number of publications concerning Al/Si interactions in hydroponic culture those investigations in soil systems are much fewer. The reasons for this are obvious. It is much more difficult to control all the variables in soil systems which have solid phases in addition to liquid phases. Adding Si fertilizers to soil not infrequently changes soil pH or adds other elements. So it is often difficult to determine the real effects of Si from the side effects of other changes due to fertilizer addition. Of course, just because this type of work is difficult to undertake, does not mean it is not important. In fact, there will be little point in all of our hydroponic investigations if we cannot eventually take this work out into the field and improve crop yields on acidic soils.

For many years, more or less the only study that attempted to investigate the mitigation of Al toxicity by Si in soil was that of Morikawa and Saigusa (2002). Working on barley growing in two andosol soils, they investigated the effects of Si fertilisation on Al toxicity. Silica gel and sodium metasilicate additions to the soils were not effective in mitigating Al toxicity, even though the treatments caused soluble soil Si concentration to increase. Mitigation of Al toxicity was observed with waste porous hydrated calcium silicate addition, but this caused an increase in soil pH which was the probable reason. In recent years there have been a few more investigations in soil systems, and they seem, at last, to be gaining in popularity. I will highlight those that I have located in my literature search, but I am aware that much of this type of work will probably never be published. Many of these types of investigations will happen in agricultural research stations, and the results will be contained in internal reports. In some cases, the effects of Si treatments on Al toxicity will be part of a much larger study.

Perhaps the most exciting of the recent papers on this topic was that by Qian et al. (2016) who investigated the effects of Si-rich biochar derived from pyrolyzed rice straw on the mitigation of Al toxicity in wheat plants. The Si in the biochar seemed to be effective in two ways: particulate Si in the biochar reduced the amount of soil exchangeable Al that was available to the plant; and the soluble Si released from the biochar formed Al/Si co-deposits in the epidermis of the wheat roots (similar to those observed by Cocker et al. (1997) for wheat grown in hydroponic culture). The authors saw the continual release of Si from the biochar as a sustainable Si source for decreasing Al toxicity in acidic soil.

Ribera-Fonseca et al. (2018) worked on ryegrass in andosol soils in greenhouse experiments, and the impacts of sodium silicate and calcium silicate fertilisation. Uptake of Si decreased the toxic effects of soil acidity by increasing total phenols and activities of antioxidant enzymes but reducing lipid peroxidation. The effects of silicon dioxide nanoparticles on Al toxicity in maize were investigated by de Sousa et al. (2019). They found that Al accumulation was not reduced, but mitigation of toxicity was. It was suggested that mitigation resulted from increased organic exudation by the roots. Baggio et al. (2021) adopted sand culture for their experiments on Xaraés Palisadegrass. Increased Si supply led to an increase in root Al content, which decreased Al transport to the shoot, and seemed to be responsible for mitigating Al toxicity.

As yet, the investigations of the mitigation of Al toxicity by Si in soil systems are quite disparate, involving different approaches and ideas. It is certainly not possible yet to assess the body of data that is available in the same way as we can for plants grown in hydroponic solutions (Table 12.2). Much more work is needed which takes more consistent approaches. Only then will we be able to assess how important Si might be in the mitigation of Al toxicity in acid soils.

12.9 Conclusion

It is now 60 years since Okuda and Takahashi (1962) first conducted their experiments on the mitigation of Al toxicity by Si in rice. We undoubtedly have a much better understanding now of the processes involved in mitigation by Si, and of the complexities of the systems we are working with, than we did then. There have been very many investigations in hydroponic culture systems (Table 12.2), and this is the most understood system overall. We also have a fairly good understanding of Al/Si co-deposition and the places it occurs, but we are still uncertain of its significance. Areas where our knowledge is weak include how Si and Al transporters are involved in mitigation, if indeed they are. We certainly have a lot more to find out about mitigation in soil systems, and that remains the area where much more focus is needed.

References

Baker G, Jones LHP, Wardrop ID (1961) Opal phytoliths and mineral particles in the rumen of the sheep. Aust J Agric Res 12:462–472. https://doi.org/10.1071/AR9610462

Baggio G, Dupas E, Galindo FS, Megda MM, Pereira NCM, Luchetta MO, Tritapepe CA, da Silva MR, Jalal A, Teixeira Filho MCM (2021) Silicon application induced alleviation of aluminum toxicity in Xaraés Palisadegrass. Agronomy 11:1938. https://doi.org/10.3390/agronomy11101938

Barceló J, Guevara P, Poschenrieder C (1993) Silicon amelioration of aluminium toxicity in teosinte (*Zea mays* L. ssp. mexicana). Plant Soil 154:249–255. https://doi.org/10.1007/BF00012530

Bartoli F (1985) Crystallochemistry and surface properties of biogenic opal. J Soil Sci 36:335–350. https://doi.org/10.1111/j.1365-2389.1985.tb00340.x

Baylis AD, Gragopoulou C, Davidson KJ, Birchall JD (1994) Effects of silicon on the toxicity of aluminium to soybean. Comm Soil Sci Plant Anal 25:537–546. https://doi.org/10.1080/00103629409369061

Birchall JD, Exley C, Chappell JS, Phillips MJ (1989) Acute toxicity of aluminium to fish eliminated in silicon-rich waters. Nature 338:146–148. https://doi.org/10.1038/338146a0

Bollard EG (1983) Involvement of unusual elements in plant growth and nutrition. In: Lauchli A, Bieleski RL (eds) Inorganic plant nutrition, Encyclopedia of plant physiology, new series, vol 15b. Springer-Verlag, pp 695–744

Britez RM, Watanabe T, Jansen S, Reissmann CB, Osaki M (2002) The relationship between aluminium and silicon accumulation in leaves of *Faramea marginata* (Rubiaceae). New Phytol 156:437–444. https://doi.org/10.1046/j.1469-8137.2002.00531.x

Carnelli AL, Madella M, Theurillat J, Ammann B (2002) Aluminum in the opal silica reticule of phytoliths: a new tool in palaeoecological studies. Am J Bot 89:346–351. https://doi.org/10.3732/ajb.89.2.346

Chen C-H, Lewin JC (1969) Silicon as a nutrient element for *Equisetum arvense*. Can J Bot 47:125–131. https://doi.org/10.1139/b69-016

Chenery EM (1949) Aluminium in the plant world. Part 1. General survey in Dicotyledons. Kew Bull 1948:173–183

Chenery EM (1950a) Aluminium in the plant world. Part 2. Monocotyledons and gymnosperms. Kew Bull 1949:463–466

Chenery EM (1950b) Aluminium in the plant world. Part 3. Cryptogams. Kew Bull 1949:466–473

Cocker KM, Hodson MJ, Evans DE (1998a) The amelioration of aluminium toxicity by silicon in wheat (*Triticum aestivum* L.): malate exudation as evidence for an *in planta* mechanism. Planta 204:318–323. https://doi.org/10.1007/s004250050262

Cocker KM, Hodson MJ, Evans DE (1998b) The amelioration of aluminium toxicity by silicon in higher plants: solution chemistry or an *in planta* mechanism? Physiol Plant 104:608–614. https://doi.org/10.1034/j.1399-3054.1998.1040413.x

Cocker KM, Hodson MJ, Evans DE, Sangster AG (1997) The interaction between silicon and aluminium in *Triticum aestivum* L. (cv. Celtic). Israel J Plant Sci 45:285–292. https://doi.org/10.1080/07929978.1997.10676691

Corrales I, Poschenrieder C, Barceló J (1997) Influence of silicon pretreatment on aluminium toxicity in maize roots. Plant Soil 190:203–209. https://doi.org/10.1023/A:1004209828791

Coskun D, Deshmukh R, Shivaraj SM, Isenring P, Bélanger RR (2021) Lsi2: a black box in plant silicon transport. Plant Soil 466:1–20. https://doi.org/10.1007/s11104-021-05061-1

de Sousa A, Saleh AM, Habeeb TH, Hassan YM, Zrieq R, Wadaan MAM, Hozzein WM, Selim S, Matos M, AbdElgawad H (2019) Silicon dioxide nanoparticles ameliorate the phytotoxic hazards of aluminum in maize grown on acidic soil. Sci Tot Environ 693:133636. https://doi.org/10.1016/j.scitotenv.2019.133636

Delavar K, Ghanati F, Zare-Maivan H, Behmanesh M (2016) The effect of the silicon and aluminum interaction on the physiological parameters of maize. Iran J Plant Physiol 6:1785–1794. https://doi.org/10.22034/ijpp.2016.532649

Dorneles AOS, Pereira AS, Sasso VM, Possebom G, Tarouco CP, Schorr MRW, Rossato L, Ferreira PAA, Tabaldi LA (2019) Aluminum stress tolerance in potato genotypes grown with silicon. Bragantia 78:12–25. https://doi.org/10.1590/1678-4499.2018007

Epstein E (1994) The anomaly of silicon in plant biology. Proc Natl Acad Sci U S A 91:11–17. https://doi.org/10.1073/pnas.91.1.11

Exley C (1998) Silicon in life: a bioinorganic solution to bioorganic essentiality. J Inorg Biochem 69:139–144. https://doi.org/10.1016/S0162-0134(97)10010-1
Exley C, Birchall JD (1992) Hydroxyaluminosilicate formation in solutions of low total aluminium concentration. Polyhedron 11:1901–1907. https://doi.org/10.1016/S0277-5387(00)83738-9
Exley C, Birchall JD (1993) A mechanism of hydroxyaluminosilicate formation. Polyhedron 12:1007–1017. https://doi.org/10.1016/S0277-5387(00)87177-6
Foy CD (1992) Soil chemical factors limiting plant root growth. Adv Soil Sci 19:97–149. https://doi.org/10.1007/978-1-4612-2894-3_5
Freitas LB, Fernandes DM, Maia SCM, Fernandes AM (2017) Effects of silicon on aluminum toxicity in upland rice plants. Plant Soil 420:263–275. https://doi.org/10.1007/s11104-017-3397-4
Galvez L, Clark RB, Gourley LM, Maranville JW (1987) Silicon interactions with manganese and aluminium toxicity in sorghum. J Plant Nutr 10:1139–1147. https://doi.org/10.1080/01904168709363642
Geis JW (1973) Biogenic silica in selected species of deciduous angiosperms. Soil Sci 116:113–130
Godde D, Homburg H, Methfessel S, Rosenkranz J (1988) Die Röntgenanalyse hilft beider Aufklärung individueller Waldschaden. LÖLF Mitteilungen 4:23–27
Grennfelt P, Engleryd A, Forsius M, Hov Ø, Rodhe H, Cowling E (2020) Acid rain and air pollution: 50 years of progress in environmental science and policy. Ambio 49:849–864. https://doi.org/10.1007/s13280-019-01244-4
Hammond KE, Evans DE, Hodson MJ (1995) Aluminium/silicon interactions in barley (*Hordeum vulgare* L.) seedlings. Plant Soil 173:89–95. https://doi.org/10.1007/BF00155521
Hara T, Gu M-H, Koyama H (1999) Ameliorative effect of silicon on aluminium injury in the rice plant. Soil Sci Plant Nutr 45:929–936. https://doi.org/10.1080/00380768.1999.10414342
Haruyama Y, Fujiwara T, Yasuda K, Saito M, Suzuki K (2019) Localization of aluminum in epidermal cells of mature tea leaves. Quantum Beam Sci 3:9. https://doi.org/10.3390/qubs3020009
Haynes RJ (2017) The nature of biogenic Si and its potential role in Si supply in agricultural soils. Agric Ecosyst Environ 245:100–111. https://doi.org/10.1016/j.agee.2017.04.021
Hodson MJ (2012) Metal toxicity and tolerance in plants. Biochemist 34:28–32. https://doi.org/10.1042/BIO03405028
Hodson MJ (2016) The development of phytoliths in plants and its influence on their chemistry and isotopic composition. Implications for palaeoecology and archaeology. J Archaeol Sci 68:62–69. https://doi.org/10.1016/j.jas.2015.09.002
Hodson MJ (2018) Phytoliths in archaeology: chemical aspects. In: Smith C (ed) Encyclopedia of global archaeology. Springer International Publishing. https://doi.org/10.1007/978-3-319-51726-1_3250-1
Hodson MJ (2020) Aluminium-silicon interactions in higher plants: a brief personal history. Botany One. https://www.botany.one/2020/03/aluminium-silicon-interactions-in-higher-plants-a-brief-personal-history/. Accessed 30 July 2022
Hodson MJ, Evans DE (1995) Aluminium/silicon interactions in higher plants. J Exp Bot 46:161–171. https://doi.org/10.1093/jxb/46.2.161
Hodson MJ, Evans DE (2020) Aluminium/silicon interactions in higher plants- an update. J Exp Bot 71:6719–6729. https://doi.org/10.1093/jxb/eraa024
Hodson MJ, Sangster AG (1993) The interaction between silicon and aluminium in *Sorghum bicolor* (L.) Moench: growth analysis and x-ray microanalysis. Ann Bot 72:389–400. https://doi.org/10.1006/anbo.1993.1124
Hodson MJ, Sangster AG (1998) Mineral deposition in the needles of white spruce [*Picea glauca* (Moench.) Voss]. Ann Bot 82:375–385. https://doi.org/10.1006/anbo.1998.0694
Hodson MJ, Sangster AG (1999) Aluminium/silicon interactions in conifers. J Inorg Biochem 76:89–98. https://doi.org/10.1016/S0162-0134(99)00119-1
Hodson MJ, Sangster AG (2000) Aluminium localization in conifers growing on highly acidic soils in Ontario, Canada. In: International symposium on impact of potential tolerance of plants on the increased productivity under aluminum stress. Research Institute for Bioresources, Okayama University, Kurashiki, Japan, pp 103–106

Hodson MJ, Sangster AG (2002) X-ray microanalytical studies of mineral localisation in the needles of white pine (*Pinus strobus* L.). Ann Bot 89:367–374. https://doi.org/10.1093/aob/mcf052
Hodson MJ, White PJ, Mead A, Broadley MR (2005) Phylogenetic variation in the silicon composition of plants. Ann Bot 96:1027–1046. https://doi.org/10.1093/aob/mci255
Hodson MJ, Wilkins DA (1991) Localization of aluminium in the roots of Norway spruce [*Picea abies* (L.) Karst.] inoculated with *Paxillus involutus* Fr. New Phytol 118:273–278. https://doi.org/10.1111/j.1469-8137.1991.tb00977.x
Jansen S, Broadley MR, Robbrecht E, Smets E (2002) Aluminum hyperaccumulation in angiosperms: a review of its phylogenetic significance. Bot Rev 68:235–269. https://doi.org/10.1663/0006-8101(2002)068[0235:AHIAAR]2.0.CO;2
Jiang D, Wu H, Cai H, Chen G (2022) Silicon confers aluminium tolerance in rice via cell wall modification in the root transition zone. Plant Cell Environ 45:1765–1778. https://doi.org/10.1111/pce.14307
Jones LHP, Handreck KA (1967) Silica in soils, plants and animals. Adv Agron 19:107–149. https://doi.org/10.1016/S0065-2113(08)60734-8
Junior GSS, Hurtado AC, Junior JPS, Prado RM, Santos DMM (2021) Nutritional and structural role of silicon in attenuating aluminum toxicity in sugarcane plants. Silicon 14:5041–5055. https://doi.org/10.1007/s12633-021-01294-y
Katz O (2014) Beyond grasses: the potential benefits of studying silicon accumulation in non-grass species. Front Plant Sci 5:376. https://doi.org/10.3389/fpls.2014.00376
Kidd PS, Llugany M, Poschenrieder C, Gunsé B, Barceló J (2001) The role of root exudates in aluminium resistance and silicon-induced amelioration of aluminium toxicity in three varieties of maize (*Zea mays* L.). J Exp Bot 52:1339–1352. https://doi.org/10.1093/jexbot/52.359.1339
Klein RL, Geis JW (1978) Biogenic silica in the Pinaceae. Soil Sci 126:145–156
Kochian LV, Piñeros MA, Liu J, Magalhaes JV (2015) Plant adaptation to acid soils: the molecular basis for crop aluminum resistance. Ann Rev Plant Biol 66:571–598. https://doi.org/10.1146/annurev-arplant-043014-114822
Kopittke PM, Gianoncelli A, Kourousias G, Green K, McKenna BA (2017) Alleviation of Al toxicity by Si is associated with the formation of Al–Si complexes in root tissues of sorghum. Front Plant Sci 89:2189. https://doi.org/10.3389/fpls.2017.02189
Li YC, Alva AK, Sumner AK (1989) Response of cotton to aluminum in solutions with varying silicon concentrations. J Plant Nutr 12:881–892. https://doi.org/10.1080/01904168909364000
Li Z, Huang F, Hu B, Qiu M (2022) Detoxification of aluminum by Ca and Si is associated to modified root cell wall properties. Theor Exp Plant Physiol 34:131–142. https://doi.org/10.1007/s40626-022-00235-3
Liang Y, Yang C, Shi H (2001) Effects of silicon on growth and mineral composition of barley grown under toxic levels of aluminum. J Plant Nutr 24:229–243. https://doi.org/10.1081/PLN-100001384
Liu W-S, Laird JS, Ryan CG, Tang Y-T, Qiu R-L, Echevarria G, Morel J-L, van der Ent A (2021) Rare earth elements, aluminium and silicon distribution in the fern *Dicranopteris linearis* revealed by μPIXE Maia analysis. Ann Bot 128:17–30. https://doi.org/10.1093/aob/mcab026
Liu W-S, Zheng H-X, Guo MN, Liu C, Huot H, Morel JL, van der Ent A, Tang Y-T, Qiu R-L (2019) Co-deposition of silicon with rare earth elements (REEs) and aluminium in the fern *Dicranopteris linearis* from China. Plant Soil 437:427–437. https://doi.org/10.1007/s11104-019-04005-0
Ma JF, Sasaki M, Matsumoto H (1997) Al-induced inhibition of root elongation in corn, *Zea mays* L. is overcome by Si addition. Plant Soil 188:171–176. https://doi.org/10.1023/A:1004274223516
Ma JF, Takahashi E (2002) Soil, fertilizer, and plant silicon research in Japan. Elsevier
Ma JF, Tamai K, Yamaji N, Mitani N, Konishi S, Katsuhara M, Ishiguro M, Murata Y, Yano M (2006) A silicon transporter in rice. Nature 440:688–691. https://doi.org/10.1038/nature04590
Ma JF, Yamaji N (2015) A cooperative system of silicon transport in plants. Trends Plant Sci 20:435–442. https://doi.org/10.1016/j.tplants.2015.04.007

Ma JF, Yamaji N, Mitani N, Tamai K, Konishi S, Fujiwara T (2007) An efflux transporter of silicon in rice. Nature 448:209–213. https://doi.org/10.1038/nature05964

Malta PG, Arcanjo-Silva S, Ribeiro C, Campos NV, Azevedo AA (2016) *Rudgea viburnoides* (Rubiaceae) overcomes the low soil fertility of the Brazilian Cerrado and hyperaccumulates aluminum in cell walls and chloroplasts. Plant Soil 408:369–384. https://doi.org/10.1007/s11104-016-2926-x

Marschner H (1991) Mechanisms of adaptation of plants to acid soils. Plant Soil 134:1–20. https://doi.org/10.1007/BF00010712

Matsumoto H, Hirasawa E, Morimura S, Takahashi E (1976) Localization of aluminium in tea leaves. Plant Cell Physiol 17:627–631. https://doi.org/10.1093/oxfordjournals.pcp.a075318

Mitani-Ueno N, Ma JF (2021) Linking transport system of silicon with its accumulation in different plant species. Soil Sci Plant Nutr 67:10–17. https://doi.org/10.1080/00380768.2020.1845972

Morikawa CK, Saigusa M (2002) Si amelioration of Al toxicity in barley (*Hordeum vulgare* L.) growing in two Andosols. Plant Soil 240:161–168. https://doi.org/10.1023/A:1015804401190

Okuda A, Takahashi E (1962) Effect of silicon supply on the injuries due to excessive amounts of Fe^{II}, Mn^{II}, Cu^{II}, As^{III}, Al^{III}, Co^{II}. J Sci Soil and Manure 33:59–64

Okuda A, Takahashi E (1965) The role of silicon. In: The mineral nutrition of the rice plant. Proceedings of a symposium of the International Rice Research Institute, Feb 1964. The Johns Hopkins Press, Baltimore, Maryland, pp 123–146

Pontigo S, Godoy K, Jiménez H, Gutiérrez-Moraga A, Mora ML, Cartes P (2017) Silicon-mediated alleviation of aluminum toxicity by modulation of Al/Si uptake and antioxidant performance in ryegrass plants. Front Plant Sci 8:642. https://doi.org/10.3389/fpls.2017.00642

Prabagar S, Hodson MJ, Evans DE (2011) Silicon amelioration of aluminium toxicity and cell death in suspension cultures of Norway spruce (*Picea abies* (L.) Karst.). Environ Exp Bot 70:266–276. https://doi.org/10.1016/j.envexpbot.2010.10.001

Pressel S, P'Ng KMY, Duckett JG (2011) An ultrastructural study of the liverwort *Mizutania riccardioides* Furuki et Iwatsuki: new insights into its systematic affinities and unique surface ornamentation. Bryologist 114:38–51. https://doi.org/10.1639/0007-2745-114.1.38

Qian L, Chen B, Chen M (2016) Novel alleviation mechanisms of aluminum phytotoxicity *via* released biosilicon from rice straw-derived biochars. Sci Rep 6:29346. https://doi.org/10.1038/srep29346

Quintal-Tun F, Munoz-Sanchez JA, Ramos-Diaz A, Escamilla-Bencomo A, Martinez-Estevez M, Exley C, Hernandez-Sotomayor SMT (2007) Aluminum-induced phospholipid signal transduction pathway in *Coffea arabica* suspension cells and its amelioration by silicic acid. J Inorg Biochem 101:362–369. https://doi.org/10.1016/j.jinorgbio.2006.10.010

Rahman MT, Kawamura K, Koyama H, Hara T (1998) Varietal differences in the growth of rice plants in response to aluminum and silicon. Soil Sci Plant Nutr 44:423–431. https://doi.org/10.1080/00380768.1998.10414464

Rahman MT, Koyama H, Hara T (1999) Effect of aluminium and silicon in solution on the growth of rice (*Oryza sativa*) suspension culture. Soil Sci Plant Nutr 45:693–700. https://doi.org/10.1080/00380768.1999.10415832

Raven JA (1983) The transport and function of silicon in plants. Biol Rev 58:179–207. https://doi.org/10.1111/j.1469-185X.1983.tb00385.x

Ribera-Fonseca A, Rumpel C, de la Luz MM, Nikolic M, Cartes P (2018) Sodium silicate and calcium silicate differentially affect silicon and aluminium uptake, antioxidant performance and phenolics metabolism of ryegrass in an acid Andisol. Crop Pasture Sci 69:205–215. https://doi.org/10.1071/CP17202

Ryder M, Gérard F, Evans DE, Hodson MJ (2003) The use of root growth and modelling data to investigate amelioration of aluminium toxicity by silicon in *Picea abies* seedlings. J Inorg Biochem 97:52–58. https://doi.org/10.1016/S0162-0134(03)00181-8

Sangster AG, Hodson MJ, Huang CX (2001) X-ray microanalytical studies of mineral composition in cell walls of needle tissues of American larch [*Larix laricina* (Du Roi) K. Koch] and

European larch [*L. decidua* (L.) Mill.]. In: Labrecque M (ed) L'arbre 2000 The Tree. 4th international symposium on the tree, Montreal Botanic Garden. Isabelle Quentin Press, pp 160–167

Sangster AG, Hodson MJ, Ling LEC (2009) Biomineralisation/environment interactions in conifers: illustrated by hemlock, *Tsuga canadensis* (L.) Carr. Quat Int 193:3–10. https://doi.org/10.1016/j.quaint.2007.08.025

Sangster AG, Ling L, Gérard F, Hodson MJ (2007) X-ray microanalysis of needles from Douglas fir growing in environments of contrasting acidity. Water Air Soil Pollut Focus 7:143–149. https://doi.org/10.1007/978-1-4020-5885-1_17

Schmitt M, Mehltreter K, Sundue M, Testo W, Watanabe T, Jansen S (2017) The evolution of aluminum accumulation in ferns and lycophytes. Am J Bot 104:573–583. https://doi.org/10.3732/ajb.1600381

Shen X, Xiao X, Dong Z, Chen Y (2014) Silicon effects on antioxidative enzymes and lipid peroxidation in leaves and roots of peanut under aluminum stress. Acta Physiol Plant 36:3063–3069. https://doi.org/10.1007/s11738-014-1676-8

Singh VP, Tripathi DK, Kumar D, Chauhan DK (2011) Influence of exogenous silicon addition on aluminium tolerance in rice seedlings. Biol Trace Elem Res 144:1260–1274. https://doi.org/10.1007/s12011-011-9118-6

Striegel (1912) Mineralstoff aufnahme verschiedener Planzenarten aus ungedungtem Boden. Deutsche Landwirtschafts-Gesellschaft Jahrbuch der Deutschen Landwirtschafts-Gesellschaft 43:349–371

Tanaka A, Navasero SA (1966) Aluminum toxicity of the rice plant under water culture conditions. Soil Sci Plant Nutr 12:55–60. https://doi.org/10.1080/00380768.1966.10431183

Tolrà R, Vogel-Mikuš K, Hajiboland R, Kump P, Pongrac P, Kaulich B, Gianoncelli A, Babin V, Barceló J, Regvar M, Poschenrieder C (2011) Localization of aluminium in tea (*Camellia sinensis*) leaves using low energy X-ray fluorescence spectro-microscopy. J Plant Res 124:165–172. https://doi.org/10.1007/s10265-010-0344-3

Tubana BS, Babu T, Datnoff LE (2016) A review of silicon in soils and plants and its role in US agriculture: history and future perspectives. Soil Sci 181:393–411. https://doi.org/10.1097/SS.0000000000000179

Turnau K, Henriques FS, Anielska T, Renker C, Buscot F (2007) Metal uptake and detoxification mechanisms in *Erica andevalensis* growing in a pyrite mine tailing. Environ Exp Bot 61:117–123. https://doi.org/10.1016/j.envexpbot.2007.05.001

Vega I, Nikolic M, Pontigo S, Godoy K, Mora MLL, Cartes P (2019) Silicon improves the production of high antioxidant or structural phenolic compounds in barley cultivars under aluminum stress. Agronomy 9:388. https://doi.org/10.3390/agronomy9070388

Vega I, Rumpel C, Ruíz A, Mora MDLL, Calderini DF, Cartes P (2020) Silicon modulates the production and composition of phenols in barley under aluminum stress. Agronomy 10:1138. https://doi.org/10.3390/agronomy10081138

Wang Y, Stass A, Horst WJ (2004) Apoplastic binding of aluminum is involved in silicon-induced amelioration of aluminum toxicity in maize. Plant Physiol 136:3762–3770. https://doi.org/10.1104/pp.104.045005

Watanabe T (2022) Basic understanding of aluminum accumulator plants. Eurasian J For Res 22:59–62. https://doi.org/10.14943/EJFR.22.59

Watanabe T, Osaki M, Tadano T (1997) Aluminum-induced growth stimulation in relation to calcium, magnesium, and silicate nutrition in *Melastoma malabathricum* L. Soil Sci Plant Nutr 43:827–837. https://doi.org/10.1080/00380768.1997.10414649

Wei Y, Han R, Xie Y, Jiang C, Yu Y (2021) Recent advances in understanding mechanisms of plant tolerance and response to aluminum toxicity. Sustainability 13:1782. https://doi.org/10.3390/su13041782

Xiao Z, Liang Y (2022) Silicon prevents aluminum from entering root tip by promoting formation of root border cells in rice. Plant Physiol Biochem 175:12–22. https://doi.org/10.1016/j.plaphy.2022.02.003

Xiao Z, Yan G, Ye M, Liang Y (2021) Silicon relieves aluminum-induced inhibition of cell elongation in rice root apex by reducing the deposition of aluminum in the cell wall. Plant Soil 462:189–205. https://doi.org/10.1007/s11104-021-04850-y

Xiao Z, Ye M, Gao Z, Jiang Y, Zhang X, Nikolic N, Liang Y (2022) Silicon reduces aluminum-induced suberization by inhibiting the uptake and transport of aluminum in rice roots and consequently promotes root growth. Plant Cell Physiol 63:340–352. https://doi.org/10.1093/pcp/pcac001

Zhang X, Long Y, Huang J, Xia J (2019) Molecular mechanisms for coping with Al toxicity in plants. Int J Mol Sci 20:1551. https://doi.org/10.3390/ijms20071551

Zsoldos F, Vashegyit Á, Bóna L, Pécsváradi A (2003a) Effects of silicon on growth and potassium transport in triticale and rye exposed to aluminium and nitrite stress. Cereal Res Comun 31:379–386. https://doi.org/10.1007/BF03543368

Zsoldos F, Vashegyi Á, Pécsváradi A, Bóna L (2003b) Influence of silicon on aluminium toxicity in common and durum wheats. Agronomie 23:349–354. https://doi.org/10.1051/agro:2003008

Chapter 13
Structural Role of Silicon-Mediated Cell Wall Stability for Ammonium Toxicity Alleviation

Mikel Rivero-Marcos, Gabriel Barbosa Silva Jr., and Idoia Ariz

13.1 Introduction

Nitrogen (N) is an essential and often limiting plant macronutrient. Discovery of the Haber-Bosch process provided agriculture with industrial-scale N fertilizers and helped feed a growing world population (Erisman et al. 2008). Today, the challenge to 'produce more food and energy with less pollution' is a social and scientific priority (Sutton et al. 2013). Increases in population and per capita energy consumption, intensive farming, excessive N fertilizer use, and relatively poor fertilizer use efficiency, have exacerbated N loss and pollution (Sutton et al. 2013). Massive application of N fertilizers to crops has resulted in N inputs that exceed crop demand. Plants are unable to absorb the full amount of applied N. Most agricultural soils have become enriched with N mainly in the form of nitrate. Consequently, plants have developed low N utilization efficiency (NUE), nutrient imbalances have occurred, and N has been lost to the environment mainly via nitrate leaching and gaseous NO_x emissions (Subbarao and Searchinger 2021). The use of NH_4^+ and NH_3 has been proposed as a viable alternative to the current nitrate-based fertilizers (Subbarao and Searchinger 2021). However, in less than a single century, plant breeders have selected crop varieties that use principally nitrate, overlooking the intolerance of most of these cultivars to alternate other N sources, such as ammonium. Prolonged application of millimolar concentrations of ammonium may

M. Rivero-Marcos · I. Ariz (✉)
Institute for Multidisciplinary Research in Applied Biology (IMAB), Universidad Pública de Navarra (UPNA), Campus de Arrosadia, Pamplona, Spain
e-mail: idoia.ariz@unavarra.es

G. B. Silva Jr.
Department of Plant Science, Research Center of Agricultural Sciences, Campus of Ministro Petrônio Portella, Federal University of Piauí, Teresina, Brazil

R. de Mello Prado (ed.), *Benefits of Silicon in the Nutrition of Plants*,
https://doi.org/10.1007/978-3-031-26673-7_13

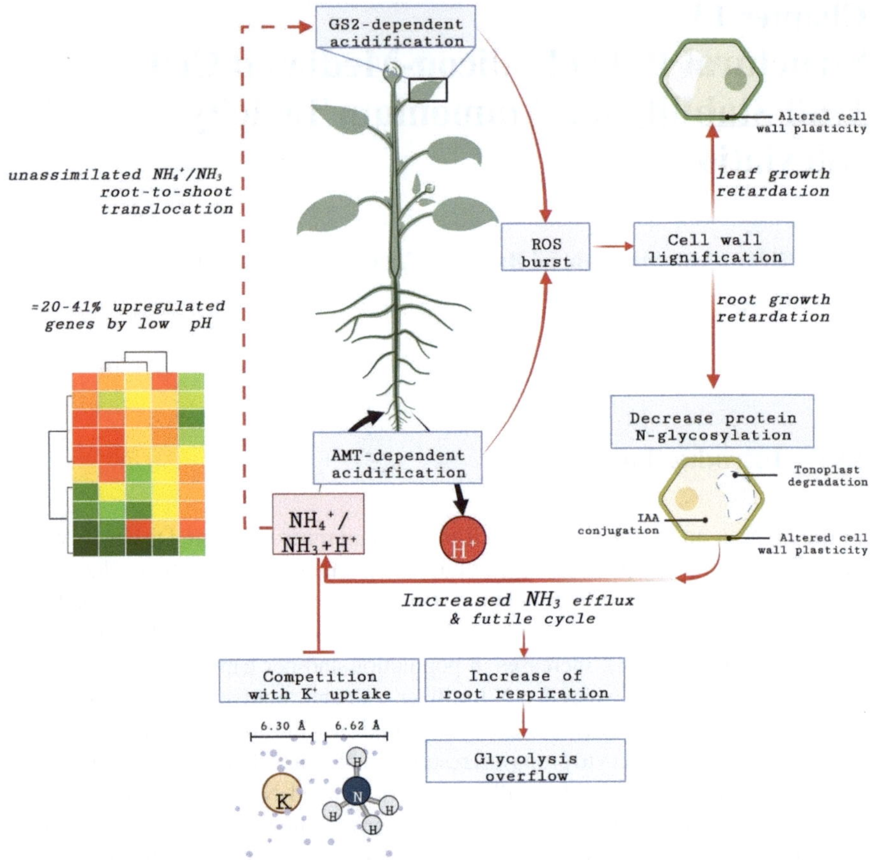

Fig. 13.1 Mechanisms of ammonium toxicity may result from the cumulative effects of divergent impairments. Excessive ammonium uptake and assimilation induce AMT- and GLN2-dependent pH acidification in the apoplasts of roots (Meier et al. 2020) and shoots (Hachiya et al. 2021), respectively. Between 20% and 41% of all genes responding to ammonium are also induced by low external pH (Patterson et al. 2010). Repression of metal cation uptake by high external ammonium may lead to K^+ depletion in plant tissues (Hoopen et al. 2010). Decreases in protein N-glycosylation (Li et al. 2010), and vacuole macrophagy (Robert et al. 2021a) may compromise membrane integrity and promote ammonium efflux and futile N cycle. These responses along with auxin (IAA) conjugation (Di et al. 2021) arrest root growth, initiate ROS burst (Yang et al. 2020), cause progressive carbohydrate shortage (Li et al. 2020), and reduce plant cell growth and energy metabolism. (This figure was created with BioRender (https://biorender.com/))

induce morphophysiological disorders, retard growth, and cause toxicity (Britto and Kronzucker 2002; Esteban et al. 2016).

The causes of 'ammonium syndrome' are unknown. Nevertheless, it has been postulated that they are related to intracellular metabolism (Fig. 13.1), including futile transmembrane ammonium cycling, deficiencies in inorganic cations and organic acids, impaired phytohormone homeostasis, pH dysregulation, and photophosphorylation uncoupling (Esteban et al. 2016). As some of these symptoms are not directly associated with ammonium toxicity-induced growth suppression, it is difficult to determine the key factors contributing to ammonium toxicity. Prior research has not fully explained the origin of ammonium sensitivity in plants. It has, however, provided vital information about the metabolic targets involved, and could facilitate the development of tools and practices to counteract ammonium toxicity. The symptoms of ammonium toxicity may be alleviated by buffering nutrient solution pH (Roosta and Schjoerring 2008a; Sarasketa et al. 2016), and supplementing potassium (K^+) (Roosta and Schjoerring 2008b; Szczerba et al. 2008b; Balkos et al. 2010; Weng et al. 2020; Fang et al. 2021), sucrose (Li et al. 2020), TCA in the Krebs' cycle such as succinate and α-ketoglutarate (Magalhães et al. 1992; Yuan et al. 2007b; Roosta and Schjoerring 2008a), GABA (Ma et al. 2016b), nitrate (Chaillou et al. 1991; Roosta and Schjoerring 2008b; Garnica et al. 2010; Hachiya et al. 2012; Zheng et al. 2015), or silicon (Ferreira Barreto et al. 2017; Viciedo et al. 2019; Song et al. 2022).

Ammonium-mediated inhibition of primary root growth has been associated with the repression of cell elongation rather than cell division (Li et al. 2010). This phenomenon has been associated with two ammonium toxicity mechanisms, namely, a futile plasma transmembrane influx/efflux cycle of ammonium through the root cells at a high energetic cost (Britto et al. 2001; Coskun et al. 2013) and abnormal GDP-mannose pyrophosphorylase (GMPase) activity in plants showing hypersensitivity to ammonium nutrition (Qin et al. 2008).

Li et al. (2010) showed that high ammonium concentrations induced ammonium efflux in the root elongation zone and inhibited root cell and primary root elongation within the same root zone of *Arabidopsis thaliana* (Arabidopsis) plants. Furthermore, they reported that GMPase activity regulated root ammonium efflux. They demonstrated that Arabidopsis GMPase mutants showed higher net ammonium efflux in their root elongation zones than wild-type plants. In fact, a reduction in pH-dependent GMPase activity in response to ammonium supply could lead to protein N-glycosylation deficiency, unfolded proteins, and cell death in the roots (Qin et al. 2008). Arabidopsis null mutants of GMPase (*cyt1-2*) showed N-glycosylation deficiency and, consequently, reduced cellulose biosynthesis, cell wall stability, and cell viability (Lukowitz et al. 2001).

The foregoing findings underscore the fact that it is necessary to elucidate the fine-tune regulation and sensing of plant ammonium transport, storage, and assimilation to determine how they can be used to identify potential toxicity mechanisms. While the original toxicity mechanisms are investigated, the study of real solutions to alleviate some of the effects/symptoms caused by ammonium toxicity may be a provisional solution of great interest and importance in today's agriculture. In this sense, the use of some elements to combat ammonium stress, such as the use of silicon (Si), could contribute to a localized repair of the possible "damage" in the structural part of the cell (plant wall), by fortification, to improve tolerance to ammonium nutrition of current crops.

Silicon is one of the nutritional elements that is gaining more and more attention due to its observed properties improving plant tolerance against biotic and abiotic stress, such as ammonium toxicity (examples cited here and extensively described in part II – Sect. 13.3) for maize (Silva Campos et al. 2015); tomato (Barreto et al. 2016, 2022); cauliflower and broccoli (Ferreira Barreto et al. 2017); yellow passionfruit (Silva Júnior et al. 2019); sugar beet (Viciedo et al. 2019); radish (Olivera Viciedo et al. 2020); and cabbage (Song et al. 2022).

Although Si is the second most abundant element in the Earth's crust, it is not considered essential to plants (Epstein 2009). In the soil solution, the available Si occurs mainly as salicylic acid (H_4SiO_4) at concentrations ranging from 0.01 to 2.0 mM, being absorbed in this non-ionic form. Consequently, the Si concentrations in plant tissues vary considerably among plant species, ranging from 0.1% to 10% of dry matter, including non-accumulating, intermediate, and accumulating Si species (Adrees et al. 2015; Bakhat et al. 2018). After its uptake, Si accumulates in the cell wall of various plant tissues as an amorphous silica polymer (He et al. 2015; Sheng et al. 2018).

The literature reports several effects related to Si-mediated abiotic stresses mediated in plants, with beneficial results on plant nutrition, physiology, and morphology (Pavlovic et al. 2021; He et al. 2015; Kim et al. 2017; Song et al. 2022). In this chapter, the role of Si as an ammonium toxicity mitigator in plants is presented by emphasizing the following topics: in part I (Sect. 13.2), we focused on understanding the mechanisms of ammonium toxicity, and in part II (Sect. 13.3), how Si might be involved in repairing them.

13.2 Metabolic Targets and Structural Vulnerability in Root Cell Membranes and Cell Walls in Response to Ammonium Toxicity

13.2.1 High Ammonium Uptake Increases AMT-Dependent Apoplastic Acidification

Ammonium is present in the ionic NH_4^+ and the neutral ammonia (NH_3) forms. The predominance of each species is pH-dependent. At pKa = 9.25, NH_4^+ accounts for ~99% of total ammonium content in most soils including alkaline ones (Marini et al. 1994; Ninnemann et al. 1994; Antonenko et al. 1997). Thus, ammonium has been considered the main form of reduced inorganic N transported across biological membranes.

Two ammonium transport systems have been identified. The first is the high-affinity ammonium transport system (HATS) operating at concentrations from micromolar to 0.5–1 mM N. HATS is saturable and energy-dependent (Yuan et al. 1993; Ninnemann et al. 1994; Marini et al. 1997). HATS includes members of the extended Mep/Amt/Rh protein superfamily. In plants, they are mainly type 1 ammonium transporters (AMT1). Above 0.5–1 mM ammonium, they are repressed and the ammonium may be absorbed by lower-affinity systems (Yuan et al. 2007a). The second ammonium transport system is the non-saturable low-affinity transport system (LATS) operating in the millimolar Km range ([N] > 0.5–1 mM) in most plant roots (Yuan et al. 1993; Kronzucker et al. 1995; Forde and Clarkson 1999). LATS is poorly defined and understood. However, it might involve putative cation channels with low substrate selectivity (Szczerba et al. 2008a; Hoopen et al. 2010; Coskun et al. 2017), as well as AMT2-type proteins (Guether et al. 2009; Giehl et al. 2017) that also serve as the main transport proteins involved in root-to-shoot ammonium translocation.

The molecular mechanism of AMTs has been extensively studied for Arabidopsis (Mayer and Ludewig 2006; Neuhäuser et al. 2007, 2009; Yuan et al. 2007a; Straub et al. 2017). It has also been investigated in tomato (Mayer et al. 2006a), bean (Ortiz-Ramirez et al. 2011), and rice (Kumar et al. 2006; Li et al. 2009). The Arabidopsis genome encodes six AMT proteins divided into the AMT1 (AMT1.1, AMT1.2, AMT1.3, AMT1.4, and AMT1.5) and AMT2 (AMT2.1) families (Loqué and von Wirén 2004; Ludewig et al. 2007; Lanquar et al. 2009). Of the five AMT proteins expressed in Arabidopsis roots, only AMT1.1, AMT1.2, AMT1.3, and, to a lesser extent, AMT1.5, mediate root ammonium uptake under HATS conditions (Loqué et al. 2006; Yuan et al. 2007a). Between 60% and 70% of the ammonium is acquired by AMT1.1 and AMT1.3 (Yuan et al. 2007a). On the other hand, the net contribution of AMT2.1 to ammonium influx is negligible as the AMT1 transporters outcompete AMT2.1 within the micromolar ammonium concentration range (Yuan et al. 2007a, 2013). In the millimolar ammonium concentration range, however, AMT2.1 makes a small but significant contribution to the net ammonium influx (Giehl et al. 2017).

AMT1.4 is specifically expressed in pollen grains and tubes (Yuan et al. 2009; Bindel and Neuhäuser 2021). AMT1.1, AMT1.3, and AMT1.5 are localized to the rhizodermis where they symplastically transport ammonium from the soil solution into the root cells (Yuan et al. 2007a). AMT1.2 is expressed mainly in the endodermis where it facilitates ammonium transport from the apoplast into the endodermal cells. In the maturation zones where the endodermal cells undergo suberization, AMT1.2 is expressed in the cortex (Duan et al. 2018). AMT2.1 localization to the pericycle cells facilitates ammonium loading into the xylem (Giehl et al. 2017).

Ammonium uptake acidifies the apoplast and alkalinizes the cytosols in response to proton export across plasma membranes (Murányi et al. 1994; Kosegarten et al. 1997). These effects depend on AMT-mediated ammonium uptake as strong acidification is absent in *amt1;1 amt1;2 amt1;3 amt2;1* quadruple mutants (*qko*) (Meier et al. 2020). AMTs are post-translationally regulated to prevent toxic ammonium accumulation levels. In Arabidopsis, the protein kinase CIPK23 phosphorylates and partially inactivates the C-terminal T460 residues in AMT1.1 and AMT1.2 but not those in AMT1.3 (Straub et al. 2017). However, T460 phosphorylation signals, but at reduced levels, were detected in *cipk23* mutants. Hence, other protein kinases or phosphatases might be implicated in equivalent T460 phosphorylation in AMTs (Straub et al. 2017). Neither ammonium assimilation products such as glutamine, nor MSX-induced increases in intracellular ammonium, induce T460 phosphorylation. Therefore, phosphorylation-dependent AMT1.1 inactivation is specifically triggered by extracellular ammonium (Loqué et al. 2007; Straub et al. 2017). Yuan et al. (2013) verified whether phosphorylation at position T464 in the C-terminus of AMT1.3 (corresponding to position 460 in the *C*-terminus of AMT1.1) also allosterically regulates ammonium transport (Yuan et al. 2013). Trimer formation is required for functional AMTs (Loqué et al. 2007).

AMTs participate in apoplastic acidification as they have high ammonium specificity. Computational (Wang et al. 2012) and experimental (Ariz et al. 2011, 2018) studies on plants and yeast revealed that NH_4^+ deprotonation is probably a major step in ammonium transport. After NH_4^+ is deprotonated at the periplasmic side of the ammonium transporter, the NH_3 traverses the membrane through the hydrophobic channel while the proton crosses through the pore at the bridge formed by the twin histidine motif (His168 and His318). This point constitutes the major NH_4^+ transport selectivity motif, preventing K^+ flow, which has a hydrated size similar to that of NH_4^+ (Williamson et al. 2020). Inside the cell, NH_3 and H^+ recombine to form NH_4^+. This mechanism might be conserved amongst the electrogenic members of the Amt/Mep/Rh family which include AMT1.1 and AMT1.2 in various plant species (Mayer et al. 2006b; Loqué et al. 2009; Ortiz-Ramirez et al. 2011; Wacker et al. 2014; Yang et al. 2015). In contrast, non-electrogenic ammonium or methylammonium transport was detected in Arabidopsis AMT2-expressing oocytes (Neuhäuser et al. 2009). Thus, the substance being transported by the AMT2s is most likely electroneutral NH_3. However, there were no significant differences in growth among yeast cells harboring AMT2 at different pH values. In this case, direct NH_3 transport is partially ruled out. NH_4^+ may be recruited by the binding site, deprotonated in AMT2, and transported across the membrane as uncharged NH_3 (Neuhäuser et al. 2009).

Despite AMT repression in the presence of high ammonium levels, ammonium nonetheless enters the plant through the aforementioned putative channels. Once ammonium is inside the cell, it may be assimilated, stored, or effluxed to the external medium.

13.2.2 Translocation of Ammonium from the Root Increases Ammonium Assimilation and Acidification in the Shoot

Most of the ammonium transported into plant cells is locally assimilated in the cytoplasm and plastids by glutamine synthetases (GS). This process generates protons. The stoichiometry of this reaction is two protons per glutamine. One proton is derived from ATP hydrolysis while the other originates from NH_4^+ deprotonation (Britto and Kronzucker 2005).

A substantial proportion of the nitrate absorbed by the roots is translocated to the shoot. It was previously assumed that ammonium is either directly absorbed by the roots from the soil solution or generated in the roots by nitrate reduction and almost exclusively assimilated there (Van Beusichem et al. 1988; Kafkafi and Ganmore-Neumann 1997). However, ammonium assimilation in the roots requires abundant carbon skeletons and reducing equivalents. Therefore, plants with limited root assimilation capacity may favor ammonium loading from the xylem to the shoot (Giehl et al. 2017). Lasa et al. (2001, 2002) compared the growth of spinach, sunflower, and snap pea plants in the presence of ammonium as the sole N source and demonstrated a close relationship between ammonium accumulation in the shoots and a reduction in shoot growth (Lasa et al. 2001, 2002). Spinach growth was most severely affected and presented with the highest shoot ammonium concentration. Sunflower was moderately affected and presented with intermediate shoot ammonium concentrations. Snap pea tolerated ammonium supply and its shoot ammonium content was similar to that for plants fed nitrate. Thus, plants that assimilate ammonium in their roots have relatively greater exogenous ammonium tolerance than those that translocate the ammonium to their shoots (Barker and Mills 2011). Moreover, effective ammonium sequestration in root cell vacuoles prevents transport to ammonium-sensitive leaves (Schjoerring et al. 2002; Cruz et al. 2006). Rice plants are highly ammonium-tolerant as they vacuolate ammonium when their root assimilation capacity is exceeded (Kronzucker et al. 1999).

Hachiya et al. (2021) demonstrated that elevated foliar ammonium assimilation by GS2 led to ammonium toxicity in Arabidopsis. They also observed that root GS1.2 attenuated foliar ammonium toxicity. Hence, root GS1.2 may act as a barrier preventing root-to-shoot ammonium transport and avoiding ammonium assimilation by shoot GS2 (Hachiya et al. 2021). The results of other studies confirmed this hypothesis. Jian et al. (2018) reported increases and decreases in root and shoot GS activity, respectively, in the ammonium-tolerant *nrt1.1* (*chl1-5*) Arabidopsis mutant compared to the more sensitive wild-type Col-0 Arabidopsis (Jian et al. 2018). Guan et al. (2016) demonstrated that Arabidopsis mutants lacking the GS1.2 isoform accumulated ammonium and were hypersensitive to ammonium stress. Magalhäes and Huber (1989) reported significantly higher root GS activity in rice than in tomato or maize (Magalhäes and Huber 1989). Future research should aim to determine how AMTs, GS1.2, and GS2 differentially regulate pH. This information could explain why 20–41% of all genes responding to ammonium are also upregulated by low external pH (Patterson et al. 2010).

Plants also avoid ammonium toxicity by increasing extracellular ammonium efflux. Nevertheless, the high energetic cost associated with this mechanism partially accounts for the growth impairment observed in response to ammonium nutrition. A futile N influx/efflux cycle consists of a root NH_3 influx accompanied by an energetically demanding NH_3 efflux, and has been linked to ammonium-induced growth suppression (Britto et al. 2001; Kronzucker et al. 2001). In sensitive plant species, these fluxes occur at lower ammonium thresholds (Chen et al. 2013; Coskun et al. 2013). Moreover, sensitive plant species may have a relatively lower capacity to store ammonium as they have smaller root vacuoles than tolerant species (Loqué et al. 2005; Bai et al. 2014; Liang et al. 2020).

Root cell vacuoles are vital for non-assimilated ammonium storage. A few studies focused on the functionality of the vacuole structure under high ammonium nutrition. Robert et al. (2021a) reported that alterations in the integrity of the root vacuole led to ammonium leakage from root cells and substantial efflux in the presence of high ammonium concentrations. Previous studies proposed that membrane flux from macroautophagy may contribute to vacuole biogenesis and enlargement (Marty 1978; Inoue et al. 2006; Yano et al. 2007a, b). At very high ammonium concentrations, this process is impaired and a global imbalance in membrane flux to and from the vacuole is created. Vacuoles with reduced reservoirs and less storage capacity have limited ability to increase in size through dynamic morphological changes in response to growth and environmental signals. Cell elongation requires spatiotemporal coordination between increases in vacuole size and cell wall loosening. Under ammonium stress conditions, then, cell expansion is limited (Robert et al. 2021a).

In conjunction with vacuole dynamics and volume, plant cell growth is also regulated and cell size is determined by the cell wall status. It was originally believed that the role of Si in plants was to create a physical barrier that 'fortifies' the cell wall (reviewed by Mandlik et al. 2020). Interestingly, it was recently shown in rice that Si bound to xyloglucan increases resistance to cellulase degradation, and thus improves the nanomechanical properties of the cell wall (Pu et al. 2021). It was implied that reduced cell elongation is the principal mechanism by which ammonium inhibits primary root growth. Future research will endeavor to clarify the roles of the cell wall structure in ammonium-induced growth suppression and determine how Si supply could alleviate this key attribute of the 'ammonium syndrome'.

13.2.3 Ammonium Nutrition Decreases Protein N-Glycosylation-Dependent Ammonium Efflux and Arrests Root Elongation

The aforementioned ammonium toxicity mechanisms are linked to ammonium cation or neutral NH_3 permeation into the cell (Mattsson and Schjoerring 2002; Coskun et al. 2013). Thus, useful information may be obtained by studying transmembrane ammonium fluxes that might closely depend on root cell wall stability. Li et al. (2010) showed that elevated ammonium efflux inhibited primary root growth in Arabidopsis mainly by affecting its elongation zone (Li et al. 2010). Britto et al. (2001) compared the relative ammonium sensitivity of barley and rice and found that the most sensitive barley plants presented with higher ammonium efflux and more severe growth arrest than rice. The authors proposed that elevated ammonium efflux leads to a futile N cycle in the root cell plasma membranes and incurs a metabolic energy cost (Britto et al. 2001).

Ammonium influx through the AMTs is well documented. However, the molecular identity of the ammonium efflux transporter or channel is unknown. It also remains to be determined how they are affected by high-ammonium stress. Extensive physiological studies revealed that ammonium efflux is dominated by NH_3 export across the plasma membrane. This process may be facilitated by (1) NH_3-transporting aquaporins (Britto et al. 2001; Coskun et al. 2013), included among intrinsic tonoplast proteins (Jahn et al. 2004; Loqué et al. 2005; Kirscht et al. 2016), or by (2) AMT2-type proteins such as AtAMT2.1 that permeate NH_3 (Neuhäuser et al. 2009) and reroute it towards the inner root cells in response to ammonium nutrition (Giehl et al. 2017). The proposed model suggests that NH_3 transporters across cell membranes do not dissipate energy and permit the NH_3 to penetrate various cell compartments based on concentration gradients and pH. Hence, the ammonium cation is trapped in the vacuole (Coskun et al. 2013). Ammonium hyperaccumulation in root cell vacuoles increased tonoplast degradation, reduced membrane flow by impairing macroautophagic flux (Robert et al. 2021a), and created vacuoles with simplified structures that increase ammonium efflux and retard root growth. Ammonium accumulation in vacuoles might explain the decreases in the concentration of cations such as K^+. This model might partially explain nitrate/ammonium co-provision. Balancing the cation/anion ratio in the vacuole combined with flow-dependent tonoplast stabilization optimizes the nutrient status (Miller and Cramer 2004; Robert et al. 2021b).

Decreases in protein *N*-glycosylation precede the transporters facilitating ammonium efflux. This phenomenon is associated with reductions in cellulose biosynthesis, cell wall stability, and, therefore, root cell viability (Jae et al. 2008; Qin et al. 2008; Barth et al. 2010; Li et al. 2010). In the endeavor to identify the genes associated with ammonium toxicity, it was found that the expression levels of those related to protein *N*-glycosylation were significantly altered (Qin et al. 2008). N-glycosylation is a common post-transcriptional protein modification in eukaryotes. It affects numerous processes including enzyme activity and protein folding, stability, and intermolecular interactions (Zeng et al. 2018). In plants, alteration of the genes encoding protein *N*-glycosylation is associated with primary root inhibition under high-ammonium stress. These genes encode the enzymes GMPase (*GMP1*, or Vitamin C1, *VTC1*), dolichol phosphate mannose synthase 1 (*DPMS1*), and GDP-mannose pyrophosphohydrolase (*NUDX9*) (Qin et al. 2008; Jadid et al. 2011; Kempinski et al. 2011). GMPase synthesizes the *N*-glycosylation substrate GDP-mannose. Hence, the lack of *N*-glycoproteins might be implicated in ammonium hypersensitivity. Ammonium-dependent inhibition of primary root growth was partially attenuated in Arabidopsis *nudx9* mutants unable to hydrolyze GDP-mannose to mannose 1-phosphate and GMP (Tanaka et al. 2015). Di et al. (2021) demonstrated that protein *N*-glycosylation could be promoted by downregulating *NUDX9* through the transcription factor WRK46 that responds to high ammonium levels (Di et al. 2021). By contrast, *gmp1* and *dpms1* mutations reduced *N*-glycosylation and increased ammonium sensitivity. Under high-ammonium stress, then, protein *N*-glycosylation prevents ammonium efflux from the roots and, by extension, the futile N cycle (Qin et al. 2008; Barth et al. 2010; Tanaka et al. 2015; Di et al. 2021).

The regulatory domains that mediate ammonium efflux might directly inhibit *N*-glycosylation, thereby preventing closure. Nevertheless, the specific transporters or channels involved have not been identified. In contrast, stabilization of the cell walls by *N*-glycosylation could improve ammonium transport efficiency and prevent efflux. Defects in cell wall pectins, hemicelluloses, and lignin (Głazowska et al. 2018) could deform membranes and cause depolarization. The transmembrane ammonium transporters and other proteins in deformed cell membranes may fuse and form oligomers and aggregates that further destabilize, increase tension, and increase the curvature of membranes (McMahon and Gallop 2005; Fig. 13.2). In this case, transport efficiency would decrease and ion efflux would increase. Membrane potential plays a key role in the transport efficiency of voltage-gated K^+ channels. Voltage-gated channels are sensitive to changes in membrane depolarization or hyperpolarization. Changes in membrane potential alter the channel protein opening (Ivashikina et al. 2001). For K^+ channels, K^+ efflux enhanced by plasma membrane depolarization might occur in response to high ammonium supply under both low and high K^+ (Ayling 1993; Wang et al. 1994; Nocito et al. 2002; Coskun et al. 2010), thereby facilitating unspecified ammonium efflux to the external medium.

Root zone and root cell type-specific responses also influence ion fluxes during elevated ammonium conditions. Li et al. (2010) demonstrated ammonium effluxes in the elongation zones and meristem but ammonium influxes in the maturation zones when roots were supplied with ammonium (Li et al. 2010). These results suggest that the growing cells of the elongation zone and meristem are substantially

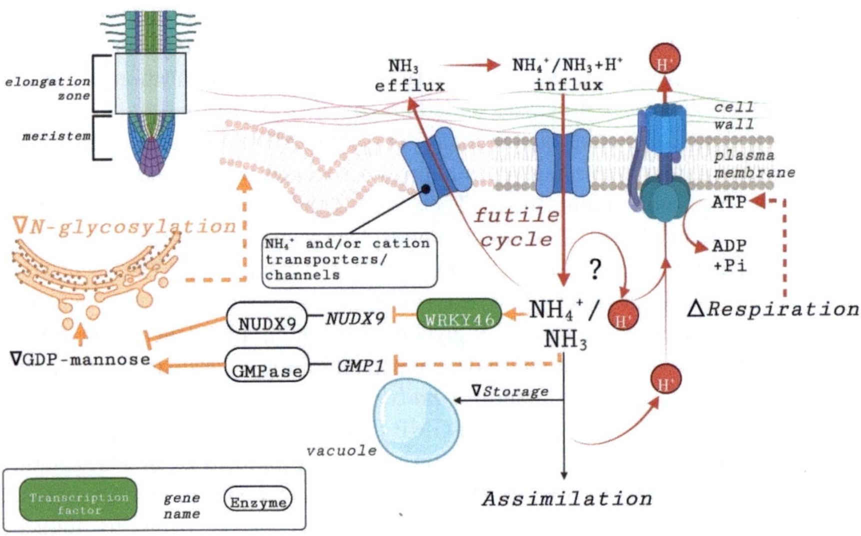

Fig. 13.2 Ammonium inhibits root elongation by decreasing protein *N*-glycosylation. In roots, ammonium downregulates GDP-mannose by repressing GMP1 and inducing NUDX9 (Qin et al. 2008; Tanaka et al. 2015). Consequently, protein *N*-glycosylation decreases, cell walls are destabilized, and membranes are deformed (McMahon and Gallop 2005). Hence, ammonium efflux increases in the root apex, vacuolar sequestration is reduced, the futile N cycle is fed, respiratory energy costs increase, and root growth is reduced (Britto et al. 2001). Elevated ammonium levels induce the transcription factor WRKY46 which inhibits NUDX9 transcription and prevents even attenuated protein *N*-glycosylation (Di et al. 2021). GMP1: GDP-mannose pyrophosphorylase; NUDX9: GDP-mannose pyrophosphohydrolase. (This figure was created with BioRender (https://biorender.com/))

more sensitive to ammonium than the cells of the maturation zone. Unlike mature basal cells, growing apical cells may rapidly send signals to promote ammonium efflux after they sense high ammonium levels. In addition, the relative differences in the characteristics of growing and mature cells may be critical. Cells that are not completely differentiated such as those in the meristem are characterized by small vacuoles, volumes, and water content (Luxová 1988; Li et al. 2010). Hence, the root apex may be relatively more sensitive to ammonium and export it even at low concentrations. By contrast, mature cells have large vacuoles that may occupy >90% of their internal volume. Thus, they possess large ammonium storage capacity and the bulk of the ammonium they absorb may be localized to their vacuoles (Loqué and von Wirén 2004). Vacuolar sequestration may be a mechanism of avoiding ammonium toxicity as it prevents transport to ammonium-sensitive leaves (Yuan et al. 1993; Miller et al. 2001; Schjoerring et al. 2002; Loqué and von Wirén 2004; Cruz et al. 2006). A comprehensive analysis of cell type-specific sensitivity to salt stress in barley (Shabala et al. 2016) and potato (Liu et al. 2019) roots revealed that K^+ efflux was highest in the root meristem and elongation zones. Therefore, ammonium efflux might be facilitated through nonspecific K^+ channels in the root apex.

It was determined that deficiency in N-glycosylation caused by high ammonium levels may result in insufficient cellulose biosynthesis and cell wall formation in root meristematic and elongation zones. Cell wall defects may cause membrane deformation and, by extension, reduce transport efficiency and increase ammonium and K^+ fluxes possibly by common non-selective cation channels. The futile N cycle that results from elevated ammonium efflux is detrimental to root growth and plant survival. Future research should aim to establish the agronomic consequences of the aforementioned mechanisms.

13.2.4 Internal Ammonium Accumulation Initiates ROS-Dependent Cell Wall Lignification and Limits Cell Growth

Foliar ammonium assimilation was cited as a major cause of ammonium toxicity (Hachiya et al. 2021). However, the immediate responses might be initiated by the induction of reactive oxygen species (ROS) bursts and their downstream reactions including photophosphorylation uncoupling in response to free radical injury (Foyer 2018; Yang et al. 2020), reduced glutathione (GSH) cycling for ROS cleavage (Yang et al. 2020), and the induction of glycolysis/glycogen breakdown-related genes to provide the metabolic energy and carbon skeletons required for ROS scavenging (Li et al. 2020; Yang et al. 2020). NADPH oxidase, laccases, peroxidase, and other enzymes catalyzing ROS play vital roles in the formation of the Casparian strip which is a lignin-based diffusion barrier in the endodermis of plant root cells (Lee et al. 2013). As ammonium-induced ROS participate in lignin formation, they directly modify primary and secondary cell wall structures and, therefore, affect root and leaf growth (Barros et al. 2015; Tsukagoshi 2016; Głazowska et al. 2019; Royo et al. 2019; Yang et al. 2022). Podgórska et al. (2017) reported that elevated peroxidase activity combined with increased cell wall phenolic content in plants grown on ammonium may indicate the occurrence of cell wall lignification and, by extension, limited leaf cell expansion. Interestingly, Radotić et al. (2022) have shown that moderate Si supply modifies lignin polymerization by binding to coniferyl alcohol dimers, and making lignin structure more compact. These results encourage further research on the Si-lignin interaction for cell wall integrity, expanding current understanding of beneficial effects against stresses such as ammonium toxicity.

13.3 Repairing Role of Si in Plant Cell Structural Components Resulting from Ammonium Nutrition

13.3.1 Silicon Decreases Oxidative Stress Caused by Excess Ammonium

One of the immediate plant responses during abiotic stresses such as ammonium toxicity is generating ROS such as singlet oxygen ($^1O_2$), superoxide (O^{-2}), H_2O_2, and hydroxyl radicals (OH), which cause severe damage to the cell structure, organelles, and, consequently, their functions (Sharma et al. 2012; Das and Roychoudhury 2014). Therefore, these effects decrease the chlorophyll and carotenoid contents (Barreto et al. 2022) and the photosynthetic rates (Viciedo et al. 2019; Silva Junior et al. 2020; Song et al. 2022). Oxidative stress occurs due to the high energy consumption for the ammonium efflux (Sheng et al. 2018), maintaining its low concentration in the cell cytosol (Britto et al. 2014).

Plants mitigate and repair this damage by activating their defense system, inducing a complex antioxidant system to maintain cell homeostasis through the activation of the antioxidant enzymes superoxide dismutase (SOD), catalase (CAT), ascorbate peroxidase (APX), polyphenol peroxidase (PPO), peroxidase (POD), guaiacol peroxidase (GPX), and glutathione reductase (GR) (Tripathi et al. 2017; Kim et al. 2017).

From this perspective, Si supply can induce plant tolerance to ammonium stress by inducing the antioxidant system through antioxidant enzymes (Kim et al. 2017) in order to balance ROS production, decreasing electrolyte leakage (Silva Júnior et al. 2019) and lipid peroxidation (Barreto et al. 2022), which corresponds to the malondialdehyde content in the plant tissue. Si can bind to free monophenol complexes, forming Si-phenol (Rogalla and Römheld 2002) and preventing the accumulation of these phenols and ROS production in cell organelles (Maksimovic et al. 2012). However, the role of Si in the metabolism of polyphenols should be further clarified, especially with regard to structural changes associated with the damage caused by oxidate stress in the plant cell wall.

Favorable results regarding the role of Si as a mitigator of abiotic stresses depend on the stress exposure time (Sharma et al. 2012; Kim et al. 2017), the stressing agent, and the level of stress caused to plants, ranging from moderate to severe (Campos et al. 2020). However, this topic will address studies that evaluated the effect of Si on the mitigation of oxidative stress caused by excess ammonium and mediated by the activity of antioxidant enzymes in tropical plant species (Table 13.1), given that this theme is still scarcely addressed in the literature.

The damage levels caused by ammonium toxicity and the mitigating effect of Si are also dependent on the genotypes within each plant species. From this perspective, Barreto et al. (2022) assessed the effects of nitrate and ammonium supply in the presence and absence of Si in plants of the tomato cultivars Micro Ton ('MT', wild type) and Micro Ton CKX2 (transgenic type) by evaluating the components of lipid peroxidation. Thus, in the MT tomato, the MDA levels were higher in plants supplied with ammonium. In contrast, in MT CKX2 plants, ammonium excess decreased the MDA levels (Fig. 13.3a). In the roots of the cv. MT, the lowest MDA content was observed in plants treated with nitrate. There was no difference between the MDA contents of MT CKX2 roots in the different treatments (Fig. 13.3b). With regard to H_2O_2, Si supply decreased the MDA levels in the leaves of the cv. MT cultivated in sole ammonium solution compared to the sole nitrate treatment. However, the lowest H_2O_2 contents decreased in the leaves of MT CKX2 plants cultivated with ammonium regardless of Si supply (Fig. 13.3c). In the roots of MT plants cultivated with ammonium, the H_2O_2 contents decreased with Si supply. On the other hand, this effect was not observed in the roots of MT CKX2 plants (Fig. 13.3d).

In cabbage plants stressed with ammonium, O_2^- and H_2O_2 accumulation and peroxidation were mitigated by Si due to the higher action of the antioxidant enzymes SOD and CAT. Complementary, the plants treated with Si showed the highest activities of the enzymes responsible for ammonium assimilation, including

Table 13.1 Antioxidant activities and lipid peroxidation mediated by Si application under ammonium stress in tropical plants

Tropical species	Si concentration	NH_4^+ toxic concentration	Si activated enzyme	Si neutralized ROS and lipid peroxidation	References
Cabbage	1.0 mmol L^{-1}	13 mmol L^{-1}	SOD and CAT	O_2^-, H_2O_2, and MDA	Song et al. (2022)
Salvia	1.0 mmol L^{-1}	13 mmol L^{-1}	APX, CAT, GPX, GR	O_2^-, H_2O_2, and MDA	Song et al. (2022)
Tomato	1.3 mmol L^{-1}	5.9 mmol L^{-1}	SOD and CAT	MDA	Barreto et al. (2022)
Cucumber	2.0 mmol L^{-1}	–	SOD, POD, CAT, and APX	MDA	Zhang et al. (2014)
Cucumber	–	–	SOD, POD, CAT, and APX	O_2^- and H_2O_2	Gao et al. (2014)

CAT catalase, *SOD* superoxide dismutase, *APX* ascorbate peroxidase, *POD* peroxidase, O_2^- superoxide, H_2O_2 hydrogen peroxide, *MDA* malondialdehyde

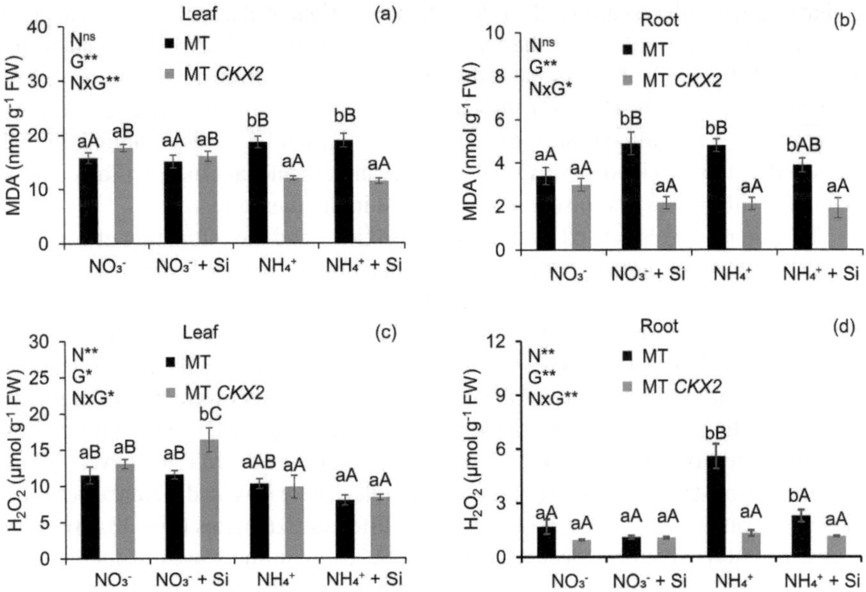

Fig. 13.3 Malondialdehyde (MDA) levels in leaves (**a**) and roots (**b**) and hydrogen peroxide (H_2O_2) levels in leaves (**c**) and roots (**d**) of the tomato cultivars MT and *MT CKX2* grown in nutrient solutions with nitrate and ammonium in the presence and absence of Si. (Adapted from Barreto et al. (2022))

GS, glutamate synthase (GOGAT), and glutamate dehydrogenase (NADH-GDH) (Song et al. 2022). There is evidence that excessive ammonium assimilation causes acid stress since high contents of this cation in chloroplasts can block the N metabolism mediated by the GS-GOGAT enzymatic complex responsible for N assimilation in plants (Bittsánszky et al. 2015). These factors indicate that new studies on the effects of Si on H^+ regulation and distribution should be conducted since the interaction between Si and the antioxidant enzyme system of plants requires deeper analyses in order to understand the mechanisms responsible for regulating plant responses to Si-mediated abiotic stresses.

13.3.2 Structural Role of Si in Cell Wall Stability Aiming at Ammonium Toxicity Alleviation

Plants develop specific strategies to deposit large amounts of Si in the intra- or extracellular region (Sheng et al. 2018) by absorbing silicic acid using Si transporters (Ma et al. 2006). This component is deposited in the cell wall as amorphous silica, especially in the hemicellulose fraction, forming organosilicon through the reticulation of hemicellulose polysaccharides (He et al. 2015).

Plant cell growth boosted by turgor pressure depends on the cell wall structure. Cell wall strengthening through the association and interaction with Si can result in better nutrient uptake and optimized plant metabolism. Therefore, under ammonium toxicity conditions, Si can contribute to regenerating cell wall components (He et al. 2015), improving cell rigidity and membrane stability, optimizing ammonium assimilation, and decreasing the cell demand for the regulation of AMTs (Sheng et al. 2018). These strategies can explain the structural role of Si in cell wall stability aiming at alleviating ammonium toxicity in plants. Confirming this hypothesis, Sheng et al. (2018) observed that rice plants grown in solutions with 1 mmol L^{-1} silicic acid showed a higher ammonium influx compared to the absence of Si.

The possible promoting mechanisms involved in capturing ammonium ions as a function of the Si bound to the cell wall include a higher cell division rate in the different cell growth phases, improvement of the synthesis and expression of AMTs in the cell membrane, and cell wall stabilization with improvements in the transport efficiency of ammonium. The higher cell wall rigidity resulting from the action of Si (Luyckx et al. 2017) enables cells to withstand the turgor pressure (Ma et al. 2016a) caused by the solutes within the cells, stabilizing the membranes and optimizing nutrient uptake under osmotic stress conditions caused by ammonium (Sheng et al. 2018).

The latest studies indicate that Si promotes the biosynthesis of cytokinins and can accelerate the cell division cycle (Markovich et al. 2017). The transport of assimilates in plants treated with Si provides greater energy storage in the cells, benefiting plant tolerance to abiotic stress (Zhu et al. 2016). In the cell wall, Si binds mainly to pectins (He et al. 2015) through direct organosilicon bonds (C-O-Si) instead of being deposited as amorphous silica (He et al. 2013), thus enabling the beneficial element to mitigate deleterious effects in cells stressed by ammonium. However, the chemical and physiological mechanisms, nutrient uptake, and the optimization of the ammonium metabolism associated with Si needs scientific elucidation in order to better understand the structural effect of Si on the cell wall.

13.3.3 Silicon Supply Mitigates Ammonium Toxicity Symptoms Related to Plant Growth and Development

As previously described, Si application under ammonium stress conditions is responsive to ROS elimination since it decreases lipid peroxidation through the activation of antioxidant enzymes of the plant defense system, including CAT, SOD, PPO, POD, APX, GPX, and GSSH (Kim et al. 2017). In the absence of Si, ammonium toxicity can cause physiological and biochemical damage, compromising plant growth. The symptoms of ammonium toxicity begin with chlorosis, leaf necrosis, and root stunting, which can be mitigated by Si supply in the growth medium (Silva Júnior et al. 2019; Song et al. 2022).

From this perspective, under ammonium toxicity (9.75 mmol L^{-1}), Si supply increased the stem diameter (Fig. 13.4a), root length (Fig. 13.4b), N-use efficiency (Fig. 13.4c), and root dry matter (Fig. 13.4d) in yellow passion fruit plants (Silva Júnior et al. 2019). The authors showed that silicon interacted with N, especially under ammonium stress conditions, mitigating toxicity with increased N and Si accumulation in the plant tissue and providing greater dry matter accumulations. Si supply under ammonium excess increased the leaf chlorophyll index (Silva Júnior et al. 2019) and decreased the leaf electrolyte leakage index, mitigating chlorotic and necrotic symptoms in the leaves and favoring root development (Fig. 13.5) (Silva Júnior 2015).

Ammonium increase from 50% to 100% of the nutrient solution (i.e., up to the concentration of 13 mmol L^{-1}) resulted in damage to the height and shoot dry matter of cabbage plants with Si application (1.0 mmol L^{-1}), improving these two parameters (Song et al. 2022). At the concentration of 2.0 mmol L^{-1}, Si also mitigated

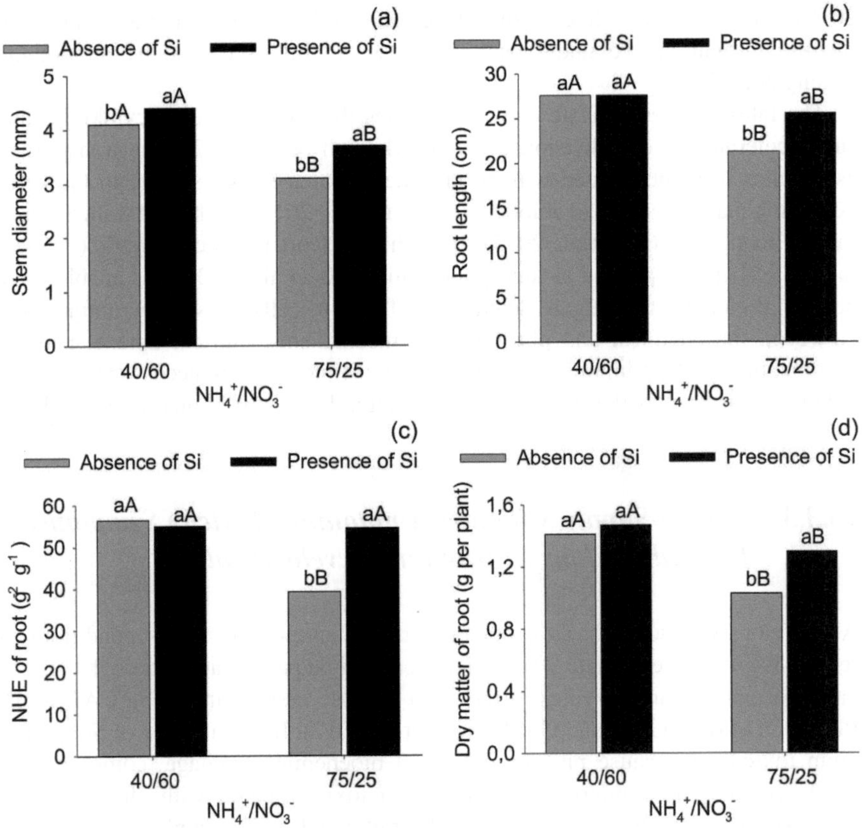

Fig. 13.4 Stem diameter (**a**), root length (**b**), nitrogen-use efficiency (**c**), and root dry matter (**d**) of passion fruit seedlings as a function of ammonium and nitrate proportions in the absence and presence of Si in the nutrient solution. (Adapted from Silva Júnior et al. (2019))

Fig. 13.5 Si effect on the mitigation of ammonium toxicity in the shoot (**a**), leaves (**b**), and roots (**c**) of passion fruit seedlings. (Adapted from Silva Júnior (2015))

ammonium toxicity in cauliflower when the total N concentration was 15 mmol L^{-1}, with half the N in the form ammonium, and in broccoli at the total concentration of 15 mmol L^{-1} of N supplied as ammonium (Barreto et al. 2017). Likewise, the supply of 1.0 mmol L^{-1} Si associated with 1.0 mmol L^{-1} ammonium in tomato plants grown in a nutrient solution increased the dry root biomass and the shoot biomass 40 days after emergence by decreasing the leaf contents of MDA and H_2O_2 (Barreto et al. 2016).

In sugar beet plants grown until 50 days after emergence, ammonium supply above 15 mmol L^{-1} decreased the shoot dry matter and, more severely, the dry root matter. However, this effect was mitigated in plants grown in a nutrient solution with 2.0 mmol L^{-1} Si by favoring shoot dry matter production, with higher Si-use efficiency (Fig. 13.6) (Viciedo et al. 2019).

Plants of the tomato cultivars Micro Ton (MT, wild type) and Micro Ton CKX2 (transgenic type) grown with ammonium showed the lowest growth rates compared to those grown with nitrate (Fig. 13.7). On the other hand, Si supply favored a higher growth in the tomato cv. MT grown with ammonium, confirming the beneficial effect of Si under ammonium stress conditions for this cultivar (Fig. 13.7a). The

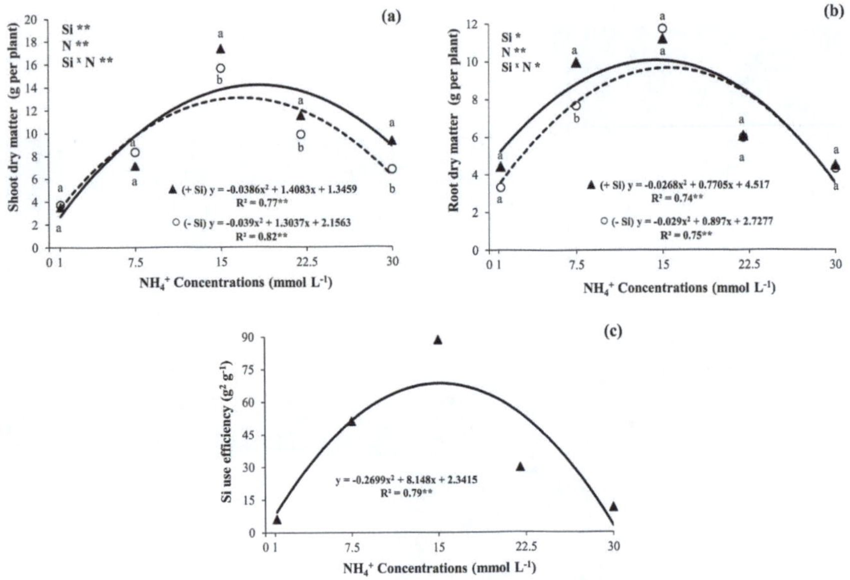

Fig. 13.6 Effects of ammonium concentrations in the presence (▲) and absence (O) of Si on the shoot dry matter (**a**), root dry matter (**b**), and Si-use efficiency (**c**) of sugar beet 50 days after emergence. (Adapted from Viciedo et al. (2019))

Fig. 13.7 Tomato cultivars MT (**a**) and *MT CKX2* (**b**) grown in a nutrient solution with nitrate or ammonium in the presence and absence of Si. (Adapted from Barreto et al. (2022))

absence of Si effect on the cv. Micro Ton CKX2 stressed with ammonium (Fig. 13.7b) confirms that the action of Si depends, among other factors, on the genotype studied (Barreto et al. 2022).

High ammonium concentrations (60 mmol L^{-1}) damaged root formation, with Si supply decreasing dry matter production in two cultivars of *Cucumis sativus,* cv. Tsubasa and cv. Hokushin (Campos et al. 2016). These findings indicate that, under severe toxicity conditions, the beneficial element did not mitigate the damage caused by ammonium. In this same study, the authors showed that, although Si favored plant growth when provided with ammonium as the sole source of N, no difference was observed between the Si concentrations of 1 and 10 mmol L^{-1}. This fact could be explained by Si polymerization in the nutrient solution with the increase in its concentration, reducing its uptake. Furthermore, most of the Si absorbed is immobilized as amorphous silica outside the cell wall (He et al. 2015; Sheng et al. 2018), with only a small amount remaining free in the plant. With that, it is evident that adjustments in the ammonium and Si concentrations in the growth medium are essential to confirming the mitigating effect of the beneficial element, justifying further investigations on the effects of Si in regulating ammonium nutrition in plants.

It is possible to use ammonium for plant growth in nutrient solutions or irrigated systems at adequate concentrations to increase growth and production. However, depending on the concentration, the nutrient could present toxicity risks. In this scenario, it was clear that Si could be a strategy for association with ammonium to maximize the N-use efficiency, minimizing ammonium toxicity and favoring sustainable cultivation. Therefore, ways are opened for future research aiming to map the interaction between ammonium concentrations and Si in the growth medium in species with different tolerances to ammonium toxicity and Si accumulation capacity, thus improving N metabolism in such plants.

13.4 Conclusions and Future Perspective

This chapter discussed the interesting and promising topic of how Si could alleviate the impact of ammonium nutrition and its toxicity on plant growth. Excess ammonium concomitant to pH-dependent ROS production can be alleviated by enhancing antioxidant systems in a Si-dependent manner. Although the interaction of Si and the plant antioxidant enzyme system remains poorly understood, the regulation of cell wall lignification might be involved, as ROS are closely associated with lignin deposition. Indeed, the alteration in the polymerization of lignin in the presence of Si could increase the sink capacity of ROS required for peroxidase and laccase activity, as well as the developmental signal in the secondary wall differentiation. Consequently, the stability of the membrane transport proteins and ions uptake could be modified, reducing the reported electrolytic leakage or the harmful net efflux of ammonium.

Acknowledgments MR was supported by a UPNa PhD Scholarship. IA was supported by a MCIN RyC Programme MCIN/AEI/10.13039/501100011033 and by "European Union Next Generation EU/PRTR" under grant No. RYC2021-032345-I. This work was supported by the AEI under grant No. PID2019-107463RJ-I00/AEI/10.13039/501100011033.

References

Adrees M, Ali S, Rizwan M, Rehman MZ, Ibrahim M, Abbas F, Farid M, Qayyum MK, Irshad MK (2015) Mechanisms of silicon-mediated alleviation of heavy metal toxicity in plants: a review. Ecotoxicol Environ Saf 119:186–197. https://doi.org/10.1016/j.ecoenv.2015.05.011

Antonenko YN, Pohl P, Denisov GA (1997) Permeation of ammonia across bilayer lipid membranes studied by ammonium ion selective microelectrodes. Biophys J 72:2187–2195. https://doi.org/10.1016/S0006-3495(97)78862-3

Ariz I, Cruz C, Moran JF, González-Moro MB, García-Olaverri C, González-Murua C, Martins-Loução MA, Aparicio-Tejo PM (2011) Depletion of the heaviest stable N isotope is associated with NH_4^+/NH_3 toxicity in NH_4^+-fed plants. BMC Plant Biol 11:83. https://doi.org/10.1186/1471-2229-11-83

Ariz I, Boeckstaens M, Gouveia C, Martins AP, Sanz-Luque E, Fernández E, Soveral G, von Wirén N, Marini AM, Aparicio-Tejo PM et al (2018) Nitrogen isotope signature evidences ammonium deprotonation as a common transport mechanism for the AMT-Mep-Rh protein superfamily. Sci Adv 4:3599. https://doi.org/10.1126/sciadv.aar3599

Ayling SM (1993) The effect of ammonium ions on membrane potential and anion flux in roots of barley and tomato. Plant Cell Environ 16:297–303. https://doi.org/10.1111/j.1365-3040.1993.tb00872.x

Bai L, Ma X, Zhang G, Song S, Zhou Y, Gao L, Miao Y, Song C-P (2014) A receptor-like kinase mediates ammonium homeostasis and is important for the polar growth of root hairs in arabidopsis. Plant Cell 26:1497–1511. https://doi.org/10.1105/tpc.114.124586

Bakhat HF, Bibi N, Zia Z, Abbas S, Hammad HM, Fahad S, Ashraf MR, Shah GM, Rabbani F, Saeed S (2018) Silicon mitigates biotic stresses in crop plants: a review. Crop Prot 104:21–34. https://doi.org/10.1016/j.cropro.2017.10.008

Balkos KD, Britto DT, Kronzucker HJ (2010) Optimization of ammonium acquisition and metabolism by potassium in rice (*Oryza sativa* L. cv. IR-72). Plant Cell Environ 33:23–34. https://doi.org/10.1111/j.1365-3040.2009.02046.x

Barker AV, Mills HA (2011) Ammonium and nitrate nutrition of horticultural crops. Wiley, Hoboken, pp 395–423. https://doi.org/10.1002/9781118060759.ch8

Barreto RF, Prado RM, Leal AJF, Troleis MJB, Junior GBS, Monteiro CC, Santos LCN, Carvalho RF (2016) Mitigation of ammonium toxicity by silicon in tomato depends on the ammonium concentration. Acta Agric Scand Sect B Soil Plant Sci 66:483–488. https://doi.org/10.1080/09064710.2016.1178324

Barreto RF, Schiavon Júnior AA, Maggio MA, de Mello Prado R (2017) Silicon alleviates ammonium toxicity in cauliflower and in broccoli. Sci Hortic (Amsterdam) 225:743–750. https://doi.org/10.1016/j.scienta.2017.08.014

Barreto RF, de Mello Prado R, Lúcio JCB, López-Díaz I, Carrera E, Carvalho RF (2022) Ammonium toxicity alleviation by silicon is dependent on cytokinins in tomato cv. micro-tom. J Plant Growth Regul 41:417–428. https://doi.org/10.1007/s00344-021-10314-5

Barros J, Serk H, Granlund I, Pesquet E (2015) The cell biology of lignification in higher plants. Ann Bot 115:1053–1074. https://doi.org/10.1093/aob/mcv046

Barth C, Gouzd ZA, Steele HP, Imperio RM (2010) A mutation in GDP-mannose pyrophosphorylase causes conditional hypersensitivity to ammonium, resulting in Arabidopsis root growth inhibition, altered ammonium metabolism, and hormone homeostasis. J Exp Bot 61:379–394. https://doi.org/10.1093/jxb/erp310

Bindel N, Neuhäuser B (2021) High-affinity ammonium transport by *Arabidopsis thaliana* AMT1;4. Acta Physiol Plant 43:69. https://doi.org/10.1007/s11738-021-03235-z

Bittsánszky A, Pilinszk K, Gyulai G, Komives T (2015) Overcoming ammonium toxicity. Plant Sci 231:184–190. https://doi.org/10.1016/j.plantsci.2014.12.005

Britto DT, Kronzucker HJ (2002) NH_4^+ toxicity in higher plants: a critical review. J Plant Physiol 159:567–584. https://doi.org/10.1078/0176-1617-0774

Britto DT, Kronzucker HJ (2005) Nitrogen acquisition, PEP carboxylase, and cellular pH homeostasis: new views on old paradigms. Plant Cell Environ 28:1396–1409. https://doi.org/10.1111/j.1365-3040.2005.01372.x

Britto DT, Siddiqi MY, Glass ADM, Kronzucker HJ (2001) Futile transmembrane NH_4^+ cycling: a cellular hypothesis to explain ammonium toxicity in plants. Proc Natl Acad Sci U S A 98:4255–4258. https://doi.org/10.1073/pnas.061034698

Britto DT, Balkos KD, Becker A, Coskun D, Huynh WQ, Kronzucker HJ (2014) Potassium and nitrogen poising: physiological changes and biomass gains in rice and barley. Can J Plant Sci 94:1085–1089. https://doi.org/10.4141/cjps2013-143

Campos CNS, Prado RM, Caione G, Neto AJL, Mingotte (2016) Silicon and excess ammonium and nitrate in cucumber plants. Afr J Agric Res 11(4):276–283. https://doi.org/10.5897/AJAR2015.10221

Campos CNS, Silva Júnior GB, Prado RM, David CHO, Souza Junior JP, Teodoro PE (2020) Silicon mitigates ammonium toxicity in plants. Agron J 11:635–647. https://doi.org/10.1002/agj2.20069

Chaillou S, Vessey JK, Morot-gaudry JF, Raper CD, Henry LT, Boutin JP (1991) Expression of characteristics of ammonium nutrition as affected by pH of the root medium. J Exp Bot 42:189–196. https://doi.org/10.1093/jxb/42.2.189

Chen G, Guo S, Kronzucker HJ, Shi W (2013) Nitrogen use efficiency (NUE) in rice links to NH_4^+ toxicity and futile NH_4^+ cycling in roots. Plant Soil 369:351–363. https://doi.org/10.1007/s11104-012-1575-y

Coskun D, Britto DT, Kronzucker HJ (2010) Regulation and mechanism of potassium release from barley roots: an in planta $^{42}K^+$ analysis. New Phytol 188:1028–1038. https://doi.org/10.1111/j.1469-8137.2010.03436.x

Coskun D, Britto DT, Li M, Becker A, Kronzucker HJ (2013) Rapid ammonia gas transport accounts for futile transmembrane cycling under NH_3/NH_4^+ toxicity in plant roots. Plant Physiol 163:1859–1867. https://doi.org/10.1104/pp.113.225961

Coskun D, Britto DT, Kronzucker HJ (2017) The nitrogen–potassium intersection: membranes, metabolism, and mechanism. Plant Cell Environ 40:2029–2041. https://doi.org/10.1111/pce.12671

Cruz C, Bio AFM, Domínguez-Valdivia MD, Aparicio-Tejo PM, Lamsfus C, Martins-Loução MA (2006) How does glutamine synthetase activity determine plant tolerance to ammonium? Planta 223:1068–1080. https://doi.org/10.1007/s00425-005-0155-2

Das K, Roychoudhury A (2014) Reactive oxygen species (ROS) and response of antioxidants as ROS-scavengers during environmental stress in plants. Front Environ Sci 2:53. https://doi.org/10.3389/fenvs.2014.00053

Di D, Sun L, Wang M, Wu J, Kronzucker HJ, Fang S, Chu J, Shi W, Li G (2021) WRKY46 promotes ammonium tolerance in Arabidopsis by repressing NUDX9 and indole-3-acetic acid-conjugating genes and by inhibiting ammonium efflux in the root elongation zone. New Phytol 232:190–207. https://doi.org/10.1111/nph.17554

Duan F, Giehl RFH, Geldner N, Salt DE, von Wirén N (2018) Root zone–specific localization of AMTs determines ammonium transport pathways and nitrogen allocation to shoots. PLoS Biol 16:2006024. https://doi.org/10.1371/journal.pbio.2006024

Epstein E (2009) Silicon: its manifold roles in plants. Ann Appl Biol 5:155–160. https://doi.org/10.1111/j.1744-7348.2009.00343.x

Erisman JW, Sutton MA, Galloway J, Klimont Z, Winiwarter W (2008) How a century of ammonia synthesis changed the world. Nat Geosci 1:636–639. https://doi.org/10.1038/ngeo325

Esteban R, Ariz I, Cruz C, Moran JF (2016) Review: mechanisms of ammonium toxicity and the quest for tolerance. Plant Sci 248:92–101. https://doi.org/10.1016/j.plantsci.2016.04.008

Fang G, Yang J, Sun T, Wang X, Li Y (2021) Evidence that synergism between potassium and nitrate enhances the alleviation of ammonium toxicity in rice seedling roots. PLoS One 16:0248796. https://doi.org/10.1371/journal.pone.0248796

Ferreira Barreto R, Schiavon Júnior AA, Maggio MA, de Mello Prado R (2017) Silicon alleviates ammonium toxicity in cauliflower and in broccoli. Sci Hortic 225:743–750. https://doi.org/10.1016/j.scienta.2017.08.014

Forde BG, Clarkson DT (1999) Nitrate and ammonium nutrition of plants: physiological and molecular perspectives. Adv Bot Res 30:1–90

Foyer CH (2018) Reactive oxygen species, oxidative signaling and the regulation of photosynthesis. Environ Exp Bot 154:134–142. https://doi.org/10.1016/j.envexpbot.2018.05.003

Gao Q, Wang Y, Lu X (2014) Effects of exogenous silicon on physiological characteristics of cucumber seedlings under ammonium stress. J Appl Ecol 25:1395–1400

Garnica M, Houdusse F, Zamarreño AM, Garcia-Mina JM (2010) Nitrate modifies the assimilation pattern of ammonium and urea in wheat seedlings. J Sci Food Agric 90:357–369. https://doi.org/10.1002/jsfa.3811

Giehl RFH, Laginha AM, Duan F, Rentsch D, Yuan L, von Wirén N (2017) A critical role of AMT2;1 in root-to-shoot translocation of ammonium in arabidopsis. Mol Plant 10:1449–1460. https://doi.org/10.1016/j.molp.2017.10.001

Głazowska S, Baldwin L, Mravec J, Bukh C, Hansen TH, Jensen MM, Fangel JU, Willats WGT, Glasius M, Felby C et al (2018) The impact of silicon on cell wall composition and enzymatic saccharification of *Brachypodium distachyon*. Biotechnol Biofuels 11:171. https://doi.org/10.1186/s13068-018-1166-0

Głazowska S, Baldwin L, Mravec J, Bukh C, Fangel JU, Willats WG, Schjoerring JK (2019) The source of inorganic nitrogen has distinct effects on cell wall composition in *Brachypodium distachyon*. J Exp Bot 70:6461–6473. https://doi.org/10.1093/jxb/erz388

Guan M, de Bang TC, Pedersen C, Schjoerring JK (2016) Cytosolic glutamine synthetase Gln1;2 is the main isozyme contributing to GS1 activity and can be up-regulated to relieve ammonium toxicity. Plant Physiol 171:1921–1933. https://doi.org/10.1104/pp.16.01195

Guether M, Neuhäuser B, Balestrini R, Dynowski M, Ludewig U, Bonfante P (2009) A mycorrhizal-specific ammonium transporter from *Lotus japonicus* acquires nitrogen released by arbuscular mycorrhizal fungi. Plant Physiol 150:73–83. https://doi.org/10.1104/pp.109.136390

Hachiya T, Watanabe CK, Fujimoto M, Ishikawa T, Takahara K, Kawai-Yamada M, Uchimiya H, Uesono Y, Terashima I, Noguchi K (2012) Nitrate addition alleviates ammonium toxicity without lessening ammonium accumulation, organic acid depletion and inorganic cation depletion in *Arabidopsis thaliana* shoots. Plant Cell Physiol 53:577–591. https://doi.org/10.1093/pcp/pcs012

Hachiya T, Inaba J, Wakazaki M, Sato M, Toyooka K, Miyagi A, Kawai-Yamada M, Sugiura D, Nakagawa T, Kiba T et al (2021) Excessive ammonium assimilation by plastidic glutamine synthetase causes ammonium toxicity in *Arabidopsis thaliana*. Nat Commun 12:1–10. https://doi.org/10.1038/s41467-021-25238-7

He CW, Wang LJ, Liu J, Liu X, Li XL, Ma J, Lin YJ, Xu FS (2013) Evidence for 'silicon' within the cell walls of suspension-cultured rice cells. New Phytol 200:700–709. https://doi.org/10.1111/nph.12401

He C, Ma J, Wang L (2015) A hemicellulose-bound form of silicon with potential to improve the mechanical properties and regeneration of the cell wall of rice. New Phytol 206:1051–1062. https://doi.org/10.1111/nph.13282

Hoopen FT, Cuin TA, Pedas P, Hegelund JN, Shabala S, Schjoerring JK, Jahn TP (2010) Competition between uptake of ammonium and potassium in barley and Arabidopsis roots: molecular mechanisms and physiological consequences. J Exp Bot 61:2303–2315. https://doi.org/10.1093/jxb/erq057

Inoue Y, Suzuki T, Hattori M, Yoshimoto K, Ohsumi Y, Moriyasu Y (2006) *AtATG* genes, homologs of yeast autophagy genes, are involved in constitutive autophagy in arabidopsis root tip cells. Plant Cell Physiol 47:1641–1652. https://doi.org/10.1093/pcp/pcl031

Ivashikina N, Becker D, Ache P, Meyerhoff O, Felle HH, Hedrich R (2001) K^+ channel profile and electrical properties of Arabidopsis root hairs. FEBS Lett 508:463–469. https://doi.org/10.1016/S0014-5793(01)03114-3

Jadid N, Mialoundama A, Heintz D, Ayoub D, Erhardt M, Mutterer J, Meyer D, Alioua A, van Dorsselaer A, Rahier A et al (2011) Dolichol phosphate mannose synthase 1 mediates the biogenesis of isoprenyl-linked glycans and influences development, stress response, and ammonium hypersensitivity in Arabidopsis. Plant Cell 23:1985–2005. https://doi.org/10.1105/2Ftpc.111.083634

Jae SK, Frank J, Chang HK, Kajiura H, Vikram M, Ueda A, Kim S, Jeong DB, Triplett B, Fujiyama K et al (2008) Salt tolerance of *Arabidopsis thaliana* requires maturation of N-glycosylated proteins in the Golgi apparatus. Proc Natl Acad Sci U S A 105:5933–5938. https://doi.org/10.1073/pnas.0800237105

Jahn TP, Møller ALB, Zeuthen T, Holm LM, Klærke DA, Mohsin B, Kühlbrandt W, Schjoerring JK (2004) Aquaporin homologues in plants and mammals transport ammonia. FEBS Lett 574:31–36. https://doi.org/10.1016/j.febslet.2004.08.004

Jian S, Liao Q, Song H, Liu Q, Lepo JE, Guan C, Zhang J, Ismail AM, Zhang Z (2018) NRT1.1-related NH_4^+ toxicity is associated with a disturbed balance between NH_4^+ uptake and assimilation. Plant Physiol 178:1473–1488. https://doi.org/10.1104/pp.18.00410

Kafkafi U, Ganmore-Neumann R (1997) Ammonium in plant tissue: real or artifact? J Plant Nutr 20:107–118. https://doi.org/10.1080/01904169709365237

Kempinski CF, Haffar R, Barth C (2011) Toward the mechanism of NH_4^+ sensitivity mediated by Arabidopsis GDP-mannose pyrophosphorylase. Plant Cell Environ 34:847–858. https://doi.org/10.1111/j.1365-3040.2011.02290.x

Kim YH, Khan AL, Waqas M, Lee IJ (2017) Silicon regulates antioxidant activities of crop plants under abiotic-induced oxidative stress: a review. Front Plant Sci 8:510. https://doi.org/10.3389/fpls.2017.00510

Kirscht A, Kaptan SS, Bienert GP, Chaumont F, Nissen P, de Groot BL, Kjellbom P, Gourdon P, Johanson U (2016) Crystal structure of an ammonia-permeable aquaporin. PLoS Biol 14:1002411. https://doi.org/10.1371/journal.pbio.1002411

Kosegarten H, Grolig F, Wieneke J, Wilson G, Hoffmann B (1997) Differential ammonia-elicited changes of cytosolic pH in root hair cells of rice and maize as monitored by 2′,7′-bis-(2-carboxyethyl)-5 (and -6)-carboxyfluorescein-fluorescence ratio. Plant Physiol 113:451–461. https://doi.org/10.1104/2Fpp.113.2.451

Kronzucker HJ, Yaeesh Siddiqi M, Glass AD (1995) Kinetics of NO_3^- influx in spruce. Plant Physiol 109:319–326. https://doi.org/10.1104/pp.109.1.319

Kronzucker HJ, Siddiqi MY, Glass ADM, Kirk GJD (1999) Nitrate-ammonium synergism in rice. A subcellular flux analysis. Plant Physiol 119:1041–1046. https://doi.org/10.1104/2Fpp.119.3.1041

Kronzucker HJ, Britto DT, Davenport RJ, Tester M (2001) Ammonium toxicity and the real cost of transport. Trends Plant Sci 6:335–337. https://doi.org/10.1016/S1360-1385(01)02022-2

Kumar A, Kaiser BN, Siddiqi MY, Glass ADM (2006) Functional characterisation of *OsAMT1.1* overexpression lines of rice, *Oryza sativa*. Funct Plant Biol 33:339–346. https://doi.org/10.1071/fp05268

Lanquar V, Loqué D, Hörmann F, Yuan L, Bohner A, Engelsberger WR, Lalonde S, Schulze WX, von Wirén N, Frommer WB (2009) Feedback inhibition of ammonium uptake by a phospho-dependent allosteric mechanism in arabidopsis. Plant Cell 21:3610–3622. https://doi.org/10.1105/2Ftpc.109.068593

Lasa B, Frechilla S, Lamsfus C, Aparicio-Tejo PM (2001) The sensitivity to ammonium nutrition is related to nitrogen accumulation. Sci Hortic 91:143–152. https://doi.org/10.1016/S0304-4238(01)00239-4

Lasa B, Frechilla S, Aparicio-Tejo PM, Lamsfus C (2002) Role of glutamate dehydrogenase and phosphoenolpyruvate carboxylase activity in ammonium nutrition tolerance in roots. Plant Physiol Biochem 40:969–976. https://doi.org/10.1016/S0981-9428(02)01451-1

Lee Y, Rubio MC, Alassimone J, Geldner N (2013) A mechanism for localized lignin deposition in the endodermis. Cell 153:402–412. https://doi.org/10.1016/j.cell.2013.02.045

Li B, Merrick M, Li S, Li H, Zhu S, Shi W, Su Y (2009) Molecular basis and regulation of ammonium transporter in rice. Rice Sci 16:314–322. https://doi.org/10.1016/S1672-6308(08)60096-7

Li Q, Li B-H, Kronzucker HJ, Shi W-M (2010) Root growth inhibition by NH_4^+ in Arabidopsis is mediated by the root tip and is linked to NH_4^+ efflux and GMPase activity. Plant Cell Environ 33:1529–1542. https://doi.org/10.1111/j.1365-3040.2010.02162.x

Li Y, Zhou J, Hao D, Yang S, Su Y (2020) Arabidopsis under ammonium over-supply: characteristics of ammonium toxicity in relation to the activity of ammonium transporters. Pedosphere 30:314–325. https://doi.org/10.1016/S1002-0160(20)60011-X

Liang G, Song H, Xiao Y, Zhang Z (2020) Ammonium accumulation caused by reduced tonoplast v-ATPase activity in *Arabidopsis thaliana*. Int J Mol Sci 22:2. https://doi.org/10.3390/ijms22010002

Liu Y, Yu Y, Sun J, Cao Q, Tang Z, Liu M, Xu T, Ma D, Li Z, Sun J (2019) Root-zone-specific sensitivity of K^+-and Ca^{2+}-permeable channels to H_2O_2 determines ion homeostasis in salinized diploid and hexaploid *Ipomoea trifida*. J Exp Bot 70:1389–1405. https://doi.org/10.1093/jxb/ery461

Loqué D, von Wirén N (2004) Regulatory levels for the transport of ammonium in plant roots. J Exp Bot 55:1293–1305. https://doi.org/10.1093/jxb/erh147

Loqué D, Ludewig U, Yuan L, Von Wirén N (2005) Tonoplast intrinsic proteins AtTIP2;1 and AtTIP2;3 facilitate NH_3 transport into the vacuole. Plant Physiol 137:671–680. https://doi.org/10.1104/pp.104.051268

Loqué D, Yuan L, Kojima S, Gojon A, Wirth J, Gazzarrini S, Ishiyama K, Takahashi H, Von Wirén N (2006) Additive contribution of AMT1;1 and AMT1;3 to high-affinity ammonium uptake across the plasma membrane of nitrogen-deficient Arabidopsis roots. Plant J 48:522–534. https://doi.org/10.1111/j.1365-313x.2006.02887.x

Loqué D, Lalonde S, Looger LL, von Wirén N, Frommer WB (2007) A cytosolic trans-activation domain essential for ammonium uptake. Nature 446:195–198. https://doi.org/10.1038/nature05579

Loqué D, Mora SI, Andrade SLA, Pantoja O, Frommer WB (2009) Pore mutations in ammonium transporter AMT1 with increased electrogenic ammonium transport activity. J Biol Chem 284:24988–24995. https://doi.org/10.1074/2Fjbc.M109.020842

Ludewig U, Neuhäuser B, Dynowski M (2007) Molecular mechanisms of ammonium transport and accumulation in plants. FEBS Lett 581:2301–2308. https://doi.org/10.1016/j.febslet.2007.03.034

Lukowitz W, Nickle TC, Meinke DW, Last RL, Conklin PL, Somerville CR (2001) Arabidopsis *cyt1* mutants are deficient in a mannose-1-phosphate guanylyltransferase and point to a requirement of N-linked glycosylation for cellulose biosynthesis. Proc Natl Acad Sci U S A 98:2262–2267. https://doi.org/10.1073/pnas.051625798

Luxová M (1988) The participation of the primary maize root on the assimilation of NH_4^+ ions. Plant Soil 111:187–189. https://www.jstor.org/stable/42937654

Luyckx M, Hausman JF, Lutts S, Guerriero G (2017) Impact of silicon in plant biomass production: focus on bast fibres, hypotheses, and perspectives. Plants 6:37. https://doi.org/10.3390/2Fplants6030037

Ma JF, Tamai K, Yamaji N, Mitani N, Konishi S, Katsuhara M, Ishiguro M, Murata Y, Yano M (2006) A silicon transporter in rice. Nature 440:688–691. https://doi.org/10.1038/nature04590

Ma J, Sheng HC, Li XL, Wang LJ (2016a) iTRAQ-based proteomic analysis reveals the mechanisms of silicon-mediated cadmium tolerance in rice (*Oryza sativa*) cells. Plant Physiol Biochem 104:71–80. https://doi.org/10.1016/j.plaphy.2016.03.024

Ma X, Zhu C, Yang N, Gan L, Xia K (2016b) γ-Aminobutyric acid addition alleviates ammonium toxicity by limiting ammonium accumulation in rice (*Oryza sativa*) seedlings. Physiol Plant 158:389–401. https://doi.org/10.1111/ppl.12473

Magalhäes JR, Huber DM (1989) Ammonium assimilation in different plant species as affected by nitrogen form and pH control in solution culture. Fertil Res 21:1–6. https://doi.org/10.1007/BF01054728

Magalhães JR, Huber DM, Tsai CY (1992) Evidence of increased ^{15}N-ammonium assimilation in tomato plants with exogenous α-ketoglutarate. Plant Sci 85:135–141. https://doi.org/10.1016/0168-9452(92)90108-X

Maksimovic JD, Mojovic M, Maksimovic V, Römheld V, Nikolic M (2012) Silicon ameliorates manganese toxicity in cucumber by decreasing hydroxyl radical accumulation in the leaf apoplast. J Exp Bot 63:2411–2420. https://doi.org/10.1093/jxb/err359

Mandlik R, Thakral V, Raturi G, Shinde S, Nikolić M, Tripathi DK, Sonah H, Deshmukh R (2020) Significance of silicon uptake, transport, and deposition in plants. J Exp Bot 71:6703–6718. https://doi.org/10.1093/jxb/eraa301

Marini A-M, Vissers S, Urrestarazu A, André B (1994) Cloning and expression of the MEP1 gene encoding an ammonium transporter in *Saccharomyces cerevisiae*. EMBO J 13:3456–3463. https://doi.org/10.1002/2Fj.1460-2075.1994.tb06651.x

Marini A-M, Soussi-Boudekou S, Vissers S, Andre B (1997) A family of ammonium transporters in *Saccharomyces cerevisiae*. Mol Cell Biol 17:4282–4293. https://doi.org/10.1128/mcb.17.8.4282

Markovich O, Steiner E, Kouril S, Tarkowski P, Aharoni A, Elbaum R (2017) Silicon promotes cytokinin biosynthesis and delays senescence in Arabidopsis and Sorghum. Plant Cell Environ 40:1189–1196. https://doi.org/10.1111/pce.12913

Marty F (1978) Cytochemical studies on GERL, provacuoles, and vacuoles in root meristematic cells of Euphorbia. Proc Natl Acad Sci U S A 75:852–856. https://doi.org/10.1073/2Fpnas.75.2.852

Mattsson M, Schjoerring JK (2002) Dynamic and steady-state responses of inorganic nitrogen pools and NH_3 exchange in leaves of *Lolium perenne* and *Bromus erectus* to changes in root nitrogen supply. Plant Physiol 128:742–750. https://doi.org/10.1104/2Fpp.010602

Mayer M, Ludewig U (2006) Role of AMT1;1 in NH_4^+ acquisition in *Arabidopsis thaliana*. Plant Biol 8:522–528. https://doi.org/10.1055/s-2006-923877

Mayer M, Dynowski M, Ludewig U (2006a) Ammonium ion transport by the AMT/Rh homologue LeAMT1;1. Biochem J 396:431–437. https://doi.org/10.1042/2FBJ20060051

Mayer M, Schaaf G, Mouro I, Lopez C, Colin Y, Neumann P, Cartron J, Ludewig U (2006b) Different transport mechanisms in plant and human amt/rh-type ammonium transporters. Currents 127:142–143. https://doi.org/10.1085/jgp.200509369

McMahon HT, Gallop JL (2005) Membrane curvature and mechanisms of dynamic cell membrane remodelling. Nature 438:590–596. https://doi.org/10.1038/nature04396

Meier M, Liu Y, Lay-Pruitt KS, Takahashi H, von Wirén N (2020) Auxin-mediated root branching is determined by the form of available nitrogen. Nat Plants 6:1136–1145. https://doi.org/10.1038/s41477-020-00756-2

Miller AJ, Cramer MD (2004) Root nitrogen acquisition and assimilation. Plant Soil 274:1–36. https://doi.org/10.1007/s11104-004-0965-1

Miller AJ, Cookson SJ, Smith SJ, Wells DM (2001) The use of microelectrodes to investigate compartmentation and the transport of metabolized inorganic ions in plants. J Exp Bot 52:541–549. https://pubmed.ncbi.nlm.nih.gov/11373303/

Murányi A, Seeling B, Ladewig E, Jungk A (1994) Acidification in the rhizosphere of rape seedlings and in bulk soil by nitrification and ammonium uptake. Z Pflanzenernähr Bodenkd 157:61–65. https://doi.org/10.1002/jpln.19941570111

Neuhäuser B, Dynowski M, Mayer M, Ludewig U (2007) Regulation of NH_4^+ transport by essential cross talk between AMT monomers through the carboxyl tails. Plant Physiol 143:1651–1659. https://doi.org/10.1104/pp.106.094243

Neuhäuser B, Dynowski M, Ludewig U (2009) Channel-like NH_3 flux by ammonium transporter AtAMT2. FEBS Lett 583:2833–2838. https://doi.org/10.1016/j.febslet.2009.07.039

Ninnemann O, Jauniaux J-C, Frommer WB (1994) Identification of a high affinity NH_4^+ transporter from plants. EMBO J 13:3464–3471. https://doi.org/10.1002/2Fj.1460-2075.1994.tb06652.x

Nocito FF, Sacchi GA, Cocucci M (2002) Membrane depolarization induces K^+ efflux from subapical maize root segments. New Phytol 154:45–51. https://doi.org/10.1046/j.1469-8137.2002.00369.x

Olivera Viciedo D, de Mello PR, Lizcano Toledo R, Salas Aguilar D, dos Santos LCN, Calero Hurtado A, Peña Calzada K, Betancourt Aguilar C (2020) Physiological role of silicon in radish seedlings under ammonium toxicity. J Sci Food Agric 100:5637–5644. https://doi.org/10.21203/rs.3.rs-1393734/v1

Ortiz-Ramirez C, Mora SI, Trejo J, Pantoja O (2011) PvAMT1;1, a highly selective ammonium transporter that functions as H^+/NH_4^+ symporter. J Biol Chem 286:31113–31122. https://doi.org/10.1074/jbc.m111.261693

Patterson K, Cakmak T, Cooper A, Lager I, Rasmusson AG, Escobar MA (2010) Distinct signalling pathways and transcriptome response signatures differentiate ammonium- and nitrate-supplied plants. Plant Cell Environ 33:1486–1501. https://doi.org/10.1111/j.1365-3040.2010.02158.x

Pavlovic J, Kostic L, Bosnic P, Kirkby EA, Nikolic M (2021) Interactions of silicon with essential and beneficial elements in plants. Front Plant Sci 12:1224. https://doi.org/10.3389/fpls.2021.697592

Podgórska A, Burian M, Gieczewska K, Ostaszewska-Bugajska M, Zebrowski J, Solecka D, Szal B (2017) Altered cell wall plasticity can restrict plant growth under ammonium nutrition. Front Plant Sci 8. https://doi.org/10.3389/fpls.2017.01344

Pu J, Wang L, Zhang W, Ma J, Zhang X, Putnis CV (2021) Organically-bound silicon enhances resistance to enzymatic degradation and nanomechanical properties of rice plant cell walls. Carbohydr Polym 266:118057. https://doi.org/10.1016/j.carbpol.2021.118057

Qin C, Qian W, Wang W, Wu Y, Yu C, Jiang X, Wang D, Wu P (2008) GDP-mannose pyrophosphorylase is a genetic determinant of ammonium sensitivity in *Arabidopsis thaliana*. Proc Natl Acad Sci U S A 105:18308–18313. https://doi.org/10.1073/pnas.0806168105

Radotić K, Djikanović D, Kalauzi A, Tanasijević G, Maksimović V, Dragišić Maksimović J (2022) Influence of silicon on polymerization process during lignin synthesis. Implications for cell wall properties. Int J Biol Macromol 198:168–174. https://doi.org/10.1016/j.ijbiomac.2021.12.143

Robert G, Yagyu M, Koizumi T, Naya L, Masclaux-Daubresse C, Yoshimoto K (2021a) Ammonium stress increases microautophagic activity while impairing macroautophagic flux in Arabidopsis roots. Plant J 105:1083–1097. https://doi.org/10.1111/tpj.15091

Robert G, Yagyu M, Lascano HR, Masclaux-Daubresse C, Yoshimoto K (2021b) A proposed role for endomembrane trafficking processes in regulating tonoplast content and vacuole dynamics under ammonium stress conditions in Arabidopsis root cells. Plant Signal Behav 16:1924977. https://doi.org/10.1080/15592324.2021.1924977

Rogalla H, Römheld V (2002) Effects of silicon on the availability of boron: possible effects on the phenol pathway and on the redox status in *Cucumis sativus* L. In: Goldbach H et al (eds) Boron nutrition in animals and plants. Kluwer Plenum Academic Publishers, pp 205–213. https://doi.org/10.1007/978-1-4615-0607-2_19

Roosta HR, Schjoerring JK (2008a) Root carbon enrichment alleviates ammonium toxicity in cucumber plants. J Plant Nutr 31:941–958. https://doi.org/10.1080/01904160802043270

Roosta HR, Schjoerring JK (2008b) Effects of nitrate and potassium on ammonium toxicity in cucumber plants. J Plant Nutr 31:1270–1283. https://doi.org/10.1080/01904160802135050

Royo B, Esteban R, Buezo J, Santamaría E, Fernández-Irigoyen J, Becker D, Moran JF (2019) The proteome of *Medicago truncatula* in response to ammonium and urea nutrition reveals the role of membrane proteins and enzymes of root lignification. Environ Exp Bot 162:168–180. https://doi.org/10.1016/j.envexpbot.2019.02.010

Sarasketa A, González-Moro MB, González-Murua C, Marino D (2016) Nitrogen source and external medium pH interaction differentially affects root and shoot metabolism in Arabidopsis. Front Plant Sci 7:29. https://doi.org/10.3389/fpls.2016.00029

Schjoerring JK, Husted S, Mäck G, Mattsson M (2002) The regulation of ammonium translocation in plants. J Exp Bot 53:883–890. https://doi.org/10.1093/jexbot/53.370.883

Shabala L, Zhang J, Pottosin I, Bose J, Zhu M, Fuglsang AT, Velarde-Buendia A, Massart A, Hill CB, Roessner U et al (2016) Cell-type-specific H^+ -ATPase activity in root tissues enables k^+ retention and mediates acclimation of barley (*Hordeum vulgare*) to salinity stress. Plant Physiol 172:2445–2458. https://doi.org/10.1104/pp.16.01347

Sharma P, Jha AB, Dubey RS, Pessarakli M (2012) Reactive oxygen species, oxidative damage, and antioxidative defense mechanism in plants under stressful conditions. J Bot 2012:217037. https://doi.org/10.1155/2012/217037

Sheng H, Ma J, Pu J, Wang L (2018) Cell wall-bound silicon optimizes ammonium uptake and metabolism in rice cells. Ann Bot 122:303–313. https://doi.org/10.1093/aob/mcy068

Silva Campos CN, de Mello Prado R, Garcia Roque C, de Lima Neto AJ, Pereira Marques LJ, Peixoto Chaves A, Cruz CA (2015) Use of silicon in mitigating ammonium toxicity in maize plants. Am J Plant Sci 06:1780–1784. https://doi.org/10.4236/ajps.2015.611178

Silva Júnior GB (2015) Ammonium and nitrate ratio, mitigation of ammonia toxicity with silicon and nutrient accumulation curve in passion fruit seedlings. Ph.D. thesis. São Paulo State University. http://hdl.handle.net/11449/136064

Silva Júnior GB, Prado RM, Campos CNS, Agostinho FB, Silva SLO, Santos LC, González LC (2019) Silicon mitigates ammonium toxicity in yellow passion fruit seedlings. Chil J Agric Res 79:425–434. https://doi.org/10.4067/S0718-58392019000300425

Silva Junior GB, Prado RM, Silva SLO, Campos CNS, Castellanos LC, Santos LCN, Barreto RF, Teodoro PE (2020) Nitrogen concentrations and proportions of ammonium and nitrate in the nutrition and growth of yellow passion fruit seedlings. J Plant Nutr 43:2533–2547. https://doi.org/10.1080/01904167.2020.1783299

Song J, Yang J, Jeong BR (2022) Silicon mitigates ammonium toxicity in cabbage (*Brassica campestris* L. ssp. pekinensis) 'Ssamchu'. Front Sustain Food Syst 6:245. https://doi.org/10.3389/fsufs.2022.922666

Straub T, Ludewig U, Neuhäuser B (2017) The Kinase CIPK23 inhibits ammonium transport in *Arabidopsis thaliana*. Plant Cell 29:409–422. https://doi.org/10.1105/tpc.16.00806

Subbarao GV, Searchinger TD (2021) Opinion: a "more ammonium solution" to mitigate nitrogen pollution and boost crop yields. Proc Natl Acad Sci 118:2107576118. https://doi.org/10.1073/pnas.2107576118

Sutton MA, Bleeker A, Howard C, Bekunda M, Grizzetti B, de Vries W, van Grinsven HJM, Abrol YP, Adhya TK, Billen G et al (2013) Our nutrient world: the challenge to produce more food and energy with less pollution. https://nora.nerc.ac.uk/id/eprint/500700/1/N500700BK.pdf

Szczerba MW, Britto DT, Ali SA, Balkos KD, Kronzucker HJ (2008a) NH_4^+-stimulated and -inhibited components of K^+ transport in rice (*Oryza sativa* L.). J Exp Bot 59:3415–3423. https://doi.org/10.1093/2Fjxb/2Fern190

Szczerba MW, Britto DT, Balkos KD, Kronzucker HJ (2008b) Alleviation of rapid, futile ammonium cycling at the plasma membrane by potassium reveals K^+-sensitive and -insensitive components of NH_4^+ transport. J Exp Bot 59:303–313. https://doi.org/10.1093/jxb/erm309

Tanaka H, Maruta T, Ogawa T, Tanabe N, Tamoi M, Yoshimura K, Shigeoka S (2015) Identification and characterization of Arabidopsis AtNUDX9 as a GDP-d-mannose pyrophosphohydrolase: its involvement in root growth inhibition in response to ammonium. J Exp Bot 16:5797–5808. https://doi.org/10.1093/2Fjxb/2Ferv281

Tripathi DK, Singh S, Singh VP, Prasad SM, Dubey NK, Chauhan DK (2017) Silicon nanoparticles more effectively alleviated UV-B stress than silicon in wheat (*Triticum aestivum*) seedlings. Plant Physiol Biochem 110:70–81. https://doi.org/10.1016/j.plaphy.2016.06.026

Tsukagoshi H (2016) Control of root growth and development by reactive oxygen species. Curr Opin Plant Biol 29:57–63. https://doi.org/10.1016/j.pbi.2015.10.012

Van Beusichem ML, Kirkby EA, Baas R (1988) Influence of nitrate and ammonium nutrition on the uptake, assimilation, and distribution of nutrients in *Ricinus communis*. Plant Physiol 86:914–921. https://doi.org/10.1104/pp.86.3.914

Viciedo DO, de Mello Prado R, Lizcano Toledo R, dos Santos LCN, Calero Hurtado A, Nedd LLT, Castellanos Gonzalez L (2019) Silicon supplementation alleviates ammonium toxicity in sugar beet (*Beta vulgaris* L.). J Soil Sci Plant Nutr 19:413–419. https://doi.org/10.1007/s42729-019-00043-w

Wacker T, Garcia-Celma JJ, Lewe P, Andrade SLA (2014) Direct observation of electrogenic NH_4^+ transport in ammonium transport (Amt) proteins. Proc Natl Acad Sci U S A 111:9995–10000. https://doi.org/10.1073/pnas.1406409111

Wang MY, Glass ADM, Shaff JE, Kochian LV (1994) Ammonium uptake by rice roots: III. Electrophysiology. Plant Physiol 104:899–906. https://doi.org/10.1104/pp.104.3.899

Wang S, Orabi EA, Baday S, Bernèche S, Lamoureux G (2012) Ammonium transporters achieve charge transfer by fragmenting their substrate. J Am Chem Soc 134:10419–10427. https://doi.org/10.1021/ja300129x

Weng L, Zhang M, Wang K, Chen G, Ding M, Yuan W, Zhu Y, Xu W, Xu F (2020) Potassium alleviates ammonium toxicity in rice by reducing its uptake through activation of plasma membrane H^+-ATPase to enhance proton extrusion. Plant Physiol Biochem 151:429–437. https://doi.org/10.1016/j.plaphy.2020.03.040

Williamson G, Tamburrino G, Bizior A, Boeckstaens M, Mirandela GD, Bage M, Pisliakov A, Ives CM, Terras E, Hoskisson PA et al (2020) A two-lane mechanism for selective biological ammonium transport. elife 9:1–41. https://doi.org/10.7554/eLife.57183

Yang S, Hao D, Cong Y, Jin M, Su Y (2015) The rice OsAMT1;1 is a proton-independent feedback regulated ammonium transporter. Plant Cell Rep 34:321–330. https://doi.org/10.1007/s00299-014-1709-1

Yang S, Hao D, Jin M, Li Y, Liu Z, Huang Y, Chen T, Su Y (2020) Internal ammonium excess induces ROS-mediated reactions and causes carbon scarcity in rice. BMC Plant Biol 20:143. https://doi.org/10.1186/s12870-020-02363-x

Yang D, Zhao J, Bi C, Li L, Wang Z (2022) Transcriptome and proteomics analysis of wheat seedling roots reveals that increasing NH_4^+/NO_3^- ratio induced root lignification and reduced nitrogen utilization. Front Plant Sci 12:3251. https://doi.org/10.3389/fpls.2021.797260

Yano K, Hattori M, Moriyasu Y (2007a) A novel type of autophagy occurs together with vacuole genesis in miniprotoplasts prepared from tobacco culture cells. Autophagy 3:215–221. https://doi.org/10.4161/auto.3739

Yano K, Suzuki T, Moriyasu Y (2007b) Constitutive autophagy in plant root cells. Autophagy 3:360–362. https://doi.org/10.4161/auto.4158

Yuan WM, Siddiqi MY, Ruth TJ, Glass ADM (1993) Ammonium uptake by rice roots. II. Kinetics of $^{13}NH_4^+$ influx across the plasmalemma. Plant Physiol 103:1259–1267. https://doi.org/10.1104/2Fpp.103.4.1259

Yuan L, Loqué D, Kojima S, Rauch S, Ishiyama K, Inoue E, Takahashi H, Von Wirén N (2007a) The organization of high-affinity ammonium uptake in Arabidopsis roots depends on the spatial arrangement and biochemical properties of AMT1-type transporters. Plant Cell 19:2636–2652. https://doi.org/10.1105/tpc.107.052134

Yuan Y, Ou J, Wang Z, Zhang C, Zhou Z, Lin Q (2007b) Regulation of carbon and nitrogen metabolisms in rice roots by 2-oxoglutarate at the level of hexokinase. Physiol Plant 129:296–306. https://doi.org/10.1111/j.1399-3054.2006.00806.x

Yuan L, Graff L, Loqué D, Kojima S, Tsuchiya YN, Takahashi H, von Wirén N (2009) AtAMT1;4, a pollen-specific high-affinity ammonium transporter of the plasma membrane in Arabidopsis. Plant Cell Physiol 50:13–25. https://doi.org/10.1093/pcp/pcn186

Yuan L, Gu R, Xuan Y, Smith-Valle E, Loqué D, Frommer WB, von Wirén N (2013) Allosteric regulation of transport activity by heterotrimerization of Arabidopsis ammonium transporter complexes in vivo. Plant Cell 25:974–984. https://doi.org/10.1105/tpc.112.108027

Zeng W, Ford KL, Bacic A, Heazlewood JL (2018) N-linked glycan micro-heterogeneity in glycoproteins of arabidopsis. Mol Cell Proteomics 17:413–421. https://doi.org/10.1074/mcp.ra117.000165

Zhang PY, Gao RG, Yang FJ, Wang XF, Wei M, Shi QH, Li Y (2014) Effects of silicon on photosynthetic characteristics and activity of antioxidant enzymes in continuous-cropped cucumber seedlings. J Appl Ecol 25:1733–1738. https://pubmed.ncbi.nlm.nih.gov/25223031

Zheng X, He K, Kleist T, Chen F, Luan S (2015) Anion channel SLAH3 functions in nitrate-dependent alleviation of ammonium toxicity in Arabidopsis. Plant Cell Environ 38:474–486. https://doi.org/10.1111/pce.12389

Zhu YX, Guo J, Feng R, Jia JH, Han WH, Gong HJ (2016) The regulatory role of silicon on carbohydrate metabolism in *Cucumis sativus* L. under salt stress. Plant Soil 406:231–249. https://doi.org/10.1007/s11104-016-2877-2

Chapter 14
Silicon Mitigates the Effects of Potentially Toxic Metals

Lilian Aparecida de Oliveira, Flávio José Rodrigues Cruz, Dalila Lopes da Silva, Cassio Hamilton Abreu-Junior, and Renato de Mello Prado

14.1 Introduction

Heavy metals (HM) occur naturally in soils, from geological and anthropic activities (Chibuike and Obiora 2014), and when released into the environment, they tend to accumulate in soils and become available to plants, then being absorbed through their roots (Kim et al. 2014; Yang et al. 2022). This release, when in high concentrations, can become toxic to plants and cause damage to the environment, promoting reduced growth and changes in physiological and biochemical processes, such as damage to cellular structures, increased oxidative stress, changes in enzymatic activity, changes in the dynamics of plant nutrients, and even decreased soil organic matter content (Hossain et al. 2012; Chibuike and Obiora 2014; Alsahli et al. 2021; Siddiqui et al. 2021; Singh et al. 2020).

HM are elements that exhibit metallic properties such as ductility, malleability, conductivity, cation stability, and ligand specificity (Duffus 2002; Tangahu et al. 2011). Some HM, such as cobalt (Co), copper (Cu), iron (Fe), manganese (Mn), molybdenum (Mo), nickel (Ni), selenium (Se), and zinc (Zn) are required in minimal amounts by organisms, but can become harmful in excessive amounts. Other HM, such as lead (Pb), cadmium (Cd), mercury (Hg), and arsenic (As) (a metalloid but generally referred to as a heavy metal) have no beneficial effect on organisms

L. A. de Oliveira · D. L. da Silva · C. H. Abreu-Junior (✉)
Universidade de São Paulo, Center for Nuclear Energy in Agriculture, Soil Fertility and Plant Nutrition Laboratory, Piracicaba, Brazil
e-mail: cahabreu@cena.usp.br

F. J. R. Cruz
Federal Institute of Education, Science and Technology of Amapá, Laranjal do Jari, Brazil

R. de Mello Prado
São Paulo State University, Jaboticabal, São Paulo, Brazil

R. de Mello Prado (ed.), *Benefits of Silicon in the Nutrition of Plants*,
https://doi.org/10.1007/978-3-031-26673-7_14

and are therefore considered the "major threats", as they are very harmful to both plants and animals (Chibuike and Obiora 2014; Aprile and De Bellis 2020).

Currently, HM pollution is increasing rapidly, which leads to many environmental problems. The metals are mainly accumulated in the soil and then transferred to the food chain through plants grown in these soils. Among these metals, Cd is a threat all over the world and its accumulation in grains causes a deleterious effect on human health, affecting several organs when in high concentration, causing kidney and bone dysfunctions, and even cancer. In addition, Cd is highly harmful to plant development by reducing growth and levels of photosynthetic pigments while increasing oxidative stress. Therefore, this element must have its concentration minimized in soils and consequently in plants.

Pb is also a contaminant that has caused damage to the environment worldwide. In the earth's crust, the average content of Pb is 15 mg kg^{-1}, while in soils its average content is 27 mg kg^{-1}, which result from the weathering process of the matrix rock (Kabata-Pendias 2010). However, the increase in Pb levels in soils results from human activities, such as industries, burning of fossil fuels, use of sewage sludge in agriculture, improper disposal of batteries, among others.

To mitigate and reduce the negative effects promoted by HM toxicity, studies have shown that the supply of Si, in addition to reducing the negative effects caused, also favors the growth and development of plants by increasing defense mechanisms (Emamverdian et al. 2018; Rahman et al. 2021; Meena et al. 2022). This allows the culture under this type of stress to have a greater tolerance to high concentrations of HM, through internal mechanisms such as increased antioxidant activity, complexation, and compartmentalization of Si ions with metal ions, in addition to changes in the structural components and also inhibition of the transport of metals from the roots to the aerial part of the plant (Liang et al. 2015; Bhat et al. 2019). Therefore, this chapter aims to address the mechanisms involved and the potential role of Si in the attenuation of the effects caused by the toxicity of heavy metals in cultivated plants and the importance of this element in agricultural production.

14.2 HM Stress Mitigation Mechanisms

There are some techniques to reduce the concentration of HM in the environment, namely the supply of silicon has shown beneficial effects, as the Si in the soil interacts with HM. For instance, with Cd, there is co-precipitation, complexation in the soil and/or by interactions that occur inside plants, and transport of Cd from the roots to the aerial part and/or compartmentalization within the plants with deposition associated with cell wall constituents, which can lead to changes in the expression of multiple genes, as well as in the metabolism (de Oliveira 2009).

Si has a mitigating effect on the toxicity of HM, because it inhibits the uptake of Cd, As, Pb, and Cr (Imtiaz et al. 2016) through changes in soil pH, metal speciation, or reduction in availability of HM due to coprecipitation with Si (Bhatti et al. 2018) and compartmentalization of HM in vacuoles or the cell wall (Meena et al. 2022). Other Si-induced mitigating mechanisms for the toxicity of HM are related to root exudation of phenolic compounds, such as catechin and quercetin, that alter the mobility of toxic metals in the rhizosphere (Vaculík et al. 2020).

Si induces exudation of organic acids (i.e., fumaric, tartaric, oxalic, acetic, and maleic) that act as ligands for toxic metals, reducing the mobility of HM in the rhizosphere (Wu et al. 2016). These effects on the mobility of toxic metals in the vicinity of the rhizosphere allow greater absorption of essential nutrients, improving plant growth under conditions of HM toxicity. However, although Si modulates the exudation of organic acids by the plant's root system, this mechanism can affect the availability of both HM and essential nutrients. For example, the supply of 1.5 g Si kg^{-1} soil increased the root and leaf contents of Ca, Mg, and Fe in ginger (*Zingiber officinale* Roscoe) under toxicity of 500 mg Pb kg^{-1} soil. But for Cu, Mn, and Zn, the effect of Si was not pronounced. These results were accompanied by a reduction in Pb contents in shoots and roots due to Si supplementation, which decreased Pb mobility in the rhizosphere (Chen et al. 2019). The effect of Si on plant nutritional improvement was also observed in the interaction between Si and Cr, Cd, and Cu, with a positive effect on Ca, Mg, K, and P contents (Rizwan et al. 2012; Keller et al. 2015; Tripathi et al. 2015).

Silicon reduces HM stress by providing the maintenance of some important processes for plants. At the physiological level, the supplementation of plants with Si allows gains in dry mass because Si modulates a set of physiological changes, such as the increase in number of grana, size of chloroplasts, and concentration of chlorophylls, which have a positive impact on the increase in the photosynthetic capacity of plants (Hussain et al. 2021). These positive changes modulated by Si are accompanied by an increase in the concentration of photosynthetic pigments (chlorophyll *a*, *b*, total *a* + *b*, and carotenoids) that improve photosynthetic yield and growth against toxic levels of Pb (Bharwana et al. 2014; Rahman et al. 2021), evidencing the protective effect of Si on the photosynthetic machinery, which results in plant growth gains. Chen et al. (2019) showed that the toxic effect of Pb (500 mg kg^{-1} soil) can also be minimized by Si (1.5 g kg^{-1} soil) by increasing chlorophyll and carotenoid concentration plus vegetative growth (height, diameter of stem, fresh weight of root, stem, and leaves).

In rice plants at different doses of cadmium, the dose of 0.2 mmol L^{-1} of silicon was sufficient to increase efficient water use, light efficiency, and increase photosynthetic and chlorophyll efficiency parameters (Nwugo and Huerta 2008). Still in rice plants, now supplied with 0.5 and 1.0 g kg^{-1} of Si, chlorophyll content and catalase activity increased, while cell membrane permeability decreased. Quantitative and qualitative analyses revealed that Cd and Si formed soluble complexes of Si-Cd (structural unit O–Si–O–Cd), and this form could not be absorbed by plants, being a mechanism to reduce the promotion of water contamination, through the migration of Cd to water due to its solubility; therefore, it is a strategy to reduce contamination with Cd (Guo et al. 2022b).

14.3 Effects of Silicon on Absorption, Transport, and Accumulation of HM

Studies show that Si can strongly influence the absorption, transport, and accumulation of HM from the soil to the roots and shoots of plants, according to Khan et al. (2018). For example, Cd, which is a non-accumulating HM, increases its concentration in the roots of plants that are not accumulators, or less tolerant to this element when they are in a growth medium with Si additions, since it relieves Cd phytotoxicity by inhibiting Cd transport from the roots to the shoot. This is because Si deposition occurs in the endoderm, which physically blocks the passage of Cd through the apoplast from the root to the shoot (Zhang et al. 2013; Wu et al. 2019; Malik et al. 2021), as in bean plants (Table 14.1), which can be a resistance factor to the damage that Cd could cause in the whole plant (de Oliveira 2009).

In tobacco plants, which also do not accumulate silicon, the application of Si successfully retained the Cd content in the root and reduced the Cd translocation from the roots to the shoot, and the Cd transfer factor is reduced in the treatments with addition of Si (Lu et al. 2018). This can be attributed to the fact that Si deposition physically increases the strength and rigidity of cell walls, provides binding sites for metals, and reduces apoplastic flux, thus decreasing the uptake and translocation of toxic metals from roots to shoots (Ma and Yamaji 2006). Cd and Si can, therefore, form an unknown complex compound or deposit on cell walls and, as a result, can decrease the toxicity of Cd (Farooq et al. 2013).

Another benefit worth mentioning is that, in addition to Si improving the retention of Cd in the roots, fixing it to the cell wall, Si also restricts the intracellular transport of Cd, as its main binding component is pectin, inducing the accumulation of Cd in the covalent soluble fraction of pectin and in hemicellulose, in which after a long-term exposure to Cd, Si can promote phytochelatin synthesis parallel to the regulation of protein expression related to efflux and limited transport of Cd from the vacuole to the cytosol (de Oliveira 2009).

Table 14.1 Cadmium concentration in the aerial part, roots, pods, and grains of bean plants, cultivated in a nutrient solution containing increasing doses of cadmium, in the presence of silicon (+Si) and in the absence of silicon (−Si)

Cadmium concentration in the solution (μmol L^{-1})	Aerial part		Roots		Beans		Grains	
	−Si	+Si	−Si	+Si	−Si	+Si	−Si	+Si
	Cd (mg kg^{-1})							
0.00	0.11a	0.10a	5.0b	16.9a	0.05a	0.03a	0.07a	0.03a
0.05	0.16a	0.11b	6.1b	34.8a	0.05a	0.03a	0.15a	0.03a
0.45	0.48a	0.33b	39.1b	104.6a	0.11a	0.09a	0.017a	0.07b
0.90	0.63a	0.48b	131.3b	142.2a	0.18a	0.17a	1.16a	0.13b
4.50	13.83a	9.62b	422.4b	463.5a	2.85a	0.90a	2.18a	1.12b

Adapted from de Oliveira (2009)
Means followed by distinct letters differ according to the doses of silicon to Tukey ($p < 0.05$)

Thus, Si limits Cd translocation through the root cell wall under short-term Cd exposure and promotes compartmentalization of vacuoles to mitigate long-term Cd toxicity, thereby decreasing Cd deposition in shoots (Wei et al. 2021). This effect was observed in rice, where about 64% of the total Si in the cell walls was bound to hemicellulose constituents, being responsible for the inhibition of Cd uptake in rice cells by a complexation mechanism of [Si-hemicellulose matrix] Cd and subsequent co-deposition (Ma et al. 2017).

As stated, the effect of Si on decreasing Cd uptake by roots in the apoplastic and symplastic pathways depends on the time of exposure to this element by plants (Wu et al. 2018). In wheat, short-term exposure to Cd did not present significant changes in the symplastic pathway by Si supply, however, Cd concentrations in the apoplastic root fluid were decreased (Wu et al. 2018). This effect can be attributed to the fact that Si delayed endodermal suberization of roots resulting in apoplastic translocation of Cd to the shoot, thus decreasing Cd in the apoplastic fluid of roots after short-term exposure.

Qi et al. (2020) demonstrated that Si deposited in the cell wall of the leaves and roots of this accumulating plant has extreme affinity for Cd, since silicon accumulating plants such this may show a reduction in the translocation of cadmium by apoplast, due to the fact that the diffusion process between cells is hampered, with possible isolation of Cd in root cell walls by the formation of highly insoluble colloidal silica. However, for a significant difference to occur in the decrease of Cd concentration in the roots of accumulating plants such as rice, doses lower than 1.8 mmol L^{-1} did not present positive effects (Table 14.2). This decrease in Cd uptake and translocation in rice plants can be attributed to the fact that rice root tissue has a natural barrier to prevent Cd uptake and translocation.

The benefits of Si supply under the effects of HM toxicity by reducing the uptake and transport of toxic metals to the aerial part of plants and indirectly improving physiological responses of plants, photosynthetic activity, transpiration, stomatal conductance, and water use efficiency (Sharma 2005; Sharma et al. 2012; Adrees

Table 14.2 Cadmium concentration (μmol L^{-1}) in the aerial part (stem + leaf), roots, husk, and seed of rice plants grown in a nutrient solution, in the presence of silicon (+Si) and in the absence of silicon (−Si)

Cadmium concentration in the solution (μmol L^{-1})	Aerial part		Roots		Beans		Grains	
	−Si	+Si	−Si	+Si	−Si	+Si	−Si	+Si
	Cd (mg kg^{-1})							
0.00	0.5a	1.0a	3.4a	3.3a	0.2a	0.6a	0.1a	0.3a
0.45	4.5a	2.5b	53.2a	38.0b	2.4a	1.2a	1.0a	0.7a
4.50	33.6a	16.3b	533.9a	57.0b	9.7a	6.7a	2.9a	2.1b
9.00	35.7a	17.4b	1344.1a	627.1b	13.9a	6.5a	5.6a	3.2b
22.50	102.9a	28.3b	1584.7a	1196.3b	15.7a	9.3a	6.8a	5.1b
45.00	104.9a	47.1	2247.7a	1309.0b	24.9	20.4	8.3a	7.1b

Adapted from de Oliveira (2009)
Means followed by distinct letters differ according to the doses of silicon to Tukey ($p < 0.05$)

et al. 2015) are also extensively studied under the effect of Pb toxicity. The toxicity effects of Pb are related to important physiological disorders in gas exchange, photosynthetic machinery, water status, and plant hormones, where photosynthetic activity is negatively affected by toxic levels of Pb due to changes in the anatomical structure of chloroplasts, inhibition of the synthesis of chlorophylls, plastoquinones and carotenoids, damage to the electron transport chain, inhibition of the activity of Calvin cycle enzymes, and reduction of CO_2 diffusion due to stomatal closure (Kumar and Prasad 2018).

Despite the differential accumulation of Si in plants of the Poaceae family and other plant families, Si is shown to be efficient in mitigating the negative impacts of HMs (Khan et al. 2021) and other types of abiotic stresses (Ahire et al. 2021; Mir et al. 2022). This characteristic of Si allows it to assign a prominent role in the management of plants under conditions of abiotic stress.

14.4 Antioxidant Defense Mechanisms

Plants have enzymatic and non-enzymatic antioxidant mechanisms to counterbalance the production of free radicals due to HM, and thus reduce lipid peroxidation and cellular damage triggered by HM in plants. In this sense, there is a set of antioxidant enzymes that act by neutralizing free radicals such as superoxide dismutase (SOD), catalase (CAT), ascorbate peroxidase (APX), glutathione reductase (GR), peroxidase (POD), and glutathione peroxidase (GPx).

As well as the production of reactive oxygen species (ROS), lipid peroxidation in plants subjected to stress by HM is high and can induce leaf chlorosis, plus oxidative damage to membranes, directly increasing the production of ROS. In the case of plants grown under Cu toxicity, this effect can be mitigated with the supply of Si, which reduces the degradation of the lipid membrane by regulating the production of α-linoleic acid, that makes up more than 75% of the malondialdehyde (MDA) content, because under Cu stress conditions, the plant significantly reduces the percentage of α-linoleic acid, promoting ROS increase. Therefore, there are high concentrations of α-linolenic acid in chloroplasts and mitochondria, via the production of small oxygenated compounds such as MDA, which signal and absorb the ROS generated in these organelles (Kim et al. 2014). In addition to the fact that α-linolenic acid is a precursor of jasmonic acid (JA), essential in signaling and stress defenses.

Another factor that may be affecting the integrity of the cell wall is the effect of oxidative stress, as in this condition there is an imbalance between the production of ROS, which at high concentrations in the cellular environment can impose oxidative damage on structures and cellular functions (Choudhury et al. 2013). Cd may not generate free radicals directly, but may increase lipid peroxidation, contributing to the process of cell damage (Gratão et al. 2006).

Si stimulates the antioxidant system in plants (Zhang et al. 2018; Kim et al. 2017), for example in plants under Cd toxicity, where the phytotoxic effect of Cd is related to the inhibition of antioxidant compounds that results in lipid peroxidation and generation of superoxide anions (Ma et al. 2022). This fact can greatly reduce the production of plant biomass by affecting the physiological processes. However, the application of Si promotes an increase in antioxidant activity (Zhou et al. 2022) and this leads to a reduction in the damage caused by Cd and a consequent increase in production.

This effect can be reduced in plants supplemented with Si. Plants under Cd toxicity stress supplemented with Si at concentrations of 0.3, 0.6, and 1.2 mg kg^{-1} of Cd showed an increase in the activity of antioxidants, such as peroxidases, but the activity of superoxide dismutase decreased (Chen et al. 2007).

Furthermore, it is also important to show that different phytohormones are involved in the growth of roots and other plant organs in the presence of Si. Cytokinin and auxins have been reported as producers in the meristem and pericycle of the root and participate in signaling mechanisms in response to the condition of stress (Tripathi et al. 2021). Besides, their levels are influenced by the nature of the stress, whether abiotic (salt, drought, heat, cold, HM, etc.) or biotic (pests or diseases). The production of abscisic acid, jasmonic acid, and salicylic acid increases by Cd stress on exposure to Si application, which are all involved in strengthening the cell membrane, reducing electrolyte leakage and lipid peroxidation and improving the defense system of plants under various abiotic stresses (Kim et al. 2014; Khan et al. 2020; Ma et al. 2022). Thus, the production and activities of different phytohormones can change the structure and response of roots to HM uptake.

Pb is also an HM that increases the production of free radicals in plant cells, causing oxidative stress and growth inhibition, and because it is an HM that does not show variation in its oxidation state, it does not alter the redox balance of cells and does not directly participate in the generation of free radicals (Zulfiqar et al. 2019). The production mechanism of oxidative stress triggered by Pb happens through the substitution of cations in biomolecules and alteration in the activity of enzymes containing metals and activating NADPH oxidase, thus increasing the production of free radicals (Chen et al. 2019; Shu et al. 2012; Singh et al. 2020). In this sense, the literature reports a close relationship between Pb toxicity in plants and the production of free radicals.

Rice plants from seeds soaked with 1 mM of Pb show an increase in the activity of the antioxidant enzymes SOD, POD and CAT, with an increase in the concentration of oxidative stress indicators (MDA, H_2O_2, O_2^-, OH^-) and in lipid peroxidation (Khan et al. 2018). Toxic levels of Pb (50 and 100 μM) induce an increase in the production of H_2O_2, greater electrolyte extravasation, and lipid peroxidation in cotton (Bharwana et al. 2014). Toxic levels of Pb (50 μM) also increase the activity of SOD, CAT, APX, and GPx in cotton leaves and roots, although the levels of oxidative stress indicators such as malonic aldehyde, H_2O_2, and electrolyte extravasation increased with the toxicity of applied Pb (Bharwana et al. 2014).

These results suggest that, depending on the species, level of Pb toxicity, and time of exposure of plants to the pollutant, antioxidant mechanisms may not be effective in efficiently neutralizing free radicals. However, the exogenous supply of Si maximizes the activity of antioxidant enzymes and reduces the concentration of oxidative stress indicators, suggesting that Si mitigates the harmful effects of Pb in plants (Bharwana et al. 2014; Khan et al. 2018). The mechanism of action of Si in modulating the activity of antioxidant enzymes is unclear, indicating the need for molecular studies to evidence the relationship between Si, antioxidant enzymes, and the respective genes that encode them.

14.5 Morphological Alterations

The supply of Si can positively affect the anatomical parameters of plants cultivated with high concentrations of HM, especially those involving root growth, since HM stress initially affects the root system of plants, which is in direct contact with the pollutant. HM trigger nutritional, physiological, biochemical, and anatomical disorders, in addition to changes in xylem diameter, mesophyll and epidermis thickness, and transversal area occupied by collenchyma and median vessels, which may confer greater plant resistance, as recorded in plants under the stress caused by Cd (Adrees et al. 2015; Riaz et al. 2021).

In rice plants grown under Cu toxicity and analyzed under microscope, it was observed that Si mainly influenced root growth, since the exodermis, epidermis, endodermis and cortex regions of plants treated with Si showed cells without major damage, lamella more suberized, and a greater proximity between the cells, with better development of the cell wall and no visible deformities, which were present in plants without Si, in addition to higher shoot biomass and lower rate of necrosis in leaf tissues (Kim et al. 2014).

Still in the roots of rice plants, but under Cd toxicity, some anatomical changes were triggered, such as formation of granules, probably of starch, plus laceration of epidermal cells and disruption of the cell wall (Fig. 14.1a, b). However, the presence of Si reduced this effect, as Si supplementation can apparently denature/dissolve the granules formed by excess cadmium in the roots (Fig. 14.1d), in addition to the fact that, when observing the image, we can see that supplementation with Si helps to preserve the epidermis (Fig. 14.1c), having a positive effect on maintaining the absorption of water and nutrients by rice, helping to mitigate Cd toxicity (de Oliveira 2009).

It is worth noting that the morphological alterations of the cells depend on factors such as metal concentration, exposure period, and degree of sensitivity of the species to the metal (Prasad 1995). Hence, the increase of the Cd concentration in the solution caused the deleterious effect, including on the turgor potential and on the cellular plasticity (Barceló et al. 1988).

Physiological, histochemical, and molecular biological data show that Si has an active impact on rice root anatomy and gene transcription. (Fleck et al. 2011) demonstrated that Si nutrition affects exodermis suberization and sclerenchyma lignification, as Si activates 12 genes related to suberin and lignin synthesis in sclerenchyma, including a gene encoding a leucine replication protein that exhibited a level of transcription 25 times higher with Si nutrition. The oxidizing power of rice roots was also enhanced, since the supply of Si promoted a decrease in the oxidation zone

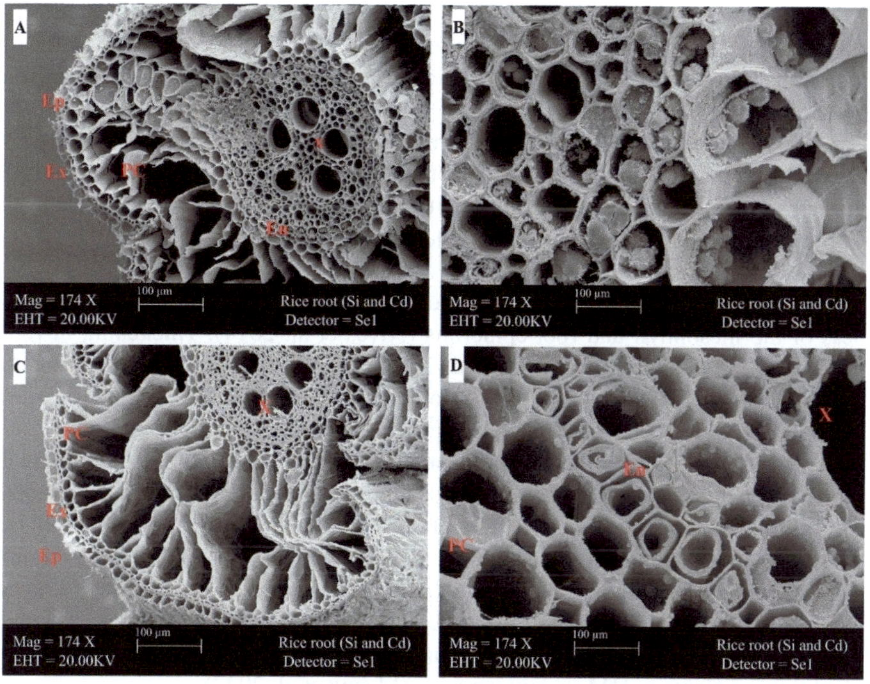

Fig. 14.1 Rice roots subjected to cross-sections by cryofracture seen in scanning electron microscopy (SEM). The roots of plants exposed to 9.0 μmol L^{-1} of cadmium (**a, b**) and of plants exposed to 9.0 μmol L^{-1} of cadmium and 1.8 mmol L^{-1} of silicon are observed (**c, d**). CP cortical parenchyma, X xylem, En endoderm, Ep epidermis, Ex exoderm

of adventitious roots from 12 to 5 cm in length along the entire length of the root (Fleck et al. 2011).

Changes mediated by the supply of Si have also been reported in the bean crop, where it was observed that Cd triggers important anatomical changes in their roots, such as changes in the number and size of xylem cells and the appearance of electron-dense granules in the xylem as a result of the HM application (Barceló et al. 1988). Changes in tomato plant anatomy have already been observed by the presence of Cd, namely changes in leaf density, cell arrangement, increase in the size of palisade parenchyma cells, decrease in the number of chloroplasts, and an increase in intercellular spaces in leaves (Gratão et al. 2006).

Cunha and Nascimento (2009), when studying the relationship between the accumulation of Si, Cd, and Zn, also verified structural changes in corn plants grown in soil enriched with cadmium and zinc, and observed that the addition of Si to the contaminated soil promoted an increase in growth parameters, indicating that phytotoxicity was reduced. The authors also verified that the addition of Si increases the xylem diameter, improving the hydraulic balance of corn plants under stress and playing a fundamental role in the production of biomass.

Pb also causes changes in both the root system and the aerial part of plants. In rice plants, Pb (200 μM) reduced root epidermis, endoderm and exoderm thickness, the aerenchyma area, root cortex thickness, and vascular cylinder diameter. The toxic effect of Pb is manifested in the leaf anatomy, although this pollutant is poorly transported from the root to the aerial part of the plant. Toxic levels of Pb (200 μM) also reduced the thickness of the adaxial and abaxial epidermis, the thickness of the mesophyll, and the diameter of the bulliform cells. All these anatomical changes were accompanied by an increase in Pb levels in the roots and leaves of the plants (Guedes et al. 2021).

In *Oxycaryum cubens* plants, toxic levels of Pb (4–32 mg L^{-1}) reduced the thickness of the mesophyll and root epidermis (Alves et al. 2014). Thickening of root tissues such as endoderm and exodermis is a strategy to reduce Pb translocation to the aerial part of plants and attenuate its toxic effects (Gomes et al. 2011). However, this strategy varies between plant genotypes and HM types.

A considerable part of the HM in plants is present in the sap of the apoplast and in the cell wall, with a smaller portion located in the cellular protoplast of plants. The ratio between the concentration of HM in the protoplast/apoplast can be decreased by Si, which contributes in reducing the toxic effects of HM in metabolically active cellular compartments such as the cytoplasm and organelles (Greger et al. 2015; Wu et al. 2018; Vaculík et al. 2020). In addition, the translocation and movement of HM between plant organs are affected by Si, which increases the survival time of plants to HM toxicity stresses (Meena et al. 2022).

Si may be strongly associated with the plant cell wall, in the form of organosilicon (Si-O-C), especially in the cross-links of the hemicellulose of the cell wall (He et al. 2015). In this sense, the modifications promoted by Si in the cell wall bindings may be a crucial mechanism to minimize the effect of toxicity by HM. For example, because of the co-localization of Si with Al in several plant species, it is suggested that the formation of insoluble Al-Si complexes may be one of the mechanisms promoted by Si to reduce the mobility of HM within the plant, thus reducing stress (Turnau et al. 1993).

14.6 Altering Gene Expression

Some works have been developed on the molecular mechanisms by which Si can reduce the toxicity of HM, its absorption and transfer from the soil to the plant roots, and how the regulation of gene expression mediated by Si involves the absorption, transport, and accumulation of HM by plants and their possible detoxification mechanisms.

Khan et al. (2021), in a recent study, show the interaction between Si and HM at the molecular level in root cells, indicating that different transporters and aquaporins present competition for the entry of different HM and Si, and that vacuolar transporters are also responsible for capturing HM in the vacuole and reducing the impact of HM inside the cell. Guo et al. (2022a), when analyzing RNA-seq, observed that the foliar application of Si decreased the expression of the metallic transporters

OsZIP7 and OsZIP8, and that the supply of Si increased the expression of genes that participate in the metabolism of glutathione and of cellulose synthesis, and also demonstrated that Cd remobilization in rice leaves is affected by the amount of Cd that was stored in non-labile subcellular compartments, which can be regulated by Si application, and that most of the Cd (69%) was distributed in the cell walls of rice leaves and in various cell wall fractions (pectin, hemicellulose 1 and residue).

In rice cultivated under Cd stress, increasing the concentration of Si in the medium upregulated the expression of the low silicon rice 1 (*Lsi1*) transporter gene, while the expression of the gene involved in Cd transport, macrophage protein 5 associated with natural resistance (Nramp5), was downregulated, demonstrating that Si can also mitigate the negative effects of HM toxicity (Ma et al. 2022). Kim et al. (2014) observed a reduction in lipid peroxidation and desaturation of fatty acids in plant tissues of cells in rice cultivated with Si, and that the supply of Si decreased the concentration of Cd and Cu, which led to a decrease of mRNA expression of enzymes encoding HM transporters (OsHMA2 and OsHMA3) and a significant increase in the genes responsible for transporting Si (OsLSi1 and OsLSi2).

14.7 Conclusions

The supply of Si under stress conditions due to metal toxicity is a viable and sustainable practice to the detriment of commonly used agrochemicals, especially for crops that accumulate this beneficial element. The use of silicon in culture not only improves growth, morphological, and physiological parameters by reducing the deleterious effects of heavy metal stress while increasing antioxidants that help combat reactive oxygen species and photosynthetic efficiency by decreasing oxidative stress, but also develops mechanisms of detoxification and remobilization of heavy metals, involving various mechanisms of prevention or tolerance.

Despite presenting numerous benefits, there are still many scientific gaps that need to be understood regarding the mechanisms of reduction of the effects caused by the toxicity of heavy metals, with the impacts of silicon on the activities of antioxidant and non-enzymatic enzymes being the most addressed aspects and considered one of the important mechanisms to reduce the negative effects of heavy metals. Nevertheless, mechanisms related to physiological alterations and changes in structures related to gene expression still require further research.

References

Adrees M, Ali S, Rizwan M, Zia-ur-Rehman M, Ibrahim M, Abbas F, Farid M, Qayyum MF, Irshad MK (2015) Mechanisms of silicon-mediated alleviation of heavy metal toxicity in plants: a review. Ecotoxicol Environ Saf 119:186–197. https://doi.org/10.1016/j.ecoenv.2015.05.011

Ahire ML, Mundada PS, Nikam TD, Bapat VA, Penna S (2021) Multifaceted roles of silicon in mitigating environmental stresses in plants. Plant Physiol Biochem 169:291–310. https://doi.org/10.1016/j.plaphy.2021.11.010

Alsahli AA, Bhat JA, Alyemeni MN, Ashraf M, Ahmad P (2021) Hydrogen sulfide (H_2S) mitigates arsenic (as)-induced toxicity in pea (*Pisum sativum* L.) plants by regulating osmoregulation, antioxidant defense system, ascorbate glutathione cycle and glyoxalase system. J Plant Growth Regul 40:2515–2531. https://doi.org/10.1007/s00344-020-10254-6

Alves LQ, de Jesus RM, de Almeida AAF, Souza VL, Mangabeira PAO (2014) Effects of lead on anatomy, ultrastructure and concentration of nutrients in plants *Oxycaryum cubense* (Poep. & Kunth) Palla: a species with phytoremediator potential in contaminated watersheds. Environ Sci Pollut Res 21:6558–6570. https://doi.org/10.1007/s11356-014-2549-9

Aprile A, De Bellis L (2020) Editorial for special issue "heavy metals accumulation, toxicity, and detoxification in plants". Int J Mol Sci 21:1–5. https://doi.org/10.3390/ijms21114103

Barceló J, Vázquez MD, Poschenrieder C (1988) Structural and ultrastructural disorders in cadmium-treated bush bean plants (*Phaseolus vulgaris* L.). New Phytol 108:37–49. https://doi.org/10.1111/j.1469-8137.1988.tb00202.x

Bharwana SA, Ali S, Farooq MA, Iqbal N, Hameed A, Abbas F, Ahmad MSA (2014) Glycine betaine-induced lead toxicity tolerance related to elevated photosynthesis, antioxidant enzymes suppressed lead uptake and oxidative stress in cotton. Turk J Bot 38:281–292. https://doi.org/10.3906/bot-1304-65

Bhat JA, Shivaraj SM, Singh P, Navadagi DB, Tripathi DK, Dash PK, Solanke AU, Sonah H, Deshmukh R (2019) Role of silicon in mitigation of heavy metal stresses in crop plants. Plan Theory 8:1–20. https://doi.org/10.3390/plants8030071

Bhatti SS, Kumar V, Sambyal V, Singh J, Nagpal AK (2018) Comparative analysis of tissue compartmentalized heavy metal uptake by common forage crop: a field experiment. Catena 160:185–193. https://doi.org/10.1016/j.catena.2017.09.015

Chen CF, Zhong JH, Li SY (2007) Effect of silicon on growth and anti-stress ability of Chinese cabbage (*Brassica pekinensis* Rupr.) in cadmium contaminated soil. Plant Physiol Commun 43:479–482

Chen Z, Xu J, Xu Y, Wang K, Cao B, Xu K (2019) Alleviating effects of silicate, selenium, and microorganism fertilization on lead toxicity in ginger (*Zingiber officinale* roscoe). Plant Physiol Biochem 145:153–163. https://doi.org/10.1016/j.plaphy.2019.10.027

Chibuike GU, Obiora SC (2014) Heavy metal polluted soils: effect on plants and bioremediation methods. Appl Environ Soil Sci 2014:2708. https://doi.org/10.1155/2014/752708

Choudhury S, Panda P, Sahoo L, Panda SK (2013) Reactive oxygen species signaling in plants under abiotic stress. Plant Signal Behav 8:23681. https://doi.org/10.4161/psb.23681

Cunha KPV, Nascimento CWA (2009) Silicon effects on metal tolerance and structural changes in maize (*Zea mays* L.) grown on a cadmium and zinc enriched soil. Water Air Soil Pollut 197:323–330. https://doi.org/10.1007/s11270-008-9814-9

de Oliveira LA (2009) Silicon in bean and rice plants: absorption, transport, redistribution and tolerance to cadmium. Dissertation. São Paulo University. https://teses.usp.br/teses/disponiveis/64/64134/tde-03122009-094223/publico/Doutorado.pdf. Accessed Oct 2022

Duffus JH (2002) "Heavy metals" – a meaningless term? (IUPAC technical report). Pure Appl Chem 74:793–807. https://doi.org/10.1351/pac200274050793

Emamverdian A, Ding Y, Xie Y, Sangari S (2018) Silicon mechanisms to ameliorate heavy metal stress in plants. Biomed Res Int 22:8492898. https://doi.org/10.1155/2018/8492898

Farooq MA, Ali S, Hameed A, Ishaque W, Mahmood K, Iqbal Z (2013) Alleviation of cadmium toxicity by silicon is related to elevated photosynthesis, antioxidant enzymes; suppressed cadmium uptake and oxidative stress in cotton. Ecotoxicol Environ Saf 96:242–249. https://doi.org/10.1016/j.ecoenv.2013.07.006

Fleck AT, Nye T, Repenning C, Stahl F, Zahn M, Schenk MK (2011) Silicon enhances suberization and lignification in roots of rice (*Oryza sativa*). J Exp Bot 62:2001–2011. https://doi.org/10.1093/jxb/erq392

Gomes MP, Marques TC, Marques TCLLSM, Nogueira MOG, Castro EM, Soares ÂM (2011) Ecophysiological and anatomical changes due to uptake and accumulation of heavy metal in *Brachiaria decumbens*. Plant Physiol Biochem 566–573:145–152. https://doi.org/10.1590/S0103-90162011000500009

Gratão PL, Pompeu GB, Azevedo RA (2006) Plant antioxidant responses to toxic elements. Curr Top Biochem Res 65:1330–1337

Greger M, Bergqvist C, Sandhi A, Landberg T (2015) Influence of silicon on arsenic uptake and toxicity in lettuce. J Appl Bot Food Qual 88:234–240. https://doi.org/10.5073/JABFQ.2015.088.034

Guedes FRCM, Maia CF, da Silva BRS, Batista BL, Alyemeni MN, Ahmad P, da Lobato AKS (2021) Exogenous 24-Epibrassinolide stimulates root protection, and leaf antioxidant enzymes in lead stressed rice plants: central roles to minimize Pb content and oxidative stress. Environ Pollut 280:116992. https://doi.org/10.1016/j.envpol.2021.116992

Guo J, Ye D, Zhang X, Huang H, Wang Y, Zheng Z, Li T, Yu H (2022a) Characterization of cadmium accumulation in the cell walls of leaves in a low-cadmium rice line and strengthening by foliar silicon application. Chemosphere 287:132374. https://doi.org/10.1016/j.chemosphere.2021.132374

Guo L, Chen A, Li C, Wang Y, Yang D, He N, Liu M (2022b) Solution chemistry mechanisms of exogenous silicon influencing the speciation and bioavailability of cadmium in alkaline paddy soil. J Hazard Mater 438:129526. https://doi.org/10.1016/j.jhazmat.2022.129526

He C, Ma J, Wang L (2015) A hemicellulose-bound form of silicon with potential to improve the mechanical properties and regeneration of the cell wall of rice. New Phytol 206:1051–1062. https://doi.org/10.1111/nph.13282

Hossain MA, Piyatida P, da Silva JAT, Fujita M (2012) Molecular mechanism of heavy metal toxicity and tolerance in plants: central role of glutathione in detoxification of reactive oxygen species and methylglyoxal and in heavy metal chelation. J Bot 2012:1–37. https://doi.org/10.1155/2012/872875

Hussain S, Shuxian L, Mumtaz M et al (2021) Foliar application of silicon improves stem strength under low light stress by regulating lignin biosynthesis genes in soybean (*Glycine max* L. Merr.). J Hazard Mater 401:123256. https://doi.org/10.1016/j.jhazmat.2020.123256

Imtiaz M, Rizwan MS, Mushtaq MA, Ashraf M, Shahzad SM, Yousaf B, Saeed DA, Rizwan M, Nawaz MA, Mehmood S, Tu S (2016) Silicon occurrence, uptake, transport and mechanisms of heavy metals, minerals and salinity enhanced tolerance in plants with future prospects: a review. J Environ Manag 183:521–529. https://doi.org/10.1016/j.jenvman.2016.09.009

Kabata-Pendias A (2010) Trace elements in soils and plants. CRC Press/Taylor & Francis Group, Boca Raton

Keller C, Rizwan M, Davidian JC, Pokrovsky OS, Bovet N, Chaurand P, Meunier JD (2015) Effect of silicon on wheat seedlings (*Triticum turgidum* L.) grown in hydroponics and exposed to 0 to 30 μM Cu. Planta 241:847–860. https://doi.org/10.1007/s00425-014-2220-1

Khan F, Hussain S, Tanveer M, Khan S, Hussain HA, Iqbal B, Geng M (2018) Coordinated effects of lead toxicity and nutrient deprivation on growth, oxidative status, and elemental composition of primed and non-primed rice seedlings. Environ Sci Pollut Res 25:21185–21194. https://doi.org/10.1007/s11356-018-2262-1

Khan A, Bilal S, Khan AL, Imran M, Al-Harrasi A, Al-Rawahi A, Lee IJ (2020) Silicon-mediated alleviation of combined salinity and cadmium stress in date palm (*Phoenix dactylifera* L.) by regulating physio-hormonal alteration. Ecotoxicol Environ Saf 188:109885. https://doi.org/10.1016/j.ecoenv.2019.109885

Khan I, Awan SA, Rizwan M, Ali S, Hassan MJ, Brestic M, Zhang X, Huang L (2021) Effects of silicon on heavy metal uptake at the soil-plant interphase: a review. Ecotoxicol Environ Saf 222:112510. https://doi.org/10.1016/j.ecoenv.2021.112510

Kim YH, Khan AL, Kim DH, Lee SY, Kim KM, Waqas M, Jung HY, Shin JH, Kim JG, Lee IJ (2014) Silicon mitigates heavy metal stress by regulating P-type heavy metal ATPases, *Oryza sativa* low silicon genes, and endogenous phytohormones. BMC Plant Biol 14:13. https://doi.org/10.1186/1471-2229-14-13

Kim YH, Khan AL, Waqas M, Lee IJ (2017) Silicon regulates antioxidant activities of crop plants under abiotic-induced oxidative stress: a review. Front Plant Sci 8:0510. https://doi.org/10.3389/fpls.2017.00510

Kumar A, Prasad MNV (2018) Plant-lead interactions: transport, toxicity, tolerance, and detoxification mechanisms. Ecotoxicol Environ Saf 166:401–418. https://doi.org/10.1016/j.ecoenv.2018.09.113

Liang Y, Nikolic M, Bélanger R, Gong H, Song A (2015) Silicon-mediated tolerance to metal toxicity. In: Liang Y, Nikolic M, Bélanger R, Gong H, Song A (eds) Silicon in agriculture: from theory to practice. Springer, pp 1–235

Lu Y, Ma J, Teng Y, He J, Christie P, Zhu L, Ren W, Zhang M, Deng S (2018) Effect of silicon on growth, physiology, and cadmium translocation of tobacco (*Nicotiana tabacum* L.) in cadmium-contaminated soil. Pedosphere 28:680–689. https://doi.org/10.1016/S1002-0160(17)60417-X

Ma J, Zhang X, Wang L (2017) Synergistic effects between [Si-hemicellulose matrix] ligands and Zn ions in inhibiting cd ion uptake in rice (*Oryza sativa*) cells. Planta 245:965–976. https://doi.org/10.1007/s00425-017-2655-2

Ma J, Hamzah M, Alsafran M, Al H, Rizwan M, Nawaz M, Ali S, Usman K (2022) Response of cauliflower (*Brassica oleracea* L.) to nitric oxide application under cadmium stress. Ecotoxicol Environ Saf 243:113969. https://doi.org/10.1016/j.ecoenv.2022.113969

Malik MA, Wani AH, Mir SH, Rehman IU, Tahir I, Ahmad P, Rashid I (2021) Elucidating the role of silicon in drought stress tolerance in plants. Plant Physiol Biochem 165:187–195. https://doi.org/10.1016/j.plaphy.2021.04.021

Meena V, Dotaniya ML, Saha JK, Patra AK (2022) Silicon potential to mitigate plant heavy metals stress for sustainable agriculture: a review. SILICON 14:4447–4462. https://doi.org/10.1007/s12633-021-01200-6

Mir RA, Bhat BA, Yousuf H, Islam ST, Raza A, Rizvi MA, Charagh S, Albaqami M, Sofi PA, Zargar SM (2022) Multidimensional role of silicon to activate resilient plant growth and to mitigate abiotic stress. Front Plant Sci 13:819658. https://doi.org/10.3389/fpls.2022.819658

Ma JF, Yamaji N (2006) Silicon uptake and accumulation in higher plants. Trends Plant Sci, 11:392–397. http://dx.doi.org/10.1016/j.tplants.2006.06.007

Nwugo CC, Huerta AJ (2008) Silicon-induced cadmium resistance in rice (*Oryza sativa*). J Plant Nutr Soil Sci 171:841–848. https://doi.org/10.1002/jpln.200800082

Prasad MNV (1995) Cadmium toxicity and tolerance in vascular plants. Environ Exp Bot 35:525–545. https://doi.org/10.1016/0098-8472(95)00024-0

Qi X, Tam NF-Y, Li WC, Ye Z (2020) The role of root apoplastic barriers in cadmium translocation and accumulation in cultivars of rice (*Oryza sativa* L.) with different Cd-accumulating characteristics. Environ Pollut 264:114736. https://doi.org/10.1016/j.envpol.2020.114736

Rahman S, Durrani MS, Ahmad I (2021) Effect of silicon and silicon nanoparticles on lead-treated wheat seedlings in hydroponic condition. Int J Biosci 19:175–180. https://doi.org/10.12692/ijb/19.4.175-180

Riaz M, Kamran M, Rizwan M, Ali S, Parveen A, Malik Z, Wang X (2021) Cadmium uptake and translocation: selenium and silicon roles in cd detoxification for the production of low cd crops: a critical review. Chemosphere 273:129690. https://doi.org/10.1016/j.chemosphere.2021.129690

Rizwan M, Meunier JD, Miche H, Keller C (2012) Effect of silicon on reducing cadmium toxicity in durum wheat (*Triticum turgidum* L. cv. Claudio W.) grown in a soil with aged contamination. J Hazard Mater 209–210:326–334. https://doi.org/10.1016/j.jhazmat.2012.01.033

Sharma P (2005) Lead toxicity im plants. Braz J Plant Physiol 17:35–52. https://doi.org/10.1590/S1677-04202005000100004

Sharma P, Jha AB, Dubey RS, Pessarakli M (2012) Reactive oxygen species, oxidative damage, and antioxidative defense mechanism in plants under stressful conditions. J Bot 2012:1–26. https://doi.org/10.1155/2012/217037

Shu X, Yin LY, Zhang QF, Wang WB (2012) Effect of Pb toxicity on leaf growth, antioxidant enzyme activities, and photosynthesis in cuttings and seedlings of *Jatropha curcas* L. Environ Sci Pollut Res 19:893–902. https://doi.org/10.1007/s11356-011-0625-y

Siddiqui MH, Alamri S, Mukherjee S, Al-Amri AA, Alsubaie QD, Al-Munqedhi BMA, Ali HM, Kalaji HM, Fahad S, Rajput VD, Narayan OP (2021) Molybdenum and hydrogen sulfide synergistically mitigate arsenic toxicity by modulating defense system, nitrogen and cysteine assimilation in faba bean (*Vicia faba* L.) seedlings. Environ Pollut 290:117953. https://doi.org/10.1016/j.envpol.2021.117953

Singh R, Parihar P, Prasad SM (2020) Interplay of calcium and nitric oxide in improvement of growth and arsenic-induced toxicity in mustard seedlings. Sci Rep 10:1–12. https://doi.org/10.1038/s41598-020-62831-0

Tangahu BV, Sheikh Abdullah SR, Basri H, Idris M, Anuar N, Mukhlisin M (2011) A review on heavy metals (as, Pb, and hg) uptake by plants through phytoremediation. Int J Chem Eng 2011:939161. https://doi.org/10.1155/2011/939161

Tripathi DK, Singh VP, Prasad SM, Chauhan DK, Kishore Dubey N, Rai AK (2015) Silicon-mediated alleviation of Cr(VI) toxicity in wheat seedlings as evidenced by chlorophyll florescence, laser induced breakdown spectroscopy and anatomical changes. Ecotoxicol Environ Saf 113:133–144. https://doi.org/10.1016/j.ecoenv.2014.09.029

Tripathi DK, Vishwakarma K, Singh VP, Prakash V, Sharma S, Muneer S, Nikolic M, Deshmukh R, Vaculík M, Corpas FJ (2021) Silicon crosstalk with reactive oxygen species, phytohormones and other signaling molecules. J Hazard Mater 408:124820. https://doi.org/10.1016/j.jhazmat.2020.124820

Turnau K, Kottke I, Oberwinkler F (1993) Paxillus involutus – Pinus sylvestris mycorrhizae from heavily polluted forest. Plant Biol 106:213–219. https://doi.org/10.1111/j.1438-8677.1993.tb00742.x

Vaculík M, Lukačová Z, Bokor B, Martinka M, Tripathi DK, Lux A (2020) Alleviation mechanisms of metal(loid) stress in plants by silicon: a review. J Exp Bot 71:6744–6757. https://doi.org/10.1093/jxb/eraa288

Wei W, Peng H, Xie Y, Wang X, Huang R, Chen H, Ji X (2021) The role of silicon in cadmium alleviation by rice root cell wall retention and vacuole compartmentalization under different durations of cd exposure. Ecotoxicol Environ Saf 226:112810. https://doi.org/10.1016/j.ecoenv.2021.112810

Wu J, Geilfus CM, Pitann B, Mühling KH (2016) Silicon-enhanced oxalate exudation contributes to alleviation of cadmium toxicity in wheat. Environ Exp Bot 131:10–18. https://doi.org/10.1016/j.envexpbot.2016.06.012

Wu Z, Xu S, Shi H, Zhao P, Liu X, Li F, Deng T, Du R, Wang X, Wang F (2018) Comparison of foliar silicon and selenium on cadmium absorption, compartmentation, translocation and the antioxidant system in Chinese flowering cabbage. Ecotoxicol Environ Saf 166:157–164. https://doi.org/10.1016/j.ecoenv.2018.09.085

Wu J, Mock HP, Giehl RFH, Pitann B, Mühling KH (2019) Silicon decreases cadmium concentrations by modulating root endodermal suberin development in wheat plants. J Hazard Mater 364:581–590. https://doi.org/10.1016/j.jhazmat.2018.10.052

Yang Z, Yang F, Liu JL, Wu HT, Yang H, Shi Y, Liu J, Zhang YF, Luo YR, Chen KM (2022) Heavy metal transporters: functional mechanisms, regulation, and application in phytoremediation. Sci Total Environ 809:151099. https://doi.org/10.1016/j.scitotenv.2021.151099

Zhang Q, Yan C, Liu J, Lu H, Wang W, Du J, Duan H (2013) Silicon alleviates cadmium toxicity in *Avicennia marina* (Forsk.) Vierh. Seedlings in relation to root anatomy and radial oxygen loss. Mar Pollut Bull 76:187–193. https://doi.org/10.1016/j.marpolbul.2013.09.005

Zhang Y, Shi Y, Gong HJ, Zhao HL, Li HL, Hu YH, Wang YC (2018) Beneficial effects of silicon on photosynthesis of tomato seedlings under water stress. J Integr Agric 17:2151–2159. https://doi.org/10.1016/S2095-3119(18)62038-6

Zhou Q, Cai Z, Xian P, Yang Y, Cheng Y, Lian T, Ma Q, Nian H (2022) Silicon-enhanced tolerance to cadmium toxicity in soybean by enhancing antioxidant defense capacity and changing cadmium distribution and transport. Ecotoxicol Environ Saf 241:113766. https://doi.org/10.1016/j.ecoenv.2022.113766

Zulfiqar U, Farooq M, Hussain S, Maqsood M, Hussain M, Ishfaq M, Ahmad M, Anjum MZ (2019) Lead toxicity in plants: impacts and remediation. J Environ Manag 250:109557. https://doi.org/10.1016/j.jenvman.2019.109557

Chapter 15
Beneficial Role of Silicon in Plant Nutrition Under Salinity Conditions

Alexander Calero Hurtado, Dilier Olivera Viciedo, and Renato de Mello Prado

15.1 Introduction

Plants are exposed during their lifetime to different environmental stresses, particularly abiotic stresses such as salinity (Calero Hurtado et al. 2020a; Hurtado et al. 2021), heat, and drought (Olivera-Viciedo et al. 2019, 2021a); nutrient imbalance (Olivera-Viciedo et al. 2021b); and heavy metal toxicity (Olivera-Viciedo et al. 2020; de Mello Prado 2021).

Salinity is one of the most severe abiotic stresses limiting crop production worldwide. It affects approximately 7% of the world's total land area, 20% of the world's cultivated land area, and nearly 50% of the world's irrigated land area, which represents one-third of food-producing land (Khan et al. 2019; Kibria et al. 2017). In Brazil, although information on salt-affected areas is not well defined, estimates show that 20–25% of irrigated areas in the northeast region face soil salinization problems (Peña Calzada et al. 2023; Pessoa et al. 2016). Part of these degraded saline areas, which are unsuitable for conventional crops, can be used to grow alternative plants to serve as a source of food, fuel, fodder, fiber, resin, essential oils, and pharmaceutical products, as well as for landscape reintegration (Bessa et al. 2017).

Under salinity conditions, plants tend to activate different physiological, biochemical, and stoichiometry mechanisms to cope with stress. This occurs by altering their morphology, anatomy, water relations, photosynthesis, and protein synthesis, as well as through biochemical adaptations such as the antioxidative

A. Calero Hurtado (✉)
University of Sancti Spiritus, Sancti Spiritus, Cuba

D. Olivera Viciedo
Institute of Agrifood, Animals and Environmental Sciences, Universidad de O'Higgins, San Fernando, Chile

R. de Mello Prado
São Paulo State University, Jaboticabal, São Paulo, Brazil

R. de Mello Prado (ed.), *Benefits of Silicon in the Nutrition of Plants*,
https://doi.org/10.1007/978-3-031-26673-7_15

metabolism response (EL Sabagh et al. 2020; Calero Hurtado et al. 2020a; Hurtado et al. 2021).

Salinity inhibits plant growth in three principle ways: ion toxicity (mainly of Na^+ and Cl^-), osmotic stress, and nutritional disruption (Munns 2011; Calero Hurtado et al. 2020b). Adverse salinity effects on plants are evident from negative growth trends from alteration or inhibition of biochemical and physiological processes (Shahid et al. 2020). To understand the physiological mechanisms responsible for salinity tolerance in a given species, it is necessary to know whether plant growth is being limited by the osmotic effect of the salt in the soil or by the toxic effect of the salt within the plant (Munns and Tester 2008). Ion imbalance is one of the major consequences, in which a high concentration of sodium (Na^+) and chloride (Cl^-) ions, as an example, can lead to harmful biochemical processes in the plant. Moreover, soil salinity prevents the plant from absorbing water from the soil, resulting in a decline in cell water that affects cell turgor (Shahid et al. 2020).

Among the key physiological and biochemical processes impaired by high salinity are transpiration, photosynthesis, protein synthesis, and stoichiometric changes of C, N, and P (Asrar et al. 2017; Calero Hurtado et al. 2020a, b). Soil salinity also encourages the production of reactive oxygen species (ROS), thus reducing plant growth (Calero Hurtado et al. 2019; Khan et al. 2019). The main reasons for oxidative stress are: (i) imbalance between ROS generation and detoxification due to disturbance of 'normal' cell physiology; (ii) ROS biosynthesis as a constituent part of stress signaling and immunity response needed for defense and adaptation (Demidchik 2015; Shahid et al. 2020).

Apart from restrictions in the uptake of mineral nutrients and hindrances in key physiological processes, greater accumulation of toxic ions leads to excessive generation of ROS, increasing oxidative damage to cells (Siddiqui et al. 2017). The most toxic ROS include singlet oxygen, superoxide ion, hydroxyl ion, and hydrogen peroxide, which can easily target important biomolecules like membrane lipids and proteins, nucleic acids, and others (Demidchik 2015; Asrar et al. 2017).

Toxic ions, predominantly Na^+ and Cl^-, accumulate in the protoplasm and cause ionic imbalance and metabolic and nutritional disorders while decreasing plant growth (Munns 2011; Shahid et al. 2020). Excess Na^+ in soil decreases K^+ content in plants, especially in roots (Agarie et al. 1998; James et al. 2011; Akhter et al. 2022). This is due to the combined action of decreasing inflow and increasing efflux of K^+ (Apse and Blumwald 2007) mediated by transporters and K^+ channels located in the plasmalemma (Voigt et al. 2009). Therefore, Na^+ accumulation in the cell is partly a result of competition between Na^+ and K^+ for the active sites of enzymes and ribosomes (Riedelsberger et al. 2019), whereby essential metabolic processes might be hampered (Munns et al. 2016). Consequently, K deficiency reduces plant tolerance to water deficit due to its influence on plant osmoregulation, playing a key role in stress avoidance and adaptation (Tränkner et al. 2018).

Salinity induces changes in the uptake, transport, and redistribution of nutrients in a plant. This condition can cause physiological inactivation of a certain nutrient and increase the internal need of the plant for that nutrient (Grattan and Grieve 1999; Munns et al. 2016). Furthermore, high cytosolic sodium (Na^+) concentration induces changes in the absorption and metabolism of calcium (Ca^{2+}). In other words,

Na^+ replaces Ca^{2+} in cell membranes (Rengel 1992) and in the cell wall (Nilsen and Orcutt 1996), hence decreasing their permeability, which in turn reduces their elasticity (Neumann et al. 1994) and worsens salinity damage to the plant. In addition, a low Ca^{2+} concentration in the plant interrupts stress-signal transduction processes that would otherwise attenuate plant damage (Türkan and Demiral 2009).

Excess accumulation of Na^+ in plant tissues tends to increase the Na^+/K^+ ratio, thereby altering cellular ionic and nutritional homeostasis (Assaha et al. 2017; Nouripour-Sisakht et al. 2022). The resulting decrease in cellular metabolic efficiency can induce symptoms that start with chlorosis and develop into early leaf senescence (Munns et al. 2016; Hanin et al. 2016), ultimately leading to plant death (Jalili et al. 2022). Overall salinity decreases the equilibrium between cations and anions, affecting osmoregulation and many other essential metabolic processes (El-Ramady et al. 2018). This disruption of homeostasis occurs at the cellular level and throughout the plant, causing molecular damage, restricting growth, and even leading to plant death (de Luma et al. 2016).

15.2 Silicon and Salt Stress Remediation

High salt concentration in the growth medium results in high Na^+ concentration. This causes nutritional imbalances that decrease plant growth, biomass production, and yield. An important strategy for reducing the adverse effects of salt stress in plants consists of applying silicon (Si) via root supplementation (Alzahrani et al. 2018; Calero Hurtado et al. 2019, 2020b), foliar spraying (Calero Hurtado et al. 2020a; Hurtado et al. 2021; Sattar et al. 2019), or even combining these two techniques (Calero Hurtado et al. 2019, 2020b). Several authors report that Si application may increase salt tolerance in many important crops such as wheat (Zia et al. 2021; Premkumar et al. 2022), rice (Flam-Shepherd et al. 2018; Akhter et al. 2022), maize (Xie et al. 2015; Bosnic et al. 2018; Khan et al. 2018), barley (Akhter et al. 2022; Premkumar et al. 2022), sorghum (Calero Hurtado et al. 2019, 2020a; Yin et al. 2016), pea (Al-Huqail et al. 2017; Zia et al. 2021), tomato (Muneer et al. 2014; Muneer and Jeong 2015; Liu et al. 2015), soybean (Farhangi-Abriz and Torabian 2018; Gonzalo et al. 2013), cucumber (Alsaeedi et al. 2018; Wang et al. 2015; Zhu et al. 2015), and sunflower (Calero Hurtado et al. 2020b; Hurtado et al. 2021; Conceição et al. 2019). Here we summarized the beneficial role of Si in plant nutrition under salt stress.

15.3 Role of Si in Decreasing Na^+ Uptake, Transport, and Accumulation

High concentrations of soluble salts, such as NaCl, increase the concentration of Na^+ and Cl^- ions in soil solution and then in the plants. The beneficial effects of Si on attenuating salt stress in plants depend on the form of application, which can be via root supplementation (growth medium or soil) or via foliar application. Many studies have reported that the different forms of Si supply via soil and via foliar spraying can decrease Na^+ uptake and transport in tropical plants. Table 15.1 describes in short, the role of Si in decreasing Na uptake.

Among the most important mechanisms of plant tolerance to salt stress is the decrease in Na^+ uptake, transport, and accumulation by plants (Rizwan et al. 2015; Etesami and Jeong 2018; Liu et al. 2019; Zhu et al. 2019). Some details of Si supplied in different forms (via foliar spraying, via medium growth, and via the combination of both) are given below.

Results from several studies show that Si application may increase salt tolerance in several important crops. Many authors reported that foliar Si spraying decreased Na^+ uptake and accumulation in plants grown under salt stress conditions (Calero Hurtado et al. 2019). In salt-stressed faba bean (*Vicia faba* L.) plants, foliar Si spraying (1000 ppm SiO_2) decreased Na^+ concentration in both shoots and roots (Hellal et al. 2012). Bybordi (2014) reported that foliar application of Si decreased Na^+ uptake in wheat plants. Foliar Si spraying (300 ppm SiO_2) with potassium silicate decreased Na^+ uptake in roots and shoots of salt-stressed moringa (*Moringa oleifera* L.) plants (Hussein and Abou-baker 2014). Similarly, foliar application of Si (4 mmol L^{-1}) as potassium silicate decreased Na^+ uptake in shoots of salt-stressed wheat (*Triticum aestivum* L.) plants (Bybordi 2014). In addition, foliar spraying with Si (1 mM) as potassium silicate on *Cynodon dactylon* (L.) Pers., *Festuca arundinacea* (Schreb.), and *Lolium perenne* (L.) ameliorated the adverse effects of salt stress by decreasing Na^+ concentration in the shoots of turfgrasses (Esmaeili et al. 2015).

In salt-stressed soybean plants, foliar Si spraying (1 mM) with silica nanoparticles (nano-SiO_2) decreased Na^+ concentration in shoots (Farhangi-Abriz and Torabian 2018). More recently, foliar Si spraying (28.6 mM) decreased Na^+ uptake and accumulation in shoots and roots of sorghum and sunflower seedlings grown under salt stress conditions (Calero Hurtado et al. 2019, 2020b). The above studies showed that foliar Si might decrease Na^+ uptake and accumulation under salt stress. Notwithstanding, detailed mechanisms still need to be explored for better understanding the role of Si in tropical plants under salt stress at early stages, especially with regard to foliar spraying under field conditions. In salt-stressed tomato plants, Si foliar spraying decreased Na^+ uptake and shoot accumulation (Hernández-Salinas et al. 2021).

Decreasing Na^+ uptake and accumulation by Si addition in the growth medium is one of the most important mechanisms for salt stress attenuation in plants under salt stress conditions. Silicon increases plant resistance to salt stress by decreasing Na^+ uptake via the Na^+-H^+ exchangers HvSOS1 in the plasma membrane and HvNHX1 in the tonoplast, respectively, and by increasing K^+ ion uptake via K^+-H^+ symporters such as HvHAK1 (Rizwan et al. 2015). The decrease in Na^+ uptake might be due to Si-mediated stimulation of the root plasma membrane H^+-ATPase activity under salt stress as previously reported for barley (Liang et al. 2005, 2006) and aloe (Xu et al. 2015). This increase in K^+ uptake and decrease in Na^+ uptake and translocation might be due to Si deposition in roots, which could decrease Na^+ bypass flow. Some

Table 15.1 Role of Si in decreasing Na^+ uptake, transport, and accumulation

Crop	Var./Cv./hybrid	Role of Si in decreasing Na uptake	Reference
Barley (*Hordeum vulgare* L.)	Jian 4 (salt-tolerant), Kepin No. 7 (salt-susceptible)	Decreased Na^+ concentration in shoots and roots; great inhibition of Na^+ uptake and transport from roots to shoots	Liang (1999)
Rice (*Oryza sativa* L.)	CSR 10 (salt-tolerant), GR 4 (salt-sensitive), IR36 (intermediate tolerant)	Reduced Na^+ transport; the action of silicate was through partial blockage of the transpirational bypass flow, an important pathway for Na^+ uptake	Yeo et al. (1999)
Rice	IR36	Decreased Na^+ uptake through a reduction in apoplastic transport across the roots	Gong et al. (2006)
Faba beans (*Vica faba* L.)	Nobaria 12	Increased K^+ concentration and K:Na ratio	Hellal et al. (2012)
Rice	IR4630 (salt-tolerant), IR15324 (salt-sensitive)	Reduced sodium (Na^+) transport and bypass flow in rice	Gurmani et al. (2013b)
Wheat (*Triticum aestivum* L.).	Kharchia-65 (salt-tolerant), Punjab-85 (salt-sensitive)	Inhibited Na^+ transport and the Na^+/K^+ ratio in both wheat genotypes by limiting Na^+ bypass flow	Gurmani et al. (2013a)
Wheat	Pishgam, Afagh, and Alvand	Decreased Na^+ uptake	Bybordi (2014)
Okra (*Abelmoschus esculentus* L.)	Okra-7080 (salt- sensitive), OH-713 (salt-tolerant)	Decreased Na^+ uptake	Abbas et al. (2015)
Sunflower (*Helianthus annuus* L.)	'Hysun-33'	Decreased Na^+ concentration in shoots and roots	Ashraf et al. (2015)
Cynodon dactylon [L.] Pers, *Festuca arundinacea* Schreb., *Lolium perenne* L.		Decreased Na^+ concentration in the shoots of all turfgrasses	Esmaeili et al. (2015)
Sorghum	Gadambalia	Reduced Na^+ uptake and accumulation from ion channel regulation	Yin et al. (2016)
Maize (*Zea mays* L.)	ZP-560	Maize plants accumulated less Na^+ in both root apex and cortex	Bosnic et al. (2018)
Soybean (*Glycine max* L.)	M7	Decreased Na^+ concentration	Farhangi-Abriz and Torabian (2018)
Wheat		Increased nitrate accumulation in shoots and roots	Sienkiewicz-Cholewa et al. (2018)
Rice	IR29 (salt-sensitive), Pokkali (salt-tolerant)	Reduced Na^+ translocation via bypass flow in some (but not all) rice cultivars	Flam-Shepherd et al. (2018)

(continued)

Table 15.1 (continued)

Crop	Var./Cv./hybrid	Role of Si in decreasing Na uptake	Reference
Sorghum (*Sorghum bicolor* L. 'Moench')	'Dekalb 540'	Decreased Na^+ concentration in shoots and roots	Calero Hurtado et al. (2019)
Maize	'Sygenta 8441', 'Pearl'	Reduced Na^+ concentration in roots, shoots, new and old leaves	Raza et al. (2019)
Cucumber	Beit Alpha	Reduced root Na^+ uptake and Na^+ translocation to stems and leaves	Alsaeedi et al. (2019)
Mung bean		Decreased Na^+ uptake	Ahmad et al. (2019)
Strawberry (*Fragaria* × *ananassa* Duch.)	'Kurdistan', 'Paros'	Inhibited Na uptake in roots and leaves	Yaghubi et al. (2019)
Rice	Kernel Basmati	Decreased Na^+ concentration in plant tissues	Ahmed et al. (2019)
Pepper (*Capsicum annum* L.)		Improved the contents of N, K, P, Si, Zn	Hussein and Abdullah (2019)
Rice	Zhonghua11	Decreased Na concentration and net Na uptake	Yan et al. (2020)
Basil (*Ocimum basilicum* L.)		Considerably decreased Na^+ uptake and the K^+/Na^+ ratio in shoots	Farouk et al. (2020)
Sweet pepper	California Wonder	Decreased Na^+ content	Abdelaal et al. (2020)
Sunflower (*Helianthus annuus* L.)	'Double Sungold'	Decreased Na^+ concentration and Na^+/K^+ in shoots and roots	Hurtado et al. (2021)
Tomato	Liberty	Led to the lowest Na content in shoots	Hernández-Salinas et al. (2021)
Maize	P1574, Hycorn 11	Reduced Na^+ ion uptake at the surface of maize roots and its translocation in plant tissues, thus significantly reducing Na^+ ion accumulation	Ali et al. (2021)
Tomato	Strain B	Increased root Na exclusion and consequently decreased shoot Na	Sayed et al. (2022)
Roselle (*Hibiscus sabdariffa* L.)		Decreased Na concentration	Abou-Sreea et al. (2022)
Licuore	*Glycyrrhiza uralensis* Fisch, *Glycyrrhiza inflata* Bat.	Decreased Na^+ uptake and Na^+ ion translocation from roots to shoots	Shen et al. (2022)
Barley (*Hordeum vulgare* L.)	B-9006, B-15,002, Jou-87, and 14,003	Reduced Na^+ toxicity	Akhter et al. (2022)

authors reported this finding for rice (Yeo et al. 1999; Gong et al. 2006; Gurmani et al. 2013b) and wheat (Gurmani et al. 2013a). Several studies described the Si-mediated increase in K^+ uptake and decrease in Na^+ translocation towards shoots in wheat plants under salt stress (Tuna et al. 2008; Tahir et al. 2011).

Silicon application (0.83 mM) in the form of metasilicic acid via the root medium attenuated salt stress effects in sorghum seedlings by decreasing Na^+ accumulation in shoots and roots (Yin et al. 2016). Also, Si supplementation (50 mg kg^{-1} soil) with sodium silicate applied to the soil decreased Na^+ accumulation in salt-stressed sunflower plants (Ashraf et al. 2015). More recently, Flam-Shepherd et al. (2018) observed that Si (1.67 mM) applied as sodium silicate via nutrient solution increased plant growth and lowered Na^+ concentration in shoots of rice cultivars under salt stress conditions. The authors concluded that, while Si can reduce Na^+ translocation via bypass flow in some (but not all) rice cultivars, it does not affect unidirectional Na^+ transport or Na^+ cycling in roots, either across root cell membranes or within the bulk root apoplast. Similarly, Si addition (4 mM) as potassium silicate in root medium limited Na^+ content in salt-stressed wheat plants (Alzahrani et al. 2018).

According to the literature, Si addition at 1 mM with nano-silicon oxide (nano-SiO_2) in nutrient solution decreased Na^+ concentration in salt-stressed soybean seedlings (Farhangi-Abriz and Torabian 2018). Similarly, Si supplementation at 2 mM with metasilicic acid in root medium attenuated both osmotic and ionic stress in salt-stressed maize plants by decreasing Na^+ accumulation (Khan et al. 2018).

Silicon (1.5 mM) supplied to the root medium in the form of orthosilicic acid decreases Na^+ accumulation in both root apex and cortex (by increased presence of ZmSOS1 and ZmSOS2 in the root apex and cortex facilitating Na^+ exclusion) and in the root stele (by enhanced Na^+ loading into the xylem). Also, Si downregulated the expression of ZmHKT1 in the root stele, which further decreased Na^+ unloading from the xylem. Consequently, Si increased leaf Na^+ accumulation but also enhanced Na^+ sequestration into the vacuoles, thereby decreasing Na^+ accumulation in the chloroplasts of maize plants under salt stress conditions (Bosnic et al. 2018). Similarly, Si addition (2 mM) in the nutrient solution as sodium silicate decreased Na^+ accumulation in salt-stressed sunflower plants (Conceição et al. 2019). More recently, Si addition to the growth medium (2 mM) as stabilized sodium and potassium silicate decreased Na^+ concentration, transport, and accumulation in shoots and roots of salt-stressed sorghum and sunflower seedlings (Calero Hurtado et al. 2019, 2020b).

15.4 Increasing Mineral Uptake by Si Under Salt Stress

Salinity restricts nutrient uptake by roots and then nutrient transport to shoots, thereby reducing nutrient availability. This affects a series of metabolic processes such as photosynthesis, respiration, and nitrogen fixation in different plant parts (e.g., leaves, stems, and roots). High Na^+ uptake by plants limits the uptake of K^+ and Ca^{2+} ions and leads to nutritional imbalance in plants (Assaha et al. 2017). All plants need an adequate amount of mineral nutrients for standard growth and development (de Mello Prado 2021).

Silicon application increased the concentration and accumulation of macronutrients (e.g., N, P, K, Ca, Mg and S) and micronutrients (e.g., Fe, Mn, Zn, B, and Cu) by salt-stressed plants (Etesami and Jeong 2018; Alzahrani et al. 2018; Calero Hurtado et al. 2019, 2020a; Liu et al. 2019; Hurtado et al. 2021) This element can play a crucial role in balancing the uptake, transport, and distribution of minerals and nutrients in plants under salt stress conditions (Etesami and Jeong 2018; Zhu et al. 2019; Liu et al. 2019).

The mechanism of Si-mediated salinity stress resistance usually leads to the following physiological improvements in plants: lodging resistance (Epstein 1999), maintenance of structural rigidity (Schoelynck et al. 2010), management of nutrients (Rizwan et al. 2015; Etesami and Jeong 2018), regulation of genes involved in various physiological processes (Muneer and Jeong 2015; Liu et al. 2019), activation of the antioxidant system (Soundararajan et al. 2017; Calero Hurtado et al. 2020b), elicitation of secondary metabolites related to stress resistance (Manivannan and Ahn 2017; Jana et al. 2019), increased photosynthesis efficiency (Alzahrani et al. 2018; Khan et al. 2018), balancing of ion homeostasis (Alsaeedi et al. 2018; Yaghubi et al. 2019; Hurtado et al. 2021), regulation of nutritional efficiency and nutrient homeostasis (Calero Hurtado et al. 2019, 2020a), and enhancement of growth and biomass (Soundararajan et al. 2018; Ahmad et al. 2019; Zhu et al. 2020). Due to these merits, Si has been introduced in several crops to increase plant yield and stress tolerance. Table 15.2 describes the beneficial effects of Si.

A key role of Si in attenuating the adverse effects of salt stress in plants is the increase in mineral uptake and accumulation (Calero Hurtado et al. 2019). Foliar application of Si (1000 ppm SiO_2) increased K^+ content and the K:Na ratio in salt-stressed faba bean plants (Hellal et al. 2012). Similarly, foliar Si spraying (300 ppm SiO_2) with potassium silicate increased P, K, Ca, and Mg concentrations under salt-stressed moringa plants (Hussein and Abou-baker 2014). In addition, foliar application of Si (4 mmol L^{-1}) as potassium silicate increased K content in whole salt-stressed wheat plants (Bybordi 2014).

According to the literature, foliar spraying with Si (1 mM) as potassium silicate on *Cynodon dactylon* (L.) Pers., *Festuca arundinacea* (Schreb.), and *Lolium perenne* (L.) ameliorated the adverse effects of salinity by increasing shoot K^+ concentration in turfgrasses (Esmaeili et al. 2015). More recently, Abbas et al. (2017) reported that the foliar application of Si (150 mg L^{-1}) as silicic acid increased Ca and Mg concentrations in leaves and roots of salt-stressed okra plants.

In salt-stressed soybean plants, foliar Si spraying (1 mM) with silica nanoparticles (nano-SiO_2) increased root and leaf K^+ contents (Farhangi-Abriz and Torabian 2018). Likewise, Amer and El-Emary (2018) reported that foliar Si spraying (nano-silica concentration of 300 mg L^{-1}) increased N uptake and N use efficiency in maize and faba bean under salinity conditions. More recently, foliar application of Si (28.6 mM) as stabilized sodium and potassium silicate increased the accumulation of N, P, K, Ca, Mg, S, Fe, Mn, Zn, and Cu and their uptake and use efficiency in sorghum and sunflower plants under salinity stress (Calero Hurtado et al. 2019).

Table 15.2 Role of Si in increasing mineral nutrition in plants under salt stress conditions

Crop	Silicon supplementation via	Silicon concentration	Silicon source	Role of Si in plant nutrition under salt stress	Reference
Faba bean	Foliar spraying	1000 ppm	Diatomite (SiO_2)	Increased K^+ concentration and K:Na ratio	Hellal et al. (2012)
Moringa (*Moringa oleifera* Lam.)	Foliar spraying	300 ppm	Potassium silicate	Increased the concentration of N, P, K, Ca, and Mg	Hussein and Abou-baker (2014)
Wheat (*Triticum aestivum* L.)	Foliar spraying	2 and 4 mmol L^{-1}	Potassium silicate (K_2SiO_3)	Increased K^+ concentration	Bybordi (2014)
Sunflower (*Helianthus annuus* L.)	Soil	50 mg kg^{-1}	Sodium silicate (Na_2SiO_3)	Increased K^+ concentration and K:Na ratio in shoots and roots	Ashraf et al. (2015)
Okra (*Abelmoschus esculentus* L. Moench)	Foliar spraying	150 mg 15.3.L^{-1}	Silicic acid [$Si(OH)_4$]	Increased Ca and Mg concentrations in leaves and roots	Abbas et al. (2017)
Maize, Faba bean	Foliar spraying	300 mg L^{-1}	Nano-SiO_2	Increased nitrogen uptake and nitrogen use efficiency	Amer and El-Emary (2018)
Soybean plants	Foliar spraying	0.5, 1, and 2 mM	Nano-SiO_2	Decreased Na^+ concentration; increased K^+ concentration	Farhangi-Abriz and Torabian (2018)
Mung bean	Growth medium	2 mM	Na_2SiO_3	Increased the accumulation of K^+ and Ca^{2+}	Ahmad et al. (2019)
Rice	Soil	50, 75, and 100 mg kg^{-1}	Calcium silicate	Increased K^+ uptake and K:Na ratio in plant tissues	Ahmed et al. (2019)
Maize (*Zea mays* L.)	Growth medium	2 mM	Silicic acid	Enhanced K^+ levels in roots, shoots, new and old leaves	Raza et al. (2019)
Cucumber	Growth medium	100, 200, and 300 ppm	Nanosilica (SiNPs)	Increased N and K in roots, stems, and leaves; K:Na ratio; and nutrient use efficiency	Alsaeedi et al. (2019)

(continued)

Table 15.2 (continued)

Crop	Silicon supplementation via	Silicon concentration	Silicon source	Role of Si in plant nutrition under salt stress	Reference
Potato (*Solanum tuberosum* L.)	Soil	45.50%	SiO_2	Increased the concentration of nutrients (N, P, K, Ca, Zn, and B)	Mahmoud et al. (2022)
Sunflower	Nutrient solution	1, 1.5, and 2.0 mM	Na_2SiO_3	Improved nitrate uptake and its accumulation in the roots; significantly increased nitrate reductase activity	Conceição et al. (2019)
Strawberry	Nutrient solution	1000 and 1500 ppm	Potassium silicate (K_2O_3Si)	Increased K^+ uptake	Yaghubi et al. (2019)
Sorghum, Sunflower	Foliar spraying Nutrient solution	28.6 mmol L^{-1} 2 mM	Stabilized sodium and potassium silicate (SiNaKE)	Increased the uptake, accumulation, and use efficiency of N, P, K, Ca, Mg, S, Fe, Zn, Mn, Cu	Calero Hurtado et al. (2019)
				Improved the nutrient homeostasis of C, N, P and their relationship	Calero Hurtado et al. (2020b)
				Improved C, N, and P homeostasis; Improved Na:K and Na:Ca ratios	Hurtado et al. (2021)
Basil (*Ocimum basilicum* L.)	Foliar spraying	100 mg L^{-1}	Sodium metasilicate	Enhanced K^+ concentration and K^+:Na^+ ratio	Farouk et al. (2020)
Sweet pepper	Foliar spraying	2.7 mmol L^{-1}	Sodium and potassium silicate	Increased the content of N, P, and K^+	Abdelaal et al. (2020)

(continued)

Table 15.2 (continued)

Crop	Silicon supplementation via	Silicon concentration	Silicon source	Role of Si in plant nutrition under salt stress	Reference
Rice	Nutrient solution	1.5 mM	Silicic acid	Increased K and decreased Na:K ratio in shoots	Yan et al. (2020)
Tomato	Nutrient solution	1.5 mM	No source mentioned	Improved K, Ca, and K:Na balance, thus increasing shoot growth	Hernández-Salinas et al. (2021)
Tomato (*Solanum lycopersicum* L.)	Foliar spraying	0.5 ppm	Nano-silicon	Increased shoot mineral (N, P, K, Ca, Mg, Fe, and Zn) concentrations	Sayed et al. (2022)
Spinach (*Spinacia oleracea* L.)	Soil	1.5 and 3.0 mM	Potassium silicate	Increased leaf mineral (N, P, and K) content	Gabr et al. (2022)
Cucumber (*Cucumis sativus* L.)	Nutrient solution	75 and 100 mg L	Sodium silicate	Improved the accumulation of P, Mg, Ca, Fe, Zn, Mn, and Cu	Mousavi et al. (2022)
Pea (*Pisum sativum* L.)	Nutrient solution	3 mM	Sodium silicate SiO_2 nanoparticles	Enhanced K^+ content in roots and shoots	Ismail et al. (2022)
Radish (*Raphanus sativus* L.)	Soil	1% w/w	SiNPs	Increased root and shoot Zn concentration	Taqdees et al. (2022)
Roselle (*Hibiscus sabdariffa* L.)	Foliar spraying	30 and 60 g L^{-1}	Potassium silicate	Increased leaf K^+ concentration	Abou-Sreea et al. (2022)

Foliar application of silicon (as sodium and potassium silicate at 2.7 mmol L^{-1}) improved N, P, and K contents in salt-stressed sweet pepper plants (Abdelaal et al. 2020). Recently, foliar Si spraying with sodium metasilicate at 100 mg L^{-1} maintained ion homeostasis in salt-affected sweet basil (6000 mg of NaCl per kg soil) (Farouk et al. 2020).

Many reports show that Si supplementation in the growth medium increases mineral nutrition in tropical plants under salt stress (Rizwan et al. 2015; Etesami and Maheshwari 2018; Liu et al. 2019). Silicon can increase mineral nutrition in salt-stressed plants by increasing K^+ uptake (via K^+-H^+ symporters such as HvHAK1) (Xu et al. 2015; Rizwan et al. 2015). This increase in K^+ uptake might be due to root

Si deposition which can decrease Na^+ bypass flow, as reported for salt-stressed rice (Yeo et al. 1999; Gong et al. 2006; Gurmani et al. 2013b; Akhter et al. 2022) and wheat plants (Gurmani et al. 2013a). Various studies described the Si-mediated increase in K^+ uptake and decrease in Na^+ translocation towards shoots in wheat plants under salt stress (Zia et al. 2021; Premkumar et al. 2022).

Moreover, Si supplementation (50 mg kg^{-1} soil) with sodium silicate applied to the soil increased K^+ concentration and the K^+:Na^+ ratio in sunflower plants. These factors improved the adaptation of sunflower to NaCl stress (Ashraf et al. 2015). Similarly, root Si application (4 mM) as potassium silicate increased leaf K^+ and Ca^{2+} contents in salt-stressed wheat plants (Alzahrani et al. 2018). In soybean seedlings, Si added at 1 mM as nano-silicon oxide (nano-SiO_2) via nutrient solution improved K^+ content in shoots and roots of legumes under salt stress (Farhangi-Abriz and Torabian 2018).

Silicon in the form of metasilicic acid supplied at 2 mM in root medium attenuated the adverse effects of salinity by improving ion accumulation in salt-stressed maize plants (Khan et al. 2018). Recently, the application of 2 mM Si in the form of sodium silicate via nutrient solution modulated N metabolism and antioxidant enzyme activities in sunflower plants, attenuating the harmful effects of salinity (Conceição et al. 2019).

In turn, silicon supplementation in the form of sodium silicate (Na_2SiO_3) to NaCl-stressed mung bean plants ameliorated the adverse effects of NaCl on the accumulation of K^+ and Ca^{2+} and decreased Na^+, enhancing the Na^+:K^+ ratio (Ahmad et al. 2019). More recently, Si addition (2 mM) as stabilized sodium and potassium silicate to the growth medium improved the mineral nutrition of both sorghum and sunflower seedlings under salt stress by increasing macronutrient and micronutrient accumulation and their uptake and use efficiency, also enhancing the equilibrium of Na^+:K^+ ratios in shoots and roots (Calero Hurtado et al. 2019; Hurtado et al. 2021).

So far, the mechanisms of Si-mediated regulation of mineral nutrition to attenuate the adverse effects of salt stress in plants are poorly understood at the molecular and genetic levels. More genetic experiments are required to determine linkage relationships between Si and salt stress to study the expression level of genes related to transport, deposition, and translocation of macro- and micronutrients and Si in different plant species.

15.5 Special Role of Si in Increasing Plant Growth, Biomass, and Yield Under Salt Stress

The use of foliar spraying with Si compounds is relatively new. Foliar Si spraying increases plant growth and yield and reduces the detrimental effects of salinity stress (Abbas et al. 2015; Calero Hurtado et al. 2019, 2020b; Moghith et al. 2020; Mahmoud et al. 2022). Foliar spraying with Si is effective in attenuating salt stress regardless of source, compound, and concentrations applied. Foliar spraying with high Si concentration should guarantee the greatest uptake and accumulation of this element by the leaves of plants such as sorghum and sunflower (Calero Hurtado et al. 2020b; Hurtado et al. 2021). Therefore, it is important to use the best source and the best concentration of Si in the foliar spraying solution. Recently, foliar spraying with stabilized sodium and potassium silicate at a concentration of ~1 g L^{-1} (28.6 mmol L^{-1}) resulted in high Si accumulation in sorghum plants (Calero Hurtado et al. 2019; Oliveira et al. 2019). Other authors report that Si application via foliar spraying attenuated salinity stress (Abbas et al. 2015, 2017; Sattar et al. 2016, 2018). Some results suggest that foliar Si application led to significant increases in leaf Si contents in relation to nontreated plants (Laane 2018).

In addition, foliar Si spraying (1000 ppm SiO_2, potassium silicate) increased pod yield, number of seeds per plant, and yield in salt-stressed faba bean plants (Hellal et al. 2012). Similarly, foliar Si spraying (300 ppm, potassium silicate) increased plant height, leaf area, and dry weight of roots, stems, and leaves in salt-stressed moringa (*Moringa oleifera* L.) plants (Hussein and Abou-baker 2014). Furthermore, foliar application of Si (4 mmol L^{-1}, potassium silicate) improved plant weight, 100-seed weight, seed yield, and ear length in salt-stressed *Triticum aestivum* L. plants (Bybordi 2014).

Foliar spraying of Si (150 mg L^{-1}, silicic acid) also increased plant biomass production in salt-stressed okra plants (Abbas et al. 2015). Similarly, foliar spraying of Si (150 mg L^{-1}, potassium silicate) increased the growth and yield of salt-stressed *Pisum sativum* L. plants (Shahid et al. 2015). Moreover, foliar application of Si (1 mM of potassium silicate) increased shoot length and shoot number in *Cynodon dactylon* L. Pers., *Festuca arundinacea* Schreb., and *Lolium perenne* L. under salinity conditions (Esmaeili et al. 2015). Recently, Abbas et al. (2017) reported that foliar application of Si (150 mg L^{-1}, silicic acid) increased leaf area and leaf area index in salt-stressed okra plants. Similarly, Yavaş et al. (2019) (Si source not mentioned) found that foliar Si spraying increased fresh weight in salt-stressed *Stevia rebaudiana* plants.

Foliar application of Si as sodium and potassium silicate at 2.7 mmol L^{-1} significantly enhanced the number of fruits per plant, fruit fresh weight per plant (g), and fruit yield (t $hectare^{-1}$) of sweet pepper plants under NaCl stress (Abdelaal et al. 2020). Recently, foliar spraying of Si as sodium metasilicate at 100 mg L^{-1} improved the growth of salt-affected sweet basil (6000 mg NaCl per kg soil) (Farouk et al. 2020).

Foliar application of Si as sodium silicate at 600 ppm increased plant height, number of leaves, and number of nodes in cotton plants grown under salinity conditions (Nisar et al. 2022). Recently, foliar spraying with Si (100 mg L^{-1} of sodium silicate) improved the accumulation of P, Mg, Ca, Fe, Zn, Mn, and Cu in salt-stressed cucumber plants (Mousavi et al. 2022).

It has been extensively reported that Si supplementation to the growth medium increases plant growth and yield under salinity conditions (Etesami and Jeong 2018;

Souri et al. 2021). Silicon addition can increase nutrient uptake by increasing root surface area and length as well as root activity (Ma et al. 2001; Moussa and Mousa 2006; Hernandez-Apaolaza 2014).

Furthermore, silicon (1.92 g kg^{-1} of SiO_2 via soil) increased dry matter accumulation in salt-stressed sorghum plants (Nabati et al. 2013). Reports also show that Si supplementation (0.83 mM, metasilicic acid) in root medium increased leaf fresh weight, stem fresh weight, root fresh weight, total fresh weight, and leaf area in salt-stressed sorghum plants (Yin et al. 2013). Silicon application (2 mM, sodium silicate) via nutrient solution significantly increased the growth of sunflower seedlings under salinity stress (Ali et al. 2013). Similarly, Si supplementation (50 mg kg^{-1}, sodium silicate) applied to the soil enhanced fresh biomass, dry biomass, plant height, head diameter, number of seeds, and seed weight in salt-stressed sunflower plants (Ashraf et al. 2015).

Recently, Si addition (1.67 mM, sodium silicate) via nutrient solution increased the growth of rice cultivars under salt stress conditions (Flam-Shepherd et al. 2018). Similarly, Si application (2 mM, sodium silicate) via root medium improved the growth of salt-stressed rice plants (Das et al. 2018). Silicon supplementation (2.0 mM as silicic acid) via nutrient solution increased vegetative growth and dry matter of leaves and stems in salt-stressed tomato seedlings (Korkmaz et al. 2018). Likewise, Si application (4 mM as potassium silicate) in root medium increased growth in salt-stressed wheat plants (Alzahrani et al. 2018). Moreover, Si addition (1 mM as nano-silicon oxide (nano-SiO_2)) in nutrition solution improved shoot and root growth in salt-stressed soybean seedlings (Farhangi-Abriz and Torabian 2018).

Silicon application at 1.0 and 1.5 mM as silicic acid via nutrient solution significantly increased shoot and root weight in salt-stressed wheat seedlings (Sienkiewicz-Cholewa et al. 2018). Finally, positive effects of Si supplementation as sodium silicate (Na_2SiO_3) ameliorated the growth and biomass production of salt-stressed mung bean plants (Ahmad et al. 2019).

15.6 Conclusions

Silicon application methods nutritionally relieved salinity-induced damage by improving mineral nutrition, ion homeostasis, and nutritional efficiency. This resulted in a considerable increase in plant growth and yield.

Future studies should perform multi-year investigations to confirm the beneficial effects of Si-mediated attenuation of salt stress in plants, as this mechanism is poorly understood at the molecular and genetic levels. More genetic experiments are required to determine linkage relationships between silicon and salt stress to study the expression level of genes related to transport, deposition, and translocation of mineral nutrients and Si in different plant species.

References

Abbas T, Balal R, Shahid M et al (2015) Silicon-induced alleviation of NaCl toxicity in okra (*Abelmoschus esculentus*) is associated with enhanced photosynthesis, osmoprotectants and antioxidant metabolism. Acta Physiol Plant 37:1–15. https://doi.org/10.1007/s11738-014-1768-5

Abbas T, Sattar A, Ijaz M et al (2017) Exogenous silicon application alleviates salt stress in okra. Hortic Environ Biotechnol 58:342–349. https://doi.org/10.1007/s13580-017-0247-5

Abdelaal KAA, Mazrou YSA, Hafez YM (2020) Silicon foliar application mitigates salt stress in sweet pepper plants by enhancing water status, photosynthesis, antioxidant enzyme activity and fruit yield. Plan Theory 9:733. https://doi.org/10.3390/plants9060733

Abou-Sreea AIB, Roby MHH, Mahdy HAA et al (2022) Improvement of selected morphological, physiological, and biochemical parameters of roselle (*Hibiscus sabdariffa* L.) grown under different salinity levels using potassium silicate and aloe saponaria extract. Plants 11:497. https://doi.org/10.3390/PLANTS11040497

Agarie S, Agata W, Kubota F, Kaufmann PB (1998) Physiological roles of silicon in photosynthesis and dry matter production in rice plants: I. effects of silicon and shading treatments. Jpn J Crop Sci 61:200–206. https://doi.org/10.1626/jcs.61.200

Ahmad P, Abass M, Alam P et al (2019) Silicon (Si) supplementation alleviates NaCl toxicity in mung bean [*Vigna radiata* (L.) Wilczek] through the modifications of physio-biochemical attributes and key antioxidant enzymes. J Plant Growth Regul 38:70–82. https://doi.org/10.1007/s00344-018-9810-2

Ahmed N, Murtaza M, Ali MA et al (2019) Silicon improves rice nutrition and productivity under salinity. Pak J Bot 51:837–843. https://doi.org/10.30848/PJB2019

Akhter MS, Noreen S, Saleem N et al (2022) Silicon can alleviate toxic effect of NaCl stress by improving K^+ and Si uptake, photosynthetic efficiency with reduced Na^+ toxicity in barley (*Hordeum vulgare* L.). SILICON 14:4991–5000. https://doi.org/10.1007/S12633-021-01270-6

Al-Huqail AA, Alqarawi AA, Hashem A et al (2017) Silicon supplementation modulates antioxidant system and osmolyte accumulation to balance salt stress in *Acacia gerrardii* Benth. Saudi J Biol Sci 30:1–9. https://doi.org/10.1016/j.sjbs.2017.11.049

Ali S, Farooq MA, Yasmeen T et al (2013) The influence of silicon on barley growth, photosynthesis and ultra-structure under chromium stress. Ecotoxicol Environ Saf 89:66–72. https://doi.org/10.1016/j.ecoenv.2012.11.015

Ali M, Afzal S, Parveen A et al (2021) Silicon mediated improvement in the growth and ion homeostasis by decreasing Na^+ uptake in maize (*Zea mays* L.) cultivars exposed to salinity stress. Plant Physiol Biochem 158:208–218. https://doi.org/10.1016/J.PLAPHY.2020.10.040

Alsaeedi A, El-Ramady H, Alshaal T et al (2018) Exogenous nanosilica improves germination and growth of cucumber by maintaining K^+/Na^+ ratio under elevated Na^+ stress. Plant Physiol Biochem 125:164–171. https://doi.org/10.1016/J.PLAPHY.2018.02.006

Alsaeedi A, El-Ramady H, Alshaal T et al (2019) Silica nanoparticles boost growth and productivity of cucumber under water deficit and salinity stresses by balancing nutrients uptake. Plant Physiol Biochem 139:1–10. https://doi.org/10.1016/J.PLAPHY.2019.03.008

Alzahrani Y, Kuşvuran A, Alharby HF et al (2018) The defensive role of silicon in wheat against stress conditions induced by drought, salinity or cadmium. Ecotoxicol Environ Saf 154:187–196. https://doi.org/10.1016/j.ecoenv.2018.02.057

Amer M, El-Emary F (2018) Impact of foliar with nano-silica in mitigation of salt stress on some soil properties, crop-water productivity and anatomical structure of maize and faba bean. Environ Biodivers Soil Secur 2:25–38. https://doi.org/10.21608/JENVBS.2018.3753.1026

Apse MP, Blumwald E (2007) Na^+ transport in plants. FEBS Lett 581:2247–2254. https://doi.org/10.1016/j.febslet.2007.04.014

Ashraf M, Abid M, Teixeira da Silva JA et al (2015) Silicon and potassium nutrition enhances salt adaptation capability of sunflower by improving plant water status and membrane stability. Commun Soil Sci Plant Anal 46:991–1005. https://doi.org/10.1080/00103624.2015.1018527

Asrar H, Hussain T, Hadi SMS et al (2017) Salinity induced changes in light harvesting and carbon assimilating complexes of *Desmostachya bipinnata* (L.) staph. Environ Exp Bot 135:86–95. https://doi.org/10.1016/j.envexpbot.2016.12.008

Assaha DVM, Ueda A, Saneoka H et al (2017) The role of Na^+ and K^+ transporters in salt stress adaptation in Glycophytes. Front Physiol 8:509. https://doi.org/10.3389/fphys.2017.00509

Bessa MC, Lacerda CF, Amorim AV et al (2017) Mechanisms of salt tolerance in seedlings of six woody native species of the Brazilian semi-arid. Rev Ciência Agron 48:157–165. https://doi.org/10.5935/1806-6690.20170018

Bosnic P, Bosnic D, Jasnic J, Nikolic M (2018) Silicon mediates sodium transport and partitioning in maize under moderate salt stress. Environ Exp Bot 155:681–687. https://doi.org/10.1016/j.envexpbot.2018.08.018

Bybordi A (2014) Interactive effects of silicon and potassium nitrate in improving salt tolerance of wheat. J Integr Agric 13:1889–1899. https://doi.org/10.1016/S2095-3119(13)60639-5

Calero Hurtado A, Chiconato DA, de Mello Prado R et al (2019) Silicon attenuates sodium toxicity by improving nutritional efficiency in sorghum and sunflower plants. Plant Physiol Biochem 142:224–233. https://doi.org/10.1016/j.plaphy.2019.07.010

Calero Hurtado A, Aparecida Chiconato D, de Mello Prado R et al (2020a) Silicon application induces changes C:N:P stoichiometry and enhances stoichiometric homeostasis of sorghum and sunflower plants under salt stress. Saudi J Biol Sci 27:3711–3719. https://doi.org/10.1016/j.sjbs.2020.08.017

Calero Hurtado A, Chiconato DA, de Mello Prado R et al (2020b) Different methods of silicon application attenuate salt stress in sorghum and sunflower by modifying the antioxidative defense mechanism. Ecotoxicol Environ Saf 203:110964. https://doi.org/10.1016/j.ecoenv.2020.110964

Conceição S, Neto C, Marques E et al (2019) Silicon modulates the activity of antioxidant enzymes and nitrogen compounds in sunflower plants under salt stress. Arch Agron Soil Sci 65:1237–1247. https://doi.org/10.1080/03650340.2018.1562272

Das P, Manna I, Biswas AK, Bandyopadhyay M (2018) Exogenous silicon alters ascorbate-glutathione cycle in two salt-stressed indica rice cultivars (MTU 1010 and Nonabokra). Environ Sci Pollut Res 25:26625–26642. https://doi.org/10.1007/s11356-018-2659-x

de Luma CS, da Ellen GSL, de Risely FA et al (2016) Nitrogen metabolism in sorghum under salinity and silicon treatments in Brazil. Afr J Agric Res 11:199–208. https://doi.org/10.5897/ajar2015.10453

de Mello Prado R (2021) Nutrition of tropical plants. Springer

Demidchik V (2015) Mechanisms of oxidative stress in plants: from classical chemistry to cell biology. Environ Exp Bot 109:212–228. https://doi.org/10.1016/j.envexpbot.2014.06.021

EL Sabagh A, Hossain A, Barutçular C et al (2020) Consequences of salinity stress on the quality of crops and its mitigation strategies for sustainable crop production: an outlook of arid and semi-arid regions. Environ Clim Plant Veg Growth:503–533. https://doi.org/10.1007/978-3-030-49732-3_20

El-Ramady H, Alshaal T, Elhawat N et al (2018) Plant nutrients and their roles under saline soil conditions. In: Hasanuzzaman M, Fujita M, Oku H, Nahar K, Hawrylak-Nowak B (eds) Plant nutrients and abiotic stress tolerance. Springer, Singapore, pp 297–324

Epstein E (1999) Silicon. Annu Rev Plant Physiol Plant Mol Biol 50:641–664. https://doi.org/10.1146/annurev.arplant.50.1.641

Esmaeili S, Salehi H, Eshghi S (2015) Silicon ameliorates the adverse effects of salinity on turfgrass growth and development. J Plant Nutr 38:1885–1901. https://doi.org/10.1080/01904167.2015.1069332

Etesami H, Jeong BR (2018) Silicon (Si): review and future prospects on the action mechanisms in alleviating biotic and abiotic stresses in plants. Ecotoxicol Environ Saf 147:881–896. https://doi.org/10.1016/j.ecoenv.2017.09.063

Etesami H, Maheshwari DK (2018) Safety use of plant growth promoting rhizobacteria (PGPRs) with multiple plant growth promoting traits in stress agriculture: action mechanisms and future prospects. Ecotoxicol Environ Saf 156:225–246. https://doi.org/10.1016/j.ecoenv.2018.03.013

Farhangi-Abriz S, Torabian S (2018) Nano-silicon alters antioxidant activities of soybean seedlings under salt toxicity. Protoplasma 255:953–962. https://doi.org/10.1007/s00709-017-1202-0

Farouk S, Elhindi KM, Alotaibi MA (2020) Silicon supplementation mitigates salinity stress on *Ocimum basilicum* L. via improving water balance, ion homeostasis, and antioxidant defense system. Ecotoxicol Environ Saf 206:111396. https://doi.org/10.1016/J.ECOENV.2020.111396

Flam-Shepherd R, Huynh WQ, Coskun D et al (2018) Membrane fluxes, bypass flows, and sodium stress in rice: the influence of silicon. J Exp Bot 69:1679–1692. https://doi.org/10.1093/jxb/erx460

Gabr SM, Abouelsaad IA, Brengi S, Gouda A (2022) Growth and yield of spinach as affected by silicon and fulvic acid under salt stress. J Adv Agric Res 27:26–42. https://doi.org/10.21608/JALEXU.2022.114829.1037

Gong HJ, Randall DP, Flowers TJ (2006) Silicon deposition in the root reduces sodium uptake in rice (*Oryza sativa* L.) seedlings by reducing bypass flow. Plant Cell Environ 29:1970–1979. https://doi.org/10.1111/j.1365-3040.2006.01572.x

Gonzalo MJ, Lucena JJ, Hernández-Apaolaza L (2013) Effect of silicon addition on soybean (*Glycine max*) and cucumber (*Cucumis sativus*) plants grown under iron deficiency. Plant Physiol Biochem 70:455–461. https://doi.org/10.1016/j.plaphy.2013.06.007

Grattan SR, Grieve CM (1999) Salinity-mineral nutrient relations in horticultural crops. Sci Hortic 78:127–157. https://doi.org/10.1016/S0304-4238(98)00192-7

Gurmani A, Bano A, Najeeb U et al (2013a) Exogenously applied silicate and abscisic acid ameliorates the growth of salinity stressed wheat ('*Triticum aestivum*' L) seedlings through Na^+ exclusion. Aust J Crop Sci 7:1123–1130

Gurmani A, Bano A, Ullah N et al (2013b) Exogenous abscisic acid (ABA) and silicon (Si) promote salinity tolerance by reducing sodium (Na^+) transport and bypass flow in rice ("*Oryza sativa*" indica). Aust J Crop Sci 7:1219–1226

Hanin M, Ebel C, Ngom M et al (2016) New insights on plant salt tolerance mechanisms and their potential use for breeding. Front Plant Sci 7:1–17. https://doi.org/10.3389/fpls.2016.01787

Hellal FA, Abdelhameid M, Abo-Basha DM, Zewainy RM (2012) Alleviation of the adverse effects of soil salinity stress by foliar application of silicon on faba bean (*Vicia faba* L.). J Appl Sci Res 8:4428–4433

Hernandez-Apaolaza L (2014) Can silicon partially alleviate micronutrient deficiency in plants? A review. Planta 240:447–458. https://doi.org/10.1007/s00425-014-2119-x

Hernández-Salinas M, Valdez-Aguilar LA, Alia-Tejacal I et al (2021) Silicon enhances the tolerance to moderate NaCl-salinity in tomato grown in a hydroponic recirculating system. J Plant Nutr 45:413–425. https://doi.org/10.1080/01904167.2021.1963772

Hurtado AC, Chiconato DA, de Mello PR et al (2021) Silicon alleviates sodium toxicity in sorghum and sunflower plants by enhancing ionic homeostasis in roots and shoots and increasing dry matter accumulation. SILICON 13:475–486. https://doi.org/10.1007/s12633-020-00449-7

Hussein MJ, Abdullah AK (2019) Exogenous of silicon and glycine betaine improves salinity tolerance of pepper plants (*Capsicum annum* L.). Plant Arch 19:664–672

Hussein MM, Abou-baker NH (2014) Growth and mineral status of moringa plants as affected by silicate and salicylic acid under salt stress. Int J Plant Soil Sci 3:163–177

Ismail LM, Soliman MI, Abd El-Aziz MH, Abdel-Aziz HMM (2022) Impact of silica ions and nano silica on growth and productivity of pea plants under salinity stress. Plan Theory 11:494. https://doi.org/10.3390/PLANTS11040494

Jalili S, Ehsanpour AA, Javadirad SM (2022) The role of melatonin on caspase-3-like activity and expression of the genes involved in programmed cell death (PCD) induced by in vitro salt stress in alfalfa (*Medicago sativa* L.) roots. Bot Stud 63:1–11. https://doi.org/10.1186/S40529-022-00348-7/FIGURES/5

James RA, Blake C, Byrt CS, Munns R (2011) Major genes for Na^+ exclusion, Nax1 and Nax2 (wheat HKT1;4 and HKT1;5), decrease Na^+ accumulation in bread wheat leaves under saline and waterlogged conditions. J Exp Bot 62:2939–2947. https://doi.org/10.1093/jxb/err003

Jana GA, Al Kharusi L, Sunkar R et al (2019) Metabolomic analysis of date palm seedlings exposed to salinity and silicon treatments. Plant Signal Behav 14:11. https://doi.org/10.1080/15592324.2019.1663112

Khan W, Aziz T, Maqsood M et al (2018) Silicon nutrition mitigates salinity stress in maize by modulating ion accumulation, photosynthesis, and antioxidants. Photosynthetica 56:1047–1057. https://doi.org/10.1007/s11099-018-0812-x

Khan A, Khan AL, Muneer S et al (2019) Silicon and salinity: crosstalk in crop-mediated stress tolerance mechanisms. Front Plant Sci 10:1429. https://doi.org/10.3389/FPLS.2019.01429/BIBTEX

Kibria MG, Hossain M, Murata Y, Hoque MA (2017) Antioxidant defense mechanisms of salinity tolerance in rice genotypes. Rice Sci 24:155–162. https://doi.org/10.1016/j.rsci.2017.05.001

Korkmaz A, Karagöl A, Akınoğlu G, Korkmaz H (2018) The effects of silicon on nutrient levels and yields of tomatoes under saline stress in artificial medium culture. J Plant Nutr 41:123–135. https://doi.org/10.1080/01904167.2017.1381975

Laane H-M (2018) The effects of foliar sprays with different silicon compounds. Plan Theory 7:45. https://doi.org/10.3390/plants7020045

Liang Y (1999) Effects of silicon on enzyme activity and sodium, potassium and calcium concentration in barley under salt stress. Plant Soil 209:217–224. https://doi.org/10.1023/A:1004526604913

Liang Y, Si J, Römheld V (2005) Silicon uptake and transport is an active process in *Cucumis sativus*. New Phytol 167:797–804. https://doi.org/10.1111/j.1469-8137.2005.01463.x

Liang Y, Zhang W, Chen Q et al (2006) Effect of exogenous silicon (Si) on H+-ATPase activity, phospholipids and fluidity of plasma membrane in leaves of salt-stressed barley (Hordeum vulgare L.). Environ Exp Bot 57:212–219. https://doi.org/10.1016/j.envexpbot.2005.05.012

Liu SL, Yang RJ, Ma MD et al (2015) Effects of exogenous NO on the growth, mineral nutrient content, antioxidant system, and ATPase activities of *Trifolium repens* L. plants under cadmium stress. Acta Physiol Plant 37:1721. https://doi.org/10.1007/s11738-014-1721-7

Liu B, Soundararajan P, Manivannan A (2019) Mechanisms of silicon-mediated amelioration of salt stress in plants. Plan Theory 8:307. https://doi.org/10.3390/plants8090307

Ma JF, Goto S, Tamai K, Ichii M (2001) Role of root hairs and lateral roots in silicon uptake by rice. Plant Physiol 127:1773–1780. https://doi.org/10.1104/pp.010271

Mahmoud LM, Shalan AM, El-Boray MS et al (2022) Application of silicon nanoparticles enhances oxidative stress tolerance in salt stressed 'Valencia' sweet orange plants. Sci Hortic 295:110856. https://doi.org/10.1016/J.SCIENTA.2021.110856

Manivannan A, Ahn YK (2017) Silicon regulates potential genes involved in major physiological processes in plants to combat stress. Front Plant Sci 8:1–13. https://doi.org/10.3389/fpls.2017.01346

Moghith WMA, Youssef ASM, El-Wahab MAA et al (2020) Effect of saline water stress in the presence of silicon foliar application on growth, productivity and chemical constituents of chia (*Salvia hispanica* L.) under egyptian conditions. Asian Plant Res J 4:28–45. https://doi.org/10.9734/aprj/2020/v4i230083

Mousavi SAA, Roosta HR, Esmaeilizadeh M, Eshghi S (2022) Silicon and selenium supplementations modulate antioxidant systems and mineral nutrition to mitigate salinity-alkalinity stresses in cucumber (*Cucumis sativus* L.) plants under hydroponic conditions. J Plant Process Funct 10:41–49

Moussa HR, Mousa HR (2006) Influence of exogenous application of silicon on physiological response of salt-stressed maize (*Zea mays* L.). Int J Agric Biol 8:293–297

Muneer S, Jeong BR (2015) Proteomic analysis of salt-stress responsive proteins in roots of tomato (*Lycopersicon esculentum* L.) plants towards silicon efficiency. Plant Growth Regul 77:133–146. https://doi.org/10.1007/s10725-015-0045-y

Muneer S, Kim E, Park J, Lee J (2014) Influence of green, red and blue light emitting diodes on multiprotein complex proteins and photosynthetic activity under different light intensities in lettuce leaves (*Lactuca sativa* L.). Int J Mol Sci 15:4657–4670. https://doi.org/10.3390/ijms15034657

Munns R (2011) Plant adaptations to salt and water stress. Differences and commonalities. In: Turkan I (ed) Advances in botanical research, 1st edn. Elsevier Ltd, Amsterdam, pp 1–32

Munns R, Tester M (2008) Mechanisms of salinity tolerance. Annu Rev Plant Biol 59:651–681. https://doi.org/10.1146/annurev.arplant.59.032607.092911

Munns R, James R, Gilliham M et al (2016) Tissue tolerance: an essential but elusive trait for salt-tolerant crops. Funct Plant Biol 43:1103–1113. https://doi.org/10.1071/FP16187

Nabati J, Kafi M, Masoumi A, Mehrjerdi MZ (2013) Effect of salinity and silicon application on photosynthetic characteristics of sorghum (*Sorghum bicolor* L.). Int J Agric Sci 3:483–492

Neumann PM, Azaizeh H, Leon D (1994) Hardening of root cell walls: a growth inhibitory response to salinity stress. Plant Cell Environ 17:303–309. https://doi.org/10.1111/j.1365-3040.1994.tb00296.x

Nilsen ET, Orcutt DM (1996) Physiology of plants under stress. Abiotic factors. Wiley, New York

Nisar S, Iqbal M, Ashraf J et al (2022) Enhancing salt tolerance in cotton by improving its morpho-physiological and antioxidant potential through foliar applied silicon. Silicon 14:11243. https://doi.org/10.1007/S12633-022-01849-7

Nouripour-Sisakht J, Ehsanzadeh P, Ehtemam MH (2022) Fennel outperforms ajwain and anise in saline environment: physiological response mechanisms in germinating seeds and mature plants. Ital J Agron. https://doi.org/10.4081/IJA.2022.2096

Oliveira RLL, de Mello Prado R, Felisberto G, Cruz FJR (2019) Different sources of silicon by foliar spraying on the growth and gas exchange in sorghum. J Soil Sci Plant Nutr 19:948–953. https://doi.org/10.1007/s42729-019-00092-1

Olivera-Viciedo D, de Mello Prado R, Martínez CA et al (2019) Short-term warming and water stress affect *Panicum maximum* Jacq. stoichiometric homeostasis and biomass production. Sci Total Environ 681:267–274. https://doi.org/10.1016/j.scitotenv.2019.05.108

Olivera-Viciedo D, Mello Prado R, Lizcano Toledo R et al (2020) Physiological role of silicon in radish seedlings under ammonium toxicity. J Sci Food Agric 100:5637–5644. https://doi.org/10.1002/jsfa.10587

Olivera-Viciedo D, de Mello Prado R, Martinez CA et al (2021a) Changes in soil water availability and air-temperature impact biomass allocation and C:N:P stoichiometry in different organs of *Stylosanthes capitata* Vogel. J Environ Manag 278:111540. https://doi.org/10.1016/j.jenvman.2020.111540

Olivera-Viciedo D, de Mello Prado R, Martinez CA et al (2021b) Are the interaction effects of warming and drought on nutritional status and biomass production in a tropical forage legume greater than their individual effects? Planta 254:1–10. https://doi.org/10.1007/S00425-021-03758-2/FIGURES/5

Peña Calzada K, Calero Hurtado A, Olivera Viciedo D et al (2023) Regulatory role of silicon on growth, potassium uptake, ionic homeostasis, Proline accumulation, and antioxidant capacity of soybean plants under salt stress. J Plant Growth Regul 1–13. https://doi.org/10.1007/S00344-023-10921-4

Pessoa LGM, Freire MBGDS, Wilcox BP et al (2016) Spectral reflectance characteristics of soils in northeastern Brazil as influenced by salinity levels. Environ Monit Assess 188:1–11. https://doi.org/10.1007/s10661-016-5631-6

Premkumar A, Javed MT, Pawlowski K, Lindberg SM (2022) Silicate inhibits the cytosolic influx of chloride in protoplasts of wheat and affects the chloride transporters, TaCLC1 and TaNPF2.4/2.5. Plants 11:1162. https://doi.org/10.3390/PLANTS11091162

Raza MM, Ullah S, Aziz T et al (2019) Alleviation of salinity stress in maize using silicon nutrition. Not Bot Horti Agrobot 47:1340–1347. https://doi.org/10.15835/NBHA47411584

Rengel Z (1992) The role of calcium in salt toxicity. Plant Cell Environ 15:625–632. https://doi.org/10.1111/j.1365-3040.1992.tb01004.x

Riedelsberger J, Vergara-Jaque A, Piñeros M et al (2019) Extracellular cation binding pocket is essential for ion conduction of OsHKT2;2. BMC Plant Biol 19:316. https://doi.org/10.1186/s12870-019-1909-5

Rizwan M, Ali S, Ibrahim M et al (2015) Mechanisms of silicon-mediated alleviation of drought and salt stress in plants: a review. Environ Sci Pollut Res 22:15416–15431. https://doi.org/10.1007/s11356-015-5305-x

Sattar A, Cheema MA, Ali H et al (2016) Silicon mediates the changes in water relations, photosynthetic pigments, enzymatic antioxidants activity and nutrient uptake in maize seedling under salt stress. Grassl Sci 62:262–269. https://doi.org/10.1111/grs.12132

Sattar A, Cheema MA, Sher A et al (2018) Foliage applied silicon alleviates the combined effects of salinity and drought stress on wheat seedlings. Int J Agric Biol 20:2537–2543. https://doi.org/10.17957/IJAB/15.0800

Sattar A, Cheema MA, Sher A et al (2019) Foliar applied silicon improves water relations, stay green and enzymatic antioxidants activity in late sown wheat. Silicon:1–8. https://doi.org/10.1007/s12633-019-00115-7

Sayed EG, Mahmoud AWM, El-Mogy MM et al (2022) The effective role of nano-silicon application in improving the productivity and quality of grafted tomato grown under salinity stress. Horticulturae 8:293. https://doi.org/10.3390/HORTICULTURAE8040293

Schoelynck J, Bal K, Backx H et al (2010) Silica uptake in aquatic and wetland macrophytes: a strategic choice between silica, lignin and cellulose? New Phytol 186:385–391. https://doi.org/10.1111/j.1469-8137.2009.03176.x

Shahid M, Balal R, Pervez M et al (2015) Foliar spray of phyto-extracts supplemented with silicon: an efficacious strategy to alleviate the salinity-induced deleterious effects in pea (*Pisum sativum* L.). Tur J Bot 39:408–419. https://doi.org/10.3906/bot-1406-84

Shahid MA, Sarkhosh A, Khan N et al (2020) Insights into the physiological and biochemical impacts of salt stress on plant growth and development. Agronomy 10:938. https://doi.org/10.3390/AGRONOMY10070938

Shen Z, Pu X, Wang S et al (2022) Silicon improves ion homeostasis and growth of liquorice under salt stress by reducing plant Na+ uptake. Sci Reports 121:1–13. https://doi.org/10.1038/s41598-022-09061-8

Siddiqui MN, Mostofa MG, Akter MM et al (2017) Impact of salt-induced toxicity on growth and yield-potential of local wheat cultivars: oxidative stress and ion toxicity are among the major determinants of salt-tolerant capacity. Chemosphere 187:385–394. https://doi.org/10.1016/J.CHEMOSPHERE.2017.08.078

Sienkiewicz-Cholewa U, Sumisławska J, Sacała E et al (2018) Influence of silicon on spring wheat seedlings under salt stress. Acta Physiol Plant 40:1–8. https://doi.org/10.1007/s11738-018-2630-y

Soundararajan P, Manivannan A, Ko CH et al (2017) Leaf physiological and proteomic analysis to elucidate silicon induced adaptive response under salt stress in Rosa hybrida 'Rock Fire'. Int J Mol Sci 18:1768. https://doi.org/10.3390/ijms18081768

Soundararajan P, Manivannan A, Ko CH, Jeong BR (2018) Silicon enhanced redox homeostasis and protein expression to mitigate the salinity stress in Rosa hybrida 'Rock Fire'. J Plant Growth Regul 37:16–34. https://doi.org/10.1007/s00344-017-9705-7

Souri Z, Khanna K, Karimi N, Ahmad P (2021) Silicon and plants: current knowledge and future prospects. J Plant Growth Regul 40:906–925. https://doi.org/10.1007/s00344-020-10172-7

Tahir M, Aziz T, Farooq M, Sarwar G (2011) Silicon-induced changes in growth, ionic composition, water relations, chlorophyll contents and membrane permeability in two salt-stressed wheat genotypes. Arch Agron Soil Sci 58:247–256. https://doi.org/10.1080/03650340.2010.518959

Taqdees Z, Khan J, Khan WUD et al (2022) Silicon and zinc nanoparticles-enriched Miscanthus biochar enhanced seed germination, antioxidant defense system, and nutrient status of radish under NaCl stress. Crop Pasture Sci 73:556–572. https://doi.org/10.1071/CP21342

Tränkner M, Tavakol E, Jákli B (2018) Functioning of potassium and magnesium in photosynthesis, photosynthate translocation and photoprotection. Physiol Plant 163:414–431. https://doi.org/10.1111/PPL.12747

Tuna AL, Kaya C, Higgs D et al (2008) Silicon improves salinity tolerance in wheat plants. Environ Exp Bot 62:10–16. https://doi.org/10.1016/j.envexpbot.2007.06.006

Türkan I, Demiral T (2009) Recent developments in understanding salinity tolerance. Environ Exp Bot 67:2–9. https://doi.org/10.1016/J.ENVEXPBOT.2009.05.008

Voigt EL, Caitano RF, Maia JM et al (2009) Involvement of cation channels and NH_4^+-sensitive K^+ transporters in Na^+ uptake by cowpea roots under salinity. Biol Plant 53:764–768. https://doi.org/10.1007/s10535-009-0140-x

Wang S, Liu P, Chen D et al (2015) Silicon enhanced salt tolerance by improving the root water uptake and decreasing the ion toxicity in cucumber. Front Plant Sci 6:759. https://doi.org/10.3389/fpls.2015.00759

Xie Z, Song R, Shao H et al (2015) Silicon improves maize photosynthesis in saline-alkaline soils. Sci World J 2015:1–6. https://doi.org/10.1155/2015/245072

Xu CX, Ma YP, Liu YL (2015) Effects of silicon (Si) on growth, quality and ionic homeostasis of aloe under salt stress. S Afr J Bot 98:26–36. https://doi.org/10.1016/j.sajb.2015.01.008

Yaghubi K, Vafaee Y, Ghaderi N, Javadi T (2019) Potassium silicate improves salinity resistant and affects fruit quality in two strawberry cultivars grown under salt stress. Commun Soil Sci Plant Anal 50:1439–1451. https://doi.org/10.1080/00103624.2019.1621333

Yan G, Fan X, Peng M et al (2020) Silicon improves rice salinity resistance by alleviating ionic toxicity and osmotic constraint in an organ-specific pattern. Front Plant Sci 11:260. https://doi.org/10.3389/FPLS.2020.00260/BIBTEX

Yavaş İ, Yılmaz F, Ünay A (2019) Promoting effect of foliar silicon on steviol glycoside contents of *Stevia rebaudiana* Bertoni under salt stress. Int J Second Metab 6:263–269

Yeo AR, Flowers SA, Rao G et al (1999) Silicon reduces sodium uptake in rice (*Oryza sativa* L.) in saline conditions and this is accounted for by a reduction in the transpirational bypass flow. Plant Cell Environ 22:559–565. https://doi.org/10.1046/j.1365-3040.1999.00418.x

Yin L, Wang S, Li J et al (2013) Application of silicon improves salt tolerance through ameliorating osmotic and ionic stresses in the seedling of *Sorghum bicolor*. Acta Physiol Plant 35:3099–3107. https://doi.org/10.1007/s11738-013-1343-5

Yin L, Wang S, Tanaka K et al (2016) Silicon-mediated changes in polyamines participate in silicon-induced salt tolerance in *Sorghum bicolor* L. Plant Cell Environ 39:245–258. https://doi.org/10.1111/pce.12521

Zhu Y, Xu X, Hu Y et al (2015) Silicon improves salt tolerance by increasing root water uptake in *Cucumis sativus* L. Plant Cell Rep 34:1629–1646. https://doi.org/10.1007/s00299-015-1814-9

Zhu YX, Gong HJ, Yin JL (2019) Role of silicon in mediating salt tolerance in plants: a review. Plan Theory 8:0147. https://doi.org/10.3390/plants8060147

Zhu Y, Jiang X, Zhang J et al (2020) Silicon confers cucumber resistance to salinity stress through regulation of proline and cytokinins. Plant Physiol Biochem 156:209–220. https://doi.org/10.1016/J.PLAPHY.2020.09.014

Zia A, Hegazy HS, Hassan NS et al (2021) Biochemical responses of wheat to silicon application under salinity. J Plant Nutr Soil Sci 184:255–262. https://doi.org/10.1002/JPLN.202000181

Chapter 16
Silicon Mitigates the Effects of Water Deficit in Tropical Plants

Gelza Carliane Marques Teixeira and Renato de Mello Prado

16.1 Introduction

Life is intimately associated with water, particularly with water in its liquid phase. Water is the form in which the H atom, an essential element of all organic molecules, is absorbed and then assimilated in the course of photosynthesis. About 500 g of water is absorbed by roots to produce 1 g of organic material and thus may be considered as a plant nutrient, in the same way as CO_2 or NO_3^- are also plant nutrients (Mengel et al. 2001).

The exponential growth of the world's population has put unprecedented pressure on the availability of freshwater resources. Food production systems consume more than 70% of the world's freshwater withdrawal. This highlights the relevance of research-backed practice efforts to increase water-use efficiency. In tropical climate regions, cyclical water scarcity is evident in different regions of cultivation. This increases the urgency of using innovative practices to increase water efficiency aiming for an economic advantage and at the same time reducing environmental impacts.

The future scenario indicates a possible worsening of the occurrence of environmental stresses, such as water deficit, which may require the frequent use of silicon (Si), a beneficial element used in agriculture with a potential to collaborate in ensuring food security. This requires an increase in the efficiency of applications of this beneficial element. The strategy is to use fertigation, which could reduce the rates of polymerization of Si in the soil and, consequently, the doses of this element per area. The aim is to ensure adequate absorption of Si by the plant and reduce environmental impacts. The decrease in Si polymerization rates in this new form of element application via fertigation occurs due to the use of soluble sources associated

G. C. M. Teixeira · R. de Mello Prado (✉)
São Paulo State University, Jaboticabal, São Paulo, Brazil
e-mail: rm.prado@unesp.br

R. de Mello Prado (ed.), *Benefits of Silicon in the Nutrition of Plants*,
https://doi.org/10.1007/978-3-031-26673-7_16

with low concentrations of Si at higher application frequencies, consequently obtaining greater beneficial effects of Si and at the same time decreasing the required doses of this element per area.

In this chapter, we show that it is possible to improve water-use efficiency by adopting fertigation with soluble Si at different doses per area and that are agronomically and economically viable. This effect may be a strategy for agricultural cultivation in areas of low water availability or could improve the future growth of irrigated agriculture, which uses large, irrigated poles. This could expand world production.

The objective of this article is to address the potentials of this new form of application of Si via fertigation to improve the Si use efficiency in irrigated agriculture aiming to mitigate damage caused by water deficit and to show the need for further studies aiming to reduce irrigation depths for different cultivated species depending on the type of soil.

16.2 Damage to Tropical Plants Caused by Water Deficit

Water deficit is one of the abiotic stresses that threaten food security and energy production the most. It is an important environmental factor that may limit plant growth and interfere with physiological characteristics, affecting plant stability, productivity, and survival (Contin et al. 2014). The effects of water deficit have increased due to the climate change, and agricultural crops are increasingly subject to limited water resources and longer periods of exposure to water deficit (Besharat et al. 2020). Such periods have occurred with greater intensity, especially in regions with sandy soils and low water retention capacity (Schaller et al. 2020).

Damage caused by water deficit has limited the production of several crops of agricultural importance, especially those grown in tropical climate conditions (Ferreira et al. 2017). They can be exposed to adverse conditions at important stages of development, considering that drought may occur during all climatic seasons and plantations are carried out at almost all times of the year (Basnayake et al. 2012). The exposure of plants to periods of low rainfall may limit their development, especially when it occurs during important growth phases, such as the initial establishment in the field and the reproductive phase. Drought during the crop establishment phase causes severe damage to plants (Jain et al. 2015), especially to sensitive plants from seeds, which have low nutrient and water reserves (Martins et al. 2015). This may cause death of seedlings, decreasing the population of plants in the field and consequently productivity.

The low availability of water in the soil may cause molecular, biochemical, physiological, and morphological changes in plants. The effects caused by water deficit in plants begin with physiological damage, with loss of water from plant tissues. It is characterized mainly by a decrease in leaf water content (Teixeira et al. 2020a) due to increase in transpiration (Verma et al. 2021), decreased leaf water potential (de Oliveira Filho et al. 2021), and osmotic adjustment imbalance (Pei et al. 2010). The low volume of water in tissues triggers harmful effects on biochemical aspects related to plant defense, with a decrease in the action of enzymatic (Bezerra et al. 2019) and non-enzymatic antioxidants (Teixeira et al. 2021a). This favors cell oxidation processes and degradation of lipid layers of membranes of these compounds as a reflection of the increase in oxidative stress.

Oxidative stress is a biological event commonly observed in plants that are subjected to abiotic stresses. It is characterized by increased production of reactive oxygen species (ROS) mainly in chloroplasts, mitochondria, and peroxisomes (Gratão et al. 2005). High concentrations of ROS, such as superoxide radicals (O_2^-), hydrogen peroxide (H_2O_2), hydroxyl radical (OH^-), and singlet oxygen ($1O_2$), can cause damage to nucleic acids, enzymatic inhibition, chlorophyll degradation, oxidation of proteins, and lipid peroxidation (Sachdev et al. 2021). Thus, oxidative stress occurs when there is an imbalance between ROS formation and the ability of the antioxidant machinery to detoxify or eliminate ROS. Oxidative damage to cellular components may eventually occur if an increased formation of ROS and other pro-oxidants is not counterbalanced (Kim et al. 2017). These combined effects cause a decrease in photosynthetic rates (Verma et al. 2021) and, consequently, in the development and production of dry mass.

Recent research has proposed new approaches to justify the decrease in dry mass production caused by water deficit. Such new approaches refer to the modification of the stoichiometry of nutrients with structural functions such as carbon (C), nitrogen (N), and phosphorus (P). Such changes cause ionic imbalance and decrease nutritional homeostasis, compromising the use efficiency of these nutrients by living systems (Elser and Hamilton 2007). There are indications that water deficit, because it changes elementary stoichiometric homeostasis, induces nutritional imbalance. This is one of the factors responsible for reducing the growth of plants exposed to water deficit (Teixeira et al. 2020b).

16.3 Plant Defense System Against Damage Caused by Water Deficit

In order to withstand water deficit, plant organisms have multiple mechanisms of action from the cellular level to the tissue level up to the entire plant level. They are activated at different stages of development (Marcos et al. 2018). Osmotic adjustment is one of the first mechanisms used by plants to maintain plant cell turgor in conditions of low water supply (Avila et al. 2021). A limited water supply in the soil leads to an increase in the accumulation of compatible solutes in leaves, such as proline and glycine betaine (Merwad et al. 2018). In roots, there is an increase in the synthesis of abscisic acid, which is transported to leaves, where it induces stomatal closure, decreases water loss through transpiration, and maintains turgor (Maillard et al. 2018). These physiological modifications protect the integrity of membranes and maintain the influx of water into cells (Fig. 16.1) (Iannucci et al. 2002; da Silva et al. 2012).

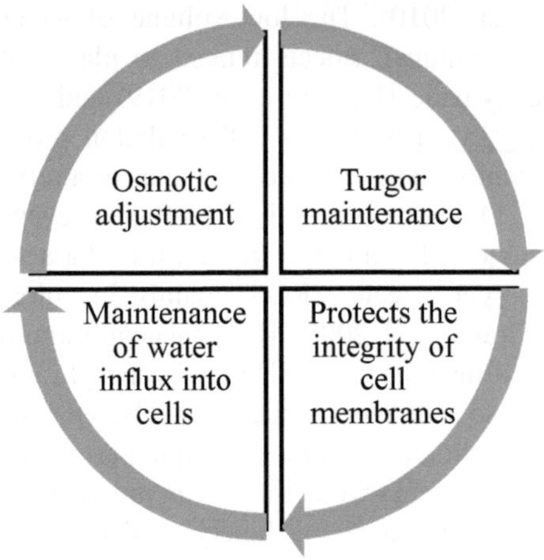

Fig. 16.1 Physiological effects associated with osmotic adjustment in plants under water deficit. (Source: Prepared by the authors)

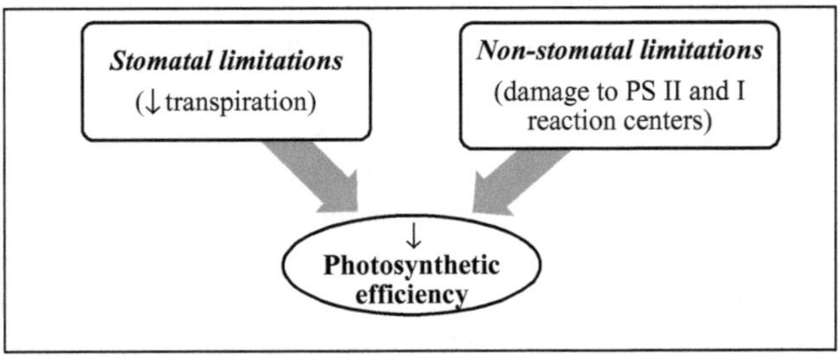

Fig. 16.2 Stomatal and non-stomatal limitations that cause a decrease in photosynthetic efficiency. (Source: Prepared by the authors)

However, when stomatal conductance decreases, there is a lower influx of CO_2 from the atmosphere into carboxylation sites. This exerts a direct effect on the reduction of photosynthetic rates under water stress conditions. In addition to stomatal limitations affecting photosynthesis, non-stomatal limitations may damage the reaction centers of the photosystems (PS) I and II and may also compromise photosynthetic efficiency (Fig. 16.2) (Verma et al. 2021).

To combat the toxic effects caused by the excessive accumulation of ROS, plants have developed an efficient antioxidant system. Antioxidant defenses include antioxidant enzymes such as superoxide dismutases (SOD), catalases (CAT), ascorbate peroxidases (APX), and class III peroxidases (POD) (Gratão et al. 2005). SOD isoforms are the first line of antioxidant defense and are identified as a family of metalloenzymes that catalyze O^{2-} into H_2O_2 and eliminate it (Fig. 16.3) (Ferreira et al. 2017).

CAT isoforms, predominantly located in peroxisomes and glyoxysomes, are heme proteins that convert H_2O_2 into O_2 and H_2O (Gratão et al. 2005). Various peroxidases, such as APX and POD, are located throughout the cell and detoxify H_2O_2. In addition, many non-enzymatic antioxidant molecules, such as ascorbate,

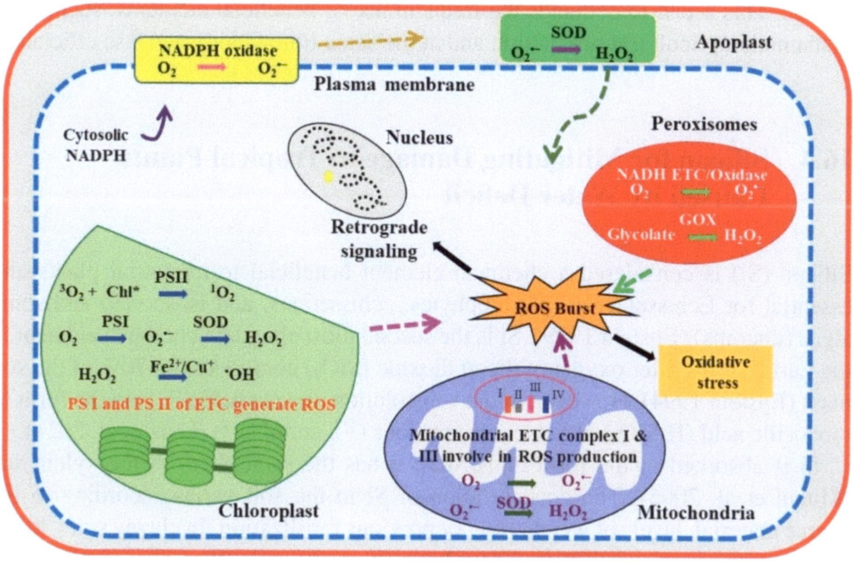

Fig. 16.3 Water deficit induces ROS production in different plant cell organelles that initiate signaling or cause oxidative stress. In the chloroplast, singlet oxygen ($1O_2$), superoxide radical ($O_2^{\bullet-}$), hydrogen peroxide (H_2O_2), and hydroxyl radical (•OH) are produced by an excited chlorophyll (Chl*) through the electron transport chain (ETC) in the PS I and II (Mehler reaction), $O_2^{\bullet-}$ dismutation by superoxide dismutase (SOD) and Fenton reaction catalyzed by reduced iron (Fe^{2+}) and copper (Cu^+), respectively. In peroxisomes, photorespiration (glycolate), enzymes, and small ETC-dependent NADH (nicotinamide adenine dinucleotide) induce the production of $O_2^{\bullet-}$ and H_2O_2. Mitochondrial ETC participates in the generation of $O_2^{\bullet-}$, which, by SOD dismutation, produces H_2O_2. Cytosolic NADPH induces the conversion of O_2 into $O_2^{\bullet-}$ by the action of the plasma membrane NADPH oxidase, which further dismutes to H_2O_2 in the apoplast by SOD. ROS produced in different cell organelles under pressure from abiotic stresses mediate signaling pathways at a low/moderate concentration or induce oxidative stress at a high concentration. (Source: Adapted from Sachdev et al. (2021))

glutathione, carotenoids, and tocopherols, are components of the plant's defense system against oxidative stresses (Fig. 16.3) (Sachdev et al. 2021).

Carotenoids and tocopherols prevent ROS formation in chloroplasts and therefore play an essential role in photoprotection (Parveen et al. 2019). Tocopherols also have antioxidant activity, inhibiting lipid peroxidation in biological membranes and eliminating ROS (Hiraga et al. 2001). Carotenoids also have antioxidant activity by eliminating and quenching ROS and triplet-state chlorophyll (Inman-Bamber and Smith 2005). In addition, the decrease in chlorophyll content through decreased synthesis or increased degradation causes a decrease in the number of photons absorbed by the leaf under unfavorable environmental conditions, thus increasing photoprotective and antioxidant capacities (Smirnoff 1993).

In this context, to promote sustainability in agriculture, the search for strategies that mitigate the damage caused by water deficit has been a concern, especially considering future scenarios with low water availability as a result of climate change. This scenario demands the frequent use of beneficial elements, such as Si, to maintain agricultural production and at the same time favor water-use efficiency.

16.4 Silicon for Mitigating Damage to Tropical Plants Caused by Water Deficit

Silicon (Si) is considered a chemical element beneficial to terrestrial plants and essential for Echissetaceae (pteridophytes, echissetines, and isospores) and some algae (diatoms) (Epstein 1999). Si is the second most abundant chemical element in the Earth's crust after oxygen. Silicon dioxide (SiO_2) comprises 50–70% of the soil mass (Epstein 1994). However, in the soil solution, the available Si is in the form of monosilic acid (H_4SiO_4) at low concentrations (~1 mmol L^{-1}) (Katz et al. 2021).

Si is absorbed in the form of H_4SiO_4; it has the same form in the xylem sap (Mitani et al. 2005). The concentration of Si in the soil varies according to the source material, levels of weathering, or previous fertilization. In clayey soils, the Si content can vary from 200 to 300 g kg^{-1}; in sandy soils, the average content is 450 g kg^{-1} (Tubana et al. 2016). However, a large part of this element present in the soil is in an inert form, not available in solution, which limits its absorption by plants (Katz et al. 2021). The low availability of assimilable Si by plants is greater in regions with a tropical climate, as climatic conditions increase the vulnerability of soils to events that favor desilication, such as leaching and high acidity (Keeping 2017; Tubana et al. 2016). These characteristics are common in Ultisols, Latosols, and Neosols, soils that are highly weathered and therefore have low levels of available Si (Tubana et al. 2016; Schaller et al. 2020).

Due to the high degree of soil weathering in tropical regions, Si represents about 5–40% of its composition (Crusciol et al. 2013). In addition, the availability of this element also depends on soil characteristics, such as pH, which proportionally increases its dissociation and consequently its availability to plants (Mandlik et al. 2020). The element can be added to the soil by applying silicate fertilizers (Camargo et al. 2021) or by decomposing plant residues (Katz 2014).

The decomposition of plant residues is relevant in areas with plantations of species that accumulate Si, removing it by up to 700 kg ha^{-1} per year (Keeping 2017). For this reason, extensive cultivation of accumulator plants exposes the soil to a decrease in available Si levels due to continuous uptake of this element by plants without proper replenishment of Si with fertilization, resulting in a decrease in the productive performance of crops worldwide (Majumdar and Prakash 2020). Recent studies have validated that a substantial amount of Si stored in plants is removed from the field at harvest time and that it does not return directly to the soil (Katz et al. 2021; Crusciol et al. 2013).

The application of Si in crops in tropical regions does not yet have an official recommendation. It can be supplied to crops using sources from the steel industry, which basically consist of calcium and magnesium silicate, and sources of phosphorus such as magnesium thermophosphates (de Sousa et al. 2010). However, the use of calcium silicate, for example, requires the application of high doses of Si (close to 1000 kg ha^{-1}) and needs to be incorporated into the soil to favor its dissolution (Camargo et al. 2021). In addition, attention should be paid to the quality of this material, which may have a high concentration of potentially toxic heavy metals and may cause pH changes in already corrected soils (Mandlik et al. 2020).

The use of Si in agricultural crops has been efficient in mitigating the harmful effects caused by environmental stresses, such as water deficit (Epstein 1994), an increasingly common phenomenon in areas destined for agricultural exploitation (Cardozo et al. 2018). The application of Si may interfere with several factors that contribute to the increase of plant tolerance to stressful conditions, such as the sedimentation of amorphous or biogenic silica on the leaf surface, which constitutes 90% of the absorbed Si (Yoshida 1965). This effect prevents the imbalance of leaf water status in plants under water deficit, as the deposits of this element in cell walls reduce water losses through transpiration (Pei et al. 2010) and help to preserve the water content in leaf tissue (Pei et al. 2010; Teixeira et al. 2020a). Less changes occur in water potential (Teixeira et al. 2021a) and water-use efficiency improves (Frew et al. 2018).

The physical changes caused by the deposition of silica associated with cellulose in the leaf epidermis and below the cuticle are more frequent in plants of the Poaceae family (Mitani-Ueno and Ma 2021) because they absorb greater amounts of Si given the activity of specialized proteins in the cell membranes (Mitani et al. 2009). In this plant group, there is a predominance of Si in cell walls as amorphous silica (Neethirajan et al. 2009) and in organic complexes with cellulose (Yoshida 1965), hemicellulose (He et al. 2015), and other components of the cell wall (Isa et al. 2010).

The formation of these silico-organic compounds in the cell wall occurs at the expense of lignin synthesis (Schoelynck et al. 2010). This substitution is of biological interest because the energy cost of incorporating Si into structural compounds is fifty times lower than that of organic compounds due to its high intrinsic permeability in lipid bilayers (Raven 1983). This induces a decrease in the C content of the cell wall, which may reflect in leaves of plants that received Si, modifying the C:Si stoichiometric ratio, especially under water deficit (Rocha et al. 2021); however, it may also occur in plants under no stress (Neu et al. 2017). Thus, the beneficial effect of Si in attenuating water deficit also occurs due to elemental stoichiometric homeostasis involving, in addition to C, the structural nutrients N and P, which favors balance of the nutritional functions of these nutrients in plants and, consequently, in nutritional efficiency.

However, a small part of Si at a nanometric level is contained in intercellular structures (Woesz et al. 2006) with a potential to induce an effect on physiological and biochemical attributes, preserving and even activating plant defense systems that cause antioxidant action (Katz 2014). These effects may initially help plants treated with Si to have a higher water content because Si contributes to the absorption of water by roots by inducing an increase in water potential through the osmoregulatory action of proline (Avila et al. 2021). Also, there may be an increase in the activity of aquaporins, which are proteins specialized in the absorption of water by plants and which may have their gene expression stimulated by Si (Abbas et al. 2015; Chen et al. 2016). The maintenance of water status provided by Si collaborates to reduce the damage caused to photosystem II reaction centers, increasing the use of solar radiation in the photochemical phase of photosynthesis (Verma et al. 2021) and preventing the excessive formation of ROS (Fig. 16.4) (Inman-Bamber and Smith 2005).

Decreased oxidative stress contributes to the preservation of cell integrity, increasing the contents of photosynthetic pigments (Camargo et al. 2019) and regulating enzymatic (Bezerra et al. 2019) and non-enzymatic antioxidant defense systems (Teixeira et al. 2021a). In addition, Si deposits in the cell walls of xylem vessels decrease water losses through transpiration (Camargo et al. 2021). Si can also exert an osmoregulatory function by modifying proline levels (Bezerra et al. 2019; Teixeira et al. 2020a) and increasing water absorption (Teixeira et al. 2021a; Camargo et al. 2021). Also, there is another effect of Si: attenuating water deficit, as it is involved in greater elemental stoichiometric homeostasis (C:N:P), favoring the efficiency of the use of C, N, and P, as recently reported (Teixeira et al. 2020b).

The first report of the supply of soluble Si to sugarcane cultivated under water deficit indicated that the application via nutrient solution during the formation of pre-sprouted seedlings is more efficient compared to Si via leaf spraying in the first

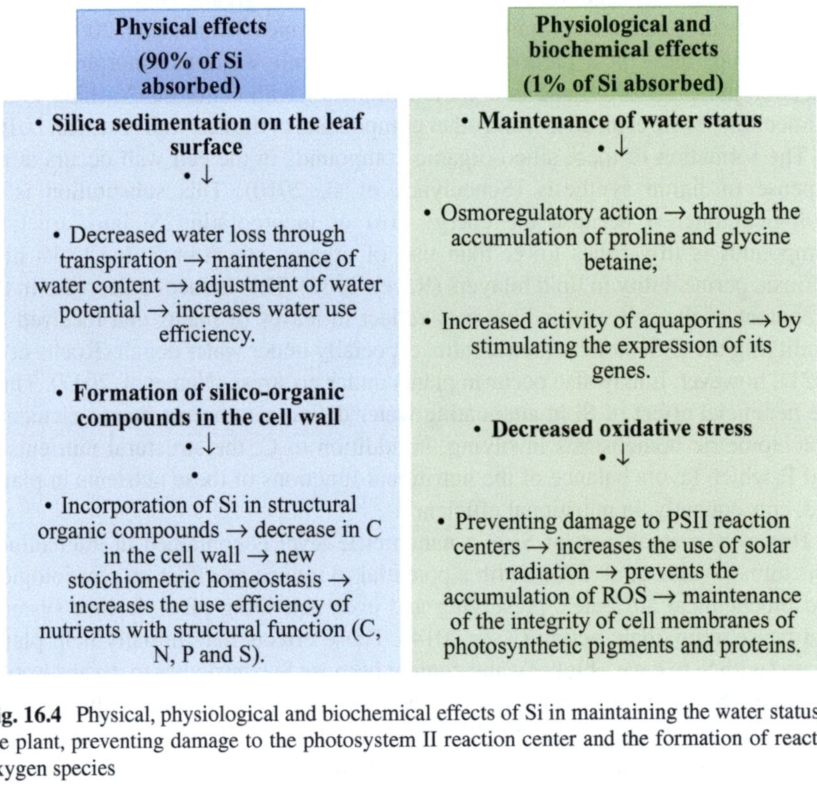

Fig. 16.4 Physical, physiological and biochemical effects of Si in maintaining the water status of the plant, preventing damage to the photosystem II reaction center and the formation of reactive oxygen species

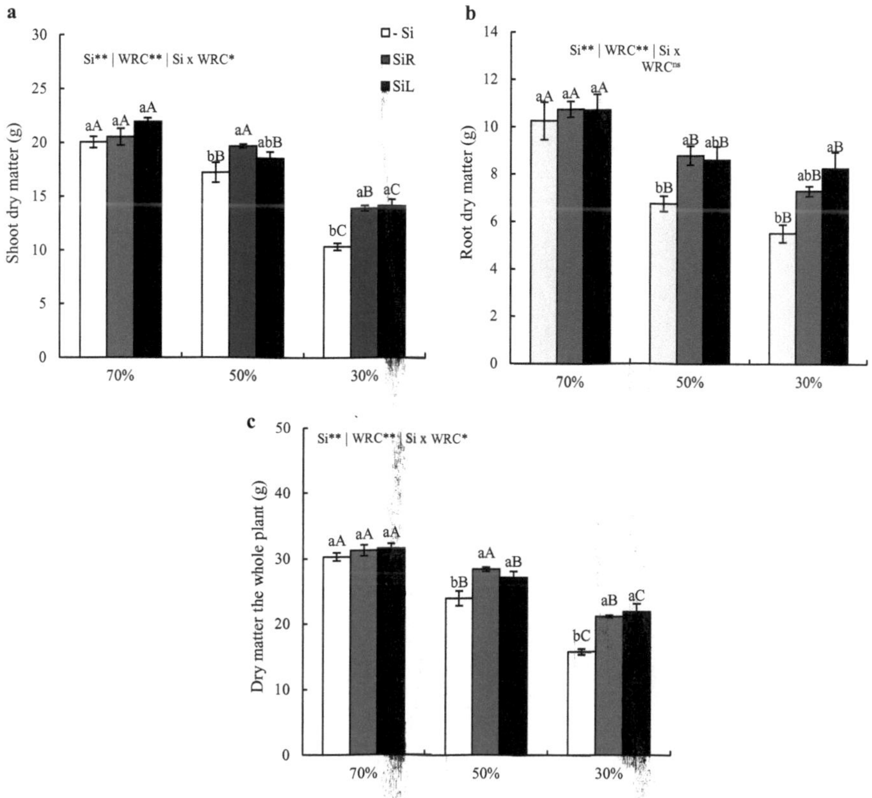

Fig. 16.5 Shoot dry mass (**a**), root dry mass (**b**), and whole plant dry mass (**c**) and images of pre-sprouted sugarcane seedlings on different soil water retention capacity levels (WRC) (70%, 50%, and 30%) and forms of Si supply (without Si: −Si; Si root: SiR; Si leaf: SiL) (**d**). ** and *: significant at 1 and 5% probability, respectively; ^{ns}: not significant by F test. Lowercase letters indicate differences in relation to Si and uppercase letters in relation to the WRC. The bars represent the mean standard error. (Source: Teixeira et al. (2020a))

30 days after transplanting. It increased the quantum efficiency of the PS II, chlorophyll content, leaf water potential, and water content; it decreased the extravasation of electrolytes in cells and free proline contents, in addition to allowing for a greater growth and biomass production (Fig. 16.5) (Teixeira et al. 2020a).

However, this effect was contested by Teixeira et al. (2021a) in an experiment developed in a hydroponic system, in which Si was initially supplied during the seedling production phase and, subsequently, water deficit was induced with PEG-6000 for a period of 72 hours. The authors found that plants under water deficit evidently increased the accumulation of Si with its supply via root and leaf by 2118% and 98%, respectively, in comparison with the absence of Si. However, although leaf Si uptake was low compared to root Si uptake, the physiological effects on relieving stress induced by water deficit were equal to or even greater than those of root application. To conclude, the amount of Si absorbed by the plant is not

Fig. 16.5 (continued)

an important factor to induce an antioxidant defense response to mitigate the immediate effects of water deficit since the two forms of Si application were similar.

The researches above were carried out to verify whether the supply of isolated Si by fertigation or by spraying to pre-sprouted sugarcane seedlings would have an immediate effect to mitigate the water deficit with short periods of three and 30 days. The results were surprising and indicated that Si could mitigate severe water deficit in the soil (30% of the soil's water-holding capacity). We also observed that the effect of Si was greater when the element was applied via fertigation (inducing absorption only by roots), as we proved that this form of application promoted greater absorption of Si in plants.

However, based on scenarios of prolonged water deficit, which are already happening and will in the future, we questioned whether it would be possible to prolong the effect of Si in mitigating the damage caused by severe water deficit for a longer period (from 60 to 160 days after transplant). In addition, we questioned whether the beneficial effects of Si can be observed in another species, energy cane, which has been shown to increase energy production and have a greater tolerance to drought.

Therefore, new studies were carried out to evaluate whether the combined application of Si via fertigation associated with leaf spraying (inducing absorption by leaves + roots) could induce a greater absorption of Si in plants. Thus, we applied Si via fertigation and leaf spraying only in the pre-sprouted seedling formation phase to evaluate the effects of mitigating severe water deficit for 60 days after transplanting.

Severe water deficit at 60 days after transplanting reduced the water content and water potential, inducing oxidative stress and impairing photosynthetic efficiency, with a consequent decrease in plant growth. Fertigation was efficient to supply Si in pre-sprouted sugarcane and energy cane seedlings. The residual effect of Si attenuated the damage caused by water deficit 60 days after transplanting in both species. The mechanisms involved are related to the increase in the activity of CAT, SOD, and APX enzymes and proline content. In addition, the water status remained stable and, consequently, there was an increase in plant growth (Fig. 16.6) (Teixeira et al. 2021b).

Furthermore, we tested whether the application of Si after transplanting to seedlings that received Si in the seedling phase would prolong the effect of mitigating the damage caused by severe water deficit for a period of up to 160 days after transplanting (Teixeira et al. 2022a). In parallel, we evaluated the structural effects provided by Si on the C:N:P nutritional balance and on the increase in nutritional efficiency in sugarcane and energy cane (Teixeira et al. 2022b).

These results are of great practical importance due to the frequent occurrence of water deficit and the expansion of irrigated sugarcane cultivation. However, it is now important to expand these studies to areas with field experimentation, increase knowledge about the benefits of Si in physiology and nutrition and in the sugarcane industry, define adequate doses for Si fertigation, and whether it would be possible to decrease the level of irrigation without loss of crop productivity, favoring efficiency in the use of water.

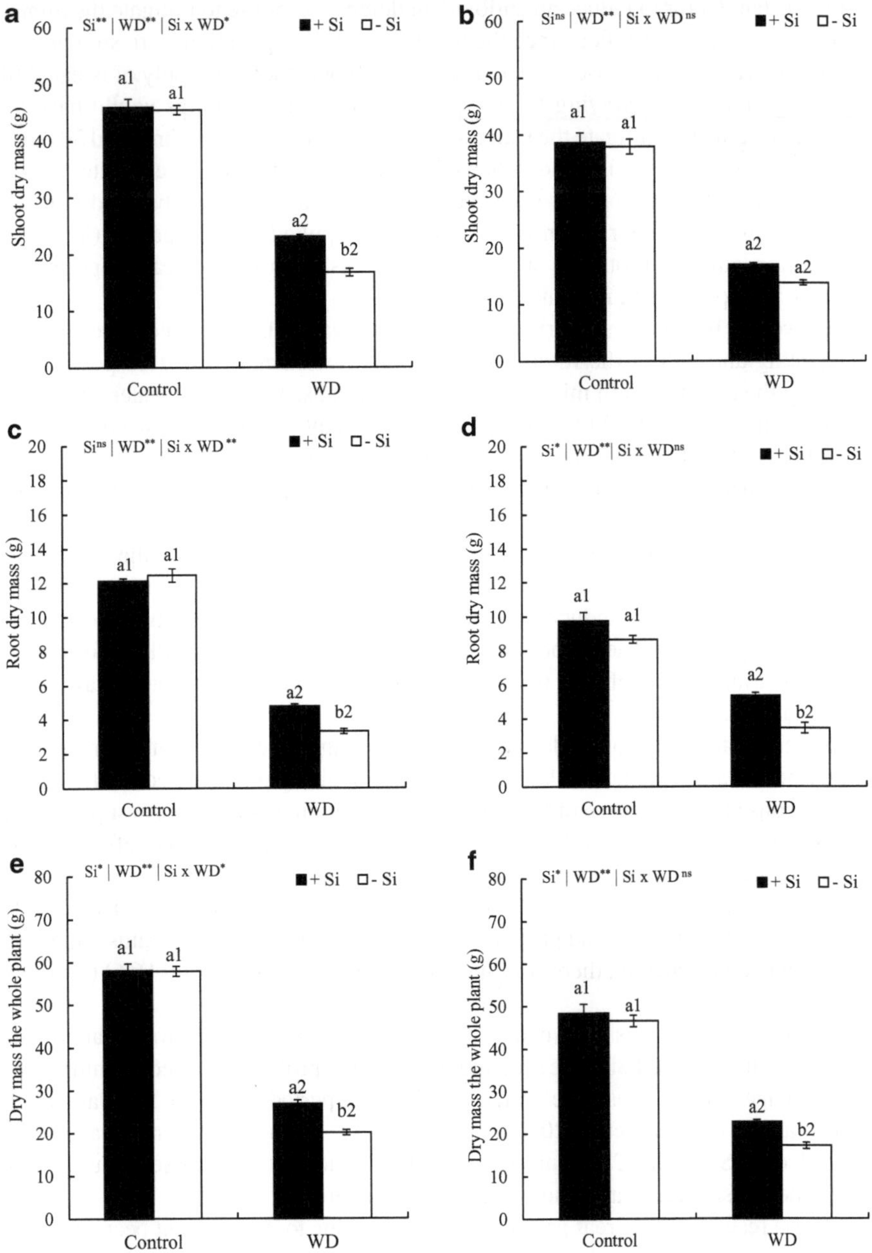

Fig. 16.6 Shoot dry mass, root dry mass, and whole plant dry mass in pre-sprouted sugarcane (*Saccharum officinarum* L.) (**a**, **c**, **e**) and energy cane (*Saccharum spontaneum*) seedlings L. (**b**, **d**, **f**) with Si applied via fertigation (+Si) and absence of Si (−Si) under adequate water regime (control) and water deficit (WD). ** and *: significant at 1 and 5% probability, respectively; ns: not significant by F test. Different letters show differences in Si in the same water regime and different numbers show differences in the water regime under the same conditions of Si supply by Tukey test. The bars represent the mean standard error; $n = 6$. (Source: Teixeira et al. (2021b))

Corn is a Si-accumulating species and one of the most cultivated annual crops in irrigated areas in tropical regions. Thus, the need to decrease consumption and increase the economic use of water is necessary to ensure adequate resources for food supply, which are prerequisites for sustainable agricultural production (Firouzabadi et al. 2021).

Initial studies on Si in corn plants under water deficit were carried out in a hydroponic medium using only a nutrient solution with polyethylene glycol (PEG) to induce stress (Gao et al. 2006) or in an inert substrate with a nutrient solution (Kaya et al. 2006). In these studies, researchers found that Si promoted the beneficial effect of mitigating water deficit stress in plants either by decreasing stomatal transpiration (Gao et al. 2006) or by maintaining the integrity of photosynthetic pigments, thus increasing leaf Ca and K levels and decreasing the electrolyte leakage rate (Kaya et al. 2006). This resulted in greater plant growth and dry mass production.

The studies mentioned above were carried out with the cultivation of plants in soil using low-solubility calcium silicate (up to 100–150 mg kg^{-1} Si), requiring incorporation into the soil to increase silicate dissolution and Si release (Amin et al. 2014, 2018). The authors found that Si attenuated a moderate water deficit (60% of soil water retention capacity) by favoring photosynthetic efficiency and reducing oxidative stress.

Recently, Teixeira et al. (2022c) showed that the beneficial effects of using soluble sources of Si applied via fertigation during the initial phase of growth of maize plants and its effects in mitigating damage caused by severe water deficit (30% of soil water retention capacity) can also be attributed to the improvement of nutritional mechanisms, which increases the efficiency of macronutrient use (Fig. 16.7). This effect is interesting because it is known that the water deficit induces a decrease in the ion-root contact and the movement of nutrients in the soil, decreasing the absorption and metabolism of nutrients in plants, that is, decreasing the performance of biological functions and the efficiency of the plant in using these nutrients in metabolism and consequently for conversion into dry matter (Ma et al. 2020).

Teixeira et al. (2022c) also evidenced that fertigation with soluble Si is an efficient option for the nutritional enrichment of plants since a solution with a low concentration of Si is used (2 mmol L^{-1}). This occurs taking into account an irrigation depth of 5 mm in four applications of Si during cultivation (Fig. 16.8), reaching an equivalent amount in an area close to 12 kg ha^{-1} of Si, that is, much lower than the other sources of low solubility of this element that require incorporation of high doses into the soil (>1 t ha^{-1} of silicate) (Keeping 2017).

Water deficiency stands out for being common in pastures that occupy a vast cultivated area. It has a global impact on agricultural production and increases the seasonality of forage production (Sbrissia et al. 2020). This fact is worrisome, since almost all beef productions are from animals fed directly on pastures, as pastures have a low production cost.

Thus, the use of Si in forages is considered as accumulators of this element (Cocker et al. 1998), and the fact that the availability of Si in the soil decreases in pasture areas as a result of degradation at different levels (Yang et al. 2019) can be

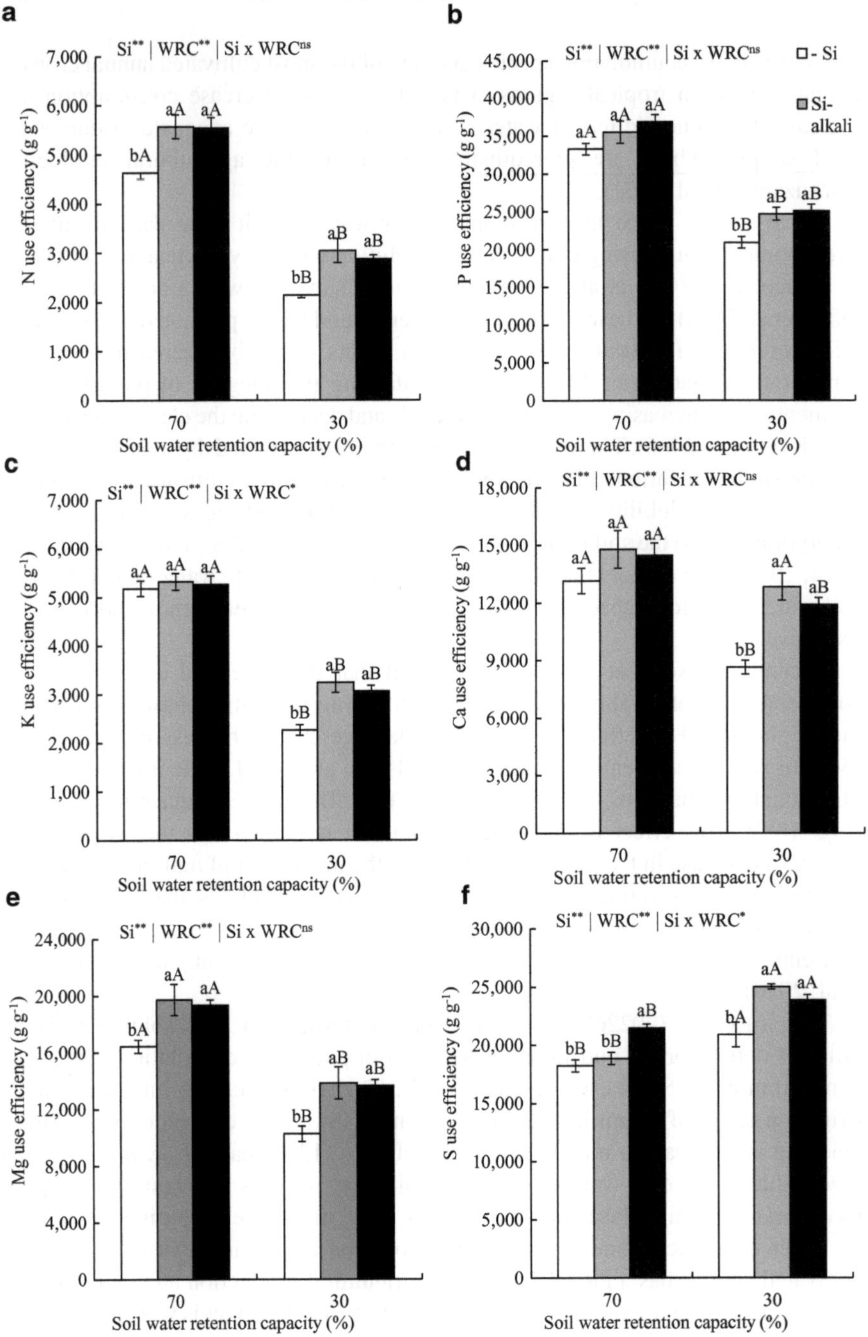

Fig. 16.7 Use efficiency of nitrogen (N) (**a**), phosphorus (P) (**b**), potassium (K) (**c**), calcium (Ca) (**d**), magnesium (Mg) (**e**), and sulfur (S) (**f**) in corn plants grown under Si supply from soluble sources applied by fertigation (–Si: control; Si-alkali: Si in the stabilized source of sodium and potassium silicate; Si-K: Si in the source of potassium silicate) in two soil water regimes: 70% (no deficit) and 30% (severe water deficit) of the soil water retention capacity (WRC). Lowercase letters show differences for Si and uppercase letters for WRC ($p < 0.05$, Tukey test). ** and *: significant at 1 and 5% probability, respectively; ^{ns}: not significant. Si x WRC: interaction. (Source: Teixeira et al. (2022c))

Fig. 16.8 Difference in the development of corn plants grown under conditions of supply of silicon (Si) from soluble sources applied by fertigation (−Si: control; Si-alkali: Si in the stabilized source of sodium and potassium silicate; Si -K: Si in the potassium source, silicate source) in two soil water regimes: 70% (no deficit) and 30% (severe water deficit) of the soil water retention capacity (WRC). (Source: Teixeira et al. (2022c))

an important strategy to mitigate water deficit. However, for forage species, there are few studies relating water deficit to Si. Melo et al. (2003) used relatively high doses of Si (up to 1452 kg ha^{-1}) in the soil in the form of calcium silicate with promising results for forage growth. However, it is possible to use relatively low doses with soluble sources via fertigation. In this sense, Rocha et al. (2021) carried out an experiment to evaluate whether the supply of Si via fertigation causes an increase in Si absorption, stoichiometric changes, improvement of physiological variables, and growth of *Brachiaria* 'BRS RB331 Ipyporã' and *Brachiaria* 'Mavuno' cultivated in two soil water regimes (70% and 40% of soil water-holding capacity) (Fig. 16.9).

The results the authors reported indicate that severe water deficit causes disturbances in C stoichiometry, with important physiological damage to forage plants. The two forages that received Si via fertigation showed a greater accumulation of Si in shoots, a decrease in C content and in C:Si ratio, an increase in the efficiency of C use, a decrease in the electrolyte extravasation rate, an increase in the water relative content, leaf water potential, chlorophyll b content, quantum efficiency of the PS II, and consequently in growth and dry matter production.

Thus, the innovative use of Si supply via fertigation with a soluble source at relatively low concentrations (2.5 mmol L^{-1}) during the forage growth cycle was proposed to efficiently increase the absorption of Si by the plant and cause a modifying effect. The C:Si stoichiometric ratio increases the efficiency of C use and consequently improves the physiological variables and forage growth under water deficit (Fig. 16.10).

It is noteworthy that the benefits of Si applied via fertigation in pastures can attenuate water stress. This management should reflect on productivity gains and water savings.

The beneficial effects of Si in dicots, such as bean plants, are still limited due to the low absorption of Si by this species. In this scenario, an unprecedented study in Brazil is being carried out to enhance the absorption of Si in bean plants using the element via drip fertigation in the soil at field level to favor daily root absorption of Si, but with intermediate doses of Si that are beneficial to favor its viability.

In parallel, this research explores the relationship of Si in two agricultural scenarios that predominate in annual crops. The first refers to the fact that it is common in field conditions to have more than one stress factor, such as water deficit or deficit irrigation associated with K deficiency. However, most studies focus on the damage of isolated stresses and even in experiments in vases. The second scenario refers to technified crops with sufficient K contents in the soil; therefore, the use of fertigation with Si could reduce the level of irrigation without crop productivity losses, favoring water-use efficiency. Unfortunately, in both scenarios, the appropriate dose for Si in fertigation, one that can meet these crops with the respective patterns of soil and leaf contents of Si in field conditions, should still be determined.

In this context, this research proposes a holistic approach to the study of Si in mitigating simultaneous stresses in plants and improving the benefits of the synergy of Si and K to increase the efficiency of water use. The initial results are surprising. There is a linear increase in the number of pods of bean plants in all water regimes studied (Fig. 16.11). Soon, the results of crop productivity and the best dose of Si will be obtained and published. It is important to emphasize that Si can enhance plant growth even in optimal water conditions, increasing the response of crops possibly because it favors other production factors that are not yet studied (Fig. 16.12).

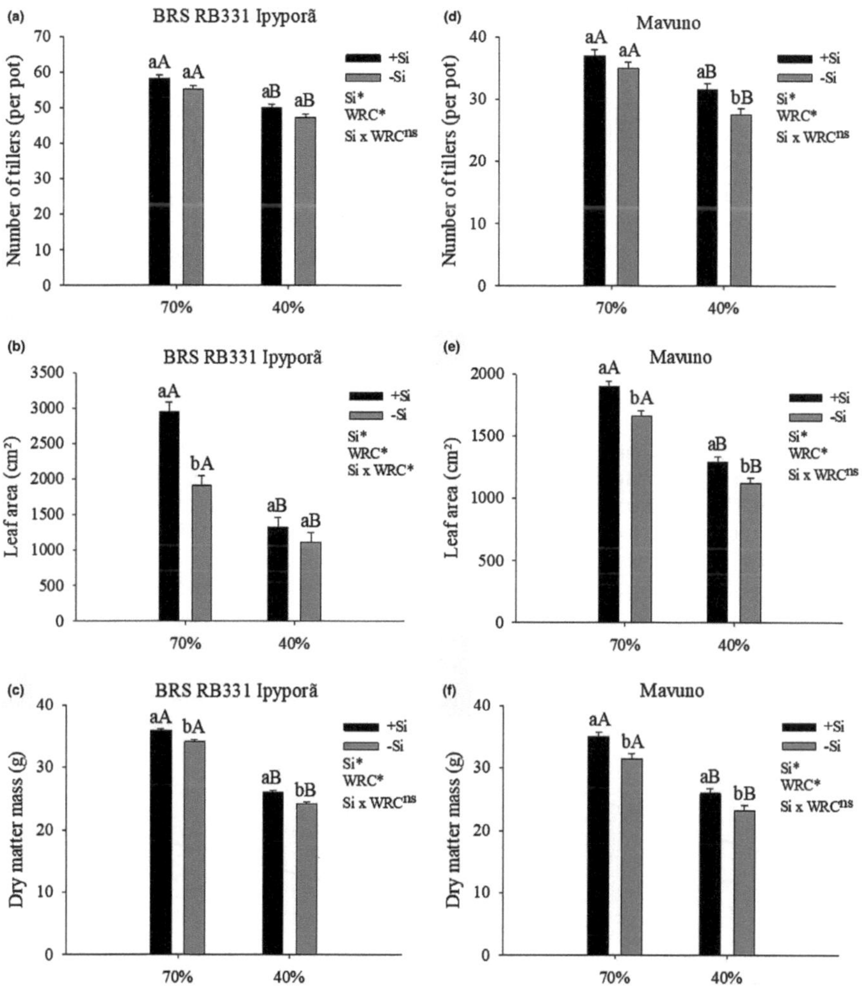

Fig. 16.9 Number of tillers (**a**, **d**), leaf area (**b**, **e**), and dry mass (**c**, **f**) of forage plants grown in soil with different soil water retention capacities (WRC) (70% and 40%) in the absence (−Si) and in the presence of fertigation with silicon (+Si). ^{ns}: not significant by F test. Lowercase letters indicate differences in relation to Si and uppercase letters in relation to WRC. The bars represent the mean standard error, $n = 6$. (Source: Rocha et al. (2021))

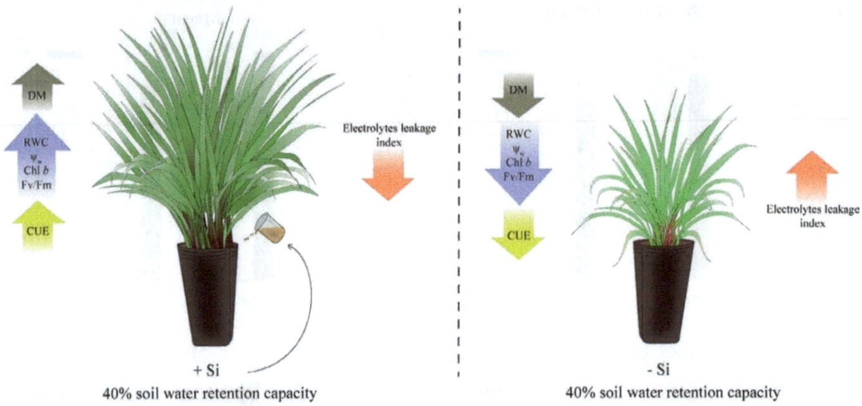

Fig. 16.10 Representation of a forage plant under water deficit condition in the absence (−Si) and in the presence of silicon fertigation (+Si) and summary of its benefits in the effects of carbon use efficiency (CUE), relative water content (RWC), leaf water potential (Ψw), chlorophyll b (Chl b), quantum efficiency of the photosystem II (Fv/Fm), and dry matter (DM). (Source: Rocha et al. (2021))

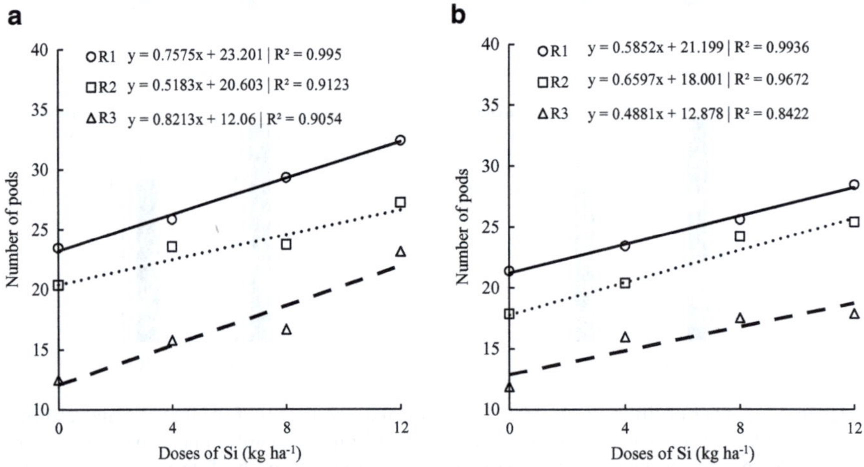

Fig. 16.11 Number of pods in bean plants under different water regimes: 80% (no water deficit – R1), 60% (moderate water deficit – R2), and 40% (severe water deficit – R3) of the water retention capacity in the soil – WRC, respectively. Two potassium fertilizations: without fertilization (K−) and with fertilization (K+). Four doses of Si supplied via fertigation: 0, 4, 8, and 12 kg ha^{-1}. (Source: Unpublished data provided by the authors)

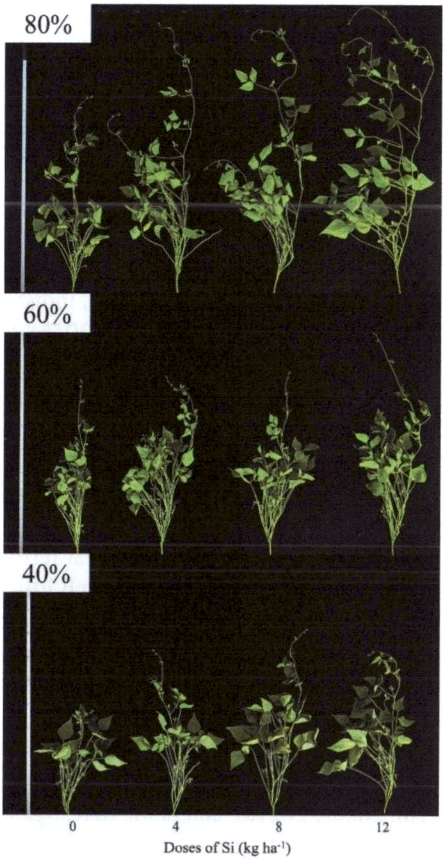

Fig. 16.12 Images of bean plants under different water regimes: 80% (no water deficit), 60% (moderate water deficit), and 40% (severe water deficit) of the soil water retention capacity – WRC, respectively. Potassium fertilization and four doses of Si supplied via fertigation: 0, 4, 8, and 12 kg ha^{-1}. (Source: Unpublished data provided by the authors)

16.5 Fertigation and Leaf Spraying with Silicon

Fertigation consists of the application of fertilizer diluted in irrigation water. It can be applied using any irrigation system, but pressurized water systems are preferable as they allow for a more uniform application of water and fertilizers and a greater water-use efficiency, which is particularly advantageous in the face of increasing water scarcity (Guido et al. 2020). Therefore, it is believed that the supply of Si via fertigation associated with leaf spraying is a potentially viable alternative because it can induce absorption of Si by organs, leaves, and roots, considerably increasing the levels of Si in plants and leading to beneficial effects.

One of the major limitations of using Si in solution is the high polymerization level. At concentrations above 3 mmol L^{-1}, it can be polymerized (Birchall 1995), forming Si dimers, trimers, and polymer chains (McKeague and Cline 1963) that are not taken up by plants. In this sense, fertigation is interesting because it allows using low doses of Si (<20 kg ha^{-1}) with Si concentrations in solution ranging from 1.5 to 3.0 mmol L^{-1}, thus avoiding polymerization. In addition, it is possible to use sources with greater stability, such as sodium and potassium silicate stabilized with sorbitol (Santos et al. 2020). Sorbitol is a polyol that helps to maintain the monomeric forms of Si-forming organic complexes (Kubicki and Heaney 2003) and decreases the droplet deliquescence point on the leaf surface (Babiker and Duncan 1974), delaying drying and contributing to its absorption by the leaf surface mainly during the phase of cuticle penetration (Fernández and Brown 2013).

The stability of monomeric forms of Si is also favored by the acidification of the solution at the time of application. It decreases the pH value of the solution (5.5 ± 0.2). Acidification of Si-containing solutions shifts the reaction to the right ($H_2SiO_4^{-2} \rightarrow H_3SiO_4^{-1} \rightarrow H_4SiO_4$), inducing a predominance of monomeric species of Si (Kudryavtsev and Figovsky 2016). In addition, with fertigation, depending on the management used, it is possible to apply it more frequently. These factors combined improve the quality of the Si solution applied to plants and the absorption of Si, making it possible to manage solutions using fertigation associated with leaf spraying.

Changes in the cultivation system by using irrigation in extensive areas, including regions with less rainfalls (Ana 2021), have contributed to the use of fertigation. Thus, fertilization is carried out via fertigation directly on the soil and sprayed on leaves. This is justified by the high efficiency provided. Therefore, it is believed that, with efficiency, Si applications in areas of plant cultivation under tropical climate conditions can use fertigation and leaf spraying after transplanting, which may contribute to increasing the amount of Si absorbed by plants.

The increase in Si absorption provided by the application of Si via fertigation combined with leaf spraying has already been tested in forage plants (Rocha et al. 2022), in sugarcane only at the production phase of pre-sprouted seedlings (Teixeira et al. 2020a, 2021a), and in the first sugarcane ratoon (de Oliveira Filho et al. 2021). However, the efficiency of Si supply via fertigation and leaf spraying may vary depending on the species and the growing environment (Ramesh 2000).

16.6 Conclusion

The practical implications obtained from the information presented in this article indicate that it is important to explore the relationship of Si in improving water use in crops. This is relevant because every agricultural year there is an increase in the frequency of water deficit, causing well-known biological damages to the plant. It is clear that the use of Si, by reducing water losses in the plant and maintaining high water content in the plant tissue, avoids oxidative stress and at the same time can reinforce the plant's antioxidant system, favoring the sustainability of crops. These benefits of Si with better water-use efficiency are possible to be achieved with fluid Si via fertigation or leaf spraying using relatively low doses of Si.

It is important to expand these studies to other species under field conditions in order to increase knowledge about the benefits of Si in irrigated areas with optimized irrigation for the physiology and nutrition of plants under tropical climate conditions.

References

Abbas T, Balal RM, Shahid MA et al (2015) Silicon-induced alleviation of NaCl toxicity in okra (*Abelmoschus esculentus*) is associated with enhanced photosynthesis, osmoprotectants and antioxidant metabolism. Acta Physiol Plant 37:1–15. https://doi.org/10.1007/s11738-014-1768-5

Amin M, Ahmad R, Basra SMA, Murtaza G (2014) Silicon induced improvement in morpho-physiological traits of maize (*Zea mays* L.) under water deficit. Pak J Agric Sci 51:187–196

Amin M, Ahmad R, Ali A et al (2018) Influence of silicon fertilization on maize performance under limited water supply. SILICON 10:177–183. https://doi.org/10.1007/s12633-015-9372-x

ANA (2021) Atlas irrigation: water use in irrigated agriculture (2a). Agência Nacional de Águas e Saneamento Básico. www.ana.gov.br

Avila RG, Magalhães PC, da Silva EM et al (2021) Application of silicon to irrigated and water deficit sorghum plants increases yield via the regulation of primary, antioxidant, and osmoregulatory metabolism. Agric Water Manag 255:107004. https://doi.org/10.1016/j.agwat.2021.107004

Babiker AGT, Duncan HJ (1974) Penetration of bracken fronds by asulam as influenced by the addition of surfactant to the spray solution and by pH. Weed Res 14:375–377. https://doi.org/10.1111/j.1365-3180.1974.tb01078.x

Basnayake J, Jackson PA, Inman-Bamber NG, Lakshmanan P (2012) Sugarcane for water-limited environments. Genetic variation in cane yield and sugar content in response to water stress. J Exp Bot 63:6023–6033. https://doi.org/10.1093/jxb/ers251

Besharat S, Barão L, Cruz C (2020) New strategies to overcome water limitation in cultivated maize: results from sub-surface irrigation and silicon fertilization. J Environ Manag 263:110398. https://doi.org/10.1016/j.jenvman.2020.110398

Bezerra BKL, Lima GPP, dos Reis AR et al (2019) Physiological and biochemical impacts of silicon against water deficit in sugarcane. Acta Physiol Plant 41:189. https://doi.org/10.1007/s11738-019-2980-0

Birchall JD (1995) The essentiality of silicon in biology. Chem Soc Rev 24:351–357. https://doi.org/10.1039/CS9952400351

Camargo MS, Bezerra BKL, Holanda LA et al (2019) Silicon fertilization improves physiological responses in sugarcane cultivars grown under water deficit. J Soil Sci Plant Nutr 19:81–91. https://doi.org/10.1007/s42729-019-0012-1

Camargo MS, Fernández Honaine M, Osterrieth M et al (2021) Silicon fertilization increases gas-exchange and biomass by silicophytolith deposition in the leaves of contrasting drought-tolerant sugarcane cultivars under well-watered conditions. Plant Soil 0123456789. https://doi.org/10.1007/s11104-021-05063-z

Cardozo NP, de Oliveira BR, la Scala N (2018) Sustainable intensification of sugarcane production under irrigation systems, considering climate interactions and agricultural efficiency. J Clean Prod 204:861–871. https://doi.org/10.1016/j.jclepro.2018.09.004

Chen D, Cao B, Wang S et al (2016) Silicon moderated the K deficiency by improving the plant-water status in sorghum. Sci Rep 6:1–14. https://doi.org/10.1038/srep22882

Cocker KM, Evans DE, Hodson MJ (1998) The amelioration of aluminium toxicity by silicon in higher plants: solution chemistry or an in planta mechanism? Physiol Plant 104:608–614. https://doi.org/10.1034/j.1399-3054.1998.1040413.x

Contin DR, Soriani HH, Hernández I et al (2014) Antioxidant and photoprotective defenses in response to gradual water stress under low and high irradiance in two Malvaceae tree species used for tropical forest restoration. Trees 28:1705–1722. https://doi.org/10.1007/s00468-014-1079-x

Crusciol CAC, Ferrari Neto J, Soratto RP, da Costa CHM (2013) Cycling of nutrients and silicon in pigeonpea and pearl millet monoculture and intercropping. Braz Soil Sci J 37:1628–1640. https://doi.org/10.1590/s0100-06832013000600019

da Silva PP, Soares L, da Costa JG et al (2012) Path analysis for selection of drought tolerant sugarcane genotypes through physiological components. Ind Crop Prod 37:11–19. https://doi.org/10.1016/j.indcrop.2011.11.015

de Oliveira Filho ASB, de Mello R, Teixeira GCM et al (2021) Silicon attenuates the effects of water deficit in sugarcane by modifying physiological aspects and C: N: P stoichiometry and its use efficiency. Agric Water Manag 255:107006. https://doi.org/10.1016/j.agwat.2021.107006
de Sousa RTX, Korndörfer GH, Wangen DRB (2010) Recovery of silicon from metallurgy slag by sugarcane cultivars. Bragantia 69:669–676. https://doi.org/10.1590/s0006-87052010000300019
Elser JJ, Hamilton A (2007) Stoichiometry and the new biology: the future is now. PLoS Biol 5:1403–1405. https://doi.org/10.1371/journal.pbio.0050181
Epstein E (1994) The anomaly of silicon in plant biology. Proc Natl Acad Sci U S A 91:11–17. https://doi.org/10.1073/pnas.91.1.11
Epstein E (1999) Silicon. Annu Rev Plant Physiol Plant Mol Biol 50:641–664
Fernández V, Brown PH (2013) From plant surface to plant metabolism: the uncertain fate of foliar-applied nutrients. Front Plant Sci 4:0289. https://doi.org/10.3389/fpls.2013.00289
Ferreira THS, Tsunada MS, Bassi D et al (2017) Sugarcane water stress tolerance mechanisms and its implications on developing biotechnology solutions. Front Plant Sci 8:1077. https://doi.org/10.3389/fpls.2017.01077
Firouzabadi AG, Baghani J, Jovzi M, Albaji M (2021) Effects of wheat row spacing layout and drip tape spacing on yield and water productivity in sandy clay loam soil in a semi-arid region. Agric Water Manag 251:106868. https://doi.org/10.1016/j.agwat.2021.106868
Frew A, Weston LA, Reynolds OL, Gurr GM (2018) The role of silicon in plant biology: a paradigm shift in research approach. Ann Bot 121:1265–1273. https://doi.org/10.1093/aob/mcy009
Gao X, Zou C, Wang L, Zhang F (2006) Silicon decreases transpiration rate and conductance from stomata of maize plants. J Plant Nutr 29:1637–1647. https://doi.org/10.1080/01904160600851494
Gratão PL, Polle A, Lea PJ, Azevedo RA (2005) Making the life of heavy metal-stressed plants a little easier. Funct Plant Biol 32:481–494. https://doi.org/10.1071/FP05016
Guido V, Finzi A, Ferrari O et al (2020) Fertigation of maize with digestate using drip irrigation and pivot systems. Agronomy 10:1453. https://doi.org/10.3390/agronomy10101453
He C, Ma J, Wang L (2015) A hemicellulose-bound form of silicon with potential to improve the mechanical properties and regeneration of the cell wall of rice. New Phytol 206:1051–1062. https://doi.org/10.1111/nph.13282
Hiraga S, Sasaki K, Ito H et al (2001) A large family of class III plant peroxidases. Plant Cell Physiol 42:462–468. https://doi.org/10.1093/pcp/pce061
Iannucci A, Russo M, Arena L et al (2002) Water deficit effects on osmotic adjustment and solute accumulation in leaves of annual clovers. Eur J Agron 16:111–122. https://doi.org/10.1016/S1161-0301(01)00121-6
Inman-Bamber NG, Smith DM (2005) Water relations in sugarcane and response to water deficits. Field Crops Res 92:185–202. https://doi.org/10.1016/j.fcr.2005.01.023
Isa M, Bai S, Yokoyama T et al (2010) Silicon enhances growth independent of silica deposition in a low-silica rice mutant, Lsi1. Plant Soil 331:361–375. https://doi.org/10.1007/s11104-009-0258-9
Jain R, Chandra A, Venugopalan V, Solomon S (2015) Physiological changes and expression of SOD and P5CS genes in response to water deficit in sugarcane. Sugar Tech 17:276–282. https://doi.org/10.1007/s12355-014-0317-2
Katz O (2014) Beyond grasses: the potential benefits of studying silicon accumulation in non-grass species. Front Plant Sci 5:1–3. https://doi.org/10.3389/fpls.2014.00376
Katz O, Puppe D, Kaczorek D et al (2021) Silicon in the soil–plant continuum: intricate feedback mechanisms within ecosystems. Plan Theory 10:1–36. https://doi.org/10.3390/plants10040652
Kaya C, Tuna L, Higgs D (2006) Effect of silicon on plant growth and mineral nutrition of maize grown under water-stress conditions. J Plant Nutr 29:1469–1480. https://doi.org/10.1080/01904160600837238
Keeping MG (2017) Uptake of silicon by sugarcane from applied sources may not reflect plant-available soil silicon and total silicon content of sources. Front Plant Sci 8:0760. https://doi.org/10.3389/fpls.2017.00760
Kim Y, Khan AL, Waqas M, Lee I (2017) Silicon regulates antioxidant activities of crop plants under abiotic-induced oxidative stress: a review. Front Plant Sci 8:0510. https://doi.org/10.3389/fpls.2017.00510

Kubicki JD, Heaney PJ (2003) Molecular orbital modeling of aqueous organosilicon complexes: implications for silica biomineralization. Geochim Cosmochim Acta 67:4113–4121. https://doi.org/10.1016/S0016-7037(03)00093-3

Kudryavtsev PG, Figovsky OL (2016) Nanocomposite organomineral hybrid materials. Part I. Nanotechnol Const 8:16–49

Ma Y, Dias MC, Freitas H (2020) Drought and salinity stress responses and microbe-induced tolerance in plants. Front Plant Sci 11:591911. https://doi.org/10.3389/fpls.2020.591911

Maillard A, Ali N, Schwarzenberg A et al (2018) Silicon transcriptionally regulates sulfur and ABA metabolism and delays leaf senescence in barley under combined sulfur deficiency and osmotic stress. Environ Exp Bot 155:394–410. https://doi.org/10.1016/j.envexpbot.2018.07.026

Majumdar S, Prakash NB (2020) An overview on the potential of silicon in promoting defence against biotic and abiotic stresses in sugarcane. J Soil Sci Plant Nutr 20:1969–1998. https://doi.org/10.1007/s42729-020-00269-z

Mandlik R, Thakral V, Raturi G et al (2020) Significance of silicon uptake, transport, and deposition in plants. J Exp Bot 71:6703–6718. https://doi.org/10.1093/jxb/eraa301

Marcos FCC, Silveira NM, Mokochinski JB et al (2018) Drought tolerance of sugarcane is improved by previous exposure to water deficit. J Plant Physiol 223:9–18. https://doi.org/10.1016/j.jplph.2018.02.001

Martins APC, Albrecht LP, Castaldo J et al (2015) New technologies in the planting of sugarcane (*Saccharum* spp). J Agron Sci 4:301–317

McKeague JA, Cline MG (1963) Silica in soils. Adv Agron 15:339–396. https://doi.org/10.1016/S0065-2113(08)60403-4

Melo SP, Korndörfer GH, Korndörfer CM et al (2003) Silicon accumulation and water deficit tolerance in *Brachiaria grasses*. Sci Agric 60:755–759. https://doi.org/10.1590/S0103-90162003000400022

Mengel K, Kirkby EA, Kosegarten H, Appel T (2001) Plant water relationships. In: Mengel K, Kirkby EA, Kosegarten H, Appel T (eds) Principles of plant nutrition. Springer. https://doi.org/10.1007/978-94-010-1009-2_4

Merwad A-RMA, Desoky E-SM, Rady MM (2018) Response of water deficit-stressed *Vigna unguiculata* performances to silicon, proline or methionine foliar application. Sci Hortic 228:132–144. https://doi.org/10.1016/j.scienta.2017.10.008

Mitani N, Jian FM, Iwashita T (2005) Identification of the silicon form in xylem sap of rice (*Oryza sativa* L.). Plant Cell Physiol 46:279–283. https://doi.org/10.1093/pcp/pci018

Mitani N, Yamaji N, Ma J (2009) Identification of maize silicon influx transporters. Plant Cell Physiol 50:5–12. https://doi.org/10.1093/pcp/pcn110

Mitani-Ueno N, Ma JF (2021) Linking transport system of silicon with its accumulation in different plant species. Soil Sci Plant Nutr 67:10–17. https://doi.org/10.1080/00380768.2020.1845972

Neethirajan S, Gordon R, Wang L (2009) Potential of silica bodies (phytoliths) for nanotechnology. Trends Biotechnol 27:461–467. https://doi.org/10.1016/j.tibtech.2009.05.002

Neu S, Schaller J, Dudel EG (2017) Silicon availability modifies nutrient use efficiency and content, C:N:P stoichiometry, and productivity of winter wheat (*Triticum aestivum* L.). Sci Rep 7:3–10. https://doi.org/10.1038/srep40829

Parveen A, Liu W, Hussain S et al (2019) Silicon priming regulates morpho-physiological growth and oxidative metabolism in maize under drought stress. Plan Theory 8:0431. https://doi.org/10.3390/plants8100431

Pei ZF, Ming DF, Liu D et al (2010) Silicon improves the tolerance to water-deficit stress induced by polyethylene glycol in wheat (*Triticum aestivum* L.) seedlings. J Plant Growth Regul 29:106–115. https://doi.org/10.1007/s00344-009-9120-9

Ramesh P (2000) Effect of different levels of drought during the formative phase on growth parameters and its relationship with dry matter accumulation in sugarcane. J Agron Crop Sci 185:83–89. https://doi.org/10.1046/j.1439-037X.2000.00404.x

Raven JA (1983) The transport and function of silicon in plants. Biol Rev 58:179–207. https://doi.org/10.1111/j.1469-185x.1983.tb00385.x

Rocha JR, de Mello PR, Teixeira GCM, de Oliveira Filho ASB (2021) Si fertigation attenuates water stress in forages by modifying carbon stoichiometry, favoring physiological aspects. J Agron Crop Sci 207:31–643. https://doi.org/10.1111/jac.12479

Rocha JR, de Mello PR, de Cássia PM (2022) New outcomes on how silicon enables the cultivation of *Panicum maximum* in soil with water restriction. Sci Rep 12:1897. https://doi.org/10.1038/s41598-022-05927-z

Sachdev S, Ansari SA, Ansari MI et al (2021) Abiotic stress and reactive oxygen species: generation, signaling, and defense mechanisms. Antioxidants 10:1–37. https://doi.org/10.3390/antiox10020277

Santos LCN, Teixeira GCM, Prado RM et al (2020) Response of pre-sprouted sugarcane seedlings to foliar spraying of potassium silicate, sodium and potassium silicate, nanosilica and monosilicic acid. Sugar Tech 22:773–781. https://doi.org/10.1007/s12355-020-00833-y

Sbrissia AF, Schmitt D, Duchini PG, da Silva SC (2020) Unravelling the relationship between a seasonal environment and the dynamics of forage growth in grazed swards. J Agron Crop Sci 206:630–639. https://doi.org/10.1111/jac.12402

Schaller J, Cramer A, Carminati A, Zarebanadkouki M (2020) Biogenic amorphous silica as main driver for plant available water in soils. Sci Rep 10:1–7. https://doi.org/10.1038/s41598-020-59437-x

Schoelynck J, Bal K, Backx H et al (2010) Silica uptake in aquatic and wetland macrophytes: a strategic choice between silica, lignin and cellulose? New Phytol 186:385–391. https://doi.org/10.1111/j.1469-8137.2009.03176.x

Smirnoff N (1993) The role of active oxygen in the response of plants to water deficit and desiccation. New Phytol 125:27–58. https://doi.org/10.1111/j.1469-8137.1993.tb03863.x

Teixeira GCM, de Mello PR, Rocha AMS et al (2020a) Silicon in pre-sprouted sugarcane seedlings mitigates the effects of water deficit after transplanting. J Soil Sci Plant Nutr 20:849–859. https://doi.org/10.1007/s42729-019-00170-4

Teixeira GCM, de Prado RM, Rocha AMS, de Piccolo MC (2020b) Root and foliar applied silicon modifies C:N:P ratio and increases the nutritional efficiency of pre-sprouted sugarcane seedlings under water deficit. PLoS One 15:e0240847. https://doi.org/10.1371/journal.pone.0240847

Teixeira GCM, de Mello PR, Rocha AMS (2021a) Low absorption of silicon via foliar in comparison to root application has an immediate antioxidant effect in mitigating water deficit damage in sugarcane. J Agron Crop Sci. https://doi.org/10.1111/jac.12511

Teixeira GCM, de Mello Prado R, Rocha AMS et al (2021b) Beneficial effect of silicon applied through fertigation attenuates damage caused by water deficit in sugarcane. J Plant Growth Regul. https://doi.org/10.1007/s00344-021-10510-3

Teixeira GCM, de Prado RM, Rocha AMS et al (2022a) Action of silicon on the activity of antioxidant enzymes and on physiological mechanisms mitigates water deficit in sugarcane and energy cane plants. Sci Rep 12:17487. https://doi.org/10.1038/s41598-022-21680-9

Teixeira GCM, de Mello PR, Rocha AMS, de Cássia PM (2022b) Silicon as a sustainable option to increase biomass with less water by inducing carbon:nitrogen:phosphorus stoichiometric homeostasis in sugarcane and energy cane. Front Plant Sci 13:826512. https://doi.org/10.3389/fpls.2022.826512

Teixeira GCM, de Mello PR, de Oliveira LT et al (2022c) Silicon fertigation with appropriate source reduces water requirement of maize under water deficit. Plant Soil 477:83–97. https://doi.org/10.1007/s11104-022-05446-w

Tubana BS, Babu T, Datnoff LE (2016) A review of silicon in soils and plants and its role in US agriculture. Soil Sci 181:393–411. https://doi.org/10.1097/SS.0000000000000179

Verma KK, Song X-P, Zeng Y et al (2021) Foliar application of silicon boosts growth, photosynthetic leaf gas exchange, antioxidative response and resistance to limited water irrigation in sugarcane (*Saccharum officinarum* L.). Plant Physiol Biochem 166:582–592. https://doi.org/10.1016/j.plaphy.2021.06.032

Woesz A, Weaver JC, Kazanci M et al (2006) Micromechanical properties of biological silica in skeletons of deep-sea sponges. J Mater Res 21:2068–2078. https://doi.org/10.1557/jmr.2006.0251

Yang R, Howe JA, Golden BR (2019) Calcium silicate slag reduces drought stress in rice (*Oryza sativa* L.). J Agron Crop Sci 205:353–361. https://doi.org/10.1111/jac.12327

Yoshida S (1965) Chemical aspects of the role of silicon in physiology of the rice plant. Bull Natl Inst Agric Sci Series B 15:1–58

Chapter 17
Association of Silicon and Soil Microorganisms Induces Stress Mitigation, Increasing Plant Productivity

Krishan K. Verma, Xiu-Peng Song, Munna Singh, Dan-Dan Tian, Vishnu D. Rajput, Tatiana Minkina, and Yang-Rui Li

17.1 Introduction

The usage pattern of synthetic fertilizers, irrigation, herbicides, and pesticides have prompted the search for alternate techniques to combat nutrient and water limitations in crops. Microorganisms may improve crop nutrition, as well as the ability of crops to withstand biotic and abiotic stresses (Verma et al. 2020a; Zhang et al. 2021a, b; Koskey et al. 2021). Consequently, the usage of microorganisms in agricultural systems may reduce the application of synthetic fertilizers, herbicides, and pesticides. Interactive demonstrations in the literature include the bacterial and mycorrhizal stabilization of soil aggregates and structure; rhizobacterial promotion of plant growth through increased nutrient availability; and the enhancement of plant development by influencing the phytohormone balance and production of stress defense components (Verma et al. 2020c). Dodd et al. (2010) demonstrated changes in the growth hormones and architecture of roots, as well as in plant

K. K. Verma (✉) · X.-P. Song · Y.-R. Li
Key Laboratory of Sugarcane Biotechnology and Genetic Improvement, Ministry of Agriculture and Rural Affairs/Guangxi Key Laboratory of Sugarcane Genetic Improvement/Sugarcane Research Institute, Guangxi Academy of Agricultural Sciences, Nanning, Guangxi, China
e-mail: liyr@gxaas.net

M. Singh
Department of Botany, University of Lucknow, Lucknow, India

D.-D. Tian
Institute of Biotechnology, Guangxi Academy of Agricultural Sciences, Nanning, Guangxi, China

V. D. Rajput · T. Minkina
Academy of Biology and Biotechnology, Southern Federal University, Rostov-on-Don, Russia

R. de Mello Prado (ed.), *Benefits of Silicon in the Nutrition of Plants*,
https://doi.org/10.1007/978-3-031-26673-7_17

performance promoted by rhizobacteria. The rhizobacterial mediation of hormones impacts root elongation and architecture, thereby mediating water and nutrient status, which influence root–to-shoot signaling, leaf growth, and the gas exchange capacity for photosynthetic carbon fixation to acquire improved biomass and yield (Andrews et al. 2012; Nazli et al. 2020).

The long-term application of fertilizers in isolation had negative consequences in the environment, soil health, and quality (Worlanyo and Jiangfeng 2021). The use of microorganisms in the rhizosphere is endophytic to reduce the detrimental effects caused by chemical fertilizers (Atieno et al. 2020; Mishra et al. 2017). Hence, beneficial communities of plant and bacteria have been investigated for precision and sustainable agriculture (Pieterse et al. 2012; Afzal et al. 2019; Verma et al. 2020b; Ke et al. 2021; Song et al. 2021; Zhang et al. 2021b). Different bacteria are closely connected with plant cells, and as a result, they may have direct beneficial effects on plants (Hallmann et al. 1997; Patel et al. 2020). Endophytic bacteria such as *Pseudomonas, Bacillus, Enterobacter, Burkholderia*, and *Azospirillum* colonize a broad range of plant cultivars (Lodewyckx et al. 2002; Weyens et al. 2009; Reis et al. 2020; Zhang et al. 2021a).

The distribution of endophytic bacteria is determined by a variety of crops and circumstances (Zinniel et al. 2002; Ulrich et al. 2008). Endophytic bacteria on plants have the potential as bio-fertilizers (Andrews et al. 2012; Afzal et al. 2019; Verma et al. 2020c). Endophytic bacteria colonize the intracellular and intercellular regions of plant tissues, promoting plant growth (Ulrich et al. 2008; Vacheron et al. 2013; Hardoim et al. 2015; Santoyo et al. 2016; Gouda et al. 2018). The processes of endophytic bacteria improved plant development, although the mechanism related to these processes seems to be inadequately explored (Rashid et al. 2012; Bruto et al. 2014; Zhang et al. 2021b).

The abilities of bacteria improve nutrient acquisition, nitrogen fixation, iron acquisition, and siderophores production in plants (Zhang et al. 2009; Hardoim et al. 2015; Zhou et al. 2016; Reis et al. 2020). The phyllosphere, endosphere, and rhizosphere are the types of plant microbiome (Lundberg et al. 2012; Berg et al. 2014). The phyllosphere refers to all the above-ground organs of plants, which may be occupied by microorganisms, while the endosphere refers to the microbial habitat. Finally, the rhizosphere is a tiny fraction of the soil that is intimately connected with the plant roots (Hiltner 1904). The number of microbial species found in the microbiome of a live plant generally outnumbers the host cells (Mendes et al. 2013; Zhang et al. 2021a). Microorganisms in the plant microbiome may play key roles in the growth and maintenance of agriculturally important crops, including roles in nutrient acquisition, stress relief, and disease suppression. Furthermore, plants and microbes may work together to reduce toxicants in atmospheric variables and in contaminated soils (Koskey et al. 2021). Plant growth and development, as well as the mixture of organisms in the plant microbiome are all affected by biotic and abiotic factors. Agricultural practices (such as synthetic fertilizers and pesticides) may also alter the composition of the plant microbiome in some instances.

The application of silicon (Si) in agriculture is considered a better option for mitigating environmental stressors in various plant species/cultivars (Verma et al.

2020a, b, 2021). Si constitutes a significant portion of soil as silicate or aluminum silicate, and most of them cannot be absorbed directly by plant roots (Zhu and Gong 2014). However, silicic acid [Si $(OH)_4$] occurs at concentrations between 0.1 and 2.0 mM (pH level <9) and is readily absorbed into root systems. The silicon level in the above-ground organs of plants ranges from 0.1–10.0% on a dry weight basis (Epstein 1994). Although it is not considered as an essential element for plant development, Si-free agro-environments find tough challenges for plant growth and development (Guntzer et al. 2012; Coskun et al. 2019). Hence, Si may be re-designated in a newly established definition of the criteria for nutrient essentiality proposed by Epstein and Bloom (2005), as it enhances biotic and abiotic stress resistance and adjusts the pH level for the acquisition of macro-and micronutrients. In addition, the beneficial effects of Si are usually found to be integrated and essential when plants face various types of stresses (Liang et al. 2015; Verma et al. 2020a).

Therefore, this chapter discusses the interaction of rhizosphere bacteria and silicon in plants with updated research findings, aimed at sustainable soil health and plant performance with improved agricultural production.

17.2 The Impact of Si and Plant Microbiome on Plants

The effects of biotic and abiotic influences on plant development may focus on plant microbiomes in the evolution of sequencing in the next generation. The impact of biotic and abiotic variables, as well as of rhizobacteria and Si have currently been demonstrated in plant microbiomes (Ambrosini et al. 2016; Thiem et al. 2018; Young et al. 2018; Verma et al. 2020c; Koskey et al. 2021).

Plants are constantly interacting with various species in natural habitats. Plants have an inherent immune system that can detect invading microorganisms and create a possibly successful response against specific phytopathogens. Unsurprisingly, the innate immune system of plants plays a key role in regulating the action of microbial species in their microbiome system (Hacquard et al. 2017; Sangiorgio et al. 2020). Salicylic acid (SA) biosynthesis and related signalings are essential for the constitution of a proper root microbiome, as it regulates root colonization using bacteria (Lebeis et al. 2015). The other two primary sphytohormones in stress responses—jasmonic acid (JA) and ethylene (ET)—have also been shown to be significantly related to the root microbiome (Lebeis et al. 2015). In this context, host genotype and age are the two main factors that influence the plant microbiome (Wagner et al. 2016). Haney et al. (2015) used *Arabidopsis* accessions with *Pseudomonas fluorescens*, while others actively prevented the development of *P. fluorescens*. As a result, the formation of the plant microbiome might vary depending on the plant genotype and presence of various bacterial strains. Chemical probes from plant hosts have altered the formation of rhizosphere microbial communities, making this a promising approach for cultivating a healthy microbiome that could enhance plant growth. Findings from prior demonstrations also revealed that the selected microbes and their interactions with plant hosts have an impact on the formation of the rhizosphere microbiome.

Bacteria are the most studied component of the plant microbiome due to their importance, widespread influence, and potential for promoting plant performance. Plant species, genotypes, life cycle stage, proximity to the roots, and soil types influence the bacterial microbiome composition (Bulgarelli et al. 2015; Sasse et al. 2018; Zhang et al. 2021a, b). As bacteria are microbiomes, they have a considerable impact on other microbiome residents, including other bacteria. Due to synergistic effects such as biofilm development, bacteria cling to the surface for protection (Buonaurio et al. 2015). Furthermore, due to additive effects, multispecies biofilms are more effective than monospecies biofilms, resulting in mutual proliferation when a new species is introduced to the microbiome (Burmolle et al. 2006). Quorum sensing signaling and metabolite exchange are two other symbiotic connections between bacterial species in the plant microbiome (Passos da Silva et al. 2014). Olive trees cultivated with knot disease are an example of a significant association between bacteria in the plant microbiome. The pathogen responsible for the olive knot is *Pseudomonas savastanoi pv. Savastanoi*, which is supplemented by the non-pathogenic bacterial species *Pantoea agglomerans* and *Erwinia toletana*, thus leading the disease to worsen (Hosni et al. 2011; Passos da Silva et al. 2014). Bacterial interactions with other bacteria in the plant microbiome can be inhibiting to fungi. Many *Pseudomonas* species have also been reported to inhibit *R. solanacearum*, reducing pathogen density and disease prevalence (Hu et al. 2016). The intricate interplay between the plant, bacteria, and other microbes is more likely to have an impact on the actual mechanisms producing changes in the bacterial population of plants by other bacteria (Ries et al. 2020; Zhang et al. 2021b).

Bacteria and fungi have been observed to extensively interact with bacterial populations influencing fungal populations and vice versa. This relationship seems to be mostly plant-independent, as bacterial-mediated modifications in the fungal community have been observed in planted and unplanted settings (Durán et al. 2018). Depending on the precise situation of the interactions, bacteria in the plant microbiome might be advantageous or inhibiting to fungi. Through symbiosis, keystone bacteria such as members of the genus *Burkholderia* may considerably enhance the abundance of arbuscular mycorrhizal fungus (AMF), except for *P. fluorescens*, which presents restricted fungal growth (Banerjee et al. 2018; Verma et al. 2020c). In plant microbiomes, symbiosis is a common association between bacteria and fungi, notably with AMF (Braga et al. 2016; Banerjee et al. 2018; Koskey et al. 2021; Nazli et al. 2020). The fungus *Rhizopus* spp. and the bacteria *Burkholderia* spp. developed an endosymbiotic relationship in which *Rhizopus* relied on *Burkholderia* activity to create spores that infected rice seedlings with blight (Braga et al. 2016). Fungi play various important roles in the plant microbiome. Fungi in the plant microbiome vary by host, are influenced by nitrogen deposition (Dean et al. 2014), and are highly dependent on soil types (Nallanchakravarthula et al. 2014). Fungi may be beneficial to plant development but may also cause plant diseases (Bonito et al. 2017). Most research on fungi and the plant microbiome focuses on their ability to promote plant growth (Aguin-Casal et al. 2004; Andrews et al. 2012; Mishra et al. 2017).

Archaea have great potential for use in plant growth promotion, despite being understudied concerning the plant microbiome. Their close interactions with environments and other microbial residents frequently form symbiotic relationships with plants (Moissl-Eichinger et al. 2018; Verma et al. 2020c). *Archaea* was revealed to interact with plants and fungi in a bog habitat, impacting nutrient delivery and the capabilities of secondary plants. *Archaea* have not been thoroughly explored regarding its possible application in agriculture. The archaeon *Methanococcoides burtonii* produces non-photosynthetic Rubisco (Ribulose-1,5-biphosphate carboxylase), which promotes a rate-limiting carbon dioxide fixation process during photosynthesis, boosting photosynthesis performance and growth capacity in tobacco plants by acquiring in vivo the biosynthesis of rubisco encoded by rbcL and rbcS genes of cpDNA and nDNA, respectively (Wilson et al. 2016). This response suggests a relationship between nematodes and the various players in the plant microbiome, which might be used in agriculture to protect plants against nematode infection (Elhady et al. 2018).

Silicon and PGPBs have shown to mitigate the adverse effects of plant diseases (Olanrewaju et al. 2017; Wang et al. 2017; Verma et al. 2020c). Some of the known mechanisms by which Si and PGPBs mitigate biotic stress in plants are shown in Table 17.1. Various suitable nutrient-rich niches inside roots attract a better diversity of microorganisms. One of the fundamental mechanisms by which PGPBs protect plants from phytopathogens is the competition for the nutrients and niches (Compant et al. 2005; Frew et al. 2018; Coskun et al. 2019). Another important mechanism through which pathogens controlling PGPBs in the rhizosphere and endorhiza stimulate the entire plant body is induced systemic resistance, which enhances the defense against an extensive range of pathogens and insect herbivores (Compant et al. 2005; Van Loon and Bakker 2005; Koskey et al. 2021). Like PGPBs, Si can also control diseases in plants through the formation of physical barriers. It has been reported that the precipitation of amorphous silica in the cell wall of plants acts as a mechanical barrier and protects the entry of pathogens (Wang et al. 2017; Song et al. 2021).

Another effect of Si on disease control through the formation of chemical barriers is the enhancement of the activity of enzymes related to defense (chitinases, β-1,3-glucanases, peroxidase, lipoxygenase, polyphenol oxidases, polyphenol oxidase, and phenylalanine ammonia-lyase) (Verma et al. 2020b). Silicon may enhance plant resistance to pathogenic infections by (i) enhancing the gene expression associated with tolerance to pests and diseases, (ii) increasing the production of phenolic compounds, callose, methylaconitate (phytoalexins), and lignin, (iii) enhancing the gene expression related to encoding the proline-rich protein (PRP1) and the main enzymes in phenylpropanoid pathways, and (iv) enhancing the concentration of polyphenol, antimicrobial flavonoids, and anthocyanin (Frew et al. 2018; Coskun et al. 2019; Verma et al. 2020b; Song et al. 2021).

Table 17.1 Impact of silicon on plant biology subjected to stress conditions

Stress condition	Crop	Impacts	Source
Water-deficit	*Saccharum officinarum* L.	Increased photosynthetic capacity, pigments, leaf water status, phytohormones, and antioxidative enzyme activities	Verma et al. (2019a, b, 2020a, b, 2021b)
	Strawberry (*Fragaria x ananassa 'Camarosa'*)	Improved leaf development, injury of membranes, SPAD units, chlorophyll fluorescence variables, leaf gas exchange, and biomass traits	Safoora et al. (2018)
	Solanum lycopersicum	Photosynthetic performance, membrane injury, enzymatic and non-enzymatic activities, and growth parameters increased	Shi et al. (2016)
	Triticum aestivum and *Zea mays*	Enhanced plant biomass and photosynthetic and productivity	Maghsoudi et al. (2015)
Saline stress	*Cucumis sativus*	Plant biomass, enzymatic and non-enzymatic activities increased with plant development	Zhu et al. (2004)
	Oryza sativa L.	Growth-biomass, photosynthetic CO_2 assimilation rate, ion distribution improved or balanced	Flam-Shepherd et al. (2018)
	Vigna radiata	Plant productivity, pigments, photosynthesis, and stress resistance capacity enhanced	Mahmood et al. (2016)
Osmotic	*Sorghum bicolor*	Leaf photosynthesis responses, root hydraulic conductivity, and biomass capacity improved	Hattori et al. (2008)
	Solanum lycopersicum	Gas exchange rate, leaf water status, MDA, hydrogen peroxide content, and enzymatic responses increased during stress with Si application	Shi et al. (2016)
Excess Cd	*Oryza sativa* L.	Increased plant weight and enzymatic activities in leaf and roots	Farooq et al. (2016)
Excess As	*Oryza sativa* L.	Improved leaf photosynthetic, fluorescence variables, and Vcmax and Jmax responses	Sanglard et al. (2014)
Excess Cu	*Spartina densiflora*	Photosynthesis, pigments, relative growth rate, and biomass upregulated	Mateos-Naranjo et al. (2015)
Excess Mn	*Cucumis sativa*	Biomass, leaf mineral content, H_2O_2 and GPx level enhanced	Rogalla and Romheld (2002), Maksimovic et al. (2012)
Excess Al	*Zea mays* L.	Root development, citrate and malate exudation, and phenol exudation increased or upgraded	Wang et al. (2004)
Low K^+	*Sorghum bicolor*	Increased whole plant biomass, photosynthesis assimilation rate of leaf, chlorophyll level, and enzymatic activities	Chen et al. (2016)

17.3 Role Played by Rhizobacteria and Si in Plants During Environmental Stress

Water and salinity stress

Water stress is a major environmental stressor in agro-ecosystems, resulting in losses of millions of dollars annually (Lesk et al. 2016; Verma et al. 2021a, b). Although water stress in plants has been extensively researched, little is known regarding the effect of drought on the plant microbiome. Water stress may cause a complete loss of plants and microbial biomass (Bastida et al. 2017; Naylor and Coleman-Derr 2018; Verma et al. 2020a). However, investigations on different plants utilizing stress treatments that resulted in an enormous enrichment in bacteria have indicated that some species of bacteria perform better during stress conditions (Naylor et al. 2017; Frew et al. 2018). Bacteria isolates *Acinetobacter* and *Pseudomonas* alleviated declines in root and shoot biomass in drought-stricken grapevines by reducing the photosynthetic inhibition caused by stress, most likely through the production of 1-aminocyclopropane-1-carboxylate (ACC) deaminase.

On the other hand, these bacterial strains did not enhance plant development in well-irrigated plants (Rolli et al. 2015). Another investigation on a variety of plants infected with *Bacillus thuringiensis* IAM 12077 in sub-tropical climates found that it improved plant growth and nutrient absorption (Armada et al. 2018). Generally, bacteria enhance plant water resistance by boosting the water intake in roots, lowering ET levels and enhancing the metabolism and nutrient synthesis (Armada et al. 2018; Gagne-Bourque et al. 2016; Koskey et al. 2021; Zhang et al. 2021a, b).

Salinity is a severe problem for the production of agricultural crops, as it has negatively affected plant health by generating nutrient absorption imbalances that change osmotic pressures in plant cells (Yaish and Kumar 2015). Due to the drying and lysis of microorganisms in the plant microbiome, salinity substantially impacts microbiome composition, since halophytic bacteria tend to resist extremely saline soils (Yan et al. 2015; Verma et al. 2020c). High saline conditions hinder mycorrhizal association, making them less effective. Some mycorrhizal fungi can supplement plants by enhancing sodium and calcium ion translocation and plant growth (Hrynkiewicz et al. 2015; Thiem et al. 2018). Salinity also reduces mycorrhizal colonization, hyphal growth, and germination, which inhibits fungal growth. The plant microbiome is a complex network of interdependent bacteria that shifts many other factors interacting with fungi (Hamdia et al. 2004). Compared to fungi, salinity appears to have a more significant effect on bacteria (Thiem et al. 2018). Since most bacteria are not acclimated to high saline conditions, rising salinity in soils appears to drastically reduce bacterial richness and diversity (Thiem et al. 2018; Yaish et al. 2016). Plant growth-promoting bacteria (PGPBs) have been identified in saline soils and show potential for increasing plant development in salt-stressed areas through an environmentally acceptable approach. The saline soil isolates *Alcaligenes* sp. AF7 had a variety of plant growth-promoting characteristics, such as the production of exopolysaccharides, indole-3-acetic acid, gibberellic acid, and siderophores that were prominent up to different salinity thresholds, and that the bacterium increased the vegetative growth phase in rice by more than twofold under salinity stress (Fatima et al. 2020).

Furthermore, the strain NCCP-11 T of *Cellulomonas pakistanensis* sp. nov. isolated from paddy rice was shown to be moderately halotolerant (Ahmed et al. 2014). *Cellulomonas* isolates have shown potential as plant growth-promoting bacteria in rice, boosting nutrient availability by decomposing organic matter via secreted enzymes such as cellulases and hemicellulases (Duy et al. 2016). More research under saline-stressed conditions could find other essential halotolerant microorganisms in the microbiome to be used in future agricultural operations under salinity stress.

The association of Si and PGPBs in plants is beneficial in salinity and water stress conditions (Kaushal and Wani 2016; Rizwan et al. 2015). Tables 17.1 and 17.2 highlight the action mechanisms through which Si and PGPBs help plants cope with salinity and water stress. Salinity and water deficit have a negative impact on plant growth, development, and yield by limiting the uptake and translocation of water and nutritional ions (Hu and Schmidhalter 2005). Root length and the region of the root surface affect plant nutrient absorption. Areas that are more exposed to the absorption of scattered ions result from an increase in root surface morphology (Barber 1995). During salinity (Ahmad et al. 2011; Kim et al. 2014; Li et al. 2015) and water (Ahmed et al. 2011; Chen et al. 2011; Verma et al. 2021a) stress, Si may improve root development and enhance nutrient uptake, also improving nutrient balance (Javid et al. 2011; Fahad et al. 2015; Verma et al. 2020a), as plant hormones regulate plant development and help plants resist to various environmental stressors (Ryu and Cho 2015; Frew et al. 2018).

Water deficit also restricts nutrient uptake and accumulation by the roots and its subsequent transfer to shoots, decreasing the nutritional availability and metabolism (Farooq et al. 2009; Verma et al. 2020a). PGPBs may directly affect plant growth by solubilizing P through the secretion of a variety of extracellular phosphatases and the production of organic acids, inorganic acids, protons, and K to improve nutrient uptake (Olanrewaju et al. 2017). PGPBs can enhance plant resistance to salinity stress and water deficit by reducing Na^+ absorption, affecting the uptake, translocation, and distribution of some major nutrient elements in plants treated with Si (Soundararajan et al. 2016). In earlier demonstrations, the exogenous use of Si increased leaf gas exchange characteristics in various plant species during salinity stress and water deficit conditions (Verma et al. 2021b). High concentrations of soluble salts and specific ions (Na^+ and Cl^-) severely influence plant growth by increasing osmotic pressures and inhibiting water absorption and translocation (Hu and Schmidhalter 2005; Coskun et al. 2019). Plants require an optimal water level for survival in salinity and water stress conditions via osmotic adjustment (Osakabe et al. 2014).

By adjusting the levels of solutes, i.e., proline, glycine betaine, carbohydrates, and polyols; and antioxidative compounds, i.e., total phenolics, total soluble sugars, and total free amino acids, as well as minimizing the osmotic shock caused by NaCl

Table 17.2 Impact of soil microbes on agricultural crops subjected to abiotic stressors

Stress	Plant	Microbe	Function	Source
High temperature	Potato	*Paraburkholderia phytofirmans*	Production of ACC deaminase	Bensalim et al. (1998)
	Tomato	Mycorrhizae	Decrease MDA and H_2O_2 content, enhance ROS activity in plant organs	Duc et al. (2018)
	Wheat	*Bacillus amyloliquefaciens, Azospirillum brasilense*	Reduce ROS, pre-activation of heat shock proteins (HSPs)	El-Daim et al. (2014)
	Soy	*Bacillus aryabhatthai* SRBO2	Production of ABA content	Park et al. (2017)
	Wheat and sorghum	*Pseudomonas putida* AKMP7, *Pseudomonas* sp. AKMP6	ROS reduce, enhance proline, photosynthetic pigments, sugar, starch, amino acid, and protein and plant hormones	Ali et al. (2009, 2011)
Cold temp.	Apple and pear	*Ps. fluorescens* A506	Interaction with INA^+ bacteria	Lindow and Brandl (2003)
	Wheat	*Pantoea dispersa* 1A, *Serratia marcescens* SRM, *Pseudomonas* spp. PGERs17, NARs9	Production of ACC deaminase	Selvakumar et al. (2008), Mishra et al. (2008, 2012)
	Grapevine	*Paraburkholderia phytofirmans*	Production of ACC deaminase	Theocharis et al. (2012), Barka et al. (2006)

(continued)

Table 17.2 (continued)

Stress	Plant	Microbe	Function	Source
Insufficient water supply	Cucumber and bean	*Phoma glomerata, Penicillium* sp., *Exophiala* sp., *Paecilomyces formosus, Glomus intraradices*	Improve soil properties and root water conductivity	Aroca et al. (2007)
	Tomato, cucumber, and orange	*Pseudomonas chlororaphis* TSAU13, *Funneliformis mosseae*	Production of IAA	Egamberdieva (2012), Liu et al. (2018)
	Cucumber	*Burkholderia, Promicromonospora, Acinetobacter, Pseudomonas* spp.	Production of gibberellin	Kang et al. (2015)
	Lettuce	*Bacillus subtilis*	Production of cytokinin	Arkhipova et al. (2007)
	Soybean	*Ps. putida* H-2-3	Production of ABA	Kang et al. (2014)
	Tomato, pepper, pea, peanut	*Achromobacter piechaudii* ARV8, *B. licheniformis* K11, *Pseudomonas* spp., *Ps. fluorescens* TDK	Production of ACC deaminase	Mayak et al. (2004), Saravanakumar and Samiyappan (2007), Zahir et al. (2008), Lim and Kim (2013)
	Carrot, soybean, lettuce, basil	*Glomus intraradices*	Increase aquaporin activities	Porcel et al. (2006), Li et al. (2013)
	Basil	*Pseudomonas* sp.	Upregulated antioxidant enzyme activities	Heidari and Golpayegani (2012)
	Sunflower, orange	*Ps. putida, Ps. aeruginosa PF23, Glomus mosseae, G. versiforme, G. diaphanum*	Production of EPS	Wu et al. (2008), Sandhya et al. (2009), Tewari and Arora (2014)
	Tomato, soybean, tangerine	*Bacillus polymyxa, Glomus intraradices, G. versiforme*	Production of osmolytes	Porcel and Ruiz-Lozano (2004), Wu et al. (2008), Shintu and Jayaram (2015), Vurukonda et al. (2016)

Stress	Plant	Microbe	Function	Source
Salinity	*Maize*	*Azospirillum*	Osmoprotection	Hamdia et al. (2004)
	Lettuce	*Azospirillum*	Increase germination rate	Barassi et al. (2006)
	Rapeseed	*Pseudomonas putida* UW 4	Upgrade plant growth, development, and yield biomass	Cheng et al. (2007)
	Peanut	*Pseudomonas fluorescens*	Increased ACC deaminase activity	Saravanakumar and Samiyappan (2007)
	Maize	*Pseudomonas syringae*, *P. fluorescens*, *Enterobacter aerogenes*	Enhanced ACC deaminase level	Nadeem et al. (2007)
	Bean	*Azospirillum brasilense*	Improve root morphological capacity and increase secretion of nod-genes	Dardanelli et al. (2008)
	Maize	*Rhizobium, Pseudomonas*	Decrease EC content and increase proline level, balance RWC status in leaves, and accumulation of K ions	Bano and Fatima (2009)
	Wheat	*Pseudomonas* sp., *Serratia* sp.	Enhance ACC deaminase activity	Zahir et al. (2007)
	Cotton	*P. putida* Rs-198	Enhance the absorption of Mg^{2+}, K^{2+}, and Ca^{2+} and reduce Na^{2+} uptake from soil	Yao et al. (2010)
	Mung bean	*P. syringae*, *P. fluorescens*, and *Rhizobium phaseoli*	Increase ACC deaminase content	Ahmad et al. (2011)

(continued)

Table 17.2 (continued)

Stress	Plant	Microbe	Function	Source
Heavy metal	Pea	*Enterobacter* sp.	Restored growth injuries by reducing ion distribution. Enhanced growth performances and chlorophyll level	Naveed et al. (2020)
	Lentil	*Bacillus* sp.	Minimize the harmful effects of Cr on plants and enhanced plant performance and development	Khan et al. (2015)
	Madagascar Periwinkle	*Bacillus megaterium* MCR−8	Increased growth characteristics, proline and total soluble protein level during metal toxicity applied media and upregulated antioxidative enzymatic activities	Hansda et al. (2017)
	Black mustard	*Kocuria* sp. CRB15	Enhanced plant growth and development and reduced contaminated ion uptake and accumulation in plant organs	Gontia-Mishra et al. (2016)
	Rice	*Klebsiella pneumoniae* MCC 3091	Enhanced seedling rate of germination and biomass capacity during stress. Upgraded the level of photosynthetic pigments, enzymatic and non-enzymatic activities and downregulated the production of ethylene during metal stress condition	Pramanik et al. (2017)
	Switchgrass	*Azospirillum*	Enhanced the growth parameters, biomass and balance the pH of soil and protect the accumulation of toxic ions uptake	Ma et al. (2015)
	Black nightshade	*Pseudomonas* sp. LK9	Upgraded the morphological traits with yield productivity. Positively enhanced or balanced the uptake of Cu, Zn, and Cd ions in plant organs. Increased the level of P and Fe in soil properties	He et al. (2013)
	Sedum	*Achromobacter* sp. E4L5, *Bacillus* sp. E4S1, *Bacillus* sp. E1S2, *Bacillus pumilus* E2S2 and *Stenotrophomonas* sp. E1L	Significantly enhanced the morphological and biomass activities of plants during stress condition	Adediran et al. (2015)
	Arugula	*Pseudomonas putida* (ATCC 39213)	Significantly enhanced the growth, biomass and development of plants and leaf photosynthetic pigments. Cd ion uptake increased	Kamran et al. (2020)

toxicity, Si improves the salinity stress tolerance of plants. Under salinity and drought stress, plants usually present water deficit, which leads to excess reactive oxygen species such as singlet oxygen, hydroxyl radical, hydrogen peroxide, and superoxide anion (Frew et al. 2018; Coskun et al. 2019; Verma et al. 2020a). Damage to the plasma membrane and endomembrane systems and disruption of the normal metabolism can result from these reactive oxygen species. Like PGPBs, Si fertilizers can reduce oxidative damage in many plant species during salinity and water stress by increasing the activities of enzymatic and non-enzymatic antioxidant activities (Kim et al. 2017). In addition, Si also maintains the permeability and stability of cell membranes in plants cultivated during abiotic stress conditions (Coskun et al. 2019). Reduced leaf relative water content is a common reaction of plants in circumstances of salinity stress (Fahad et al. 2015, Rizwan et al. 2015; Verma et al. 2021a). Silicon could lead to efficient water use through increased root water uptake and stress resistance in plants during limited water irrigation (Verma et al. 2020a, 2021a).

Plants develop a systemic resistance to PGPBs (Egamberdieva and Lugtenberg 2014). Inoculating stressed plants with PGPBs may activate signaling pathways that increase the disease resistance of the host, which is a phenomenon known as induced systemic resistance (Van Loon et al. 1998; Song et al. 2021). PGPB-mediated physical and chemical changes that result in improved tolerance to abiotic stress have also been proposed as induced systemic tolerance (Sarma et al. 2012; Nazli et al. 2020). In stressed plants, the role played by Si in activating key genes linked with salt and water stress responses has also been documented (Kim et al. 2014; Muneer and Jeong 2015; Muneer et al. 2014; Zhu et al. 2015; Coskun et al. 2019).

Heavy metal toxicity

As proven in numerous instances, heavy metal toxicity in the plant microbiome reduces microbial diversity and operational taxonomic unit (OTU) counts (Jiao et al. 2019). Zinc is considered the most effective for inducing changes in diversity and OTU counts in the microbiome, followed by cadmium and lead (Hur et al. 2011). Although the effects of metals on different members of the microbiome are not fully understood, the use of plant growth-promoting microbes to help reduce heavy metals through phytoextraction has the potential to become widely used in agriculture due to its demonstrated effectiveness (Mnasri et al. 2017; Gil-Martnez et al. 2018). Moreover, plant growth-promoting microorganisms have shown more resistance to metal toxicities (Mishra et al. 2017; El-Meihy et al. 2019). These strains could be employed to promote plant growth in heavy metal-contaminated environments. The effects of heavy metals on the microbiota may identify new players who could help with bioremediation efforts in agricultural systems (Nazli et al. 2020).

Si and PGPBs have been shown to reduce the adverse effects of heavy metal toxicity in contaminated soils (Adrees et al. 2015; Frew et al. 2018). Metal-resistant PGPBs that produce siderophores might reduce metal toxicity by providing nutrients to plants and binding metals other than iron. Phytohormones such as auxins, gibberellins, and cytokinin-producing PGPBs may mitigate metal-induced stress in plants and facilitate adaptation and resistance strategies by triggering physiological changes and boosting the absorption of essential nutrients as a result of expanding plant root systems (Nazli et al. 2020). The interactive use of Si and PGPBs may reduce toxic effects in plants exposed to metal toxicity (Verma et al. 2020c). Adrees et al. (2015) and Imtiaz et al. (2016) also revealed that Si reduced the harmful effects of metals in plants by increasing the synthesis of root exudates, which chelate metals, flavonoid-phenolics, organic acids, citrate, malate, and aconitate contents. Many plant species were shown to have increased resistance to harmful toxic ions by lowering metal uptake and translocation (Adrees et al. 2015; Imtiaz et al. 2016).

Under various stresses, PGPBs stimulate heavy metal resistance genes, as well as the development of more general stress response mechanisms, such as ethylene and stress proteins (Dimkpa et al. 2009; Rajkumar et al. 2012; Sessitsch et al. 2013; Zhang et al. 2021a, b), and genes associated with toxic ions responsible for heavy metal tolerance in rice during Cu/Cd stress (Nwugo and Huerta 2010; Adrees et al. 2015).

Nutrient deficiency

In plants, Si and PGPBs reduce the negative impacts of nutritional imbalance (Hernandez-Apaolaza 2014). Tables 17.1 and 17.2 highlight some known processes by which Si and PGPBs restore the nutritional (macro-and micro-nutrient) imbalance in plants. Si also helps plants to cope with N deficiency by (i) increasing nutrient absorption, (ii) improving nodulation and N_2 fixation in leguminous crops, (iii) increasing N use efficiency (NUE), and (iv) changing primary metabolism by driving amino acid remobilization (Frew et al. 2018; Coskun et al. 2019; Verma et al. 2020a). Si fertilization boosts P availability in graminaceous species (Soundararajan et al. 2016), while Si may inhibit P uptake and the formation of chlorosis when excessive P levels are applied, presumably by lowering the transpiration rate. Si also reduces the stress caused by K shortage by regulating soil K availability and nutrient concentrations in plants and altering antioxidant enzymes to reduce K-deficiency-induced MDA content and oxidative stress (Kaya et al. 2006; Miao et al. 2010; Pei et al. 2010; Frew et al. 2018).

PGPBs enhance micronutrient availability to plants (Miransari 2013; Zhang et al. 2021a), reduce soil pH, and generate chelating agents. Iron (Fe) is a key micronutrient that plants require in dry, calcareous, and alkaline environments. In this context, indole acetic acid (IAA) and PGPBs produce siderophores for Fe nutrition (Ramos-Solano et al. 2010; Koskey et al. 2021). PGPBs also mitigate the adverse effects of excessive micronutrient concentrations in the soil and atmospheric environment on plant growth and development through a variety of methods (Zhuang et al. 2007; Reis et al. 2020; Sangiorgio et al. 2020; Koskey et al. 2021).

Si has been shown to alleviate manganese (Mn) toxicity in some plants (Li et al. 1999; Verma et al. 2020c) by increasing the oxidizing capacity of the roots, resulting in a higher Mn oxidation rate in the rhizosphere and increased precipitation outside the plants (Li et al. 1999; Verma et al. 2020c). It has also been observed that Si quickly alleviates Mn deficiency symptoms. It also affects copper (Cu) plant nutrition in Cu-stressed plants by changing the distribution or bioavailability of Cu inside the leaves (Frantz et al. 2011), also reducing Cu stress in plants by forming Si deposits on the cell wall through the increase in the number of Cu binding sites (Frantz et al. 2011; Hernandez-Apaolaza 2014; Liang et al. 2007). The positive effects of PGPBs inoculation have been reported under unfavorable conditions, such as waterlogging, wounding, freezing stress, UV radiation, ozone stress, and high/low temperatures (Glick 2014; Frew et al. 2018; Coskun et al. 2019).

Potential of hydrogen (pH)
The pH of the soil is a key component in defining the composition of the plant microbiome and its impact on heavy metal uptake. The soil pH is one of the most important environmental factors of rhizobacterial composition; bacterial species have been shown to significantly fluctuate with small pH changes, but fungus types are very slightly associated with pH changes (Rousk et al. 2010; Deng et al. 2018). Bacterial diversity decreases when the pH of the environment becomes more acidic (Yun et al. 2016). Most bacteria have restricted optimal pH ranges, and changes in nutrient availability are likely to cause this severe sensitivity to pH shift in bacteria (Yun et al. 2016; Qi et al. 2018; Zhang et al. 2021a). Highly alkaline soils may hamper plant growth due to poor mineral uptake, while very acidic soils may hamper growth due to the excessive uptake of harmful heavy metals (George et al. 2012; Tozser et al. 2017). Some plant growth-promoting microorganisms have shown to be promising in reducing alkaline stress in plants. Dixit et al. (2020) isolated three plant growth-promoting bacteria (*Alcaligenes* sp. NBRI NB2.5, *Bacillus* sp. NBRI YE1.3, and *Bacillus* sp. NBRI YN4.4) from alkaline soils and discovered that they improved maize biomass and germination rate in vitro under alkaline conditions, with *Bacillus* strain NBRI YN4.4 showing the most remarkable enhancement in plant growth. Soil inoculation with NBRI NB2.5, NBRI YE1.3, and NBRI YN4.4 increased soil enzymatic activities such as *alkaline phosphatase*, *beta-glucosidase*, and *dehydrogenase* during alkaline stress, demonstrating the importance of these alkalo-tolerant strains for plant growth and soil fertility/health in alkaline conditions (Dixit et al. 2020).

17.4 Role Played by Plant Hormones with the Application of Plant Microbes and Silicon

In addition, endophytic bacteria promote plant growth by producing several phytohormones. Plant nutrient uptake and biomass are enhanced by phytohormones (Phetcharat and Duangpaeng 2012; Shi et al. 2014). Endophytic bacteria create plant growth regulators such as cytokines (CYT), ethylene (ETH), abscisic acid (ABA), gibberellins, and indole-3-acetic acid (IAA) during interactions with plants (Vacheron et al. 2013). Surprisingly, IAA has a significant impact on plant growth and development. Modulating IAA is one of the most unusual ways that endophytic bacteria can help plants to grow. Ethylene is also an important plant hormone that regulates a variety of developmental and physiological processes (Sun et al. 2016).

Endophytic bacteria produce the ethylene precursor enzyme 1-aminocyclopropane-1-carboxylate (ACC) deaminase (Sun et al. 2009; Mishra et al. 2017). Under stressful conditions, ACC hydrolysis improved plant development (Santoyo et al. 2016). Furthermore, some endophytic bacterial strains may emit volatile organic substances, such as acetoin, pentadecane, 2,3-butanediol, 1-hexanol, and indole, promoting plant development (Ryu et al. 2003; Gutierrez-Luna et al. 2010; Hernandez-Calderon et al. 2018).

Under water stress, phytohormone-producing PGPBs have shown to enhance the plant resistance against various environmental stresses by changing phytohormone contents such as auxins, ethylene, gibberellins, cytokinins, and abscisic acid (Kaushal and Wani 2016; Tsukanova et al. 2017; Coskun et al. 2019; Verma et al. 2020b). The generation of IAA by PGPBs alters the root architecture by increasing the number of root tips and the root surface area, enhancing water and nutrient acquisition, and ultimately assisting plants in coping with water deficits (Kaushal and Wani 2016; Verma et al. 2021a).

17.5 Crop Rotation and Fertilizer Use

Crop rotation, applying natural and synthetic fertilizers, and using helpful microbes or transgenic plants are some of the advanced agricultural practices utilized by farmers to boost the productivity and quality of crops. Crop rotation has long been recognized for its benefits, including creating disease-resistant soils (Peters et al. 2003) and increasing nutrients in the soil (Stevenson and Kessel 1996), which improve plant performance. However, the impact of crop rotation on microbial populations in the soil has been poorly known for a long time. Recent research has observed how crop rotation alters the microbial population in the rhizosphere and the bulk soil surrounding plants (Breidenbach et al. 2017; Maarastawi et al. 2018). According to Hamel et al. (2018), using a large number of leguminous species (pulse) within a wheat crop rotation boosted soil nitrogen levels but did not affect the rhizobacterial community composition. According to Venter et al. (2016), increased microbial richness and variety in general (with longer rotation trials) have a more substantially positive influence on microbial richness. To further understand the underlying mechanism(s) through which crop rotation promotes stress resistance, more research is needed on the co-occurrence of keystone beneficial and harmful microorganisms, as well as on functional gene clusters related to these microbial communities.

For many years, farmers have used both natural and chemical fertilizers in their crops. On the other hand, the impact of these fertilizers on the microbial composition of the soil was not well understood. Due to the chemical components identified in fertilizers, recent studies have focused on the influence of synthetic fertilizers on the microbiome of a given crop. Reduced nitrogen fertilization did not affect the sugarcane microbiome (Yeoh et al. 2016), but it did result in decreased plant biomass and nitrogen content. On the other hand, high nitrogen fertilization enhanced the abundance of nitrification and denitrification genes in the surrounding soil, improving crop biomass and nitrogen content. Lang et al. (2018) found that when phosphorus fertilizer application increased, the overall community richness of AMF and bacteria decreased, while both fungal and bacterial gene copies increased. As a result, heavy fertilizer application reduces microbial community diversity while favoring microorganisms that may withstand high nitrogen and phosphorus levels. Similarly, other researchers have found that seaweed components and extracts improve the plant's innate immunity (Chandia et al. 2004; Cook et al. 2018). Few studies have been conducted on the influence of seaweed extracts on plant microbiomes.

17.6 Concluding Remarks, Limitations and Future Research

As a result of advancements in DNA sequencing methods, the microbiome has become an important research topic. This breakthrough may result in a better understanding of complicated microbial communities (Kuczynski et al. 2010). It may unearth knowledge regarding micro-ecosystems that occur inside soils through the study of the community of culturable and unculturable bacteria in their natural surroundings (Fierer 2017). Targeted amplicon sequencing or metagenomic shotgun sequencing are commonly used to investigate these communities through DNA sequencing (Knight et al. 2018), with constraints to the soil environment. This method is low-cost and effective to investigate different types of microorganisms. We discussed how the plant microbiome and Si are important for plant growth and development, suppressing possible pathogens and facilitating environmental cleanup. The interplay between Si and the plant microbiome is a fascinating field of study that is rapidly increasing. There are several contributing aspects linked to microbiome components and functional groups; however, the causality of these links must be further investigated. There may not be a general answer due to the variety of biotic and abiotic pressures present in the soil and their complex interconnections. Instead, each case must be examined in its context, taking specific plants, microorganisms, and environmental factors into account. As a result, it is critical to create streamlined methodologies for the characterization of microbiomes under varied situations, which may be used to guide knowledge-based plant growth promotion. Plant growth-promoting bacteria (PGPB) are living microbes sprayed on crops that may pose a health risk to humans, animals, plants, and the environment. There is currently no internationally harmonized regulatory framework to evaluate the possible dangers related to the use of PGPB. Other key topics include exploiting the beneficial action mechanisms of Si and bacteria, their plant growth-promoting properties, and the best way to recruit or employ a healthy microbiome consortium to improve crop yield with new possibilities of intensive agro-environmental cropping systems in near future.

Acknowledgments We are very grateful to the Guangxi Academy of Agricultural Sciences, Nanning, Guangxi, China, for providing the necessary facilities. This study was financially supported by the Guangxi Innovation Teams of Modern Agriculture Technology (nycytxgxcxtd-2021-03), Youth Program of National Natural Science Foundation of China (31901594), The National Natural Science Foundation of China (31760415), Guangxi Natural Science Foundation (2021GXNSFAA220022), Fund of Guangxi Academy of Agricultural Sciences (2021YT011) and Guangxi Key Laboratory of Sugarcane Genetic Improvement Project (21-238-16-K-04-02).

References

Adediran GA, Ngwenya BT, Mosselmans JFW, Heal KV, Harvie BA (2015) Mechanism behind bacteria induced plant growth promotion and Zn accumulation in Brassica juncea. J Hazard Mater 283:490–499

Adrees M, Ali S, Rizwan M, Zia-ur-Rehman M, Ibrahim M, Abbas F, Farid M, Qayyum MF, Irshad MK (2015) Mechanisms of silicon-mediated alleviation of heavy metal toxicity in plants: a review. Ecotoxicol Environ Saf 119:186–197. https://doi.org/10.1016/j.ecoenv.2015.05.011

Afzal I, Shinwari ZK, Sikandar S, Shahzad S (2019) Plant beneficial endophytic bacteria: mechanisms, diversity, host range and genetic determinants. Microbiol Res 221:36–49. https://doi.org/10.1016/j.micres.2019.02.001

Aguin-Casal O, S'ainz-Os'es MJ, Pedro Mansilla-V'azquez J (2004) *Armillaria* species infesting vineyards in Northwestern Spain. Eur J Plant Pathol 110:683–687. https://doi.org/10.1023/B:EJPP.0000041553.98879.51

Ahmad M, Zahir ZA, Asghar HN, Asghar M (2011) Inducing salt tolerance in mung bean through coinoculation with rhizobia and plant-growth-promoting rhizobacteria containing 1-aminocyc lopropane-1-carboxylate deaminase. Can J Microbiol 57:578–589. https://doi.org/10.1139/w11-044

Ahmed I, Kudo T, Abbas S, Ehsan M, Iino T, Fujiwara T, Ohkuma M (2014) *Cellulomonas pakistanensis* sp. nov., a moderately halotolerant Actinobacteria. Int J Syst Evol Micrbiol 64:2305–2311. https://doi.org/10.1099/ijs.0.059063-0

Ahmed M, Qadeer U, Aslam MA (2011) Silicon application and drought tolerance mechanism of sorghum. Afr J Agric Res 6:594–607. https://doi.org/10.5897/AJAR10.626

Ali SZ, Sandhya V, Grover M, Kishore N, Rao LV, Venkateswarlu B (2009) Pseudomonas sp. strain AKM-P6 enhances tolerance of sorghum seedlings to elevated temperatures. Biol Fert Soils 46:45–55

Ambrosini A, de Souza R, Passaglia LMP (2016) Ecological role of bacterial inoculants and their potential impact on soil microbial diversity. Plant Soil 400:193–207. https://doi.org/10.1007/s11104-015-2727-7

Andrews M, Cripps MG, Edwards GR (2012) The potential of beneficial microorganisms in agricultural systems. Ann Appl Biol 160:1–5. https://doi.org/10.1111/j.1744-7348.2011.00519.x

Arkhipova TN, Prinsen E, Veselov SU, Martinenko EV, Melentiev AI, Kudoyarova GR (2007) Cytokinin producing bacteria enhance plant growth in drying soil. Plant Soil 292:305–315. https://doi.org/10.1007/s11104-007-9233-5

Armada E, Leite MFA, Medina A, Azc'on R, Kuramae EE (2018) Native bacteria promote plant growth under drought stress condition without impacting the rhizomicrobiome. FEMS Microbiol Ecol 94:092. https://doi.org/10.1093/femsec/fiy092fiy092

Aroca R, Rosa P, Ruiz-Lozano JM (2007) How does arbuscular mycorrhizal symbiosis regulate root hydraulic properties and plasma membrane aquaporins in Phaseolus vulgaris under drought, cold or salinity stresses? New Phytol 173:808–816

Atieno M, Herrmann L, Nguyen HT, Phan HT, Nguyen NK, Srean P et al (2020) Assessment of biofertilizer use for sustainable agriculture in the Great Mekong Region. J Environ Manag 275:111300. https://doi.org/10.1016/j.jenvman.2020.111300

Banerjee S, Schlaeppi K, van der Heijden MG (2018) Keystone taxa as drivers of microbiome structure and functioning. Nat Rev Microbiol 16:567–576. https://doi.org/10.1038/s41579-018-0024-1

Bano A, Fatima M (2009) Salt tolerance in Zea mays (L.) following inoculation with rhizobium and pseudomonas. Biol Fertil Soils 45:405–413

Barassi CA, Ayrault G, Creus CM, Sueldo RJ, Sobrero MT (2006) Seed inoculation with Azospirillum mitigates NaCl effects on lettuce. Sci Hortic 109:8–14

Barber SA (1995) Soil nutrient bioavailability: a mechanistic approach. Wiley

Barka EA, Nowak J, Clément C (2006) Enhancement of chilling resistance of inoculated grapevine plantlets with a plant growth-promoting rhizobacterium, Burkholderia phytofirmans strain PsJN. Appl Environ Microbiol 72:7246–7252

Bastida F, Torres IF, Andr'es-Abell'an M, Baldrian P, L'opez-Mond'ejar R, V˘etrovský T, Richnow HH, Starke R, Ondo˜no S, García C, L'opez-Serrano FR, Jehmlich N (2017) Differential sensitivity of total and active soil microbial communities to drought and forest management. Glob Chang Biol 23:4185–4203. https://doi.org/10.1111/gcb.13790

Berg G, Grube M, Schloter M, Smalla K (2014) Unraveling the plant microbiome: looking back and future perspectives. Front Microbiol 5:1–7

Bensalim S, Nowak J, Asiedu SK (1998) A plant growth promoting rhizobacterium and temperature effects on performance of 18 clones of potato. Am J Potato Res 75:145–152

Bonito G, Hameed K, Toome-Heller M, Healy R, Reid C, Liao HL, Aime MC, Schadt C, Vilgalys R (2017) *Atractiella rhizophila*, sp. nov., an endorrhizal fungus isolated from the *Populus* root microbiome. Mycologia 109:18–26. https://doi.org/10.1080/00275514.2016.1271689

Braga LP, Yoshiura CA, Borges CD, Horn MA, Brown GG, Drake HL, Tsai SM (2016) Disentangling the influence of earthworms in sugarcane rhizosphere. Sci Rep 15:38923. https://doi.org/10.1038/srep38923

Breidenbach B, Brenzinger K, Brandt FB, Blaser MB, Conrad R (2017) The effect of crop rotation between wetland rice and upland maize on the microbial communities associated with roots. Plant Soil 419:435–445. https://doi.org/10.1007/s11104-017-3351-5

Bruto M, Prigent-Combaret C, Muller D, Moenne-Loccoz Y (2014) Analysis of genes contributing to plant-beneficial functions in plant growthpromoting rhizobacteria and related proteobacteria. Sci Rep 4:6261. https://doi.org/10.1038/srep06261

Bulgarelli D, Garrido-Oter R, Münch PC, Weiman A, Droge J, Pan Y, McHardy AC, Schulze-Lefert P (2015) Structure and function of the bacterial root microbiota in wild and domesticated barley. Cell Host Microbe 17:392–403. https://doi.org/10.1016/j.chom.2015.01.011

Buonaurio R, Moretti C, da Silva DP, Cortese C, Ramos C, Venturi V (2015) The olive knot disease as a model to study the role of interspecies bacterial communities in plant disease. Front Plant Sci 6:434. https://doi.org/10.3389/fpls.2015.00434

Burmolle M, Webb JS, Rao D, Hansen LH, Sørensen SJ, Kjelleberg S (2006) Enhanced biofilm formation and increased resistance to antimicrobial agents and bacterial invasion are caused by synergistic interactions in multispecies biofilms. Appl Environ Microbiol 72:3916–3923. https://doi.org/10.1128/AEM.03022-05

Chandía NP, Matsuhiro B, Mejías E, Moenne A (2004) Alginic acids in *Lessonia vadosa*: partial hydrolysis and elicitor properties of the polymannuronic acid fraction. J Appl Phycol 16:127–133. https://doi.org/10.1023/B:JAPH.0000044778.44193.a8

Chen DQ, Cao BB, Qi LY, Yin LN, Wang SW, Deng XP (2016) Silicon-moderated K-deficiency-induced leaf chlorosis by decreasing putrescine accumulation in sorghum. Ann Bot 118:305–315. https://doi.org/10.1093/aob/mcw111

Chen W, Yao X, Cai K, Chen J (2011) Silicon alleviates drought stress of rice plants by improving plant water status, photosynthesis and mineral nutrient absorption. Biol Trace Elem Res 142:67–76. https://doi.org/10.1007/s12011-010-8742-x

Cheng Z, Park E, Glick BR (2007) 1-Aminocyclopropane-1-carboxylate deaminase from pseudomonas putida UW4 facilitates the growth of canola in the presence of salt. Can J Microbiol 53:912–918

Compant S, Duffy B, Nowak J, Clément C, Barka EA (2005) Use of plant growth-promoting bacteria for biocontrol of plant diseases: principles, mechanisms of action, and future prospects. Appl Environ Microbiol 71:4951–4959. https://doi.org/10.1128/AEM.71.9.4951-4959.2005

Cook J, Zhang J, Norrie J, Blal B, Cheng Z (2018) Seaweed extract (Stella maris®) activates innate immune responses in *Arabidopsis thaliana* and protects host against bacterial pathogens. Mar Drugs 16:221. https://doi.org/10.3390/md16070221

Coskun D, Deshmukh R, Sonah H, Menzies JG, Reynolds O, Ma JF, Kronzucker HJ, Belanger RR (2019) The controversies of silicon's role in plant biology. New Phytol 221:67–85. https://doi.org/10.1111/nph.15343

Dardanelli MS, Fernández de Córdoba FJ, Rosario Espuny M, Rodríguez Carvajal MA, Soria Díaz ME, Gil Serrano AM, Okon Y, Megías M (2008) Effect of Azospirillum brasilense coinoculated with rhizobium on Phaseolus vulgaris flavonoids and nod factor production under salt stress. Soil Biol Biochem 40:2713–2721

Dean SL, Farrer EC, Taylor DL, Porras-Alfaro A, Suding KN, Sinsabaugh RL (2014) Nitrogen deposition alters plant–fungal relationships: linking belowground dynamics to aboveground vegetation change. Mol Ecol 23:1364–1378. https://doi.org/10.1111/mec.12541

Deng S, Ke T, Li L, Cai S, Zhou Y, Liu Y, Guo L, Chen L, Zhang D (2018) Impacts of environmental factors on the whole microbial communities in the rhizosphere of a metal-tolerant plant: *elsholtzia haichowensis*. Sun Environ Pollut 237:1088–1097. https://doi.org/10.1016/j.envpol.2017.11.037

Dimkpa C, Weinand T, Asch F (2009) Plant-rhizobacteria interactions alleviate abiotic stress conditions. Plant Cell Environ 32:1682–1694. https://doi.org/10.1111/j.1365-3040.2009.02028.x

Dixit VK, Misra S, Mishra SK, Tewari SK, Joshi N, Chauhan PS (2020) Characterization of plant growth-promoting alkalo tolerant *Alcaligenes* and *Bacillus* strains for mitigating the alkaline stress in *Zea mays*. Antonie Van Leeuwenhoek 113:889–905. https://doi.org/10.1007/s10482-020-01399-1

Dodd I, Zinovkina N, Safronova V, Belimov A (2010) Rhizobacterial mediation of plant hormone status. Ann Appl Biol 157:361379. https://doi.org/10.1111/j.1744-7348.2010.00439.x

Duc NH, Csintalan Z, Posta K (2018) Arbuscular mycorrhizal fungi mitigate negative effects of combined drought and heat stress on tomato plants. Plant Physiol Biochem 13:297–307

Durán P, Thiergart T, Garrido-Oter R, Agler M, Kemen E, Schulze-Lefert P, Hacquard S (2018) Microbial interkingdom interactions in roots promote *Arabidopsis* survival. Cell 175:973–983. https://doi.org/10.1016/j.cell.2018.10.020

Duy MV, Hoi NT, Ve NB, Thuc LV, Trang NQ (2016) Influence of *Cellulomonas flavigena*, *Azospirillum* sp. and *Pseudomonas* sp. on rice growth and yield grown in submerged soil amended with rice straw. In: Sayyed RZ, Reddy MS, Al-Turki AI (eds) Recent trends in PGPR research for sustainable crop productivity. Scientific Publishers, India, pp 238–242

Egamberdieva D (2012) Pseudomonas chlororaphis: a salt-tolerant bacterial inoculant for plant growth stimulation under saline soil conditions. Acta Physiol Plant 34:751–756

Egamberdieva D, Lugtenberg B (2014) Use of plant growth-promoting rhizobacteria to alleviate salinity stress in plants. Use Microb Allev Soil Stress 1:73–96

El-Daim IA, Bejai S, Meijer J (2014) Improved heat stress tolerance of wheat seedlings by bacterial seed treatment. Plant Soil 379:337–350

El-Meihy RM, Abou-Aly HE, Youssef AM, Tewfike TA, El-Alkshar EA (2019) Efficiency of heavy metals-tolerant plant growth promoting bacteria for alleviating heavy metals toxicity on sorghum. Environ Exp Bot 162:295–301. https://doi.org/10.1016/j.envexpbot.2019.03.005

Elhady A, Adss S, Hallmann J, Heuer H (2018) Rhizosphere microbiomes modulated by pre-crops assisted plants in defense against plant-parasitic nematodes. Front Microbiol 9:1133. https://doi.org/10.3389/fmicb.2018.01133

Epstein E (1994) The anomaly of silicon in plant biology. Proc Natl Acad Sci U S A 91:11–17

Epstein E, Bloom AJ (2005) Mineral nutrition of plants: principles and perspectives, 2nd edn. Sinauer Assoc Inc, Sunderland, UK

Fahad S, Hussain S, Matloob A, Khan FA, Khaliq A, Saud S, Hassan S, Shan D, Khan F, Ullah N (2015) Phytohormones and plant responses to salinity stress: a review. Plant Growth Regul 75:391–404. https://doi.org/10.1007/s10725-014-0013-y

Farooq M, Wahid A, Kobayashi N, Fujita D, Basra SMA (2009) Plant drought stress: effects, mechanisms and management, sustainable agriculture. Springer, pp 153–188

Farooq MA, Detterbeck A, Clemens S, Dietz KJ (2016) Silicon-induced reversibility of cadmium toxicity in rice. J Exp Bot 67:3573–3585

Fatima T, Mishra I, Verma R, Arora NK (2020) Mechanisms of halotolerant plant growth promoting *Alcaligenes* sp. involved in salt tolerance and enhancement of the growth of rice under salinity stress. 3 Biotech 10:361. https://doi.org/10.1007/s13205-020-02348-5

Fierer N (2017) Embracing the unknown: disentangling the complexities of the soil microbiome. Nat Rev Microbiol 15:579. https://doi.org/10.1038/nrmicro.2017.87

Flam-Shepherd R, Huynh WQ, Coskun D, Hamam AM, Britto DT, Kronzucker HJ (2018) Membrane fluxes, bypass flows, and sodium stress in rice: the influence of silicon. J Exp Bot 69(7):1679–1692. https://doi.org/10.1093/jxb/erx460

Frantz JM, Khandekar S, Leisner S (2011) Silicon differentially influences copper toxicity response in silicon-accumulator and non-accumulator species. J Am Soc Hortic Sci 136:329–338. https://doi.org/10.21273/JASHS.136.5.329

Frew A, Weston LA, Reynolds OL, Gurr GM (2018) The role of silicon in plant biology: a paradigm shift in research approach. Ann Bot 121:1265–1273. https://doi.org/10.1093/aob/mcy009

Gagne-Bourque F, Bertrand A, Claessens A, Aliferis KA, Jabaji S (2016) Alleviation of drought stress and metabolic changes in timothy (*Phleum pratense* L.) colonized with *Bacillus subtilis* B26. Front Plant Sci 7:584. https://doi.org/10.3389/fpls.2016.00584

George E, Horst WJ, Neumann E (2012) In: Marschner E (ed) Adaptation of plants to adverse chemical soil conditions. https://doi.org/10.1016/B978-0-12-384905-2.00017-0

Gil-Martínez M, López-García A, Domínguez MT, Navarro-Fernández CM, Kjøller R, Tibbett M, Marañnón T (2018) Ectomycorrhizal fungal communities and their functional traits mediate plant-soil interactions in trace element contaminated soils. Front Plant Sci 9:1682. https://doi.org/10.3389/fpls.2018.01682

Glick BR (2014) Bacteria with ACC deaminase can promote plant growth and help to feed the world. Microbiol Res 169:30–39. https://doi.org/10.1016/j.micres.2013.09.009

Gontia-Mishra I, Sapre S, Sharma A, Tiwari S (2016) Alleviation of mercury toxicity in wheat by the interaction of mercury-tolerant plant growth-promoting rhizobacteria. J Plant Growth Regul 35:1000–1012

Gouda S, Kerry RG, Das G, Paramithiotis S, Shin HS, Patra JK (2018) Revitalization of plant growth promoting rhizobacteria for sustainable development in agriculture. Microbiol Res 206:131–140. https://doi.org/10.1016/j.micres.2017.08.016

Guntzer F, Keller C, Meunier J-D (2012) Benefits of plant silicon for crops: a review. Agron Sustain Dev 32:201–213. https://doi.org/10.1007/s13593-011-0039-8

Gutiérrez-Luna FM, López-Bucio J, Altamirano-Hernández J, Valencia-Cantero E, Reyes de la Cruz H, Macías-Rodríguez (2010) Plant growth-promoting rhizobacteria modulate root-system architecture in *Arabidopsis thaliana* through volatile organic compound emission. Symbiosis 51:75–83. https://doi.org/10.1007/s13199-010-0066-2

Hacquard S, Spaepen S, Garrido-Oter R, Schulze-Lefert P (2017) Interplay between innate immunity and the plant microbiota. Annu Rev Phytopathol 55:565–589. https://doi.org/10.1146/annurev-phyto-080516-035623

Hallmann J, Quadt-Hallmann A, Mahaffee WF, Kloepper JW (1997) Bacterial endophytes in agricultural crops. Can J Microbiol 43:895–914. https://doi.org/10.1139/m97-131

Hamdia MA, Shaddad MAK, Doaa MM (2004) Mechanism of salt tolerance and interactive effect of *Azospirillum bransilense* inoculation on maize cultivars grown under salt stress conditions. Plant Growth Regul 44:165–174. https://doi.org/10.1023/B:GROW.0000049414.03099.9b

Hamel C, Gan Y, Sokolski S, Bainard LD (2018) High frequency cropping of pulses modifies soil nitrogen level and the rhizosphere bacterial microbiome in 4-year rotation systems of the semi-arid prairie. Appl Soil Ecol 126:47–56. https://doi.org/10.1016/j.apsoil.2018.01.003

Haney CH, Samuel BS, Bush J, Ausubel FM (2015) Associations with rhizosphere bacteria can confer an adaptive advantage to plants. Nat Plants 1:15051. https://doi.org/10.1038/nplants.2015.51

Hansda A, Kumar V, Anshumali (2017) Cu-resistant Kocuria sp CRB15: A potential PGPR isolated from the dry tailing of Rakha copper mine. 3 Biotech 7:132

Hardoim PR, Van Overbeek LS, Berg G, Pirttila AM, Compant S, Campisano A et al (2015) The hidden world within plants: ecological and evolutionary considerations for defining functioning of microbial endophytes. Microbiol Mol Biol Rev 79:293–320. https://doi.org/10.1128/MMBR.00050-14

Hattori T, Sonobe K, Araki H, Inanaga S, AnP MS (2008) Silicon application by sorghum through the alleviation of stress-induced increase in hydraulic resistance. J Plant Nutr 31:1482–1495

He H, Ye Z, Yang D, Yan J, Xiao L, Zhong T, Yuan M, Cai X, Fang Z, Jing Y (2013) Characterization of endophytic Rahnella sp. JN6 from Polygonum pubescens and its potential in promoting growth and cd, Pb, Zn uptake by Brassica napus. Chemosphere 90:1960–1965

Heidari M, Golpayegani A (2012) Effects of water stress and inoculation with plant growth promoting rhizobacteria (PGPR) on antioxidant status and photosynthetic pigments in basil (Ocimum basilicum L.). J Saudi Soc Agr Sci 11:57–61

Hernandez-Apaolaza L (2014) Can silicon partially alleviate micronutrient deficiency in plants? A review. Planta 240:447–458. https://doi.org/10.1007/s00425-014-2119-x

Hernández-Calderón E, Aviles-Garcia ME, Castulo-Rubio DY, Macías-Rodríguez L, Ramírez VM, Santoyo G et al (2018) Volatile compounds from beneficial or pathogenic bacteria differentially regulate root exudation, transcription of iron transporters, and defense signaling pathways in *Sorghum bicolor*. Plant Mol Biol 96:291–304. https://doi.org/10.1007/s11103-017-0694-5

Hiltner L (1904) Über neuere erfahrungen und probleme auf dem gebiete der bodenbakteriologie unter besonderer berücksichtigung der gründüngung und brache. Arb DLG 98:59–78

Hosni T, Moretti C, Devescovi G, Suarez-Moreno ZR, Fatmi MB, Guarnaccia C, Pongor S, Onofri A, Buonaurio R, Venturi V (2011) Sharing of quorum-sensing signals and role of interspecies communities in a bacterial plant disease. ISME J 5:1857–1870

Hrynkiewicz K, Szyma'nska S, Piernik A, Thiem D (2015) Ectomycorrhizal community structure of *Salix* and *Betula* spp. At a saline site in Central Poland in relation to the seasons and soil parameters. Water Air Soil Pollut 226:99. https://doi.org/10.1007/s11270-015-2308-7

Hu J, Wei Z, Friman VP, Gu SH, Wang XF, Eisenhauer N, Yang TJ, Ma J, Shen QR, Xu YC, Jousset A (2016) Probiotic diversity enhances rhizosphere microbiome function and plant disease suppression. MBio 7:e01790–e01716. https://doi.org/10.1128/mBio.01790-16

Hu Y, Schmidhalter U (2005) Drought and salinity: a comparison of their effects on mineral nutrition of plants. J Plant Nutr Soil Sci 168:541–549. https://doi.org/10.1002/jpln.200420516

Hur M, Kim Y, Song H-R, Kim JM, Choi YI, Yi H (2011) Effect of genetically modified poplars on soil microbial communities during the phytoremediation of waste mine tailings. Appl Environ Microbiol 77:7611–7619. https://doi.org/10.1128/AEM.06102-11

Imtiaz M, Rizwan MS, Mushtaq MA, Ashraf M, Shahzad SM, Yousaf B, Saeed DA, Rizwan M, Nawaz MA, Mehmood S, Tu S (2016) Silicon occurrence, uptake, transport and mechanisms of heavy metals, minerals and salinity enhanced tolerance in plants with future prospects: a review. J Environ Manag 183:521–529. https://doi.org/10.1016/j.jenvman.2016.09.009

Javid MG, Sorooshzadeh A, Moradi F, Sanavy SAMM, Allahdadi I (2011) The role of phytohormones in alleviating salt stress in crop plants. Aust J Crop Sci 5:726–734

Jiao S, Chen W, Wei G (2019) Resilience and assemblage of soil microbiome in response to chemical contamination combined with plant growth. Appl Environ Microbiol 85:e02523–e02518. https://doi.org/10.1128/AEM.02523-18

Kamran M, Malik Z, Parveen A, Huang L, Riaz M, Bashir S, Mustafa A, Abbasi GH, Xue B, Ali U (2020) Ameliorative effects of biochar on rapeseed (Brassica napus L.) growth and heavy metal immobilization in soil irrigated with untreated wastewater. J Plant Growth Regul 39:266–281

Kang SM, Radhakrishnan R, You YH, Khan AL, Park JM, Lee SM, Lee IJ (2015) Cucumber performance is improved by inoculation with plant growth-promoting microorganisms. Acta Agr Scand B-SP 65:36–44

Kang SM, Radhakrishnan R, Khan AL, Kim MJ, Park JM, Kim BR, Shin DH, Lee IJ (2014) Gibberellin secreting rhizobacterium, pseudomonas putida H-2-3 modulates the hormonal and stress physiology of soybean to improve the plant growth under saline and drought conditions. Plant Physiol Biochem 84:115–124

Kaushal M, Wani SP (2016) Rhizobacterial-plant interactions: strategies ensuring plant growth promotion under drought and salinity stress. Agric Ecosyst Environ 231:68–78. https://doi.org/10.1016/j.agee.2016.06.031

Kaya C, Tuna L, Higgs D (2006) Effect of silicon on plant growth and mineral nutrition of maize grown under water-stress conditions. J Plant Nutr 29:1469–1480. https://doi.org/10.1080/01904160600837238

Ke J, Wang B, Yoshikuni Y (2021) Microbiome engineering: synthetic biology of plant-associated microbiomes in sustainable agriculture. Trends Biotechnol 39:244–261. https://doi.org/10.1016/j.tibtech.2020.07.008

Khan AL, Hussain J, Al-Harrasi A, Al-Rawahi A, Lee IJ (2015) Endophytic fungi: resource for gibberellins and crop abiotic stress resistance. Crit Rev Biotechnol 35:62–74

Khan WU, Ahmad SR, Yasin NA, Ali A, Ahmad A, Akram W (2017) Application of bacillus megaterium MCR-8 improved phytoextraction and stress alleviation of nickel in Vinca rosea. Int J Phytoremediation 19:813–824

Kim Y-H, Khan AL, Waqas M, Lee I-J (2017) Silicon regulates antioxidant activities of crop plants under abiotic-induced oxidative stress: a review. Front Plant Sci 8:00510. https://doi.org/10.3389/fpls.2017.00510

Kim YH, Khan AL, Waqas M, Shim JK, Kim DH, Lee KY, Lee IJ (2014) Silicon application to rice root zone influenced the phytohormonal and antioxidant responses under salinity stress. J Plant Growth Regul 33:137–149. https://doi.org/10.1007/s00344-013-9356-2

Knight R, Vrbanac A, Taylor BC, Aksenov A, Callewaert C, Debelius J, Gonzalez A, Kosciolek T, McCall LI, McDonald D, Melnik AV, Morton JT, Navas J, Quinn RA, Sanders JG, Swafford AD, Thompson LR, Tripathi A, Xu ZZ, Zaneveld JR, Zhu Q, Caporaso JG, Dorrestein PC (2018) Best practices for analysing microbiomes. Nat Rev Microbiol 16:410–422. https://doi.org/10.1038/s41579-018-0029-9

Koskey G, Mburu SW, Awino R, Njeru EM, Maingi JM (2021) Potential use of beneficial microorganisms for soil amelioration, phytopathogen biocontrol, and sustainable crop production in smallholder agroecosystems. Front Sustain Food Syst 5:606308. https://doi.org/10.3389/fsufs.2021.606308

Kuczynski J, Costello EK, Nemergut DR, Zaneveld J, Lauber CL, Knights D, Koren O, Fierer N, Kelley ST, Ley RE, Gordon JI, Knight R (2010) Direct sequencing of the human microbiome readily reveals community differences. Genome Biol 11:210. https://doi.org/10.1186/gb-2010-11-5-210

Lang M, Christie P, Zhang J, Li X (2018) Long-term phosphorus application to a maize monoculture influences the soil microbial community and its feedback effects on maize seedling biomass. Appl Soil Ecol 128:12–22. https://doi.org/10.1016/j.apsoil.2018.01.005

Lebeis SL, Paredes SH, Lundberg DS, Breakfield N, Gehring J, McDonald M, Malfatti S, Glavina del Rio T, Jones CD, Tringe SG, Dangl JL (2015) Salicylic acid modulates colonization of the root microbiome by specific bacterial taxa. Science 349:860–864. https://doi.org/10.1126/science.aaa8764

Lesk C, Rowhani P, Ramankutty N (2016) Influence of extreme weather disasters on global crop production. Nature 529:84–87. https://doi.org/10.1038/nature16467

Li CH, Chu TD, Liu XB, Yang Q (1999) Silicon nutrition effects and its study and application development in China, pp 329–333

Li T, Hu YJ, Hao ZP, Li H, Chen BD (2013) Aquaporin genes GintAQPF1 and GintAQPF2 from glomus intraradices contribute to plant drought tolerance. Plant Signal Behav 8:e24030

Li H, Zhu Y, Hu Y, Han W, Gong H (2015) Beneficial effects of silicon in alleviating salinity stress of tomato seedlings grown under sand culture. Acta Physiol Plant 37:1–9. https://doi.org/10.1007/s11738-015-1818-7

Liang Y, Nikolic M, Bélanger R, Gong H, Song A (2015) Silicon-mediated tolerance to metal toxicity, silicon in agriculture. Springer, pp 83–122

Liang Y, Sun W, Zhu Y-G, Christie P (2007) Mechanisms of silicon-mediated alleviation of abiotic stresses in higher plants: a review. Environ Pollut 147:422–428. https://doi.org/10.1016/j.envpol.2006.06.008

Lim JH, Kim SD (2013) Induction of drought stress resistance by multi-functional PGPR bacillus licheniformis K11 in pepper. Plant Pathol J 29:201–208

Lindow SE, Brandl MT (2003) Microbiology of the phyllosphere. Appl Environ Microbiol 69:1875–1883

Liu CY, Zhang F, Zhang DJ, Srivastava AK, Wu QS, Zou YN (2018) Mycorrhiza stimulates root-hair growth and IAA synthesis and transport in trifoliate orange under drought stress. Sci Rep 8:1–9

Lodewyckx C, Vangronsveld J, Porteous F, Moore ERB, Taghavi S (2002) Endophytic bacteria and their potential applications. Crit Rev Plant Sci 21:583–606. https://doi.org/10.1080/0735-260291044377

Lundberg DS, Lebeis SL, Paredes SH, Yourstone S, Gehring J, Malfatti S, Tremblay J, Engelbrektson A, Kunin V, Del Rio TG, Edgar RC, Eickhorst T, Ley RE, Hugenholtz P, Tringe SG, Dangl JL (2012) Defining the core *Arabidopsis thaliana* root microbiome. Nature 488:86–90. https://doi.org/10.1038/nature11237

Ma Y, Oliveira RS, Wu L, Luo Y, Rajkumar M, Rocha I, Freitas H (2015) Inoculation with metal-mobilizing plant-growth-promoting rhizobacterium bacillus sp. SC2b and its role in rhizoremediation. J Toxicol Environ Health Part A 78:931–944

Maarastawi SA, Frindte K, Linnartz M, Knief C (2018) Crop rotation and straw application impact microbial communities in Italian and Philippine soils and the rhizosphere of *Zea mays*. Front Microbiol 9:1295. https://doi.org/10.3389/fmicb.2018.01295

Maghsoudi K, Emam Y, Ashraf M (2015) Influence of foliar application of silicon on chlorophyll fluorescence, photosynthetic pigments, and growth in water-stressed wheat cultivars differing in drought tolerance. Turk J Bot 39:625–634

Mahmood S, Daur I, Al-Solaimani SG, Ahmad S, Madkour MH, Yasir M, Hirt H, Ali S, Ali Z (2016) Plant growth promoting rhizobacteria and silicon synergistically enhance salinity tolerance of mung bean. Front Plant Sci 7:876. https://doi.org/10.3389/fpls.2016.00876

Maksimovic JD, Mojovic M, Maksimovic V, Romheld V, Nikolic M (2012) Silicon ameliorates manganese toxicity in cucumber by decreasing hydroxyl radical accumulation in the leaf apoplast. J Exp Bot 63:2411–2420

Mateos-Naranjo E, Galle A, Florez-Sarasa I, Perdomo JA, Galmes J, Ribas-Carbo M, Flexas J (2015) Assessment of the role of silicon in the cu-tolerance of the C4 grass Spartina densiflora. J Plant Physiol 178:74–83

Mayak S, Tirosh T, Glick BR (2004) Plant growth-promoting bacteria that confer resistance to water stress in tomatoes and peppers. Plant Sci 166:525–530

Mendes R, Garbeva P, Raaijmakers JM (2013) The rhizosphere microbiome: significance of plant beneficial, plant pathogenic, and human pathogenic microorganisms. FEMS Microbiol Rev 37:634–663. https://doi.org/10.1111/1574-6976.12028

Miao B-H, Han X-G, Zhang W-H (2010) The ameliorative effect of silicon on soybean seedlings grown in potassium-deficient medium. Ann Bot 105:967–973. https://doi.org/10.1093/aob/mcq063

Miransari M (2013) Soil microbes and the availability of soil nutrients. Acta Physiol Plant 35:3075–3084. https://doi.org/10.1007/s11738-013-1338-2
Mishra PK, Bisht SC, Bisht JK, Bhatt JC (2012) Cold-tolerant PGPRs as bioinoculants for stress management. In: Maheshwari DK (ed) Bacteria in agrobiology: stress management. Springer, Berlin/Heidelberg, Germany, pp 95–118
Mishra PK, Mishra S, Selvakumar G, Bisht SC, Bisht JK, Kundu S, Gupta HS (2008) Characterisation of a psychrotolerant plant growth promoting pseudomonas sp. strain PGERs17 (MTCC 9000) isolated from North Western Indian Himalayas. Ann Microbiol 58:561–568
Mishra J, Singh R, Arora NK (2017) Plant growth-promoting microbes: diverse roles in agriculture and environmental sustainability. In: Kumar V et al (eds) Probiotics and plant health, pp 71–111. https://doi.org/10.1007/978-981-10-3473-2_4
Mnasri M, Janouˇskov'a M, Rydlov'a J, Abdelly C, Ghnaya T (2017) Comparison of arbuscular mycorrhizal fungal effects on the heavy metal uptake of a host and a non-host plant species in contact with extraradical mycelial network. Chemosphere 171:476–484. https://doi.org/10.1016/j.chemosphere.2016.12.093
Moissl-Eichinger C, Pausan M, Taffner J, Berg G, Bang C, Schmitz RA (2018) Archaea are interactive components of complex microbiomes. Trend Microbiol 26:70–85. https://doi.org/10.1016/j.tim.2017.07.004
Muneer S, Jeong BR (2015) Proteomic analysis of salt-stress responsive proteins in roots of tomato (*Lycopersicon esculentum* L.) plants towards silicon efficiency. Plant Growth Regul 77:133–146. https://doi.org/10.1007/s10725-015-0045-y
Muneer S, Park YG, Manivannan A, Soundararajan P, Jeong BR (2014) Physiological and proteomic analysis in chloroplasts of *Solanum lycopersicum* L. under silicon efficiency and salinity stress. Int J Mol Sci 15:21803–21824. https://doi.org/10.3390/ijms151221803
Nadeem SM, Zahir ZA, Naveed M, Arshad M (2007) Preliminary investigations on inducing salt tolerance in maize through inoculation with rhizobacteria containing ACC deaminase activity. Can J Microbiol 53:1141–1149
Nallanchakravarthula S, Mahmood S, Alstrom S, Finlay RD (2014) Influence of soil type, cultivar and *Verticillium dahliae* on the structure of the root and rhizosphere soil fungal microbiome of strawberry. PLoS One 9:e111455. https://doi.org/10.1371/journal.pone.0111455
Naveed M, Mustafa A, Majeed S, Naseem Z, Saeed Q, Khan A, Nawaz A, Baig KS, Chen JT (2020) Enhancing cadmium tolerance and pea plant health through Enterobacter sp. MN17 inoculation together with biochar and gravel sand. Plan Theory 9:530
Naylor D, Coleman-Derr D (2018) Drought stress and root-associated bacterial communities. Front Plant Sci 8:2223. https://doi.org/10.3389/fpls.2017.02223
Naylor D, DeGraaf S, Purdom E, Coleman-Derr D (2017) Drought and host selection influence bacterial community dynamics in the grass root microbiome. ISME J 11:2691–2704. https://doi.org/10.1038/ismej.2017.118
Nazli F, Mustafa A, Ahmad M, Hussain A, Jamil M, Wang X, Shakeel Q, Imtiaz M, El-Esawi MA (2020) A review on practical application and potentials of phytohormone-producing plant growth-promoting rhizobacteria for inducing heavy metal tolerance in crops. Sustainability 12:9056. https://doi.org/10.3390/su12219056
Nwugo CC, Huerta AJ (2010) The effect of silicon on the leaf proteome of rice (*Oryza sativa* L.) plants under cadmium-stress. J Proteome Res 10:518–528. https://doi.org/10.1021/pr100716h
Olanrewaju OS, Glick BR, Babalola OO (2017) Mechanisms of action of plant growth promoting bacteria. World J Microbiol Biotechnol 33:197. https://doi.org/10.1007/s11274-017-2364-9
Osakabe Y, Osakabe K, Shinozaki K, Tran LSP (2014) Response of plants to water stress. Front Plant Sci 5:86. https://doi.org/10.3389/fpls.2014.00086
Park YG, Mun BG, Kang SM, Hussain A, Shahzad R, Seo CW, Kim AY, Lee SU, Oh KY, Lee DY et al (2017) Bacillus aryabhattai SRB02 tolerates oxidative and nitrosative stress and promotes the growth of soybean by modulating the production of phytohormones. PLoS One 12:e0173203

Passos da Silva D, Casta˜neda-Ojeda MP, Moretti C, Buonaurio R, Ramos C, Venturi V (2014) Bacterial multispecies studies and microbiome analysis of a plant disease. Microbiology 160:556–566. https://doi.org/10.1099/mic.0.074468-0

Patel JS, Yadav SK, Bajpai R, Teli B, Rashid MM (2020) PGPR secondary metabolites: an active syrup for improvement of plant health. Molecular aspects of plant beneficial microbes in agriculture, pp 195–208. https://doi.org/10.1016/B978-0-12-818469-1.00017-1

Pei ZF, Ming DF, Liu D, Wan GL, Geng XX, Gong HJ, Zhou WJ (2010) Silicon improves the tolerance to water-deficit stress induced by polyethylene glycol in wheat (*Triticum aestivum* L.) seedlings. J Plant Growth Regul 29:106–115. https://doi.org/10.1007/s00344-009-9120-9

Peters RD, Sturz AV, Carter MR, Sanderson JB (2003) Developing disease-suppressive soils through crop rotation and tillage management practices. Soil Till Res 72:181–192. https://doi.org/10.1016/S0167-1987(03)00087-4

Phetcharat P, Duangpaeng A (2012) Screening of endophytic bacteria from organic rice tissue for indole acetic acid production. Process Eng 32:177–183. https://doi.org/10.1016/j.proeng.2012.01.1254

Pieterse CM, Van der Does D, Zamioudis C, León-Reyes A, Van Wees SC (2012) Hormonal modulation of plant immunity. Annu Rev Cell Dev Biol 28:489–521. https://doi.org/10.1146/annurev-cellbio-092910-154055

Porcel R, Ruiz-Lozano JM (2004) Arbuscular mycorrhizal influence on leaf water potential, solute accumulation, and oxidative stress in soybean plants subjected to drought stress. J Exp Bot 55:1743–1750

Porcel R, Aroca R, Azcon R, Ruiz-Lozano JM (2006) PIP aquaporin gene expression in arbuscular mycorrhizal Glycine max and Lactuca sativa plants in relation to drought stress tolerance. Plant Mol Biol 60:389–404

Pramanik K, Mitra S, Sarkar A, Soren T, Maiti TK (2017) Characterization of cadmium-resistant Klebsiella pneumoniae MCC 3091 promoted rice seedling growth by alleviating phytotoxicity of cadmium. Environ Sci Pollut Res 24:24419–24437

Qi D, Wieneke X, Tao J, Zhou X, Desilva U (2018) Soil pH is the primary factor correlating with soil microbiome in karst rocky desertification regions in the Wushan County, Chongqing, China. Front Microbiol 9:1027. https://doi.org/10.3389/fmicb.2018.01027

Rajkumar M, Sandhya S, Prasad MNV, Freitas H (2012) Perspectives of plant-associated microbes in heavy metal phytoremediation. Biotechnol Adv 30:1562–1574. https://doi.org/10.1016/j.biotechadv.2012.04.011

Ramos-Solano B, García JAL, Garcia-Villaraco A, Algar E, Garcia-Cristobal J, Mañero FJG (2010) Siderophore and chitinase producing isolates from the rhizosphere of Nicotiana glauca Graham enhance growth and induce systemic resistance in *Solanum lycopersicum* L. Plant Soil 334:189–197. https://doi.org/10.1007/s11104-010-0371-9

Rashid S, Charles TC, Glick BR (2012) Isolation and characterization of new plant growth-promoting bacterial endophytes. Appl Soil Ecol 61:217–224. https://doi.org/10.1016/j.apsoil.2011.09.011

Reis VM, Alves BJR, Hartmann A, James EK, Zilli JE (2020) Beneficial microorganisms in agriculture: the future of plant growth-promoting rhizobacteria. Plant Soil 451:1–3. https://doi.org/10.1007/s11104-020-04482-8

Rizwan M, Ali S, Ibrahim M, Farid M, Adrees M, Bharwana SA, Zia-ur-Rehman M, Qayyum MF, Abbas F (2015) Mechanisms of silicon-mediated alleviation of drought and salt stress in plants: a review. Environ Sci Pollut Res 22:15416–15431. https://doi.org/10.1007/s11356-015-5305-x

Rogalla H, Romheld V (2002) Role of leaf apoplast in silicon-mediated manganese tolerance of Cucumis sativus L. Plant Cell Environ 25:549–555

Rolli E, Marasco R, Vigani G, Ettoumi B, Mapelli F, Deangelis ML, Gandolfi C, Casati E, Previtali F, Gerbino R, Pierotti Cei F, Borin S, Sorlini C, Zocchi G, Daffonchio D (2015) Improved plant resistance to drought is promoted by the root-associated microbiome as a water stress-dependent trait. Environ Microbiol 17:316–331. https://doi.org/10.1111/1462-2920.12439

Rousk J, Bååth E, Brookes PC, Lauber CL, Lozupone C, Caporaso JG, Knight R, Fierer N (2010) Soil bacterial and fungal communities across a pH gradient in an arable soil. ISME J 4:1340–1351. https://doi.org/10.1038/ismej.2010.58

Ryu CM, Farag MA, Hu CH, Reddy MS, Wei HX, Pare PW et al (2003) Bacterial volatiles promote growth in Arabidopsis. Proc Natl Acad Sci U S A 100:4927–4932

Ryu H, Cho Y-G (2015) Plant hormones in salt stress tolerance. J Plant Biol 58:147–155. https://doi.org/10.1007/s12374-015-0103-z

Safoora D, Cyrus G, Bahram B, Mahdi G, Siamak S (2018) Effect of silicon on growth and development of strawberry under water deficit conditions. Hort Plant J 4(6):226–232

Sandhya VZAS, Grover M, Reddy G, Venkateswarlu B (2009) Alleviation of drought stress effects in sunflower seedlings by the exopolysaccharides producing pseudomonas putida strain GAP-P45. Biol Fert Soils 46:17–26

Sangiorgio D, Cellini A, Donati I, Pastore C, Onofrietti C, Spinelli F (2020) Facing climate change: application of microbial biostimulants to mitigate stress in horticultural crops. Agronomy 10:794. https://doi.org/10.3390/agronomy10060794

Sanglard LMVP, Martins SCV, Detmann KC, Silva PEM, Lavinsky AO, Silva MM, Detmann E, Araújo WL, DaMatta FM (2014) Silicon nutrition alleviates the negative impacts of arsenic on the photosynthetic apparatus of rice leaves: an analysis of the key limitations of photosynthesis. Physiol Plantarum 152:355–366. https://doi.org/10.1111/ppl.12178

Santoyo G, Moreno-Hagelsieb G, Orozco-Mosqueda Mdel C, Glick BR (2016) Plant growth-promoting bacterial endophytes. Microbiol Res 183:92–99. https://doi.org/10.1016/j.micres.2015.11.008

Saravanakumar D, Samiyappan R (2007) ACC deaminase from Pseudomonas fluorescens mediated saline resistance in groundnut (Arachis hypogea) plants. J Appl Microbiol 102(5):1283–1292

Sarma BK, Yadav SK, Singh DP, Singh HB (2012) Rhizobacteria mediated induced systemic tolerance in plants: prospects for abiotic stress management, bacteria in agrobiology: stress management. Springer, pp 225–238

Sasse J, Martinoia E, Northen T (2018) Feed your friends: do plant exudates shape the root microbiome? Trends Plant Sci 23:25–41. https://doi.org/10.1016/j.tplants.2017.09.003

Selvakumar G, Kundu S, Joshi P, Nazim S, Gupta AD, Mishra PK, Gupta HS (2008) Characterization of a cold-tolerant plant growth-promoting bacterium Pantoea dispersa 1A isolated from a sub-alpine soil in the North Western Indian Himalayas. World J Microb Biot 24:955–960

Sessitsch A, Kuffner M, Kidd P, Vangronsveld J, Wenzel WW, Fallmann K, Puschenreiter M (2013) The role of plant-associated bacteria in the mobilization and phytoextraction of trace elements in contaminated soils. Soil Biol Biochem 60:182–194. https://doi.org/10.1016/j.soilbio.2013.01.012

Shi Y, Yang H, Zhang T, Sun J, Lou K (2014) Illumina-based analysis of endophytic bacterial diversity and space-time dynamics in sugar beet on the north slope of Tianshan Mountain. Appl Microbiol Biotechnol 98:6375–6385. https://doi.org/10.1007/s00253-014-5720-9

Shi Y, Zhang Y, Han W, Feng R, Hu Y, Guo J, Gong H (2016) Silicon enhances water stress tolerance by improving root hydraulic conductance in Solanum lycopersicum L. Front Plant Sci 7:196

Shintu PV, Jayaram KM (2015) Phosphate solubilising bacteria (bacillus polymyxa)-an effective approach to mitigate drought in tomato (Lycopersicon esculentum mill.). Trop Plant Res 2:17–22

Song XP, Verma KK, Tian DD, Zhang XQ, Liang YJ, Huang X, Li CN, Li YR (2021) Exploration of silicon functions to integrate with biotic stress tolerance and crop improvement. Biol Res 54:19. https://doi.org/10.1186/s40659-021-00344-4

Soundararajan P, Manivannan A, Jeong BR (2016) Regulatory mechanisms by silicon to overcome the salinity-induced imbalance of essential nutrient elements, silicon in plants: advances and future prospects. CRC Press, pp 47–66

Stevenson FC, Kessel CV (1996) The nitrogen and non-nitrogen rotation benefits of pea to succeeding crops. Can J Plant Sci 76:735–745. https://doi.org/10.4141/cjps96-126

Sun L, Wang X, Li Y (2016) Increased plant growth and copper uptake of host and nonhost plants by metal-resistant and plant growth-promoting endophytic bacteria. Int J Phytoremediation 18:494–501. https://doi.org/10.1080/15226514.2015.1115962

Sun Y, Cheng Z, Glick BR (2009) The presence of a1-aminocyclopropane-1-carboxylate (ACC) deaminase deletion mutation alters the physiology of the endophytic plant growth-promoting bacterium Burkholderia phytofirmans PsJN. FEMS Microbiol Lett 296:131–136

Tewari S, Arora NK (2014) Multifunctional exopolysaccharides from Pseudomonas aeruginosa PF23 involved in plant growth stimulation, biocontrol and stress amelioration in sunflower under saline conditions. Curr Microbiol 69:484–494

Tozser D, Magura T, Simon E (2017) Heavy metal uptake by plant parts of willow species: a meta-analysis. J Hazard Mater 336:101–109. https://doi.org/10.1016/j.jhazmat.2017.03.068

Theocharis A, Bordiec S, Fernandez O, Paquis S, Dhondt-Cordelier S, Baillieul F, Barka EA (2012) Burkholderia phytofirmans PsJN primes Vitis vinifera L. and confers a better tolerance to low nonfreezing temperatures. MPMI 25:241–249

Thiem D, Gołębiewski M, Hulisz P, Piernik A, Hrynkiewicz K (2018) How does salinity shape bacterial and fungal microbiomes of *Alnus glutinosa* roots? Front Microbiol 9:651. https://doi.org/10.3389/fmicb.2018.00651

Tsukanova KA, Chebotar V, Meyer JJM, Bibikova TN (2017) Effect of plant growth-promoting Rhizobacteria on plant hormone homeostasis. S Afr J Bot 113:91–102

Ulrich K, Ulrich A, Ewald D (2008) Diversity of endophytic bacterial communities in poplar grown under field conditions. FEMS Microbiol Ecol 63:169–180. https://doi.org/10.1111/j.1574-6941.2007.00419.x

Vacheron J, Desbrosses G, Bouffaud ML, Touraine B, Moenne-Loccoz Y, Muller D et al (2013) Plant growth-promoting rhizobacteria and root system functioning. Front Plant Sci 4:356. https://doi.org/10.3389/fpls.2013.00356

Van Loon LC, Bakker P (2005) Induced systemic resistance as a mechanism of disease suppression by rhizobacteria, PGPR: biocontrol and biofertilization. Springer, pp 39–66

Van Loon LC, Bakker P, Pieterse CMJ (1998) Systemic resistance induced by rhizosphere bacteria. Annu Rev Phytopathol 36:453–483. https://doi.org/10.1146/annurev.phyto.36.1.453

Venter ZS, Jacobs K, Hawkins HJ (2016) The impact of crop rotation on soil microbial diversity: a meta-analysis. Pedobiologia 59:215–223. https://doi.org/10.1016/j.pedobi.2016.04.001

Verma KK, Li DM, Singh M, Rajput VD, Malviya MK, Minkina T, Singh RK, Singh P, Song XP, Li YR (2020c) Interactive role of silicon and plant-rhizobacteria mitigating abiotic stresses: a new approach for sustainable agriculture and climate change. Plan Theory 9:1055. https://doi.org/10.3390/plants9091055

Verma KK, Liu XH, Wu KC, Singh RK, Song QQ, Malviya MK, Song XP, Singh P, Verma CL, Li YR (2020b) The impact of silicon on photosynthetic and biochemical responses of sugarcane under different soil moisture levels. SILICON 12:1355–1367. https://doi.org/10.1007/s12633-019-00228-z

Verma KK, Singh P, Song XP, Malviya MK, Singh RK, Chen GL, Solomon S, Li YR (2020a) Mitigating climate change for sugarcane improvement: role of silicon in alleviating abiotic stresses. Sugar Tech 22:741–749. https://doi.org/10.1007/s12355-020-00831-0

Verma KK, Singh RK, Song QQ, Singh P, Zhang BQ, Song XP, Chen GL, Li YR (2019a) Silicon alleviates drought stress of sugarcane plants by improving antioxidant responses. Biomed J Sci Tech Res 17:002957. https://doi.org/10.26717/BJSTR.2019.17.002957

Verma KK, Song XP, Lin B, Guo DJ, Singh M, Rajput VD, Singh RK, Singh P, Sharma A, Malviya MK, Chen GL, Li YR (2021a) Silicon induced drought tolerance in crop plants: physiological adaptation strategies. SILICON 14:2473–2487. https://doi.org/10.1007/s12633-021-01071-x

Verma KK, Song XP, Zeng Y, Guo DJ, Singh M, Rajput VD, Malviya MK, Wei KJ, Sharma A, Li DP, Chen GL, Li YR (2021b) Foliar application of silicon boosts growth, photosynthetic leaf gas exchange, antioxidative response and resistance to limited water irrigation in sugarcanc (*Saccharum officinarum* L.). Plant Physiol Biochem 166:582–592. https://doi.org/10.1016/j.plaphy.2021.06.032

Verma KK, Wu K-C, Singh P, Malviya MK, Singh RK, Song X-P, Li YR (2019b) The protective role of silicon in sugarcane under water stress: photosynthesis and antioxidant enzymes. Biomed J Sci Tech Res 15:002685. https://doi.org/10.26717/BJSTR.2019.15.002685

Vurukonda SSKP, Vardharajula S, Shrivastava M, SkZ A (2016) Enhancement of drought stress tolerance in crops by plant growth promoting rhizobacteria. Microbiol Res 184:13–24

Wagner MR, Lundberg DS, Del Rio TG, Tringe SG, Dangl JL, Mitchell-Olds T (2016) Host genotype and age shape the leaf and root microbiomes of a wild perennial plant. Nat Commun 7:12151. https://doi.org/10.1038/ncomms12151

Wang YX, Stass A, Horst WJ (2004) Apoplastic binding of aluminum is involved in silicon-induced amelioration of aluminum toxicity in maize. Plant Physiol 136:3762–3770

Wang M, Gao L, Dong S, Sun Y, Shen Q, Guo S (2017) Role of silicon on plant pathogen interactions. Front Plant Sci 8:701. https://doi.org/10.3389/fpls.2017.00701

wor Weyens N, van der Lelie D, Taghavi S, Vangronsveld J (2009) Phytoremediation: plant-endophyte partnerships take the challenge. Curr Opin Biotech 20:248–254. https://doi.org/10.1016/j.copbio.2009.02.012

Wilson RH, Alonso H, Whitney SM (2016) Evolving *Methanococcoides burtonii* archaeal Rubisco for improved photosynthesis and plant growth. Sci Rep 6:22284. https://doi.org/10.1038/srep22284

Worlanyo AS, Jiangfeng L (2021) Evaluating the environmental and economic impact of mining for post-mined land restoration and land-use. J Environ Manag 279:111623. https://doi.org/10.1016/j.jenvman.2020.111623

Wu QS, Xia RX, Zou YN (2008) Improved soil structure and citrus growth after inoculation with three arbuscular mycorrhizal fungi under drought stress. Eur J Soil Biol 44:122–128

Yaish MW, Al-Lawati A, Jana GA, Vishwas Patankar H, Glick BR (2016) Impact of soil salinity on the structure of the bacterial endophytic community identified from the roots of caliph medic (*Medicago truncatula*). PLoS One 11:e0159007. https://doi.org/10.1371/journal.pone.0159007

Yaish MW, Kumar PP (2015) Salt tolerance research in date palm tree (*Phoenix dactylifera* L.), past, present, and future perspectives. Front Plant Sci 6:348. https://doi.org/10.3389/fpls.2015.00348

Yan N, Marschner P, Cao W, Zuo C, Qin W (2015) Influence of salinity and water content on soil microorganisms. Int Soil Water Cons Res 3:316–323. https://doi.org/10.1016/j.iswcr.2015.11.003

Yao L, Wu Z, Zheng Y, Kaleem I, Li C (2010) Growth promotion and protection against salt stress by pseudomonas putida Rs–198 on cotton. Eur J Soil Biol 46:49–54

Yeoh YK, Paungfoo-Lonhienne C, Dennis PG, Robinson N, Ragan MA, Schmidt S, Hugenholtz P (2016) The core root microbiome of sugarcanes cultivated under varying nitrogen fertilizer application. Environ Microbiol 18:1338–1351. https://doi.org/10.1111/1462-2920.12925

Young E, Carey M, Meharg AA, Meharg C (2018) Microbiome and ecotypic adaption of *Holcus lanatus* (L.) to extremes of its soil pH range, investigated through transcriptome sequencing. Microbiome 6:48. https://doi.org/10.1186/s40168-018-0434-3

Yun Y, Wang H, Man B, Xiang X, Zhou J, Qiu X, Duan Y, Engel AS (2016) The relationship between pH and bacterial communities in a single karst ecosystem and its implication for soil acidification. Front Microbiol 7:1955. https://doi.org/10.3389/fmicb.2016.01955

Zahir AZ, Ghani U, Naveed M, Nadeem SM, Asghar HN (2007) Comparative effectiveness of pseudomonas and Serratia sp. containing ACC deaminase for improving growth and yield of wheat (Triticum aestivum L.) under salt-stressed conditions. Arch Microbiol 191(5):415–424

Zahir ZA, Munir A, Asghar HN, Shaharoona B, Arshad M (2008) Effectiveness of rhizobacteria containing ACC deaminase for growth promotion of peas (Pisum sativum) under drought conditions. J Microbiol Biotechnol 18:958–963

Zhang D, Xu H, Gao J, Portieles R, Du L, Gao X, Borroto Nordelo C, Borrás-Hidalgo O (2021a) Endophytic *Bacillus altitudinis* strain uses different novelty molecular pathways to enhance plant growth. Front Microbiol 12:692313. https://doi.org/10.3389/fmicb.2021.692313

Zhang J, Cook J, Nearing JT, Zhang J, Raudonis R, Glick BR, Langille MGI, Cheng Z (2021b) Harnessing the plant microbiome to promote the growth of agricultural crops. Microbiol Res 245:126690. https://doi.org/10.1016/j.micres.2020.126690

Zhang H, Sun Y, Xie X, Kim MS, Dowd SE, Pare PW (2009) A soil bacterium regulates plant acquisition of iron via deficiency inducible mechanisms. Plant J 58:568–577. https://doi.org/10.1111/j.1365-313x.2009.03803.x

Zhou C, Guo J, Zhu L, Xiao X, Xie Y, Zhu J et al (2016) *Paenibacillus polymyxa* BFKC01 enhances plant iron absorption via improved root systems and activated iron acquisition mechanisms. Plant Physiol Biochem 105:162–173. https://doi.org/10.1016/j.plaphy.2016.04.025

Zhu Y, Gong H (2014) Beneficial effects of silicon on salt and drought tolerance in plants. Agron Sustain Dev 34:455–472. https://doi.org/10.1007/s13593-013-0194-1

Zhu Z, Wei G, Li J, Qian Q, Yu J (2004) Silicon alleviates salt stress and increases antioxidant enzymes activity in leaves of salt-stressed cucumber (*Cucumis sativus* L.). Plant Sci 167:527–533. https://doi.org/10.1016/j.plantsci.2004.04.020

Zhu Y-X, Xu X-B, Hu Y-H, Han W-H, Yin J-L, Li H-L, Gong H-J (2015) Silicon improves salt tolerance by increasing root water uptake in *Cucumis sativus* L. Plant Cell Rep 34:1629–1646. https://doi.org/10.1007/s00299-015-1814-9

Zhuang X, Chen J, Shim H, Bai Z (2007) New advances in plant growth-promoting rhizobacteria for bioremediation. Environ Int 33:406–413. https://doi.org/10.1016/j.envint.2006.12.005

Zinniel DK, Lambrecht P, Harris NB, Feng Z, Kuczmarski D, Higley P et al (2002) Isolation and characterization of endophytic colonizing bacteria from agronomic crops and prairie plants. Appl Environ Microbiol 68:2198–2208. https://doi.org/10.1128/aem.68.5.2198-2208.2002

Chapter 18
Heat Stress Mitigation by Silicon Nutrition in Plants: A Comprehensive Overview

Jayabalan Shilpha, Abinaya Manivannan, Prabhakaran Soundararajan, and Byoung Ryong Jeong

18.1 Introduction

Heat stress (HS) is one of the main abiotic factors limiting agricultural output and productivity worldwide. With urbanization, degradation of the natural environment, and climate change, heat stress has become a major issue impacting crop development. An increase in the mean world temperature will significantly impact agricultural productivity (Han et al. 2009). Due to global warming, the world air temperature is expected to increase by 0.2 °C every decade, and by 2100, temperatures will be 1.8–4.0 °C higher than current levels (IPCC 2007). Thus, this change in the global temperature may alter the geographical distribution and growing season of agricultural crops (Porter 2005). More crucially, estimates based on research of global climate models indicate that the world's tropical and subtropical

J. Shilpha
Department of Biotechnology, School of Life Sciences, Pondicherry University, Puducherry, India

Institute of Agriculture and Life Science, Gyeongsang National University, Jinju, Korea

A. Manivannan
National Institute of Plant Genome Research, Aruna Asaf Ali Marg, New Delhi, India

P. Soundararajan
Evolutionary Developmental Biology Laboratory, National Institute of Plant Genome Research, Aruna Asaf Ali Marg, New Delhi, India

B. R. Jeong (✉)
Horticulture Major, Division of Applied Life Science (BK21 Four), Graduate School, Gyeongsang National University, Jinju, Korea

Institute of Agriculture and Life Science, Gyeongsang National University, Jinju, Korea

Research Institute of Life Science, Gyeongsang National University, Jinju, Korea
e-mail: brjeong@gnu.ac.kr

R. de Mello Prado (ed.), *Benefits of Silicon in the Nutrition of Plants*,
https://doi.org/10.1007/978-3-031-26673-7_18

regions will be the most affected by the impending heat stress catastrophe (Battisti and Naylor 2009). The rise in the global mean temperature would result in increasingly intense and frequent heat waves (Perkins-Kirkpatrick and Lewis 2020). Therefore, many tropical plants might undergo prolonged, severe heat stress. Regardless of the variations in the heat stress vulnerability of certain plant life stages, heat stress impacts almost all stages. Heat stress impacts all plant process dimensions, including germination, growth, maturity, fertilization, and productivity (Hasanuzzaman et al. 2013; Firmansyah and Argosubekti 2019). The reproductive stage is highly susceptible to HS injuries, which result in severe yield losses (Telfer et al. 2013). A brief HS episode can significantly reduce crop production due to the reduced resource use efficiency caused by the compromised physiological, biochemical, and molecular processes (Wang et al. 2018). As a result, the rise in HS in different crop species is becoming a serious problem, especially regarding the security of the world's food supply.

Silicon (Si) is a subtle element with promising roles in biological organisms, mostly plants. Silicon is abundantly available on the earth's surface, which encompasses 27.7% of the lithosphere. Silicon occurs in different forms in the environment, such as sand, quartz, and sandstone. However, in biological organisms, Si occurs as soluble silicic acid and amorphous silica (Das and Chattopadhyay 2000). Silicon is the second-most common element in the earth's crust, and it has been recognized for almost a century that Si has favorable effects on plant growth (Zargar et al. 2019; Atta et al. 2019). Although Si is ubiquitous, it is exclusively available to plants as monosilicic acid [$Si(OH)_4$], which has been identified as the limiting determinant of Si-bioavailability (Deshmukh et al. 2020). Silicon, formally designated as the non-essential element among inorganic compounds, has lately been found to protect crop plants from the negative impacts of abiotic stresses (Luyckx et al. 2017). Silicon has been linked to many positive impacts, including higher photosynthetic activity, improved nutrient imbalance, greater abiotic stress tolerance, increased pest and disease resistance, and decreased mineral toxicity. Silicon is frequently added to many different crops, but mostly to rice and sugarcane to boost yield (Savant et al. 1999). Silicon advantages are typically more noticeable in stressed plants than in unstressed plants (Etesami and Jeong 2018). When boosting the lignin content, Si gives the stem mechanical strength to endure harsh weather conditions (Hussain et al. 2021). Since drought is sometimes coupled with high temperatures, Si treatment may be an alternative to minimizing the damages caused by drought and heat stresses (Halford 2011; Etesami and Jeong 2018). Silicon leaf concentration increases heat stress endurance by regulating transpiration and the intake of other micro and macronutrients, promoting cell homeostasis (Shalaby et al. 2021). According to Saha et al. (2021), Si concentration in leaf tissue directly correlates with heat stress tolerance. Furthermore, they claimed that Si reduces heat stress by boosting the expression of genes sensitive to heat stress, improving photosynthesis, membrane integrity, and water balance. Although the potential applications of Si in alleviating several abiotic stresses have been extensively reported, only a few researchers have examined the molecular mechanisms of heat tolerance brought on by Si applications. In light of recent reports, this chapter primarily aims to present an overview of the underlying molecular dynamics of Si-mediated heat tolerance in plants.

18.2 Heat Stress Impact on Plants

Heat stress results in burning upper plant tissues, primarily leaves, burned twigs, senescence, and leaf discoloration (Fahad et al. 2017). A temperature rise may cause weak seed germination, leaf senescence and abscission, slow growth of roots and shoots, and flower loss (Hussain et al. 2019). High-temperature stress harms many crucial physiological functions of plants, including photosynthesis, respiration, and transpiration, by preventing carbon assimilation (Jha et al. 2014). The loss of thylakoid membrane stability and the inactivity of the photosystem II (PSII) water-splitting complex result in a reduced photosynthetic electron transport capability, which is a warning indication of severe heat stress (Teskey et al. 2015; Niinemets 2018). Along with affecting the ability of RuBP to regenerate, heat stress also damages PSII (Wise et al. 2004). In addition, high temperatures alter thylakoids' structural makeup, reducing photosynthesis (Wahid et al. 2007; Peverelli and Rogers 2013).

Generally, high temperatures significantly impact anatomical structures, not only on the cells and tissues levels but also on a sub-cellular level. The direct or indirect consequences of heat stress on plants, including denaturation of proteins and lipids, mitochondrial death, and membrane breakdown, result in plant mortality. These immediate and delayed consequences eventually result in cell depletion, decreased ion flux, hazardous chemical formation, and more (Howarth 2005). Moreover, proteins under high-temperature stress may suffer significant damage. Their production may be disrupted, key enzymes may become inactive, and membranes may be damaged. Furthermore, heat stress impacts microtubule organization. It breaks and/or lengthens spindles, promotes the development of microtubule asters in mitotic cells, and lengthens phragmoplast microtubules (Smertenko et al. 1997). Additionally, excessive heat impairs the production of osmoprotectants such as soluble sugars, proline, and heat-shock proteins (HSPs), which are known to protect cellular organelles from heat-induced dehydration (Sharma et al. 2020).

Another clear result of heat stress is the increased generation of reactive oxygen species (ROS). Reactive oxygen species (ROS) are produced quickly and build up under high-temperature stress (Mittler 2002; Waraich et al. 2012). Heat stress inactivates major antioxidant enzymes such as superoxide dismutase (SOD), catalase (CAT), ascorbate peroxidase (APX), glutathione S-transferase (GST), and glutathione peroxidase (GPX). It also reduces the production of non-enzymatic antioxidants, such as phenolic compounds, glutathione (GSH), and ascorbic acid (AsA) (Hasanuzzaman et al. 2020; Saha et al. 2021). In addition, high ROS concentrations damage all cellular molecules and negatively impact cellular metabolic activities (Van Breusegem et al. 2001). In the end, heat stress causes molecular, transcriptional, and morpho-physiological changes necessary for plant growth and survival. Ineffective plant growth and production could result from the combined effects of all these changes (Wahid et al. 2007).

18.3 Versatile Functions of Silicon in Mitigating Stress

Silicon plays many roles in numerous organisms. For example, it is essential for diatoms and algae (Martin-Jézéquel et al. 2000; Currie and Perry 2007). Similarly, it influences the formation of connective tissues, bones, cartilages, enzyme activity, and lymphocyte proliferation (Carlisle 1997). Plants absorb Si as silicic acid (Knight and Kinrade 2001). Moreover, Ingri (1978) illustrated that monosilicic acid could interact with organic molecules with hydroxyl functional groups. Thus, Si in bioavailable form can bind with macromolecules, such as carbohydrates and proteins, especially in its hypervalent form (Kinrade et al. 2001). In plants, Si tends to couple with cellular macromolecules, such as proteins, lipids, and polyphenols, especially for lignification (Kolesnikov and Gins 2001). Numerous reports evidenced the beneficial roles of Si in plants, especially in stressed environments. Increased plant silicon improves physiological functions, growth, yield, nutrient management, and stress tolerance (Manivannan and Ahn 2017). Although several studies reported the beneficial effects of Si, the molecular mechanism behind Si-mediated growth improvement and stress tolerance is still under study. For example, reports suggested that Si-induced resistance led to the accumulation and polymerization of silicon in the cell walls and increased rigidity. Thus, it acted as a mechanical barrier for fungal penetration in host tissues (Yoshida et al. 1962). Similarly, Si also can elicit chemical resistance by inducing antifungal compounds (Fawe et al. 1998). Furthermore, Si nutrition can significantly increase the activities of antioxidant enzymes, such as superoxide dismutase (SOD), catalase (CAT), ascorbate peroxidase (APX), and glutathione peroxidase (GPX), in stressed conditions (Manivannan et al. 2016; Soundararajan et al. 2017b). Moreover, Si-mediated accumulation of phenolic compounds was identified in the microscopic studies on the cells affected by *P. ultimum* (Chérif et al. 1992).

Silicon application under abiotic stress alleviated stress and improved plant growth (Liu et al. 2019). In carnation, Si supplementation decreased the negative effects of salinity and improved the in vitro growth conditions (Soundararajan et al. 2015). Upon salinity stress, Si can proceed with the following mechanisms for stress alleviation: (1) optimal water balance maintenance, (2) improved photosynthesis and reduced transpiration, (3) controlling the oxidative stress by the antioxidant system, and (4) regulation of plant hormones and nutrient management (Zhu et al. 2019). Similarly, Al-Aghabary et al. (2005) reported the stimulation of antioxidant enzymes and the increased photochemical efficiency of PSII under salinity stress. Moreover, Si supplementation mitigated metal toxicity such as the toxicity of aluminum (Al), boron (B), cadmium (Cd), chromium (Cr), copper (Cu), and zinc (Zn) (Neumann and zur Nieden 2001; Wang et al. 2004; Liang et al. 2005; Gunes et al. 2007; Li et al. 2008; Tripathi et al. 2012). In addition, using fertilizers enriched with Si decreased the electrolyte leakage and enhanced photosynthesis in rice under drought stress conditions (Chen et al. 2011). Therefore, adding Si to plants is a vital solution for abiotic and biotic stresses.

Tailoring abiotic stress tolerance in agricultural and horticultural crops can progress growth and yield in the long term. Silicon is evidenced to provide abiotic stress resistance by regulating diverse physiological, biochemical, and molecular mechanisms in plants. According to Verma et al. (2019), Si augmentation significantly increased the photosynthesis and mechanical rigidity of leaves in sugarcane. Concordantly, Si upregulated the expression of photosynthesis-related genes, aquaporins, genes involved in secondary metabolism, defense mechanism, transcription regulation, phytohormones, and polyamine biosynthesis upon abiotic and biotic stress conditions (Manivannan and Ahn 2017). Furthermore, Si tends to alleviate Zn toxicity stress by upregulating PsbY (Os08g02630), a vital manganese interacting protein in PSII. Thus, Si supplementation can trigger the Mn-binding ability and water oxidation, which invariably increases the PSII activity and the electron transfer chain rate (Song et al. 2014). Moreover, Si also increased the abundance of PsaH involved in PSI, denoting that Si can regulate both PSI and PSII upon Zn stress (Pfannschmidt and Yang 2012).

18.4 Silicon in ROS Homeostasis

Under normal conditions, plants release reactive oxygen species through photosynthesis, respiration, and developmental processes. The mitochondria, chloroplast, and peroxisomes are major organelles involved in the ROS as a by-product of cellular functions. Usually, ROS act as secondary messengers in the signaling pathway. However, excessive ROS generated could lead to lethal effects under stress conditions. Under uncontrolled conditions, ROS can damage macromolecules, such as nucleic acid, proteins, lipids, and others. For example, the effect of Si under various stress conditions, such as salt, temperature, and hyperhydricity, has been reported (Soundararajan et al. 2017a, b; Liu et al. 2019; Mustafa et al. 2021). Furthermore, Si supplementation induces antioxidant enzymes, such as superoxide dismutase (SOD), which is involved in the reduction of superoxide (O_2^-) into hydrogen peroxide (H_2O_2). Moreover, Si induces other antioxidant enzymes, such as catalase (CAT), ascorbate peroxidase (APX), guaiacol peroxidase (GPX), glutathione reductase (GR), efficiently reducing H_2O_2 into water (Soundararajan et al. 2015; Manivannan et al. 2016; Dhakate et al. 2022). In addition to the enzymatic activity, Si also enhanced the content of non-enzymatic antioxidants, such as glutathione, ascorbate, and proline (Altaf et al. 2022). In some of the horticulturally important plants, researchers have studied the effect of Si on increasing tolerance by inducing the antioxidant enzymes to scavenge excessive ROS accumulation. Examples include salvia, petunia, carnation, rose, chrysanthemum, tomato, poinsettia, and capsicum (Sivanesan et al. 2013; Soundararajan et al. 2014; Manivannan et al. 2018; Al Murad et al. 2020; Hu et al.

2020b; Alam et al. 2021). Its effect has also been studied in agronomically important crops, such as rice (Etesami et al. 2022), maize (Saleem et al. 2022), sorghum (Avila et al. 2021), sugarcane (Verma et al. 2021), and cotton (Barros et al. 2019). Nevertheless, Si effects vary between plant species and stress conditions.

Besides protecting the electrolytic leakage and the lipid peroxidation level and controlling the H_2O_2, Si modulates vital proteins involved in nucleosome pathways and carbohydrate metabolism (Manivannan et al. 2015). A salt stress study on Zinnia showed that Si enhances membrane integrity and modulates the antioxidant enzymes (Manivannan et al. 2015). Similar effects were observed in carnation and salvia on in vitro salt exposure with Si treatment (Soundararajan et al. 2013, 2015). Regarding the Si influence on the photosystem, activities of antioxidant enzymes enhance the axillary shoot multiplication and alleviate hyperhydricity in carnation (Muneer et al. 2017; Manivannan et al. 2018). Soundararajan et al. (2017a) demonstrated overcoming hyperhydricity under Si supplementation in carnation. Proteomics data showed that Si highly influenced ribosomal binding, oxide reduction, hormone/cell signaling, metal/ion binding, defense, and photosynthesis (Soundararajan et al. 2017a). Photosynthetic protein expressions were improved with Si application in strawberry propagation (Muneer et al. 2017; Park et al. 2018). Drought tolerance improved by influencing the abundance of proteins involved in transportation, circadian rhythm, and other essential pathways in chrysanthemum (Sahithi et al. 2021). Modulating antioxidant enzymes under Si treatment during the pre-/post-harvest period improved the vase life (Song et al. 2021). Recently it has been reported that Si hinders ammonium toxicity in cabbage (Song et al. 2022). When increasing nutrient content and osmolytes, such as proline, and suppressing ROS, Si application showed resistance against powdery mildew in strawberries (Xiao et al. 2022a).

Detailed studies on roses under Si treatment reveal several molecular insights into the physiological and antioxidant enhancement (Soundararajan et al. 2017a, 2018). Regarding root tissues, Si improved root hair growth, which was severely affected in salinity conditions. Moreover, Si deposition inhibits the influx transportation of the excess Na ion. In contrast, Si increased the K ion uptake. Compared with the control, NaCl decreased the abundance of 53 proteins. At the same time, Si treatment improved the abundance of 50 proteins. These proteins are mostly involved in transcription/translation, lipid metabolism, signaling, energy and carbohydrate metabolism, transportation, cell wall, and terpene-related processes. In aerial tissues, Si supplementation prevents stomatal damage, pigment degradation, and LPO. Synergistically, it improved net photosynthetic assimilation rates, stomatal conductance, internal CO_2 concentrations, and the Fv/Fm ratio. Silicon-treated salt conditions significantly upregulated the abundance of major proteins involved in photosynthesis, carbohydrate/energy metabolism, transcription/translation, stress/redox homeostasis, ion binding, and ubiquitination (Soundararajan et al. 2018). In both leaf and root tissues, the Si treatment greatly reduced the accumulation of O_2 and H_2O_2. Antioxidant enzyme mechanism induction was observed in activity, native-PAGE, and western blot analysis (Soundararajan et al. 2017b, 2018). When considering the abovementioned studies, it was clearly shown that Si has a prominent effect on ROS homeostasis in stress alleviation. More detailed studies must be conducted to unveil its mode of action on improving tolerance against stress conditions.

18.5 Si-Mediated Regulation of Heat Stress Tolerance in Plants

The plants supplemented with Si have demonstrated improved resistance to many abiotic stresses, including drought, temperature stress, salinity, heavy metal toxicity, and radiation. However, compared to research on salinity and drought, only some reports on Si-induced heat tolerance in plants have been found. The beneficial role of Si in mitigating temperature stress and their underlying mechanisms of heat tolerance in some of the plants are discussed below and depicted in Fig. 18.1.

18.5.1 Rice

Rice is a major Si accumulator and has benefited greatly from Si nutrition regarding growth and yield, especially under biotic and abiotic stress (Chen et al. 2011; Jinger et al. 2017, 2020; Ramírez-Olvera et al. 2021; Thorne et al. 2022; Ghouri et al. 2022).

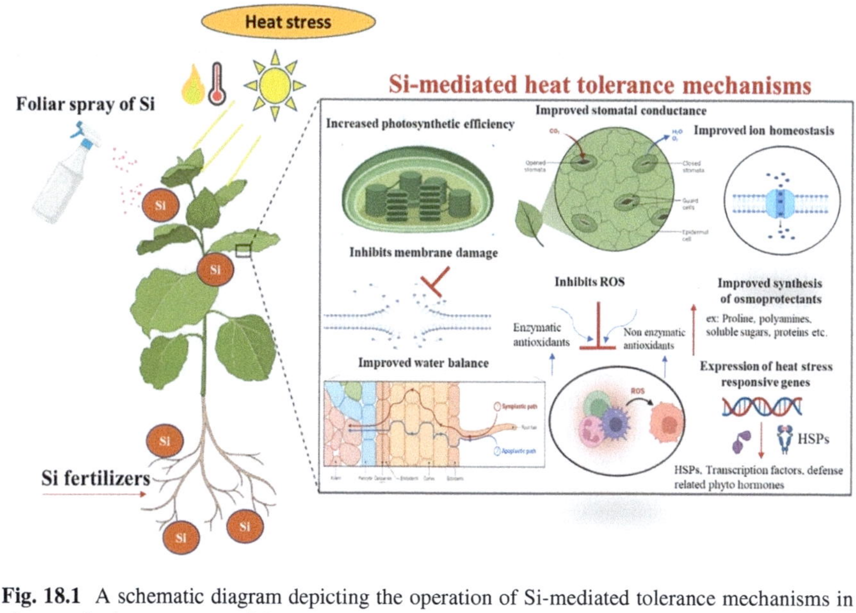

Fig. 18.1 A schematic diagram depicting the operation of Si-mediated tolerance mechanisms in plants under heat stress conditions

However, research on the effect of Si on the temperature stress of rice is limited to date. Earlier, Agarie et al. (1992) demonstrated that the electrolyte leakage in leaves of Si-treated plants was significantly lower than that of Si-untreated plants during heat stress. Moreover, they found that the leaves of Si-supplemented rice plants presented higher quantities of cell wall polysaccharides than those of Si-deficient plants. In addition, they demonstrated the Si involvement in cell wall structure strengthening. Thus, these findings show that Si in rice leaves may play critical functions in preventing electrolyte leakage through cell wall formation and mechanical support.

18.5.2 Wheat

Wheat is a significant cereal crop worldwide. It is vulnerable to heat stress, especially at the reproductive stage. Heat stress decreases wheat grain yield by shortening the filling period, damaging the photosynthetic system through oxidation, and decreasing dry matter accumulation in grains (Farooq et al. 2011). Younis et al. (2020) demonstrated the ameliorative effect of Si and silicon nanoparticles (SiNPs) on the heat stress of wheat seedlings. They observed that the administration of Si or SiNPs has dramatically reversed the ultrastructural abnormalities of cellular organelles by heat stress, especially in chloroplasts and the nucleus. These findings suggest that Si also actively contributes to the heat stability of cell membranes, most likely by preventing high-temperature-induced damage to lipids and proteins bound to the membrane. Furthermore, Si and SiNPs have improved the photosynthetic capacity by improving the PSII photochemical efficacy, performance indexes, and the number of photosynthetic pigments. Moreover, plants exposed to heat stress had higher endogenous levels of MDA, whereas plants exposed to heat stress after Si treatment had much lower endogenous MDA levels. Heat-stressed leaves treated with Si and SiNPs had considerably higher DPPH scavenging activity than non-Si-treated leaves. Compared to the untreated stressed leaves, the Si treatment dramatically enhanced RWC two-fold, further corroborated by the upregulation of aquaporin genes (*TaPIP1* and Ta*NIP2*).

In another independent study, Mustafa et al. (2021) found that foliar application of Si has significantly mitigated the heat stress at the heading stage of wheat. Furthermore, exogenous application of 4 mM Si has improved the individual (chl a, chl b) and combined chlorophyll contents (chl a+b) and carotenoid content. In addition, the activity of antioxidant enzymes, such as *SOD, CAT*, and *POD*, and the accumulation of osmoprotectants, such as proline, proteins, and soluble sugars, significantly increased upon Si treatment at terminal heat stress.

18.5.3 Barley

Hussain et al. (2019) investigated the influences of the foliar application of Si on growth, physiochemical parameters, and antioxidant system of barley plants exposed to heat stress. Barley plants treated with Si at high-temperature stress showed superior growth regarding increased shoot and root biomasses, leaf area, chlorophyll a and b, carotenoids, flavonoids, phenolics, anthocyanins, and total soluble proteins. Moreover, the Si treatment also favorably relieved the oxidative stress caused by heat stress by reducing the concentrations of MDA and H_2O_2. Meanwhile, it increased the activity of antioxidant enzymes, such as catalase, peroxidase, and superoxide dismutase.

18.5.4 Date Palm

The date palm (*Phoenix dactylifera* L.) is a key and essential fruit crop that greatly boosts the economies of arid and semi-arid areas of the world. However, the productivity and quality of date palm fruits can be significantly reduced by high and extended environmental challenges, such as heat and drought. Recently, Khan et al. (2020b) have investigated the cumulative effects of applying Si and GA_3 to date palms exogenously to reduce heat stress. The interaction between GA_3 and Si greatly decreased the negative effects of heat stress, resulting in the maximum root and shoot biomass and lengths. The combined application of GA_3 and Si under heat stress led to a significantly greater amount of chlorophyll a and b and a maximum relative water content (RWC). In addition, the formation of reactive oxygen species (O_2^-) and MDA has been greatly reduced by the combined application of GA_3 and Si. In contrast, enzymes' antioxidant activities have improved under heat stress. Interestingly, the combined application of Si and GA_3 has differentially modulated endogenous hormones, such as abscisic acid (ABA) and salicylic acid (SA). The ABA content was downregulated, while SA was more upregulated at heat stress than at control conditions. Moreover, the expression of antioxidant-related genes, such as *GPX2, Cyt-Cu/Zn SOD, GAPDH,* and *CAT*, and heat stress transcription factors and heat shock protein (HSP30-like) were significantly upregulated. The decrease in ABA content under heat stress was further substantiated by the downregulation of Abscisic acid receptor genes.

18.5.5 Tomatoes

Tomatoes (*Solanum lycopersicum* L.) are a highly valued economic crop worldwide. However, they are extremely vulnerable to heat stress. Air temperatures above 35 °C hinder tomato fruit development, flowering, pollen viability, and vegetative growth are hindered (Alsamir et al. 2021). In tomato leaves, rising temperatures reduce the efficiency of photosynthetic apparatus, alter membrane fluidity, and affect the general stability of metabolic processes. Thus, they lead to excess reactive oxygen species and oxidative stress, resulting in cell death (Camejo et al. 2005; Haque et al. 2021). In order to fight heat stress, tomatoes display a variety of morpho-physiological, biochemical, and molecular responses.

Khan et al. (2020a) investigated the underlying mechanisms of silicon-induced thermotolerance in tomatoes. The exogenous administration of Si has improved growth metrics, such as shoot and root lengths, dry weights, and stem diameter. In addition, it improved the overall phenotype of tomato plants both in the presence and absence of heat stress. Silicon mitigated the negative effects of heat stress, partly through increasing photosynthetic pigments (chlorophyll a, chlorophyll b, and carotenoids) and relative water content while lowering the levels of superoxide anions and malondialdehyde. A significant enhancement in the upregulation of mRNA expression levels of antioxidant enzymes such as *CAT, APX, POD, GR, SOD, GPX*, and *GST* was more present in Si+ heat-stressed tomato plants than in Si− heat-stressed tomato plants. Additionally, under heat stress, Si favored the expression of heat shock proteins by elevating heat transcription factors (*Hsfs*) viz., *SlHsfA1a-b*, *SlHsfA2-A3*, and *SlHsfA7*. Furthermore, Si application negatively regulated the accumulation of defense-related endogenous phytohormones (SA and ABA) by down-regulating their biosynthetic genes (*SlR1b1, SlPR-P2, SlICS, SlPAL*, and *SlNCEDI*) under heat stress. In conclusion, the upregulation of the genes encoding the Si transporters, such as *SlLsi1* and *SlLsi2*, was correlated with higher Si absorption and increased heat tolerance mechanisms in tomato plants.

18.5.6 Strawberry

Strawberry (*Fragaria × ananassa*) is a significant fruit crop worldwide. High-temperature stress hinders the establishment of leaves, flowers, and the overall photosynthesis in strawberries. Temperatures above 30 °C decrease fruit weight and fruit size in strawberries (Darrow 1966). Foliar or sub-irrigation of Si in the form of K_2SiO_3 has alleviated the high-temperature stress of strawberries at 41 °C by maintaining the expression of photosynthetic and stress response proteins (Muneer et al. 2017). The expression of two significant photosynthetic proteins, such as PsaA and PsbA, decreased in Si− plants, whereas it was retained in Si+ plants. Moreover, the activities of antioxidant enzymes, such as APX, SOD, and CAT, significantly increased in Si-supplemented plants.

Another study subjected strawberry mother and daughter plants to silicon treatment in silicate fertilizer, substrate drench, and foliar spray and found that Si treatment improved the high-temperature resistance of strawberries regardless of the application method (Xiao et al. 2022b). The above results were supported by increased shoot weight, antioxidant enzymes (APX, POD, CAT, and SOD), anthocyanin, and protein contents upon Si treatment at high-temperature stress.

18.5.7 Cucumber

Cucumber (*Cucumis sativus* L.) is one of the most important horticultural vegetable crops worldwide. It is primarily grown in greenhouses, making it susceptible to heat stress in the summer (Fan et al. 2020). In a greenhouse, Shalaby et al. (2021) investigated the role of Si in increasing the growth and yield of cucumber by mitigating the combined effect of salinity and heat stress and other anti-stressors such as nano-selenium and H_2O_2. The foliar application of Si greatly improved the overall nutrient uptake, especially by increasing the Si content from 11% to 22% in fruit. The potential involvement of Si in reducing soil salinity during heat stress could be attributed to the high Si concentration present in leaves, which modulates water losses through transpiration in addition to the high nutritional intake of other elements, such as N, P, K, Mg, and Se. Furthermore, the foliar Si application enhanced the vegetation growth parameters, such as plant height, leaf area, and the photosynthetic efficiency of cucumber, during combined heat and salinity stress. Silicon showed superior antioxidant capacity, high relative water content, and increased proline accumulation among the three anti-stressors used. Overall, Si has drastically increased the fruit quality by means of firmness, total soluble solids, and fruit yield.

Recently, in an independent study, foliar application or soaking seeds with Si was found beneficial in improving soil fertility, nutrient uptake, growth and development, chlorophyll content, antioxidant activities, and overall yield of snake cucumber under heat stress (Hassan Abd-Elrahman et al. 2022). Moreover, Si applications have increased total soluble solids and vitamin C as a protective measure to maintain osmotic balance during heat stress.

18.5.8 Poinsettia

Poinsettias (*Euphorbia pulcherrima* Willd.) are cultivated in summer and produce flowers in winter. Thus they are likely to be exposed to hot and cold temperature stresses, respectively. Under high-temperature conditions in poinsettia, higher activities of APX in Si treatment improved the maximum quantum yield of photosystem II (Fv/Fm) and photosynthetic rate (Son et al. 2015; Hu et al. 2020a). Furthermore, it was observed that nearly all of the stomata in cuttings not supplemented with Si were closed under high-temperature stress, but some of them remained open in the case of Si-treated cuttings. Moreover, the concentrations of MDA and H_2O_2 were significantly lower in Si-treated seedlings subjected to high-temperature stress. In a previous study, Si induced the rooting in poinsettia cuttings by alleviating the inhibitory effect of N-1 naphthylphthalamic acid (NPA) and modulating the tissue mineral content (Hu et al. 2020b).

18.5.9 Salvia

Salvia splendens is one of the most popular ornamental plants propagated worldwide as a bedding plant. In warm temperate continental areas, high temperatures impact this plant's development. Soundararajan et al. (2014) have investigated the positive effect of Si on the growth and antioxidant enzymes of *Salvia splendens* at high temperatures. The addition of Si has significantly improved all the aspects of growth and the activity of major antioxidant enzymes (GPX, APX, SOD, and POD). Furthermore, they observed that between 0 and 4 h of heat treatment at 35 °C, salvia plants treated with Si showed distinct protein expression patterns. Interestingly, after 4 hours of heat exposure, a protein of 46 kDa was identified to be strongly expressed, which was likely an HSP that served as a molecular chaperone to counteract heat stress.

18.6 Conclusions

Ultimately, Si has several beneficial roles in protecting plants from heat-induced detrimental effects. More crucially, Si modulates heat stress tolerance by increasing photosynthetic efficiency, tissue water potential, leaf gas exchange capacity, stomatal conductance, and epicuticular wax production. In addition, Si supplementation has a pronounced effect mitigating heat-induced oxidative damage by activating both the enzymatic and non-enzymatic antioxidant defense systems. Furthermore, the impact of Si in regulating heat resilience in plants is linked to variations in endogenous phytohormone levels as well as changes in heat stress-responsive gene and protein expression. Thus, elucidating Si-mediated heat tolerance in terms of physiological, biochemical, and molecular responses of plants using integrative OMICS techniques between Si-treated and Si-deprived plants is highly essential. Additionally, the development of heat-stress tolerant plants by manipulating Si transporter gene(s) using overexpression and gene editing techniques, such as CRISPR-Cas9, could open the way for a long-term solution to counteract the negative impacts of heat stress on crop production during climate change problems.

References

Agarie S, Agata W, Kubota F, Kaufman PB (1992) Physiological roles of silicon in photosynthesis and dry matter production in rice plants: I. Effects of silicon and shading treatments. Japanese J Crop Sci 61:200–206. https://doi.org/10.1626/jcs.61.200

Al Murad M, Khan AL, Muneer S (2020) Silicon in horticultural crops: cross-talk, signaling, and tolerance mechanism under salinity stress. Plants 9:460. https://doi.org/10.3390/plants9040460

Al-Aghabary K, Zhu Z, Shi Q (2005) Influence of silicon supply on chlorophyll content, chlorophyll fluorescence, and antioxidative enzyme activities in tomato plants under salt stress. J Plant Nutr 27:2101–2115. https://doi.org/10.1081/PLN-200034641

Alam P, Balawi TH, Altalayan FH, Hatamleh AA, Ashraf M, Ahmad P (2021) Silicon attenuates the negative effects of chromium stress in tomato plants by modifying antioxidant enzyme activities, ascorbate–glutathione cycle and glyoxalase system. Acta Physiol Plant 43:1–17. https://doi.org/10.1007/s11738-021-03276-4

Alsamir M, Mahmood T, Trethowan R, Ahmad N (2021) An overview of heat stress in tomato (*Solanum lycopersicum* L.). Saudi J Biol Sci 28:1654–1663. https://doi.org/10.1016/j.sjbs.2020.11.088

Altaf MM, Diao X, Altaf MA, ur Rehman A, Shakoor A, Khan LU, Jan BL, Ahmad P (2022) Silicon-mediated metabolic upregulation of ascorbate glutathione (AsA-GSH) and glyoxalase reduces the toxic effects of vanadium in rice. J Hazard Mater 436:129145. https://doi.org/10.1016/j.jhazmat.2022.129145

Atta B, Rizwan M, Sabir AM, Gogi MD, Ayub MA (2019) Silicon mediated induced resistance in plants for the management of agricultural insect pests: a review. World J Biol Biotechnol 4:19–27. https://doi.org/10.33865/wjb.004.01.0192

Avila RG, Magalhães PC, da Silva EM, de Souza KRD, Campos CN, de Alvarenga AA, de Souza TC (2021) Application of silicon to irrigated and water deficit sorghum plants increases yield via the regulation of primary, antioxidant, and osmoregulatory metabolism. Agric Water Manag 255:107004. https://doi.org/10.1016/j.agwat.2021.107004

Barros TC, de Mello PR, Roque CG, Arf MV, Vilela RG (2019) Silicon and salicylic acid in the physiology and yield of cotton. J Plant Nutr 42:458–465. https://doi.org/10.1080/01904167.2019.1567765

Battisti DS, Naylor RL (2009) Historical warnings of future food insecurity with unprecedented seasonal heat. Science 323:240–244. https://doi.org/10.1126/science.1164363

Camejo D, Rodríguez P, Morales MA, Dell'Amico JM, Torrecillas A, Alarcón JJ (2005) High temperature effects on photosynthetic activity of two tomato cultivars with different heat susceptibility. J Plant Physiol 162:281–289. https://doi.org/10.1016/j.jplph.2004.07.014

Carlisle EM (1997) Silicon. In: O'Dell BL, Sunde RA (eds) Handbook of nutritionally essential mineral elements. Marcel Dekker, New York, pp 603–608

Chen W, Yao X, Cai K, Chen J (2011) Silicon alleviates drought stress of rice plants by improving plant water status, photosynthesis and mineral nutrient absorption. Biol Trace Elem Res 142:67–76. https://doi.org/10.1007/s12011-010-8742-x

Chérif M, Benhamou N, Menzies JG, Bélanger RR (1992) Silicon induced resistance in cucumber plants against *Pythium ultimum*. Physiol Mol Plant Pathol 41:411–425. https://doi.org/10.1016/0885-5765(92)90053-X

Currie HA, Perry CC (2007) Silica in plants: biological, biochemical and chemical studies. Ann Bot 100:1383–1389. https://doi.org/10.1093/aob/mcm247

Darrow GM (1966) The strawberry. History, breeding and physiology. Strawberry. Hist Breed Physiol 447 pp.

Das S, Chattopadhyay UK (2000) Role of silicon in modulating the internal morphology and growth of *Mycobacterium tuberculosis*. Indian J Tuberc 47:87–91

Deshmukh R, Sonah H, Belanger RR (2020) New evidence defining the evolutionary path of aquaporins regulating silicon uptake in land plants. J Exp Bot 71:6775–6788. https://doi.org/10.1093/jxb/eraa342

Dhakate P, Kandhol N, Raturi G, Ray P, Bhardwaj A, Srivastava A, Kaushal L, Singh A, Pandey S, Chauhan DK (2022) Silicon nanoforms in crop improvement and stress management. Chemosphere 305:135165. https://doi.org/10.1016/j.chemosphere.2022.135165

Etesami H, Jeong BR (2018) Silicon (Si): review and future prospects on the action mechanisms in alleviating biotic and abiotic stresses in plants. Ecotoxicol Environ Saf 147:881–896. https://doi.org/10.1016/j.ecoenv.2017.09.063

Etesami H, Li Z, Maathuis FJM, Cooke J (2022) The combined use of silicon and arbuscular mycorrhizas to mitigate salinity and drought stress in rice. Environ Exp Bot 104955. https://doi.org/10.1016/j.envexpbot.2022.104955

Fahad S, Bajwa AA, Nazir U, Anjum SA, Farooq A, Zohaib A, Sadia S, Nasim W, Adkins S, Saud S (2017) Crop production under drought and heat stress: plant responses and management options. Front Plant Sci 8:1147. https://doi.org/10.3389/fpls.2017.01147

Fan Y, Zhang Y, Hess F, Huang B, Chen Z (2020) Nutrient balance and soil changes in plastic greenhouse vegetable production. Nutr Cycl Agroecosyst 117:77–92. https://doi.org/10.1007/s10705-020-10057-x

Farooq M, Bramley H, Palta JA, Siddique KHM (2011) Heat stress in wheat during reproductive and grain-filling phases. CRC Crit Rev Plant Sci 30:491–507. https://doi.org/10.1080/07352689.2011.615687

Fawe A, Abou-Zaid M, Menzies JG, Bélanger RR (1998) Silicon-mediated accumulation of flavonoid phytoalexins in cucumber. Phytopathology 88:396–401. https://doi.org/10.1094/PHYTO.1998.88.5.396

Firmansyah AN, Argosubekti N (2019) A review of heat stress signaling in plants. IOP Conf Ser Earth Environ Sci 484:23–25

Ghouri F, Ali Z, Naeem M, Ul-Allah S, Babar M, Baloch FS, Chattah WS, Shahid MQ (2022) Effects of silicon and selenium in alleviation of drought stress in rice. Silicon 14:5453–5461. https://doi.org/10.1007/s12633-021-01277-z

Gunes A, Inal A, Bagci EG, Coban S, Sahin O (2007) Silicon increases boron tolerance and reduces oxidative damage of wheat grown in soil with excess boron. Biol Plant 51:571–574. https://doi.org/10.1007/s10535-007-0125-6

Halford NG (2011) The role of plant breeding and biotechnology in meeting the challenge of global warming. Planet Earth 2011-Global Warm. Challenges Oppor. Policy Pract

Han F, Chen H, Li X-J, Yang M-F, Liu G-S, Shen S-H (2009) A comparative proteomic analysis of rice seedlings under various high-temperature stresses. Biochim Biophys Acta Proteins Proteomics 1794:1625–1634. https://doi.org/10.1016/j.bbapap.2009.07.013

Haque MS, Husna MT, Uddin MN, Hossain MA, Sarwar AKMG, Ali OM, Abdel Latef AAH, Hossain A (2021) Heat stress at early reproductive stage differentially alters several physiological and biochemical traits of three tomato cultivars. Horticulturae. https://doi.org/10.3390/horticulturae7100330

Hasanuzzaman M, Bhuyan MHM, Zulfiqar F, Raza A, Mohsin SM, Al MJ, Fujita M, Fotopoulos V (2020) Reactive oxygen species and antioxidant defense in plants under abiotic stress: Revisiting the crucial role of a universal defense regulator. Antioxidants 9:681. https://doi.org/10.3390/antiox9080681

Hasanuzzaman M, Nahar K, Fujita M (2013) Extreme temperature responses, oxidative stress and antioxidant defense in plants: abiotic stress responses. Appl Agric 13:169–205

Hassan Abd-Elrahman S, Abd-Elhamid Hashem F, Awny Farag A (2022) Mitigation of heat stress effects on snake cucumber plants by silicon additions and rice straw soil cover. Commun Soil Sci Plant Anal 53:759–772. https://doi.org/10.1080/00103624.2022.2028814

Howarth CJ (2005) Genetic improvements of tolerance to high temperature. Abiotic stress plant resist through breed mol approaches. Howarth Press Inc, New York, pp 277–300

Hu J, Li Y, Jeong BR (2020a) Putative silicon transporters and effect of temperature stresses and silicon supplementation on their expressions and tissue silicon content in poinsettia. Plants 9:569. https://doi.org/10.3390/plants9050569

Hu J, Li Y, Jeong BR (2020b) Silicon alleviates temperature stresses in poinsettia by regulating stomata, photosynthesis, and oxidative damages. Agronomy 10:1419. https://doi.org/10.3390/agronomy10091419

Hussain I, Parveen A, Rasheed R, Ashraf MA, Ibrahim M, Riaz S, Afzaal Z, Iqbal M (2019) Exogenous silicon modulates growth, physio-chemicals and antioxidants in barley (*Hordeum vulgare* L.) exposed to different temperature regimes. Silicon 11:2753–2762. https://doi.org/10.1007/s12633-019-0067-6

Hussain S, Shuxian L, Mumtaz M, Shafiq I, Iqbal N, Brestic M, Shoaib M, Sisi Q, Li W, Mei X (2021) Foliar application of silicon improves stem strength under low light stress by regulating lignin biosynthesis genes in soybean [*Glycine max* (L.) Merr.]. J Hazard Mater 401:123256. https://doi.org/10.1016/j.jhazmat.2020.123256

Ingri N (1978) Aqueous silicic acid, silicates and silicate complexes. In: Biochem silicon Relat Probl. Springer, pp 3–51

IPCC CC (2007) The physical science basis. Contribution of working group I to the fourth assessment report of the Intergovernmental panel on climate change, vol 996. Cambridge Univ Press, Cambridge, UK/ New York, pp 113–119

Jha UC, Bohra A, Singh NP (2014) Heat stress in crop plants: its nature, impacts and integrated breeding strategies to improve heat tolerance. Plant Breed 133:679–701

Jinger D, Devi MT, Dhar S, Dass A, Rajanna GA, Upadhaya P, Raj R (2017) Silicon in mitigating biotic stresses in rice (*Oryza sativa* L.) – a review. Ann Agric Res 38:70004

Jinger D, Devi MT, Dhar S, Dass A, Sharma VK, Vijayakumar S, Joshi E, Jatav HS, Singh N (2020) Silicon application mitigates abiotic stresses in rice: a review. Indian J Agric Sci 90:2043–2050

Khan A, Bilal S, Khan AL, Imran M, Shahzad R, Al-Harrasi A, Al-Rawahi A, Al-Azhri M, Mohanta TK, Lee I-J (2020b) Silicon and gibberellins: synergistic function in harnessing ABA signaling and heat stress tolerance in date palm (*Phoenix dactylifera* L.). Plants 9:620. https://doi.org/10.3390/plants9050620

Khan A, Khan AL, Imran M, Asaf S, Kim YH, Bilal S, Numan M, Al-Harrasi A, Al-Rawahi A, Lee IJ (2020a) Silicon-induced thermotolerance in *Solanum lycopersicum* L. via activation of antioxidant system, heat shock proteins, and endogenous phytohormones. BMC Plant Biol 20:1–18. https://doi.org/10.1186/s12870-020-02456-7

Kinrade SD, Hamilton RJ, Schach AS, Knight CTG (2001) Aqueous hypervalent silicon complexes with aliphatic sugar acids. J Chem Soc Dalton Trans 2:961–963

Knight CTG, Kinrade SD (2001) A primer on the aqueous chemistry of silicon. Stud plant Sci Elsevier 8:57–84

Kolesnikov MP, Gins VK (2001) Forms of silicon in medicinal plants. Appl Biochem Microbiol 37:524–527. https://doi.org/10.1023/A:1010262527643

Li J, Leisner SM, Frantz J (2008) Alleviation of copper toxicity in *Arabidopsis thaliana* by silicon addition to hydroponic solutions. J Am Soc Hort Sci 133:670–677. https://doi.org/10.21273/JASHS.133.5.670

Liang Y, Wong JWC, Wei L (2005) Silicon-mediated enhancement of cadmium tolerance in maize (*Zea mays* L.) grown in cadmium contaminated soil. Chemosphere 58:475–483. https://doi.org/10.1016/j.chemosphere.2004.09.034

Liu B, Soundararajan P, Manivannan A (2019) Mechanisms of silicon-mediated amelioration of salt stress in plants. Plants 8:307. https://doi.org/10.3390/plants8090307

Luyckx M, Hausman J-F, Lutts S, Guerriero G (2017) Silicon and plants: current knowledge and technological perspectives. Front Plant Sci 8:411. https://doi.org/10.3389/fpls.2017.00411

Manivannan A, Ahn YK (2017) Silicon regulates potential genes involved in major physiological processes in plants to combat stress. Front Plant Sci 8:1346. https://doi.org/10.3389/fpls.2017.01346

Manivannan A, Soundararajan P, Arum LS, Ko CH, Muneer S, Jeong BR (2015) Silicon-mediated enhancement of physiological and biochemical characteristics of *Zinnia elegans* 'Dreamland Yellow' grown under salinity stress. Hortic Environ Biotechnol 56:721–731. https://doi.org/10.1007/s13580-015-1081-2

Manivannan A, Soundararajan P, Cho YS, Park JE, Jeong BR (2018) Sources of silicon influence photosystem and redox homeostasis-related proteins during the axillary shoot multiplication of *Dianthus caryophyllus*. Plant Biosyst Int J Deal with all Asp Plant Biol 152:704–710. https://doi.org/10.1080/11263504.2017.1320312

Manivannan A, Soundararajan P, Muneer S, Ko CH, Jeong BR (2016) Silicon mitigates salinity stress by regulating the physiology, antioxidant enzyme activities, and protein expression in *Capsicum annuum* 'Bugwang'. Biomed Res Int 2016:6357. https://doi.org/10.1155/2016/3076357

Martin-Jézéquel V, Hildebrand M, Brzezinski MA (2000) Silicon metabolism in diatoms: implications for growth. J Phycol 36:821–840. https://doi.org/10.1046/j.1529-8817.2000.00019.x

Mittler R (2002) Oxidative stress, antioxidants and stress tolerance. Trends Plant Sci 7:405–410. https://doi.org/10.1016/S1360-1385(02)02312-9

Muneer S, Park YG, Kim S, Jeong BR (2017) Foliar or subirrigation silicon supply mitigates high temperature stress in strawberry by maintaining photosynthetic and stress-responsive proteins. J Plant Growth Regul 36:836–845. https://doi.org/10.1007/s00344-017-9687-5

Mustafa T, Sattar A, Sher A, Ul-Allah S, Ijaz M, Irfan M, Butt M, Cheema M (2021) Exogenous application of silicon improves the performance of wheat under terminal heat stress by triggering physio-biochemical mechanisms. Sci Rep 11:1–12. https://doi.org/10.1038/s41598-021-02594-4

Neumann D, Zur Nieden U (2001) Silicon and heavy metal tolerance of higher plants. Phytochemistry 56:685–692. https://doi.org/10.1016/S0031-9422(00)00472-6

Niinemets Ü (2018) When leaves go over the thermal edge. Plant Cell Environ 41:1247–1250. https://doi.org/10.1111/pce.13184

Park YG, Muneer S, Kim S, Hwang SJ, Jeong BR (2018) Silicon application during vegetative propagation affects photosynthetic protein expression in strawberry. Hortic Environ Biotechnol 59:167–177. https://doi.org/10.1007/s13580-018-0022-2

Perkins-Kirkpatrick SE, Lewis SC (2020) Increasing trends in regional heatwaves. Nat Commun 11:1–8. https://doi.org/10.1038/s41467-020-16970-7

Peverelli MC, Rogers WJ (2013) Heat stress effects on crop performance and tools for tolerance breeding. Rev Fac Ciencias Agrar 45:349–368

Pfannschmidt T, Yang C (2012) The hidden function of photosynthesis: a sensing system for environmental conditions that regulates plant acclimation responses. Protoplasma 249:125–136. https://doi.org/10.1007/s00709-012-0398-2

Porter JR (2005) Rising temperatures are likely to reduce crop yields. Nature 436:174. https://doi.org/10.1038/436174b

Ramírez-Olvera SM, Trejo-Téllez LI, Gómez-Merino FC, del Ruíz-Posadas LM, Alcántar-González EG, Saucedo-Veloz C (2021) Silicon stimulates plant growth and metabolism in rice plants under conventional and osmotic stress conditions. Plants 10:777. https://doi.org/10.3390/plants10040777

Saha G, Mostofa MG, Rahman MM, Tran L-SP (2021) Silicon-mediated heat tolerance in higher plants: a mechanistic outlook. Plant Physiol Biochem 166:341–347. https://doi.org/10.1016/j.plaphy.2021.05.051

Sahithi BM, Razi K, Al Murad M, Vinothkumar A, Jagadeesan S, Benjamin LK, Jeong BR, Muneer S (2021) Comparative physiological and proteomic analysis deciphering tolerance and homeostatic signaling pathways in chrysanthemum under drought stress. Physiol Plant 172:289–303. https://doi.org/10.1111/ppl.13142

Saleem MH, Parveen A, Khan SU, Hussain I, Wang X, Alshaya H, El-Sheikh MA, Ali S (2022) Silicon fertigation regimes attenuates cadmium toxicity and phytoremediation potential in two maize (*Zea mays* L.) cultivars by minimizing its uptake and oxidative stress. Sustainability 14:1462. https://doi.org/10.3390/su14031462

Savant NK, Korndörfer GH, Datnoff LE, Snyder GH (1999) Silicon nutrition and sugarcane production: a review. J Plant Nutr 22:1853–1903. https://doi.org/10.1080/01904169909365761

Shalaby TA, Abd-Alkarim E, El-Aidy F, Hamed E-S, Sharaf-Eldin M, Taha N, El-Ramady H, Bayoumi Y, Dos Reis AR (2021) Nano-selenium, silicon and H_2O_2 boost growth and productivity of cucumber under combined salinity and heat stress. Ecotoxicol Environ Saf 212:111962. https://doi.org/10.1016/j.ecoenv.2021.111962

Sharma L, Priya M, Kaushal N, Bhandhari K, Chaudhary S, Dhankher OP, Prasad PVV, Siddique KHM, Nayyar H (2020) Plant growth-regulating molecules as thermoprotectants: functional relevance and prospects for improving heat tolerance in food crops. J Exp Bot 71:569–594. https://doi.org/10.1093/jxb/erz333

Sivanesan I, Son MS, Song JY, Jeong BR (2013) Silicon supply through the subirrigation system affects growth of three chrysanthemum cultivars. Hortic Environ Biotechnol 54:14–19. https://doi.org/10.1007/s13580-013-0120-0

Smertenko A, Dráber P, Viklický V, Opatrný Z (1997) Heat stress affects the organization of microtubules and cell division in *Nicotiana tabacum* cells. Plant Cell Environ 20:1534–1542. https://doi.org/10.1046/j.1365-3040.1997.d01-44.x

Son MS, Park YG, Sivanesan I, Ko CH, Jeong BR (2015) Silicon supply through subirrigation system alleviates high temperature stress in poinsettia by enhancing photosynthetic rate. Hortic Sci Technol 33:860–868. https://doi.org/10.7235/hort.2015.15099

Song A, Li P, Fan F, Li Z, Liang Y (2014) The effect of silicon on photosynthesis and expression of its relevant genes in rice (*Oryza sativa* L.) under high-zinc stress. PloS One 9:e113782. https://doi.org/10.1371/journal.pone.0113782

Song J, Li Y, Hu J, Lee J, Jeong BR (2021) Pre-and/or postharvest silicon application prolongs the vase life and enhances the quality of cut peony (*Paeonia lactiflora* Pall.) flowers. Plants 10:1742. https://doi.org/10.3390/plants10081742

Song J, Yang J, Jeong BR (2022) Silicon mitigates ammonium toxicity in cabbage (*Brassica campestris* L. ssp. pekinensis) 'Ssamchu'. Front Sustain Food Syst 6:922666. https://doi.org/10.3389/fsufs.2022.922666

Soundararajan P, Manivannan A, Cho YS, Jeong BR (2017a) Exogenous supplementation of silicon improved the recovery of hyperhydric shoots in *Dianthus caryophyllus* L. by stabilizing the physiology and protein expression. Front Plant Sci 8:738. https://doi.org/10.3389/fpls.2017.00738

Soundararajan P, Manivannan A, Ko CH, Jeong BR (2018) Silicon enhanced redox homeostasis and protein expression to mitigate the salinity stress in *Rosa hybrida* 'Rock Fire'. J Plant Growth Regul 37:16–34. https://doi.org/10.1007/s00344-017-9705-7

Soundararajan P, Manivannan A, Ko CH, Muneer S, Jeong BR (2017b) Leaf physiological and proteomic analysis to elucidate silicon induced adaptive response under salt stress in *Rosa hybrida* 'Rock Fire'. Int J Mol Sci 18:1768. https://doi.org/10.3390/ijms18081768

Soundararajan P, Manivannan A, Park YG, Muneer S, Jeong BR (2015) Silicon alleviates salt stress by modulating antioxidant enzyme activities in *Dianthus caryophyllus* 'Tula'. Hortic Environ Biotechnol 56:233–239. https://doi.org/10.1007/s13580-015-0111-4

Soundararajan P, Sivanesan I, Jana S, Jeong BR (2014) Influence of silicon supplementation on the growth and tolerance to high temperature in *Salvia splendens*. Hortic Environ Biotechnol 55:271–279. https://doi.org/10.1007/s13580-014-0023-8

Soundararajan P, Sivanesan I, Jo EH, Jeong BR (2013) Silicon promotes shoot proliferation and shoot growth of *Salvia splendens* under salt stress in vitro. Hortic Environ Biotechnol 54:311–318. https://doi.org/10.1007/s13580-013-0118-7

Telfer P, Edwards J, Kuchel H, Reinheimer J, Bennett D (2013) Heat stress tolerance of wheat. Grains Res Dev Corp Barton, Available https://www.grdc.com.au/Research-and-Development/GRDC-Update-Papers/2013/02/Heat-stress-tolerance-of-wheat (Verified 16 May 2016)

Teskey R, Wertin T, Bauweraerts I, Ameye M, McGuire MA, Steppe K (2015) Responses of tree species to heat waves and extreme heat events. Plant Cell Environ 38:1699–1712. https://doi.org/10.1111/pce.12417

Thorne SJ, Stirnberg PM, Hartley SE, Maathuis FJM (2022) The ability of silicon fertilisation to alleviate salinity stress in rice is critically dependent on cultivar. Rice 15:1–10. https://doi.org/10.1186/s12284-022-00555-7

Tripathi DK, Singh VP, Kumar D, Chauhan DK (2012) Impact of exogenous silicon addition on chromium uptake, growth, mineral elements, oxidative stress, antioxidant capacity, and leaf and root structures in rice seedlings exposed to hexavalent chromium. Acta Physiol Plant 34:279–289. https://doi.org/10.1007/s11738-011-0826-5

Van Breusegem F, Vranová E, Dat JF, Inzé D (2001) The role of active oxygen species in plant signal transduction. Plant Sci 161:405–414. https://doi.org/10.1016/S0168-9452(01)00452-6

Verma KK, Song X-P, Zeng Y, Guo D-J, Singh M, Rajput VD, Malviya MK, Wei K-J, Sharma A, Li D-P (2021) Foliar application of silicon boosts growth, photosynthetic leaf gas exchange, antioxidative response and resistance to limited water irrigation in sugarcane (*Saccharum officinarum* L.). Plant Physiol Biochem 166:582–592b. https://doi.org/10.1016/j.plaphy.2021.06.032

Verma KK, Wu K-C, Singh P, Malviya MK, Singh RK, Song X-P, Li Y-R (2019) The protective role of silicon in sugarcane under water stress: photosynthesis and antioxidant enzymes. Biomed J Sci Tech Res 15:26717. https://doi.org/10.26717/BJSTR.2019.15.002685

Wahid A, Gelani S, Ashraf M, Foolad MR (2007) Heat tolerance in plants: an overview. Environ Exp Bot 61:199–223. https://doi.org/10.1016/j.envexpbot.2007.05.011

Wang X, Hou L, Lu Y, Wu B, Gong X, Liu M, Wang J, Sun Q, Vierling E, Xu S (2018) Metabolic adaptation of wheat grain contributes to a stable filling rate under heat stress. J Exp Bot 69:5531–5545. https://doi.org/10.1093/jxb/ery303

Wang Y, Stass A, Horst WJ (2004) Apoplastic binding of aluminum is involved in silicon-induced amelioration of aluminum toxicity in maize. Plant Physiol 136:3762–3770. https://doi.org/10.1104/pp.104.045005

Waraich EA, Ahmad R, Halim A, Aziz T (2012) Alleviation of temperature stress by nutrient management in crop plants: a review. J Soil Sci Plant Nutr 12:221–244. https://doi.org/10.4067/S0718-95162012000200003

Wise RR, Olson AJ, Schrader SM, Sharkey TD (2004) Electron transport is the functional limitation of photosynthesis in field-grown Pima cotton plants at high temperature. Plant Cell Environ 27:717–724. https://doi.org/10.1111/j.1365-3040.2004.01171.x

Xiao J, Li Y, Jeong BR (2022a) Foliar silicon spray before summer cutting propagation enhances resistance to powdery mildew of daughter plants. Int J Mol Sci 23:3803. https://doi.org/10.3390/ijms23073803

Xiao J, Li Y, Jeong BR (2022b) Foliar silicon spray to strawberry plants during summer cutting propagation enhances resistance of transplants to high temperature stresses. Front Sustain Food Syst 6:938128. https://doi.org/10.3389/fsufs

Yoshida S, Ohnishi Y, Kitagishi K (1962) Histochemistry of silicon in rice plant: III the presence of cuticle-silica double layer in the epidermal tissue. Soil Sci Plant Nutr 8:1–5. https://doi.org/10.1080/00380768.1962.10430982

Younis AA, Khattab H, Emam MM (2020) Impacts of silicon and silicon nanoparticles on leaf ultrastructure and TaPIP1 and TaNIP2 gene expressions in heat stressed wheat seedlings. Biol Plant 64:343–352. https://doi.org/10.32615/bp.2020.030

Zargar SM, Mahajan R, Bhat JA, Nazir M, Deshmukh R (2019) Role of silicon in plant stress tolerance: opportunities to achieve a sustainable cropping system. 3 Biotech 9:1–16. https://doi.org/10.1007/s13205-019-1613-z

Zhu Z, Wei G, Li J, Qian Q, Yu J (2004) Silicon alleviates salt stress and increases antioxidant enzymes activity in leaves of salt-stressed cucumber (*Cucumis sativus* L.). Plant Sci 167:527–533. https://doi.org/10.1016/j.plantsci.2004.04.020

Zhu YX, Gong HJ, Yin JL (2019) Role of silicon in mediating salt tolerance in plants: a review. Plants 8:147. https://doi.org/10.3390/plants8060147

Chapter 19
Silicon in Plants Mitigates Damage Against Pathogens and Insect Pests

Waqar Islam, Arfa Tauqeer, Abdul Waheed, Habib Ali, and Fanjiang Zeng

19.1 Introduction

Insect pests and diseases are major restricting elements in agriculture all over the world as they cause severe damage to crops. Globally, annual crop losses resulting from diseases are estimated to be at around 25% (Khakimov et al. 2022), while

W. Islam (✉)
Xinjiang Key Laboratory of Desert Plant Roots Ecology and Vegetation Restoration, Xinjiang Institute of Ecology and Geography, Chinese Academy of Sciences, Urumqi, China

State Key Laboratory of Desert and Oasis Ecology, Xinjiang Institute of Ecology and Geography, Chinese Academy of Sciences, Urumqi, China

Cele National Station of Observation and Research for Desert-Grassland Ecosystems, Cele, China

University of Chinese Academy of Sciences, Beijing, China
e-mail: waqarislam@ms.xjb.ac.cn

A. Tauqeer
College of Horticulture, FAFU-UCR Joint Center and Fujian Provincial Key Laboratory of Haixia Applied Plant System Biology, Fujian Agriculture and Forestry University, Fuzhou, China

A. Waheed · F. Zeng
Xinjiang Key Laboratory of Desert Plant Roots Ecology and Vegetation Restoration, Xinjiang Institute of Ecology and Geography, Chinese Academy of Sciences, Urumqi, China

State Key Laboratory of Desert and Oasis Ecology, Xinjiang Institute of Ecology and Geography, Chinese Academy of Sciences, Urumqi, China
e-mail: drwaheed@ms.xjb.ac.cn; zengfj@ms.xjb.ac.cn

H. Ali
Department of Agricultural Sciences/Engineering, Khwaja Fareed University of Engineering and Information Technology, Rahim Yar Khan, Pakistan
e-mail: habib.ali@kfueit.edu.pk

R. de Mello Prado (ed.), *Benefits of Silicon in the Nutrition of Plants*,
https://doi.org/10.1007/978-3-031-26673-7_19

productivity losses due to insects pests at around 10.8% (Mateos Fernández et al. 2022). These losses may increase in severe epidemic disease situations or due to sudden increases in insect pest population, which may have a major financial impact on the economy, possibly leading to a financial disaster. Therefore, it is essential to manage plant diseases and insect pests. Conventional management strategies include chemicals and different soil practices and are considered important to improve plant tolerance (Ghorbani et al. 2008; Noman et al. 2021). They include using several fertilizers, such as NPK, and micronutrients such as zinc, boron, magnesium, and silicon. Initially, silicon was categorized as a non-essential element to plant growth and development (Richmond and Sussman 2003). Recently, scientists discovered its role in reducing biotic and abiotic stresses in plants (Etesami and Jeong 2018; Seal et al. 2018; Bhat et al. 2019; Vasudev Meena et al. 2021). However, the fundamental process by which silicon improves plant tolerance to biotic and abiotic stress is still largely unknown. This may be due to numerous factors such as silicon properties or composition, its availability in soils, and the efficiency of plants to absorb it (Islam et al. 2020b).

Silicon is the second most abundant element in the soil. It is usually found in the form of mono-silicic acid (H_4SiO_4), and its appropriate absorption by plants from the soil solution occurs when the pH is <9 (Pang et al. 2019). Later, the absorbed silicic acid hydrolyzes in root cell gaps, cell walls, and leaf and stem epidermis in the form of silica gel or crystal. The assimilation of silicon varies according to plant genotype and variety (Tayyab et al. 2018). In addition, the ability to assimilate silicon is affected by its absorption in the soil and by ecological adaptations (Etesami and Jeong 2018). Several plants have a large quantity of silicon in their cells; however, the amounts vary among different species. Plants are silicon accumulators or non-accumulators (Hoffmann et al. 2020). Basically, cotyledon plants such as rice, rye, wheat, barley, and sugarcane absorb more silicon from the soil compared to dicotyledonous plants such as tomatoes and legumes (Sahebi et al. 2015). Numerous advantages in agriculture are related to the presence of silicon in the soil and its availability to plants, which may improve crop yield, productivity, and growth as well as the soil structure (Rastogi et al. 2019). Silicon decreases the rate of transpiration and enzyme activities, further reducing the abiotic stress on plants (Zhang et al. 2017).

Plant pathogens and insect pests are a serious problem in agricultural production. It may lead to significant reductions in crop quality and yield. Economically, silicon improves plant tolerance against viral, bacterial, and fungal diseases and insect pests (Table 19.1), thus playing a vital role in improving crop growth and yield. In this chapter, we focus on the mechanisms of silicon in mitigating plant pathogens and insect pests by exploring different physical, biochemical, enzymatic, molecular, and combined pathways. Moreover, the application of silicon in the management of different plant diseases and insect pests is also discussed.

Table 19.1 Role of silicon in plant defense against pathogens and insect pests

Host	Pest		Herbivore name/pathosystem	Type of defense			References
	Insect	Pathogen		Physical	Biochemical	Molecular	
Arabidopsis		√	Powdery mildew/*Erysiphe cichoracearum*	√	√	√	Fauteux et al. (2006) and Vivancos et al. (2015)
Banana		√	Black sigatoka/*Mycosphaerella fijiensis*	√	√		Kablan et al. (2012)
		√	Fusarium wilt/*Fusarium oxysporum*	√	√		Fortunato et al. (2012)
		√	Root rot/*Cylindrocladium spathiphylli*		√		Fortunato et al. (2012)
		√	Xanthomonas wilt/*Xanthomonas campestris*	√	√		Mburu et al. (2016)
Barrel medic	√		Beet armyworm	√			Korth et al. (2006)
Barley		√	Powdery mildew/*Blumeria graminis*	√			Wiese et al. (2005)
Bean		√	Angular leaf spot/*Pseudocercospora griseola*	√			Rodrigues et al. (2010)
Belle pepper		√	Phytophthora blight/*Phytophthora capsici*	√			French-Monar et al. (2010)
Bent grass		√	Dollar spot/*Sclerotinia homoeocarpa*	√	√		Zhang et al. (2006)
Bitter gourd		√	Powdery mildew/*Erysiphe* spp.		√		Ratnayake et al. (2016)
Cabbage	√		*Plutella xylostella*	√			Shoaib et al. (2018)
Capsicum		√	Anthracnose/*Colletotrichum gloeosporioides*	√	√		Jayawardana et al. (2016)
Cherry		√	Fruit decay/*Penicillium expansum, Monilinia fructicola*		√		Qin and Tian (2005)
Chestnut		√	Toink disease/*P. cinnamomi*	√			Carneiro-Carvalho et al. (2017)
		√	Chestnut blight/*Cryphonectria parasitica*	√	√		Carneiro-Carvalho et al. (2020)
Chinese cantaloupe		√	Fusarium root rot/*Fusarium* spp.	√	√		Liu et al. (2009)
		√	Postharvest pink rot/*Trichothecium roseum*	√	√		Guo et al. (2007)
Coffee		√	Leaf rust/*Hemileia vastatrix*	√			Carré-Missio et al. (2014)
		√	Root-knot Nematode/*Meloidogyne exigua*		√		Silva et al. (2010b)

(continued)

Table 19.1 (continued)

Host	Pest		Herbivore name/pathosystem	Type of defense			References
	Insect	Pathogen		Physical	Biochemical	Molecular	
Common bean		√	Anthracnose/*C. lindemuthianum*		√		Rodrigues et al. (2015)
	√		*Tetranychus urticae*, Koh	√			Gatarayiha et al. (2010)
Corn	√		*Rhopalosiphum maidis*	√			Moraes et al. (2005)
	√		*Busseola fusca*	√			Juma et al. (2015)
	√		*R. maidis*	√			Boer et al. (2019)
Cotton		√	Fusarium wilt/*Fusarium oxysporum*	√	√		Whan et al. (2016)
		√	Bacterial blight/*X. citri* subsp. *malvacearum*	√	√		Oliveira et al. (2012)
Creeping turf grass		√	Brown patch/*Rhizoctonia solani*	√	√		Zhang et al. (2006)
Cucumber		√	Crown and root rot/*Pythium ultimum*		√		Cherif et al. (1994)
		√	Powdery mildew/*Sphaerotheca fuliginea, Podosphaera xanthii*	√	√		Liang et al. (2005b)
		√	Cucumber mosaic disease/*Cucumber mosaic virus*			√	Holz et al. (2015)
	√		(*Gennadius* sp.) (Hemiptera: Aleyrodidae)	√			Correa et al. (2005)
	√		*T. urticae*, Koh	√			Gatarayiha et al. (2010)
Dragon fruit		√	Brown spot/*Neoscytalidium dimidiatum*		√		Tuan et al. (2019)
Egg plant	√		Thysanoptera: Thripidae	√			De Almeida et al. (2008)
Finger Millet	√		*Sesamia inferens* Walker	√		√	Jadhao et al. (2020)
Grass species	√		*Sitobion avenae*	√			Massey et al. (2006)
	√		*Spodoptera exempta*	√			Massey et al. (2006)
	√		*Schistocerca gregaria*	√			Massey et al. (2006)
Lettuce		√	Downy mildew/*Bremia lactucae*	√	√		Garibaldi et al. (2012)
Mango		√	Apical necrosis/*P. syringae* pv. *syringae*			√	Gutiérrez-Barranquero et al. (2012)

Melon		√	Bacterial fruit blotch/*Acidovorax citrulli*		√		Conceição et al. (2014)
		√	Powdery mildew/*P. xanthii*		√		Dallagnol et al. (2013)
		√	Bacterial fruit blotch/*Acidovorax citrulli*	√		√	Ferreira Preston et al. (2021)
Muskmelon		√	Pink rot disease/*T. roseum*		√		Li et al. (2009)
		√	Powdery mildew/*S. fuliginea*	√	√		Faure et al. (2012)
Oil palm		√	Basal stem rot/*Ganoderma boninense*	√			Najihah et al. (2015)
Pea		√	Brown rot/*M. pinodes*		√		Dann and Muir (2002)
Pearl Millet		√	Downy mildew/*Sclerospora graminicola*	√	√		Deepak et al. (2008)
Perennial ryegrass		√	Grey leaf spot/*Magnaporthe oryzae*		√		Rahman et al. (2015)
Potato		√	Dry rot/*Fusarium sulphureum*		√		Li et al. (2009)
			Late blight/*Phytophthora infestans*		√		Xue et al. (2021)

(continued)

Table 19.1 (continued)

Host	Pest		Herbivore name/pathosystem	Type of defense			References
	Insect	Pathogen		Physical	Biochemical	Molecular	
Rice		√	Blast/*Pyricularia oryzae, Magnaporthe oryzae*	√	√	√	Hayasaka et al. (2008) and Domiciano et al. (2015)
		√	Brown spot/*Bipolaris oryzae, Cochliobolus miyabeanus*	√	√	√	Dallagnol et al. (2011) and Van Bockhaven et al. (2015b)
		√	Grain discoloration/*Bipolaris oryzae*			√	Prabhu et al. (2012)
		√	Leaf scald/*Monographella albescens, Microdochium oryzae*	√	√		Tatagiba et al. (2016)
		√	Sheath blight/*R. solani*	√	√		Schurt et al. (2014)
		√	Bacterial blight/*X. oryzae* pv. *oryza*			√	Song et al. (2016)
		√	Root-knot Nematode/*M. graminicola*			√	Zhan et al. (2018)
	√		*Sogatella furcifera*	√			Jia et al. (2021)
	√		*Scirpophaga incertulas*	√			Han et al. (2018)
	√		*C. medinalis*	√			Han et al. (2018)
	√		*Nilaparvata lugens*	√			Han et al. (2018)
	√		*C. medinalis*	√			Liu et al. (2017)
	√		*N. lugens*	√			Yang et al. (2018)
	√		*S. incertulas* (Walker)	√			Jeer et al. (2017)
	√		*Chilo suppressalis* (Walker)	√			Hou and Han (2010)
	√		*Cnaphalocrocis medinalis* Guenée	√			Ye et al. (2013)
	√		*Nilaparvata lugens* (Stål)	√		√	Lin et al. (2022)
Rose		√	Powdery mildew/*P. pannosa*	√			Shetty et al. (2012)
Sorghum		√	Anthracnose/*C. sublineolum*	√	√		Resende et al. (2013a)

Soybean		√	Phytophthora stem and root rot/*P. sojae*			√	Guérin et al. (2014) and Rasoolizadeh et al. (2018)
		√	Rust/*Phakopsora pachyrhizi*		√		Cruz et al. (2014)
	√		*Bemisia tabaci* Genn. biotype B	√			Ferreira et al. (2011)
	√		*Helicoverpa armigera*	√			Alves et al. (2018)
Strawberry		√	Powdery mildew/*S. aphanis*	√	√		Kanto et al. (2004)
Sugarcane		√	Brown rust/*Puccinia melanocephala*	√	√		Ramouthar et al. (2016)
	√		*Diatraea saccharalis*	√			Meyer and Keeping (2001)
	√		*Eldana saccharina*	√			Keeping and Meyer (2002)
	√		*E. saccharina*	√			Keeping and Meyer (2006)
Sweet pepper		√	Bacterial wilt/*R. solanacearum*			√	Ramouthar et al. (2016)
Tahiti lime	√		*Diaphorina citri*	√			Ramírez-Godoy et al. (2018)
Tobacco		√	Viral infection/*Tobacco ringspot virus*			√	Zellner et al. (2011)
Tomato		√	Bacterial speck/*Pseudomonas syringae*		√		Andrade et al. (2013)
		√	Bacterial wilt/*Ralstonia solanacearum*			√	Ghareeb et al. (2011), Chen et al. (2014), and Jiang et al. (2019)
		√	Fusarium crown and root rot/*Fusarium oxysporum*	√			Huang et al. (2011)
	√		Whitefly (Homoptera: Aleyrodidae)	√			Inbar and Gerling (2008)
Tomato, Bitter guard		√	Root rot/*Pythium aphanidermatum*	√	√		Heine et al. (2007)

(continued)

Table 19.1 (continued)

Host	Pest		Herbivore name/pathosystem	Type of defense			References
	Insect	Pathogen		Physical	Biochemical	Molecular	
Wheat		√	Blast/*P. grisea*	√	√		Xavier Filha et al. (2011)
		√	Leaf blast/*P. oryzae*		√		da Silva et al. (2015)
		√	Leaf streak/*X. translucens*	√	√		Silva et al. (2010a)
		√	Powdery mildew/*B. graminis*	√	√	√	Moldes et al. (2016)
		√	Spot blotch/*Bipolaris sorokiniana*	√	√		Domiciano et al. (2010, 2013)
		√	Tan spot/*Pyrenophora tritici-repentis*	√	√		Dorneles et al. (2017) and Pazdiora et al. (2018)
	√		*S. graminum* Rond.	√			Basagli et al. (2003)
	√		*S. graminum* Rond.	√			Gomes et al. (2005)
	√		*Sitobion avenae* (F.) (Hemiptera: Aphididae)	√			Dias et al. (2014)
Zucchini squash		√	Powdery mildew/*E. cichoracearum, P. xanthii*	√	√		Savvas et al. (2009)

19.2 Mechanisms of Silicon Against Insect Pests and Pathogens

19.2.1 Formation of Physical Barrier

The concentration of silicon in plants results in the formation of a physical barrier. Through the apoplast movement, silicon is transported upward to leaf epidermis and xylem, where it accumulates in extracellular walls. Silver ions are absorbed by roots and actively transferred through evaporation, generating a second layer of epidermal cells under the cuticles (Shivaraj et al. 2022). Because of these two layers, *Pyricularia grisea* was limited and delayed in reaching rice leaves (Kim et al. 2002). Furthermore, spraying vine seedlings with potassium silicate (K_2SiO_3) at a rate of 17 mmol/L resulted in a considerable decrease in *Uncinula necator* absorption due to the formation of a silicon layer on the leaf epidermis (Bowen et al. 2019). The polymerization of K_2SiO_3 on the surface of watermelon, cucumber, and squash leaves decreased the penetration of *Sphaerotheca fuliginea* due to the formation of a physical barrier (Liang et al. 2005a, b; Menzies et al. 2019). Furthermore, a wax coating was discovered on the apical surface of coffee and soybean plants when grown in plastic tubes supplemented by silicon (Hammer et al. 2010). *Cercospora* leaf lesion was reduced and pathogen transmission was inhibited by the formation of a thick wax layer when plants were kept in silicon tubes for 7 months (Ziv et al. 2018).

In the early 1990s, studies reported the presence of silicon layers on the outermost epidermis. Research showed that silicon develops interactions with aromatic chemicals in epidermis walls, improving cell resistance and therefore reducing lesions and disease severity (Carver et al. 1987). Later, this same phenomenon was observed using a hydroponics system treated with silicon, which improved the quantity of phenolic acid in coffee leaves, resulting in a greater resistance to rust induced by *Hemileia vastatrix* (Rodrigues et al. 2011).

Silicon plays an important role in insect pest resistance by accumulating in the leaf cuticle, which helps to enhance the hardness and roughness of plant cells, thus reducing their edibility and assimilation by insect pests due to the formation of physical and mechanical barriers (Alhousari and Greger 2018). Silicon mediates leaf brashness and leads to a decrease in insect feeding quality; as a result, the insect's mouth parts may be severally damaged, thus reducing insect feeding efficiency, reproduction, and growth. Silicon mediates the hardness of cells in plant and decreases leaf degradation, enabling plants to resist insect pest attacks (Zargar et al. 2019). Also, silicon inhibits chlorophyll absorption in plants, which causes harmful effects on locusts (Leroy et al. 2019). The concentration of silicon increases the formation of a defensive mechanism related to morphological characters in coffee plants by producing a thicker midrib layer, which increases the resistance to insect attacks (Pozza et al. 2004). X-ray mapping and energy-dispersive X-rays (EDX) were used to examine the distribution of silicon deposits in sugarcane. The deposition of silicon in the root meristem and lateral root zone may have boosted plant tolerance to *Eldana saccharina* (Keeping et al. 2009). Furthermore, using EDX and electron microscopy (EM), researchers evaluated four grass varieties for their ability to absorb silicon and their efficiency against insect defense. The authors observed that a structure of spikes and phytoliths differs between and among species and is more valuable than silicon concentration in leaves in terms of leaf hardness. The authors later explained that when the supply of silicon is high, all types of grasses efficiently show an accumulation of silicon, evidencing that the physical hardness of leaves was more severe as the supply of silicon increased (Hartley et al. 2015). In silicon-treated rice rhizomes, EM showed dumbbell-shaped silicon structures with a ladder-like appearance that served as a physical barrier, providing protection against insect pests and plant hoppers (Dorairaj and Ismail 2017).

19.2.2 Biochemical Mechanisms

Plant Hormone Signaling Mechanisms
Plants developed a complex defensive system that consists of several barrier levels, including induced defense regulated by complex signaling channels, to prevent it from the actions of pathogens and insect pests. Through their ability to boost plant resistance, jasmonic acid (JA), salicylic acid (SA), and ethylene (ET) play important roles in plant defense responses and immunity (Islam et al. 2019; Noman et al. 2021). Salicylic acid is commonly used to treat hemi-biotrophic and biotrophic diseases. However, when it comes to treating necrotrophic pathogens, ET and JA are usually used (Zhang and Sonnewald 2017). Silicon has been shown to modulate plant stress responses by modulating transcription factors (TFs) through plant hormonal regulations (Manivannan and Ahn 2017; Wang et al. 2021). In response to injuries caused by plant pathogens and thrips, silicon treatment accumulates several phytohormones (Kim et al. 2014; Murali-Baskaran et al. 2021). In *Arabidopsis*, silicon-treated plant leaves infected by powdery mildew (*Erysiphe cichoracearum*) had an increased production of Et, JA, and SA, which induced tolerance against this pathogen (Ghanmi et al. 2004; Fauteux et al. 2006). Similar results occurred when tomatoes infected with *Ralstonia solanacearum* and rice blast caused by *Magnaporthe* oryzae were exposed to silicon, increasing transcription of ET and JA (Brunings et al. 2009; Ghareeb et al. 2011). As a result of silicon improving JA and SA production, rice became more resistant to *Cochliobolus miyabeanus* (Van Bockhaven et al. 2015a).

In tolerant phenotypes, silicon increases the transcription of gene encoding of enzymes involved in resistance to powdery mildew. However, the transcription of defense-related genes and SA synthesis were significantly reduced, indicating that silicon-induced resistance may not be mediated by SA-dependent mechanisms alone (Vivancos et al. 2015). *Peptidyl Arginine Deiminase 4* (*Pad4*) and *Enhanced Disease Susceptibility 1* (*EDS1*) genes are involved in the regulation of SA synthesis pathways (Cui et al. 2017). Silicon-treated *Arabidopsis* plants showed more resistance to *Golovinomyce scichoracearum.* These plants also expressed a higher level of *APR1*, *PAD4*, and *EDS1*, as well as three SA-induced resistance genes

(*PR1, PR2,* and *PR5*). In addition, *sid2* and *pad4 Arabidopsis* mutants showed a reduced disease progression curve in response to silicon treatment (Vivancos et al. 2015). As a result of SA/*WRKY* module activation, *NPR1* protein must be present for the *PR* gene expression to occur. To illustrate this, silicon increased the expression of *WRKY1* in tomato plants infected with *R. solanacearum* (Ghareeb et al. 2011).

Silicon can change JA-mediated transcriptional defense responses by using a wide range of priming chemicals (Wang et al. 2017). This could be related to an increased silicon buildup caused by JA. Silicon modulates proteolysis ligase increases in response to pathogen infections, indicating that Si may play a role in defense response signaling (Carvalhais et al. 2017; Noman et al. 2019). Ethylene and JA signaling activates ET marker genes and TF-*ACC Oxidase 1* (*ACCO1*), which is required for ET biosynthesis, and the interaction with *Tomato stress-responsive factor 1* (*TSRF1*) (Pirrello et al. 2012). A significant upregulation of genes such as *ACCO* and *TSRF1* occurred in tomato plants attacked by *R. solanacearum* when silicon treatment was applied. Ethylene and SA played an active role in this upregulation (Ghareeb et al. 2011). Similarly, silicon enhanced the expression of *Plant Defensin 1.2* (*PDF1.2*) in *Arabidopsis* in response to *Botrytis cineraria* invasion under cadmium stress conditions (Cabot et al. 2013). Moreover, *C. miyabeanus*-infected plants showed a better resistance to brown spot disease compared to the wild type (WT) when treated with *OsEIN2* (*Ethylene-insensitive protein 2*) (Van Bockhaven et al. 2015a).

Plant defense responses to sucking insects are heavily dependent upon the SA signaling pathway; however, the pathway's role in silicon-enhanced rice resistance is yet to be explored. Rice resistance to brown plant hopper (BPH), *Nilaparvata lugens* (*Stål*), was mediated by SA accumulation, defense-related enzyme activity, and gene expression. Host preference for BPH was significantly influenced by silicon application; honeydew secretion was significantly reduced, and oviposition and hatching were significantly inhibited. A significant increase in SA content and transcription levels of *BPH3, ICS1*, and *PAL4* occurred in silicon-treated plants compared with untreated plants when BPH infestation was present. There was also an increase in defense responses. *OsLsi1* mutants, however, showed a greater susceptibility to BPH and a minimal defense response. Salicylic acid and silicon applied simultaneously to WT plants led to resistance to BPH but did not exert any noticeable effect on *OsLsi1* mutants, antisense, and *as-npr1* plants. As a result of a defense priming that involves SA signaling, silicon improves rice defenses against BPH (Lin et al. 2022). Furthermore, researchers found a rapid increase in ET content 15 minutes after spraying with silicon on plants infected with potato late blight (*Phytophthora infestans*). A decrease in SA and an increase in JA occurred in plants following spraying with silicon. A pretreatment with Me-JA enhanced the resistance to *P. infestans*, while a pretreatment with DIECA, an inhibitor of JA synthesis, decreased resistance to late blight via silicon-mediated mechanisms (Xue et al. 2021).

Enzyme Synthesis Mechanisms

Silicon increases the activity of enzymes {Peroxidases (PODs), chitinase (CHT), β-1, 3-glucanase, polyphenol oxidases (PPOs), ascorbate POD (aPOD), phenylalanine ammonia-lyase (PAL), glutathione reductase (GR), superoxide dismutase

(SOD), catalase, lipoxygenase, and glucanase)}, which are linked to disease tolerance during symbioses (Tripathi et al. 2021). For example, PAL is involved in the production of secondary metabolites in plants and is thought to be necessary for resistance to diseases (Islam et al. 2018a, e). A higher PAL movement induced by silicon results in a greater accumulation of total soluble phenol derivatives and lignin-thioglycolic acid in coffee (Silva et al. 2010b) and banana (Fortunato et al. 2012) plant leaves, resulting in lower pathogen infection rates. A major enzyme involved in the oxidation process, PPO is found in chloroplasts, mitochondria, and other sub-cellular vesicles in both bound and unbound forms. The role of this enzyme in disease resistance has also been established (Prasannath 2017). The production of silicon-mediated lignins could improve interactions between hosts and pathogens, which is important for CHT and POD (Wang et al. 2017).

Inhibition of defense-associated enzyme-mediated gene expression is achieved through silicon application. When silicon was applied to perennial ryegrass plants (*Lolium perenne* L.) infected with gray spot (*M. oryzae*), lipoxygenase (LOXa), PALa, and PALb were significantly up-regulated (Rahman et al. 2015). Jasmonic acid-induced responses to herbivores can be stimulated or primed by silicon in rice plants, resulting in increased PPO and POD activity (Ye et al. 2013). The function of silicon in pathogen infection reduction by increasing defense enzyme activity has been demonstrated for a variety of pathosystems (Islam et al. 2020b). Defense-associated enzymes are responsible for the increase in antimicrobial compounds. Silicon promotes the activation of defensive enzyme mechanisms in plants, thereby reducing disease severity through the formation of antimicrobial substances and their storage in plants following pathogen invasion (Wang et al. 2017; Islam et al. 2020b). As a result of the application of silicon during *Cercospora sojina* contamination, the bottom leaves became less susceptible to the disease, as the oxidative activity of enzymes in the leaves reduced (Telles Nascimento et al. 2016).

In higher plants, antimicrobial chemicals are important components for disease control. Several studies showed that in microbial infections silicon increases the formation of antimicrobial molecules (Etesami 2018; Ahanger et al. 2020; Verma et al. 2021). Moreover, these antimicrobial compounds cause an increase in lignin-associated polyphenols, as well as PAL and PPO, which become active in response to bacteria (Ye et al. 2013; Rahman et al. 2015). PAL is responsible for the conversion of L-phenylalanine into trans-cinnamic acid, which contributes to disease resistance (Hyun et al. 2011). In rice, silicon increases the concentration of lignin and carbohydrate compounds in the epidermal cell wall (Ng et al. 2019). Similarly, treatment with silicon enhances plant disease resistance by increasing the amount of double phenolic compounds (Ahammed and Yang 2021). Silicon also promotes the production of flavonoids and improves the resistance of wheat and rose plants to diseases (Shetty et al. 2012; da Silva et al. 2015). In addition, silicon enhances the accumulation of lignin and phenolics in different pathosystems in order to increase disease resistance, such as in cucumber/dumping off/*Pythium ultimum*, rice blast/*P. oryzae*, wheat/powdery mildew/*E. cichoracearum*, *Arabidopsis*/powdery mildew/*E. cichoracearum*, etc. (Cherif et al. 1994; Bélanger et al. 2003; Ghanmi et al. 2004; Xavier Filha et al. 2011).

The role of phytoalexins in plant protection during infection by pathogens is vital. In the cucumber/*P. xanthii* pathosystem, pathogen attack is reduced by the production of higher phytoalexins with a silicon treatment (Fawe et al. 1998). Similarly, silicon increases rice resistance to blast disease by triggering the production of phytoalexins (Rodrigues et al. 2005). In the perennial ryegrass/*M. oryzae*-pathosystem, silicon boosts acid levels (chlorogenic and phenolic) and flavonoids. It also enhances the lipoxygenase and related PAL-encoding gene levels, consequently developing a strong resistance against gray leaf spot (Rahman et al. 2015).

McNaughton and Tarrants (1983) initially found enhanced defenses against insect and pest attacks in African grasses. Their study has now focused on more advanced findings (i.e., defense mechanism through microbial and biological levels). Further research showed that silicon increases plant resistance through an increased production of allelochemicals in response to feeding by insects, thereby reducing the insect's ability to feed and grow (Costa et al. 2009, 2011; Gajger and Dar 2021). Likewise, silicon use improves genes transcription levels and enzyme activities related to defense mechanisms, thus boosting the production of defense-related metabolites (phytoalexins and momilactones phenolics) (Fawe et al. 1998; Manivannan and Ahn 2017). Similarly, the resistance of wheat plants to greenbug (*Schizaphis graminum*) improved because of silicon-mediated production of PAL and PPO enzymes (Gomes et al. 2005; De Moraes 2005; Costa et al. 2007).

In response to insect attacks, plants release some volatiles, which are an important part of defense system/pathway (Hussain et al. 2019; Qasim et al. 2020). These volatiles can be enhanced by silicon supplementation. The volatiles of rice treated with silicon improved, for example, α-bergamotene and β- sesquiphellandrene, hexanal 2-ethyl, and cedrol emission in reaction to *C. medinalis* attacks (Liu et al. 2017). Thus, high *Microplitis* and *Trathala flavoorbitalis* extraction attracted more female parasitoids. Furthermore, in cucumber, silicon treatments enhanced the production of secondary metabolites that improved the attraction of a predator of *Epiphyas postvittana* called *Dicranolaius bellulus* (Kvedaras et al. 2010).

Silicon fertilization enhanced the accumulation of callose in rice plant tissues based on EPG findings (electrical penetration graphs); it reduced the plant hopper (*N. lugens*) phloem ingestion period, thus improving resistance in rice plants against insects. Silicon fertilization also increased malondialdehyde (MDA) levels. Due to scavenging reactive oxygen species (ROS), activating PAL and PPO activities, and catalyzing the conversion of phenolics into quinines, it inhibited insect growth by reducing the plant hopper's ability to digest plants (Yang et al. 2017).

19.2.3 Biochemical Mechanism and Physical Barrier: A Joint Action

Several studies have reported a joint action of physical obstacles and defense-related enzyme production in plants. There is a complex mechanism responsible for disease reduction in silicon-mediated rice plants. It is not only limited to physical barriers (deposition of silicon below the cuticle), but also includes biochemical reactions (production of phenolic compounds) (Cai et al. 2008). Another study reported that silicon supplementation to sorghum leaves inoculated with *Colletotrichum sublineolum* showed silicon deposits on the infection spot and minute acervuline. Moreover, silicon-treated leaves showed a higher concentration of PODs, PPO, and anthocyanin. As a result of a joint physical and biochemical action, the infectivity of *C. sublineolum* decreased due to silicon-stimulated joint actions (Resende et al. 2013b).

Molecular Mechanisms

Through a series of physiological and biochemical reactions, including signal transduction, silicon promotes genes related to protection and defense in host plants. During the primary response and after elicitation, silicon regulates intracellular signaling systems related to cell structural changes, hormone production, hypersensitivity, and synthesis of antimicrobial compounds (Yang et al. 2022). Many studies have investigated proteomics and transcriptomics to reveal silicon resistance response to numerous pathosystems (Kurabachew et al. 2013; Wang et al. 2013; Chen et al. 2014; Liu et al. 2009,2014; Han et al. 2015).

Silicon use can protect tomatoes from *R. solanacearum* by triggering TFs and genes related to resistance and stress, such as *ferrite*, *WRKY1*, which are proteins abundant in trehalose phosphatase in late embryogenesis (Ghareeb et al. 2011). Similar effects occurred in rhizobacteria and silicon-treated tomato stems inoculated with *R. solanacearum*. Upregulated expression happened in most of the genes involved in the synthesis of defense proteins and metabolism. The majority of genes that were downregulated were related to photosynthesis and lipid metabolism (Kurabachew et al. 2013). In cucumber, increased *proline-rich protein* (*PRP*) expressing genetic material was detected while encoding silicon-mediated systemic acquired resistance. Within epidermal cells, *PRP* enhances cell wall strength to combat invading fungi (Arana et al. 2009). The expression of *citinase-II*, *polygalacturonase-inhibiting protein,* POD, and glutamate is related to pathogen-released virulence factors that disrupt the host defenses and assist in host invasion. In tomato plants, their expression is suppressed when the pathogen (*R. solanacearum*) that attacked plants is treated with silicon (Ghareeb et al. 2011). The presence of silicon affected the levels of 26 amino acid proteins in tomatoes infected with *R. solanacearum*, implying that silicon-mediated defense is linked to changes in amino acid levels (Chen et al. 2014). Silicon may be able to reverse several transcribed changes caused by disease infection. For example, in *Arabidopsis*, a fungal infection by *E. cichoracearum* causes a change in the expression of almost 4000 genes. With a silicon treatment, the magnitude of downregulated genes related to primary metabolism was low (Ghanmi et al. 2004; Fauteux et al. 2006). About 900 genes in wheat leaves transformed in reaction to a disease caused by *Blumeria graminis f.* sp. *tritici*. Pathogens changed several genes in silicon-treated plants, implying that silicon plays a substantial role in reducing disease infestation (Chain et al. 2009). Another transcriptomic study found that silicon treatment reduced the impact of *M. oryzae* disease on rice (Brunings et al. 2009).

Recent studies indicate that induced production of reactive oxygen species (ROS) and pathogenesis-related gene expression in potato plants treated with silicon enhance resistance to late blight (*P. infestans*). The silencing of *StETR1, StEIN2, StAOS, StOPR3*, and *StNPR1* genes significantly reduced the incidence of late blight resistance, while *StCOI1* and *StSID2* genes did not modify the incidence of disease resistance. As a result of *StETR1, StEIN2, StOPR3*, and *StHSP90-VIGS* plants being attenuated by silicon, the accumulation of JA/JA-Ile was significantly low, while *StCOI1, StSID2*, and *StNPR1-VIGS* plants were unaffected. In conclusion, these results suggest that silicon may be an alternative to fungicides to control late blight and that silicon-mediated resistance requires *StHSP90-* and *StNPR1*-dependent changes (Xue et al. 2021).

In dwarf coffee seedlings in plant growth chambers silicon significantly improved their growth and development. In addition to suppressing disease and insect pests, silicon enhanced physiology (e.g., stronger root, stem, and leaf structure, higher photosynthetic efficiency, abundant organic matter accumulation, improved root activity, efficient mineral absorption, and greater activity of various enzymes) and regulated defense genes (*CaERFTF11* and *CaERF13*). Because it enhances crop varieties for an optimal production and mitigates plant diseases and insect pests, silicon could potentially contribute to global food security and safety. The use of silicon in agricultural production is therefore sustainable (Yang et al. 2022).

Silicon nanoparticles may also potentially play a role in activating the root's defensive response. Silicon nanoparticles increased the lignification of root cells in fenugreek (*Trigonellafoenum graecum*). Also, the resistance-associated gene (*tfgd 1*) is expressed as a result of this (Nazaralian et al. 2017). Furthermore, an increased expression of lignin synthesizing genes increases the girth and breaking strength of roots treated with silicon in rapeseed (*Brassica napus*) (Kuai et al. 2017). Furthermore, rice plants treated with silicon also show a better defense against the phloem-feeding brown plant hopper (*N. lugens*) through callose deposition regulation. Silicon-treated plants also showed the highest level of upregulation of the cellulose synthesis gene, *OsGSL1*, and the lowest level of upregulation of the cellulose hydrolyses gene, *Gns5* (Yang et al. 2018). To determine the effects of silicon on rice resistance to leaf folder (LF), *Cnaphalocrocis medinalis* (*Guenée*) (Lepidoptera: Pyralidae), a silicon transporter mutant *OsLsi1* and WT rice were treated with or without silicon. Plant growth significantly improved due to the application of silicon on WT plants, the level of expression of *OsLsi1* increased, and the leaves and roots accumulated more silicon, whereas mutant plants showed only marginal or no effects. Silicon-treated WT plants produced significantly higher levels of *OsLOX, OsAOS2, OsCOI1a, OsCOI1b*, and *OsBBPI* transcripts, as well as catalase, superoxide dismutase, peroxidase, and polyphenol oxidase activities, than untreated WT plants during LF infestation. In contrast, mutant *OsLsi1* plants showed an increased susceptibility to LF and minimal responses to silicon application despite higher levels of defense-related enzymes and JA-dependent genes. Rice plants lacking the silicon transporter *Lsi1* cannot induce anti-herbivore defenses and are susceptible to silicon-enhanced resistance, as results have suggested (Lin et al. 2019). Moreover, researchers recently discovered three genes (*EcLsi1s, EcLsi2s,* and *EcLsi6s*) encoding silicon transporters that are involved in silicon uptake. A transcriptome assembly method and a gene identification and characterization method were used to identify and characterize ten different Si transporter genes of finger millet. As the phylogenetic analysis revealed, *EcLsi1* and *EcLsi6* are homologs and *EcLsi2* and *EcLsi3* are also homologous. As silicon transporters, *EcLsi1* and *EcLsi6* belong to the family of *NIP2s* (*Nod26-like major intrinsic proteins*), whereas *EcLsi2* and *EcLsi3*, which are efflux silicon transporters, belong to the uncharacterized *arsB-like anion* transporter family. In addition, phylogenetic and topological analyses indicate that *EcLsi2* and *EcLsi3* evolved either by fusion or duplication from *EcLsi1* and/or *EcLsi6*. The structural differences between these silicon transporters suggest that they might have different abilities to absorb silicon. They are localized in the plasma membrane. When non-silicon-treated finger millet plants are infected with pink stem borer at 72 hpi (hours of post infestation), silicon application induces a synergistic defense mechanism by enhancing transcription levels of silicon transporter genes (*EcLsi1, EcLsi2,* and *EcLsi6*). For the purpose of developing strategies for insect pest management, this study provides insights into how molecular defense mechanisms work (Jadhao et al. 2020).

19.3 In Vivo and In Vitro Application of Silicon for Disease and Insect Pest Management

19.3.1 Role of Silicon in Viral Disease Management

Viruses are believed to be responsible for more than one-third of plant diseases (Islam et al. 2017, 2020a). Virus populations and virus spread can be slowed down by controlling insect vectors (Islam et al. 2018b, c, d) with a variety of chemicals (Jaleel et al. 2020) and natural metabolites (Islam et al. 2018a, e). Such chemical treatments will not completely stop plant virus diseases from spreading. The use of chemicals has serious harmful consequences for humans and the environment (Islam et al. 2021; Noman et al. 2021; Siddiqui et al. 2022). As a result, finding a different way to reduce viral threats is critical. The development of *tobacco ring spot virus* (*TRSV*) systemic symptoms was stopped in plants treated with 1.0 mM K_2SiO_3. Furthermore, it was proposed that tobacco plants may benefit from a leaf application of silicon to boost its resistance to *TRSV* (Zellner et al. 2011). Cucumber plants treated with silicon to combat the *cucumber mosaic virus* (*CMV*) experienced changes in the expression of defense-related genes (Holz et al. 2015). Moreover, the use of silicon considerably reduced the intensity of *papaya ring spot virus* (*PRSV*). The treatment with silicon mediated the expression of genes that were largely connected to virus infections (Elsharkawy and Mousa 2015). A recent study claimed that *Sogatella furcifera*, a white-backed plant hopper, could be prevented from transmitting *southern rice black-streaked dwarf virus* (*SRBSDV*) by silicon treatments applied to rice plants (Jia et al. 2021).

19.3.2 Role of Silicon in Bacterial Disease Management

In both dicotyledonous and monocotyledonous host plants, positive associations between bacterial disease severity and silicon have been discovered. Plant resistance to bacterial pathogens improved with silicon fertilization (both liquid and solid). In the soil, solid calcium silicate ($CaSiO_3$) fertilizers are utilized, whereas liquid K_2SiO_3 or sodium silicate (Na_2SiO_3) fertilizers are applied on leaves or via irrigation (Zellner and Datnoff 2020). The combination of biological agents (biocompatible compounds) and Si can produce a high resistance to bacterial infections. Tobacco and tomato plants are less likely to be infected by *R. solanacearum* when *Bacillus pumilis* and chitosan are present (Kiirika et al. 2013; Kurabachew and Wydra 2014). Silicon does not affect the growth of bacterial pathogens in vitro. Silicon ($CaSiO_2$) did not affect *A. citrulli* growth when tested in vitro (Ferreira 2009). Similarly, $CaSiO_3$ did not prevent the growth of *Xanthomonas citri sub* sp. *malvacearum* in growth media with any dilution of silicon (Oliveira et al. 2012). Four rice cultivars supplemented with silicon slag were resistant to bacterial blight in vivo to varying degrees (*X. oryzae pv. oryza*). Also, silicon significantly reduced the length of bacterial lesions (Rodrigues and Datnoff 2005). Furthermore, silicon-supplemented rice plants had a less severe bacterial blight (Song et al. 2016). Moderately-resistant tomatoes grown in a K_2SiO_3 solution showed a reduction in disease incidence when exposed to *R. solanacearum* (Diogo and Wydra 2007).

Silicon as peat can improve the tolerance to bacterial wilt diseases in moderately resistant plants. In an in-vivo experiment, 15 kg per 100 mL of silicon fertilizer significantly reduced the prevalence of *R. solanacearum*, thereby reducing wilt disease effects (Ayana et al. 2011). Bacterial leaf spot caused by *Xanthomonas* spp. was efficiently reduced by using K_2SiO_3 at 40 and 50 g L^{-1} (Ji et al. 2006). When silicon (2.95 g kg^{-1}) was applied to sweet pepper infected by bacterial wilt (*R. solanacearum*), it boosted the retention time (33.6%) and suppressed the bacterial wilt (98%). The area under the disease progress curve was 93.7% (Alves et al. 2015). To control bacterial blotch, researchers used different quantities of SiO_2 on melon plants affected by *A. citrulli*. SiO_2 was added to the soil (3.0 g kg^{-1}) and reduced the disease curve and disease index while also enhancing the maturation phase (Ferreira et al. 2015). A substrate with 1.41 g kg^{-1} Ca_2SiO_4 and leaf crop-dusting with 17 mM silicon (Ca_2SiO_4) reduced the disease curve and bacterial spots. It also prevented *A. citrulli* infection of melon plants for 29 days (Conceição et al. 2014). The mango pathosystem was substantially less affected by trees supplemented with silica gel (Gutiérrez-Barranquero et al. 2012). Furthermore, silicon at a dilution of 2 ml/L was applied (leaf) to tomato plants to reduce the symptoms caused by the bacteria *P. syringae* pv. tomato (Andrade et al. 2013). By fertilizing melons with silicon slag, the bacterial fruit blotch disease curve area of melons (*A. citrulli*), the disease index, and disease incidences on fruits were significantly reduced by up to 10%, 14%, and 12%, respectively, at harvesting (Ferreira Preston et al. 2021). A 25% reduction in disease index occurred when silicon was applied to the susceptible tomato genotype HYT, which showed a significant suppression of bacterial wilt (*R. solanacearum*) (Fan et al. 2018).

19.3.3 Role of Silicon in Fungal Disease Management

As a result of silicon use in a variety of crops, fungal infections in soil and air can be reduced. When a silicon-rich soil was applied to wheat fields, the incidence rate of *E. graminis* (powdery mildew) and *Septoria nodorum* (leaf and glume blotch) decreased significantly (Leusch and Buchenauer 1989). Air spraying of K_2SiO_3 at 1.7 mM on grape leaves reduced powdery mildew (*U. necator*) by more than 60% (Bowen et al. 2019). The addition of silicon to soil supplementation increased the resistance of wheat to *B. graminis* (powdery mildew), *Phaeosphaeria nodorum* (leaf spot), *M. graminicola* (septoria leaf spot), and *Puccinia recondita* (brown rust) (Rodgers-Gray and Shaw 2004). Applying silicon to rice plants reduced the incidence of neck blast (*P. oryzae*) by 40% (Seebold et al. 2004). When sweet cherry plants were sprayed with a 1% sodium metasilicate solution, *Penicillium expansum* (blue mold decay) and *Monilinia fructicola* (brown mold decay) were reduced by 63% and 87%, respectively (Qin and Tian 2005).

On a soil fertilized with K_2SiO_3, strawberry powdery mildew (*S. aphanis*) decreased by 86% and 60% in the first and second year, respectively (Kanto et al. 2004). The incidence and severity of bean anthracnose were reduced by silicon (Moraes et al. 2006). By treating the soil with silicon (1.7 mM), the powdery mildew disease (*B. graminis f.* sp. *tritici*) in wheat was reduced by 80% (Guével et al. 2007). Na_2SiO_3 significantly reduced the incidence of disease incidence of angular leaf spot (*Pseudocerco sporagriseola*) by 42 and 30% in bean plants (Rodrigues et al. 2010). Silicon spraying on wheat/*P. oryzae* plants reduced not only the lesion number but also the disease progress curve of blast disease in the wheat/*P. oryzae* interaction (Xavier Filha et al. 2011; da Silva et al. 2015). Rice plants affected by *B. oryzae* were less likely to have brown spots and a low disease progress curve when a solution of 2 mM Si was applied (Dallagnol et al. 2011). Also, silicon almost eradicated Asian rust symptoms caused by *Phakopsora pachyrhizi* to soybean plants (Arsenault-Labrecque et al. 2012). Common bean plants were sprayed with 2 mM K_2SiO_3 leaf spray to reduce the severity of *C. lindemuthianum* (anthracnose) disease by 34% (Polanco et al. 2014). A similar application of $CaSiO_3$ at a rate of five tons per hectare inhibited *M. oryzae* activity in ryegrass, decreasing disease prevalence and brutality, respectively, by 39.5% and 47.3% (Rahman et al. 2015).

Watermelon fusarium wilt was suppressed by silica nanoparticles coated with chitosan or without it. Disease progress curve measurements showed that mesoporous silica nanoparticles improved the watermelon's innate defense mechanisms by 40% and 27%, respectively (Buchman et al. 2019). A similar result was observed in the laboratory with the application of 4 and 6 g/L silicon for the inhibition of fungal linear growth (black root rot caused by *Rhizoctonia solani* and *F. solani*, respectively). There was a significant reduction in disease incidence and severity under field conditions at all concentrations tested. K_2SiO_3 and $CaSiO_3$ combined as a treatment for soil and leaf spray resulted in the greatest reduction of disease incidence and severity: 92.7% and 91.9% and 91.7% and 91.1%, respectively (Abd-El-Kareem et al. 2019).

Incorporating silicon into soil and leaf-spraying plants, a recent study examined the ability of silicon to provide resistance against four fusarium head blight-infected cultivars. With 1.50 g/kg of soil and 1.7 mM, there was a reduction in spike and spikelet bleaching (19.4% and 17.5%, respectively; on bread: 14.6% and 12.3%, respectively; on durum at 28 dpi), and the effect increased with time until 28 dpi. According to the results, silica gel-treated plants suffered from lower levels of disease incidence and severity than those sprayed with leaf spray (1.7 mM). Species-specific silicon effects were observed at 21 and 28 dpi. At 14, 21, and 28 days, both susceptible and moderately resistant cultivars treated with silicon were resistant (Sakr 2021).

19.3.4 Role of Silicon in Insect Pest Management

Silicon improves resistance not only to insect pests such as white-backed plant hoppers, green leafhoppers, and brown planthoppers, but also to other pests such as spider mites. According to a review, silicon improves crop resistance against yellow rice borer, shot borer, and stalk borer in rice, wheat, maize, and sugarcane (Islam et al. 2020b; Verma et al. 2021). In addition, silicon affects insects biologically, behaviorally, and physiologically. In *Arabidopsis,* sharp trichomes contain silica, which affects insect movement, settlement, and food intake (Handley et al. 2005). When rice leaf folders (*Cnaphalocrocis medinalis*) were fed on silicon-supplied rice plants, they experienced high mortality and population reduction (Han et al. 2015). Silicon-treated rice plants, on the other hand, induced phloem-feeding insects to shift their behavior and physiological state. Brown plant hoppers (*N. lugens*) displayed poor fertility, intensity, and amount of honeydew excretion, as well as a lower settlement rate when fed on silicon-treated plants (Yang et al. 2017). When *Spodoptera frugiperda* was fed on silicon-supplied leaves, its fecundity rate significantly reduced (Alvarenga et al. 2017). Furthermore, a high concentration of silicon in roots significantly inhibited the development and feeding rate of the sugarcane root-eating bug *Dermolepida albohirtum* (Frew et al. 2017). On the other hand, extensive nitrogen use in corn hybrid plants leads to a reduction in silicon levels and an increased crop damage by the European corn borer *Ostrinia nubilalis* (*Hübner*) larvae (Disi et al. 2018).

Plants inoculated with silicon become hard, making it difficult for insects to chew, absorb, and digest food, posing a hazard to their survival. When *Tuta absoluta* was fed and reared on silicon-supplied tomato plants, it showed a decrease in body weight and pupae and larval survival. Furthermore, silicon-mediated morphological alterations in *T. absoluta* larvae mandibles and midgut were identified (dos Santos et al. 2015). Using two application methods (leaf and soil drenching), two sources of silicon were applied to maize plants (SiO_2 and K_2SiO_3). Fall armyworm (*S. frugiperda*) mortality percentage and development period were significantly increased by leaf applications of SiO_2 and K_2SiO_3, but larval and pupal biomass decreased. Likewise, both silicon sources reduced lipase activity and fecundity of larvae, while prolonging the lifespan of adults (Haq et al. 2021). Also, biological assays found that silicon-treated plants fed less frequently on BPH species (*N. lugens*) after infestation than non-treated plants. Compared with non-treated plants, silicon-treated plants show a fast and strong callose deposition following BPH infestation (Yang et al. 2018).

19.4 Concluding Remarks

Plant growth and development depend on silicon, a pseudo-essential nutrient that is found mostly in mineral deposits and rocks, and that therefore is not easily available to plants in their natural conditions. Recent studies of silicon accessibility in soil and plant uptake, transport, and accretion have paved the way for a deeper understanding of silicon's beneficial effects in plants, including its potential to improve disease tolerance and leniency. Even though thousands of studies have investigated the physical, metabolic, and molecular pathways of silicon-mediated resistance, there is still much to learn about these processes. Further research on growth regulator signaling and transcriptome regulations in plant defense against disorders and insects is needed, with an emphasis on silicon use. Likewise, it is not known how silicon carriers react to infections, pest attacks, or other biotic stimuli. As a result, further research is essential to find and characterize novel silicon transporters and pathways. Because the plants' capacity to absorb silicon varies greatly in response to diseases and insect attacks, scientists must rely on protracted field surveys rather than greenhouse and laboratory testing to better understand individual defensive mechanisms. Moreover, there is limited information on the use of silicon, insecticides, silicon-based nanomaterials, inhibitors, and other antimicrobial agents in combination to manage insect pests and infections. In our opinion, silicon treatment should be incorporated into integrated pest management programs rather than being the primary strategy for managing diseases and insects.

References

Abd-El-Kareem F, Elshahawy IE, Abd-Elgawad MMM (2019) Effectiveness of silicon and silicate salts for controlling black root rot and induced pathogenesis-related protein of strawberry plants. Bull Natl Res Cent 43:91. https://doi.org/10.1186/s42269-019-0139-1

Ahammed GJ, Yang Y (2021) Mechanisms of silicon-induced fungal disease resistance in plants. Plant Physiol Biochem 165:200–206. https://doi.org/10.1016/j.plaphy.2021.05.031

Ahanger MA, Bhat JA, Siddiqui MH et al (2020) Integration of silicon and secondary metabolites in plants: a significant association in stress tolerance. J Exp Bot 71:6758–6774. https://doi.org/10.1093/jxb/eraa291

Alhousari F, Greger M (2018) Silicon and mechanisms of plant resistance to insect pests. Plants 7:33. https://doi.org/10.3390/plants7020033

Alvarenga R, Moraes JC, Auad AM et al (2017) Induction of resistance of corn plants to *Spodoptera frugiperda* (J E Smith, 1797) (Lepidoptera: Noctuidae) by application of silicon and gibberellic acid. Bull Entomol Res 107:527–533. https://doi.org/10.1017/S0007485316001176

Alves AL, da Costa ACT, Mottin MC, Pietrowiski V, Júnior JBD, Nunes M (2018) Evaluation of the insect Pest population dynamics in common bean cultivars in relation to the foliar fertilisation with potassium silicate. J Exp Agric Int 27:1–12. https://doi.org/10.9734/jeai/2018/44031

Alves AO, Santos MMB, Souza LJN et al (2015) Use of silicon for reducing the severity of bacterial wilt of sweet pepper. J Plant Pathol 97:419–429. https://doi.org/10.4454/JPP.V97I3.002

Andrade CCL, Resende RS, Rodrigues FÁ et al (2013) Silicon reduces bacterial speck development on tomato leaves. Trop Plant Pathol 38:436–442. https://doi.org/10.1590/S1982-56762013005000021

Arana DM, Prieto D, Román E et al (2009) The role of the cell wall in fungal pathogenesis. Microb Biotechnol 2:308–320

Arsenault-Labrecque G, Menzies JG, Bélanger RR (2012) Effect of silicon absorption on soybean resistance to *Phakopsora pachyrhizi* in different cultivars. Plant Dis 96:37–42. https://doi.org/10.1094/PDIS-05-11-0376

Ayana G, Fininsa C, Ahmed S, Wydra K (2011) Effects of soil amendment on bacterial wilt caused by ralstonia solanacerum and tomato yields in Ethiopia. J Plant Prot Res 51:72–76. https://doi.org/10.2478/v10045-011-0013-0

Basagli MAB, Moraes JC, Carvalho GA, Ecole CC, de Gonçalves-Gervásio RCR (2003) Efeito da aplicação de silicato de sódio na resistência de plantas de trigo ao pulgãoverde schizaphis graminum (Rond.) (Hemiptera: Aphididae). Neotrop Entomol 32:659–663. https://doi.org/10.1590/S1519-566X2003000400017

Bélanger RR, Benhamou N, Menzies JG (2003) Cytological evidence of an active role of silicon in wheat resistance to powdery mildew (*Blumeria graminis* f. sp. tritici). Phytopathology 93:402–412. https://doi.org/10.1094/PHYTO.2003.93.4.402

Bhat JA, Shivaraj SM, Singh P et al (2019) Role of silicon in mitigation of heavy metal stresses in crop plants. Plants 8:71. https://doi.org/10.3390/plants8030071

Boer CA, Sampaio MV, Pereira HS (2019) Silicon-mediated and constitutive resistance to Rhopalosiphum maidis (Hemiptera: Aphididae) in corn hybrids. Bull Entomol Res 109:356–364. https://doi.org/10.1017/S0007485318000585

Bowen P, Menzies J, Ehret D et al (2019) Soluble silicon sprays inhibit powdery mildew development on grape leaves. J Am Soc Hortic Sci 117:906–912. https://doi.org/10.21273/jashs.117.6.906

Brunings AM, Datnoff LE, Ma JF et al (2009) Differential gene expression of rice in response to silicon and rice blast fungus *Magnaporthe oryzae*. Ann Appl Biol 155:161–170. https://doi.org/10.1111/j.1744-7348.2009.00347.x

Buchman JT, Elmer WH, Ma C et al (2019) Chitosan-coated mesoporous silica nanoparticle treatment of *Citrullus lanatus* (watermelon): enhanced fungal disease suppression and modulated expression of stress-related genes. ACS Sustain Chem Eng 7:19649–19659. https://doi.org/10.1021/acssuschemeng.9b04800

Cabot C, Gallego B, Martos S et al (2013) Signal cross talk in Arabidopsis exposed to cadmium, silicon, and *Botrytis cinerea*. Planta 237:337–349. https://doi.org/10.1007/s00425-012-1779-7

Cai K, Gao D, Luo S et al (2008) Physiological and cytological mechanisms of silicon-induced resistance in rice against blast disease. Physiol Plant 134:324–333. https://doi.org/10.1111/j.1399-3054.2008.01140.x

Carneiro-Carvalho A, Pereira C, Marques T, Martins L, Anjos R, Pinto T, Lousada J, Gomes-Laranjo J (2017) Potential of silicon fertilization in the resistance of chestnut plants toink disease(Phytophthora cinnamomi). Int J Environ Agric Biotechnol 2:2740–2753. https://doi.org/10.22161/ijeab/2.5.60

Carré-Missio V, Rodrigues FA, Schurt DA, Resende RS, Souza NFA, Rezende DC, Moreira WR, Zambolim L (2014) Effect of foliar-applied potassium silicate on coffee leaf infection by Hemileia vastatrix. Ann Appl Biol 164:396–403. https://doi.org/10.1111/aab.12109

Carneiro-Carvalho A, Pinto T, Ferreira H, Martins L, Pereira C, Gomes-Laranjo J, Anjos R (2020) Effect of silicon fertilization on the tolerance of Castanea sativa mill. Seedlings against Cryphonectria parasitica Barr. J Plant Dis Prot. https://doi.org/10.1007/s41348-019-00283-z

Carvalhais LC, Schenk PM, Dennis PG (2017) Jasmonic acid signalling and the plant holobiont. Curr Opin Microbiol 37:42–47. https://doi.org/10.1016/j.mib.2017.03.009

Carver TLW, Zeyen RJ, Ahlstrand GG (1987) The relationship between insoluble silicon and success or failure of attempted primary penetration by powdery mildew (*Erysiphe graminis*) germlings on barley. Physiol Mol Plant Pathol 31:133–148. https://doi.org/10.1016/0885-5765(87)90012-9

Chain F, Côté-Beaulieu C, Belzile F et al (2009) A comprehensive transcriptomic analysis of the effect of silicon on wheat plants under control and pathogen stress conditions. Mol Plant-Microbe Interact 22:1323–1330. https://doi.org/10.1094/MPMI-22-11-1323

Chen Y, Liu M, Wang L et al (2014) Proteomic characterization of silicon-mediated resistance against Ralstonia solanacearum in tomato. Plant Soil 387:425–440. https://doi.org/10.1007/s11104-014-2293-4

Cherif M, Asselin A, Belanger RR (1994) Defense responses induced by soluble silicon in cucumber roots infected by *Pythium* spp. Phytopathology 84:236–242

Conceição CS, Felix KCS, Mariano RLR et al (2014) Combined effect of yeast and silicon on the control of bacterial fruit blotch in melon. Sci Hortic 174:164–170. https://doi.org/10.1016/j.scienta.2014.05.027

Correa RSB, Moraes JC, Auad AM, Carvalho GA (2005) Silício e acibenzolar-s-methyl como indutores de resistência em pepino, à mosca-branca Bemisia tabaci (Gennadius) (Hemiptera: Aleyrodidae) Biótipo B. Neotrop Entomol 34:429–433. https://doi.org/10.1590/S1519-566X2005000300011

Costa RR, Moraes JC, Antunes CS (2007) Induced resistance on wheat to *Schizaphis graminum* (Rondani, 1852) (Hemiptera: Aphididae) by silicon and acibenzolar-S-methyl. Cienc Agrotec 31:393–397. https://doi.org/10.1590/s1413-70542007000200019

Costa RR, Moraes JC, da Costa RR (2009) Silicon-imidachloprid interaction on the biological and feed behavior of *Schizaphis graminum* (Rond.) (Hemiptera: Aphididae) on wheat plants. Cienc Agrotec 33:455–460. https://doi.org/10.1590/s1413-70542009000200014

Costa RR, Moraes JC, DaCosta RR (2011) Feeding behaviour of the greenbug *Schizaphis graminum* on wheat plants treated with imidacloprid and/or silicon. J Appl Entomol 135:115–120. https://doi.org/10.1111/j.1439-0418.2010.01526.x

Cruz MFA, Rodrigues FÁ, Diniz APC, Alves Moreira M, Barros EG (2014) Soybean resistance to Phakopsora pachyrhizi as affected by Acibenzolar-S-methyl, Jasmonic acid and silicon. J Phytopathol 162:133–136. https://doi.org/10.1111/jph.12170

Cui H, Gobbato E, Kracher B et al (2017) A core function of EDS1 with PAD4 is to protect the salicylic acid defense sector in *Arabidopsis immunity*. New Phytol 213:1802–1817. https://doi.org/10.1111/nph.14302

da Silva WL, Cruz MFA, Fortunato AA, Rodrigues FÁ (2015) Histochemical aspects of wheat resistance to leaf blast mediated by silicon. Sci Agric 72:322–327. https://doi.org/10.1590/0103-9016-2014-0221

Dallagnol LJ, Rodrigues FA, Damatta FM et al (2011) Deficiency in silicon uptake affects cytological, physiological, and biochemical events in the rice-*Bipolaris oryzae* interaction. Phytopathology 101:92–104. https://doi.org/10.1094/PHYTO-04-10-0105

Dallagnol LJ, Rodrigues FA, Mielli MVB (2013) Silicon improves the emergence and sanity of rice seedlings obtained from seeds infected with Bipolaris oryzae. Trop Plant Pathol 38:478–484. https://doi.org/10.1590/S1982-56762013000600003

Dann EK, Muir S (2002) Peas grown in media with elevated plant-available silicon levels have higher activities of chitinase and β-1,3- glucanase, are less susceptible to a fungal leaf spot pathogen and accumulate more foliar silicon. Australas Plant Pathol 31:9–13. https://doi.org/10.1071/AP01047

De Almeida GD, Pratissoli D, Zanuncio JC, Vicentini VB, Holtz AM, Serrão JE (2008) Calcium silicate and organic mineral fertilizer applications reduce phytophagy by Thrips palmi Karny (Thysanoptera: Thripidae) on eggplants (Solanum melongena L.). Interciencia 33:835–838

Deepak S, Manjunath G, Manjula S, Niranjan-Raj S, Geetha NP, Shetty HS (2008) Involvement of silicon in pearl millet resistance to downy mildew disease and its interplay with cell wall

proline/hydroxyproline-rich glycoproteins. Australas Plant Pathol 37:498–504. https://doi.org/10.1071/AP08047
De Moraes JC (2005) Resistance induction in wheat plants. Sci Agric 62:547–551. https://doi.org/10.1590/S0103-90162005000600006
Dias PAS, Sampaio MV, Rodrigues MP, Korndörfer AP, Oliveira RS, Ferreira SE, Korndörfer GH (2014) Induction of resistance by silicon in wheat plants to Alate and Apterous morphs of Sitobion avenae (Hemiptera: Aphididae). Environ Entomol 43:949–956. https://doi.org/10.1603/en13234
Diogo RVC, Wydra K (2007) Silicon-induced basal resistance in tomato against *Ralstonia solanacearum* is related to modification of pectic cell wall polysaccharide structure. Physiol Mol Plant Pathol 70:120–129. https://doi.org/10.1016/j.pmpp.2007.07.008
Disi JO, Zebelo S, Kloepper JW, Fadamiro H (2018) Seed inoculation with beneficial rhizobacteria affects European corn borer (Lepidoptera: Pyralidae) oviposition on maize plants. Entomol Sci 21:48–58. https://doi.org/10.1111/ens.12280
Domiciano GP, Cacique IS, Freitas CC, Filippi MCC, DaMatta FM, Do Vale FXR, Rodrigues FÁ (2015) Alterations in gas exchange and oxidative metabolism in rice leaves infected by Pyricularia oryzae are attenuated by silicon. Phytopathology 105:738–747. https://doi.org/10.1094/PHYTO-10-14-0280-R
Domiciano GP, Rodrigues FA, Guerra AMN, Vale FXR (2013) Infection process of Bipolaris sorokiniana on wheat leaves is affected by silicon. Trop Plant Pathol 38:258–263. https://doi.org/10.1590/S1982-56762013005000006
Domiciano GP, Rodrigues FA, Vale FXR, Filha MSX, Moreira WR, Andrade CCL, Pereira SC (2010) Wheat resistance to spot blotch potentiated by silicon. J Phytopathol 158:334–343. https://doi.org/10.1111/j.1439-0434.2009.01623.x
Dorairaj D, Ismail MR (2017) Distribution of silicified microstructures, regulation of cinnamyl alcohol dehydrogenase and lodging resistance in silicon and paclobutrazol mediated *Oryza sativa*. Front Physiol 8:491. https://doi.org/10.3389/fphys.2017.00491
Dorneles KR, Dallagnol LJ, Pazdiora PC, Rodrigues FA, Deuner S (2017) Silicon potentiates biochemical defense responses of wheat against tan spot. Physiol Mol Plant Pathol 97:69–78. https://doi.org/10.1016/j.pmpp.2017.01.001
dos Santos MC, Resende Junqueira AM, Mendes de Sá VG et al (2015) Effect of silicon on the morphology of the midgut and mandible of tomato leafminer Tuta absoluta (Lepidoptera: Gelechiidae) larvae. Invertebr Surviv J 12:158–165
Elsharkawy MM, Mousa KM (2015) Induction of systemic resistance against Papaya ring spot virus (PRSV) and its vector Myzus persicae by *Penicillium simplicissimum* GP17-2 and silica (SiO_2) nanopowder. Int J Pest Manag 61:353–358. https://doi.org/10.1080/09670874.2015.1070930
Etesami H (2018) Can interaction between silicon and plant growth promoting rhizobacteria benefit in alleviating abiotic and biotic stresses in crop plants? Agric Ecosyst Environ 253:98–112. https://doi.org/10.1016/j.agee.2017.11.007
Etesami H, Jeong BR (2018) Silicon (Si): review and future prospects on the action mechanisms in alleviating biotic and abiotic stresses in plants. Ecotoxicol Environ Saf 147:881–896. https://doi.org/10.1016/j.ecoenv.2017.09.063
Fan X-Y, Lin W-P, Liu R et al (2018) Physiological response and phenolic metabolism in tomato (*Solanum lycopersicum*) mediated by silicon under Ralstonia solanacearum infection. J Integr Agric 17:2160–2171. https://doi.org/10.1016/S2095-3119(18)62036-2
Faure SP, Mercier L, Didier P et al (2012) Laser sintering process analysis: application to chromium-cobalt alloys for dental prosthesis production. In: ASME 2012 11th Bienn Conf Eng Syst Des Anal ESDA 2012, vol 4, pp 9–15. https://doi.org/10.1115/ESDA2012-82108
Fauteux F, Chain F, Belzile F et al (2006) The protective role of silicon in the Arabidopsis-powdery mildew pathosystem. Proc Natl Acad Sci U S A 103:17554–17559. https://doi.org/10.1073/pnas.0606330103
Fawe A, Abou-Zaid M, Menzies JG, Bélanger RR (1998) Silicon-mediated accumulation of flavonoid phytoalexins in cucumber. Phytopathology 88:396–401. https://doi.org/10.1094/PHYTO.1998.88.5.396

Ferreira HA (2009) Silicon in the control of water blotch in melon (*Cucumis melo* L.). Dissertation (Master's) Federal Rural University of Pernambuco, Brazil, 82p

Ferreira RS, Moraes JC, Antunes CS (2011) Silicon influence on resistance induction against bemisia tabaci biotype b (genn.) (hemiptera: Aleyrodidae) and on vegetati ve development in two soybean cultivars. Neotrop Entomol 40:495–500. https://doi.org/10.1590/S1519-566X2011000400014

Ferreira HA, do Nascimento CWA, Datnoff LE et al (2015) Effects of silicon on resistance to bacterial fruit blotch and growth of melon. Crop Prot 78:277–283. https://doi.org/10.1016/j.cropro.2015.09.025

Ferreira Preston HA, de Sousa H, Nunes G, Preston W et al (2021) Slag-based silicon fertilizer improves the resistance to bacterial fruit blotch and fruit quality of melon grown under field conditions. Crop Prot 147. https://doi.org/10.1016/j.cropro.2020.105460

Fortunato AA, Rodrigues FÁ, Do Nascimento KJT (2012) Physiological and biochemical aspects of the resistance of banana plants to Fusarium wilt potentiated by silicon. Phytopathology 102:957–966. https://doi.org/10.1094/PHYTO-02-12-0037-R

French-Monar RD, Rodrigues FA, Korndörfer GH, Datnoff LE (2010) Silicon suppresses Phytophthora blight development on bell pepper. J Phytopathol 158:554–560. https://doi.org/10.1111/j.1439-0434.2009.01665.x

Frew A, Powell JR, Hiltpold I et al (2017) Host plant colonisation by arbuscular mycorrhizal fungi stimulates immune function whereas high root silicon concentrations diminish growth in a soil-dwelling herbivore. Soil Biol Biochem 112:117–126. https://doi.org/10.1016/j.soilbio.2017.05.008

Gajger IT, Dar SA (2021) Plant allelochemicals as sources of insecticides. Insects 12:189. https://doi.org/10.3390/insects12030189

Garibaldi A, Gilardi G, Cogliati EE, Gullino ML (2012) Silicon and increased electrical conductivity reduce downy mildew of soilless grown lettuce. Eur J Plant Pathol 132:123–132. https://doi.org/10.1007/s10658-011-9855-6

Gatarayiha MC, Laing MD, Miller RM (2010) Combining applications of potassium silicate and Beauveria bassiana to four crops to control two spotted spider mite, Tetranychus urticae Koch. Int J Pest Manag 56:291–297. https://doi.org/10.1080/09670874.2010.495794

Ghanmi D, McNally DJ, Benhamou N et al (2004) Powdery mildew of *Arabidopsis thaliana*: a pathosystem for exploring the role of silicon in plant-microbe interactions. Physiol Mol Plant Pathol 64:189–199. https://doi.org/10.1016/j.pmpp.2004.07.005

Ghareeb H, Bozsó Z, Ott PG et al (2011) Transcriptome of silicon-induced resistance against *Ralstonia solanacearum* in the silicon non-accumulator tomato implicates priming effect. Physiol Mol Plant Pathol 75:83–89. https://doi.org/10.1016/j.pmpp.2010.11.004

Ghorbani R, Wilcockson S, Koocheki A, Leifert C (2008) Soil management for sustainable crop disease control: a review. Environ Chem Lett 6:149–162

Gomes FB, de Moraes JC, dos Santos CD, Goussain MM (2005) Resistance induction in wheat plants by silicon and aphids. Sci Agric 62:547–551. https://doi.org/10.1590/S0103-90162005000600006

Guérin V, Lebreton A, Cogliati EE, Hartley SE, Belzile F, Menzies JG, Bélanger RR (2014) A zoospore inoculation method with Phytophthora Sojae to assess the prophylactic role of silicon on soybean cultivars. Plant Dis 98:1632–1638. https://doi.org/10.1094/PDIS-01-14-0102-RE

Guével MH, Menzies JG, Bélanger RR (2007) Effect of root and foliar applications of soluble silicon on powdery mildew control and growth of wheat plants. Eur J Plant Pathol 119:429–436. https://doi.org/10.1007/s10658-007-9181-1

Guo Y, Liu L, Zhao J, Bi Y (2007) Use of silicon oxide and sodium silicate for controlling Trichothecium roseum postharvest rot in Chinese cantaloupe (Cucumis melo L.). Int J Food Sci Technol 42:1012–1018. https://doi.org/10.1111/j.1365-2621.2006.01464.x

Gutiérrez-Barranquero JA, Arrebola E, Bonilla N et al (2012) Environmentally friendly treatment alternatives to Bordeaux mixture for controlling bacterial apical necrosis (BAN) of mango. Plant Pathol 61:665–676. https://doi.org/10.1111/j.1365-3059.2011.02559.x

Hammer GP, Windisch G, Prodinger PM et al (2010) The cricothyroid joint-functional aspects with regard to different types of its structure. J Voice 24:140–145
Han Y, Lei W, Wen L, Hou M (2015) Silicon-mediated resistance in a susceptible rice variety to the rice leaf folder, Cnaphalocrocis medinalis Guenée (Lepidoptera: Pyralidae). PLoS One 10:557. https://doi.org/10.1371/journal.pone.0120557
Han YQ, Wen JH, Peng ZP, Zhang DY, Hou ML (2018) Effects of silicon amendment on the occurrence of rice insect pests and diseases in a field test. J Integr Agric 17:2172–2181. https://doi.org/10.1016/S2095-3119(18)62035-0
Handley R, Ekbom B, Ågren J (2005) Variation in trichome density and resistance against a specialist insect herbivore in natural populations of *Arabidopsis thaliana*. Ecol Entomol 30:284–292. https://doi.org/10.1111/j.0307-6946.2005.00699.x
Haq IU, Khurshid A, Inayat R et al (2021) Silicon-based induced resistance in maize against fall armyworm [*Spodoptera frugiperda* (Lepidoptera: Noctuidae)]. PLoS One 16:9749. https://doi.org/10.1371/journal.pone.0259749
Hartley SE, Fitt RN, McLarnon EL, Wade RN (2015) Defending the leaf surface: intra- and interspecific differences in silicon deposition in grasses in response to damage and silicon supply. Front Plant Sci 6:0035. https://doi.org/10.3389/fpls.2015.00035
Hayasaka T, Fujii H, Ishiguro K (2008) The role of silicon in preventing appressorial penetration by the rice blast fungus. Phytopathology 98:1038–1044. https://doi.org/10.1094/PHYTO-98-9-1038
Heine G, Tikum G, Horst WJ (2007) The effect of silicon on the infection by and spread of Pythium aphanidermatum in single roots of tomato and bitter gourd. J Exp Bot 58:569–577. https://doi.org/10.1093/jxb/erl
Hoffmann J, Berni R, Hausman JF, Guerriero G (2020) A review on the beneficial role of silicon against salinity in non-accumulator crops: tomato as a model. Biomol Ther 10:1–15
Holz S, Kube M, Bartoszewski G, Büttner C (2015) The impact of silicon on transcripts related to cucumber mosaic virus infection in cucumber. Commun Agric Appl Biol Sci 80:411–420
Hou M, Han Y (2010) Silicon-mediated Rice Plant resistance to the Asiatic Rice borer (Lepidoptera: Crambidae): effects of silicon amendment and Rice varietal resistance. J Econ Entomol 103:1412–1419. https://doi.org/10.1603/ec09341
Huang CH, Roberts PD, Datnoff LE (2011) Silicon suppresses fusarium crown and root rot of tomato. J Phytopathol 159:546–554. https://doi.org/10.1111/j.1439-0434.2011.01803.x
Hussain M, Debnath B, Qasim M et al (2019) Role of saponins in plant defense against specialist herbivores. Molecules 24:2067
Hyun MW, Yun YH, Kim JY, Kim SH (2011) Fungal and plant phenylalanine ammonia-lyase. Mycobiology 39:257–265. https://doi.org/10.5941/MYCO.2011.39.4.257u
Inbar M, Gerling D (2008) Plant-mediated interactions between whiteflies, herbivores, and natural enemies. Annu Rev Entomol 53:431–448. https://doi.org/10.1146/annurev.ento.53.032107.122456
Islam W, Zhang J, Adnan M et al (2017) Plant virus ecology: a glimpse of recent accomplishments. Appl Ecol Environ Res 15:691–705. https://doi.org/10.15666/aeer/1501_691705
Islam W, Adnan M, Tayyab M et al (2018a) Phyto-metabolites; an impregnable shield against plant viruses. Nat Prod Commun 13:105–112. https://doi.org/10.1177/1934578x1801300131
Islam W, Akutse KS, Qasim M et al (2018b) *Bemisia tabaci*-mediated facilitation in diversity of begomoviruses: evidence from recent molecular studies. Microb Pathog 123:162–168. https://doi.org/10.1016/j.micpath.2018.07.008
Islam W, Lin W, Islam SU et al (2018c) Genetic diversity of begomoviruses in Pakistan captured through a vector based survey. Microb Pathog 118:91–97. https://doi.org/10.1016/j.micpath.2018.03.019
Islam W, Lin W, Qasim M et al (2018d) A nation-wide genetic survey revealed a complex population structure of Bemisia tabaci in Pakistan. Acta Trop 183:119–125. https://doi.org/10.1016/j.actatropica.2018.04.015
Islam W, Qasim M, Noman A et al (2018e) Management of tobacco mosaic virus through natural metabolites. Rec Nat Prod 12:403–415. https://doi.org/10.25135/rnp.49.17.10.178

Islam W, Naveed H, Zaynab M et al (2019) Plant defense against virus diseases; growth hormones in highlights. Plant Signal Behav 14:6719. https://doi.org/10.1080/15592324.2019.1596719

Islam W, Noman A, Naveed H et al (2020a) Plant-insect vector-virus interactions under environmental change. Sci Total Environ 701:135044. https://doi.org/10.1016/j.scitotenv.2019.135044

Islam W, Tayyab M, Khalil F et al (2020b) Silicon-mediated plant defense against pathogens and insect pests. Pestic Biochem Physiol. https://doi.org/10.1016/j.pestbp.2020.104641

Islam W, Adnan M, Shabbir A et al (2021) Insect-fungal-interactions: a detailed review on entomopathogenic fungi pathogenicity to combat insect pests. Microb Pathog 159:105122. https://doi.org/10.1016/j.micpath.2021.105122

Jadhao KR, Bansal A, Rout GR (2020) Silicon amendment induces synergistic plant defense mechanism against pink stem borer (*Sesamia inferens* Walker.) in finger millet (*Eleusine coracana* Gaertn.). Sci Rep 10. https://doi.org/10.1038/s41598-020-61182-0

Jaleel W, Saeed S, Naqqash MN et al (2020) Effects of temperature on baseline susceptibility and stability of insecticide resistance against *Plutella xylostella* (Lepidoptera: Plutellidae) in the absence of selection pressure. Saudi J Biol Sci 27:1–5. https://doi.org/10.1016/j.sjbs.2019.03.004

Jayawardana HARK, Weerahewa HLD, Saparamadu MDJS (2016) The mechanisms underlying the anthracnose disease reduction by Rice Hull as a silicon source in capsicum (capsicum Annuum L.) grown in simplified hydroponics. Procedia Food Sci 6:147–150. https://doi.org/10.1016/j.profoo.2016.02.035

Jeer M, Telugu UM, Voleti SR, Padmakumari AP (2017) Soil application of silicon reduces yellow stem borer, Scirpophaga incertulas (Walker) damage in rice. J Appl Entomol 141:189–201. https://doi.org/10.1111/jen.12324

Ji P, Campbell HL, Kloepper JW et al (2006) Integrated biological control of bacterial speck and spot of tomato under field conditions using foliar biological control agents and plant growth-promoting rhizobacteria. Biol Control 36:358–367. https://doi.org/10.1016/j.biocontrol.2005.09.003

Jia L, Han Y, Hou M (2021) Silicon amendment to rice plants reduces the transmission of southern rice black-streaked dwarf virus by *Sogatella furcifera*. Pest Manag Sci 77:3233–3240. https://doi.org/10.1002/ps.6365

Jiang N, Fan X, Lin W, Wang G, Cai K (2019) Transcriptome analysis reveals new insights into the bacterial wilt resistance mechanism mediated by silicon in tomato. Int J Mol Sci 20. https://doi.org/10.3390/ijms2003

Juma G, Ahuya PO, Ong'Amo G, Le Ru B, Magoma G, Silvain JF, Calatayud PA (2015) Influence of plant silicon in Busseola fusca (Lepidoptera: Noctuidae) larvae -Poaceae interactions. Bull Entomol Res 105:253–258. https://doi.org/10.1017/S000748531500005X

Kablan L, Lagauche A, Delvaux B, Legrève A (2012) Silicon reduces black Sigatoka development in banana. Plant Dis 96:273–278. https://doi.org/10.1094/PDIS-04-11-0274

Kanto T, Miyoshi A, Ogawa T et al (2004) Suppressive effect of potassium silicate on powdery mildew of strawberry in hydroponics. J Gen Plant Pathol 70:207–211. https://doi.org/10.1007/s10327-004-0117-8

Keeping MG, Meyer JH (2002) Calcium silicate enhances resistance of sugarcane to the African stalk borer Eldana saccharina Walker (Lepidoptera: Pyralidae). Agric For Entomol 4:265–274. https://doi.org/10.1046/j.1461-9563.2002.00150.x

Keeping MG, Meyer JH (2006) Silicon-mediated resistance of sugarcane to Eldana saccharina Walker (Lepidoptera: Pyralidae): effects of silicon source and cultivar. J Appl Entomol 130:410–420. https://doi.org/10.1111/j.1439-0418.2006.01081.x

Keeping MG, Kvedaras OL, Bruton AG (2009) Epidermal silicon in sugarcane: cultivar differences and role in resistance to sugarcane borer *Eldana saccharina*. Environ Exp Bot 66:54–60. https://doi.org/10.1016/j.envexpbot.2008.12.012

Khakimov A, Salakhutdinov I, Omolikov A, Utaganov S (2022) Traditional and current-prospective methods of agricultural plant diseases detection: a review. IOP Conf Ser: Earth Environ Sci 951:012002

Kiirika LM, Stahl F, Wydra K (2013) Phenotypic and molecular characterization of resistance induction by single and combined application of chitosan and silicon in tomato against *Ralstonia solanacearum*. Physiol Mol Plant Pathol 81:1–12. https://doi.org/10.1016/j.pmpp.2012.11.002

Kim SG, Kim KW, Park EW, Choi D (2002) Silicon-induced cell wall fortification of rice leaves: a possible cellular mechanism of enhanced host resistance to blast. Phytopathology 92:1095–1103. https://doi.org/10.1094/PHYTO.2002.92.10.1095

Kim YH, Khan AL, Kim DH et al (2014) Silicon mitigates heavy metal stress by regulating P-type heavy metal ATPases, *Oryza sativa* low silicon genes, and endogenous phytohormones. BMC Plant Biol 14. https://doi.org/10.1186/1471-2229-14-13

Korth KL, Doege SJ, Park SH, Goggin FL, Wang Q, Gomez SK, Liu G, Jia L, Nakata PA (2006) Medicago truncatula mutants demonstrate the role of plant calcium oxalate crystals as an effective defense against chewing insects. Plant Physiol 141:188–195. https://doi.org/10.1104/pp.106.076737

Kuai J, Sun Y, Guo C et al (2017) Root-applied silicon in the early bud stage increases the rapeseed yield and optimizes the mechanical harvesting characteristics. Field Crop Res 200:88–97. https://doi.org/10.1016/j.fcr.2016.10.007

Kurabachew H, Wydra K (2014) Induction of systemic resistance and defense-related enzymes after elicitation of resistance by rhizobacteria and silicon application against *Ralstonia solanacearum* in tomato (*Solanum lycopersicum*). Crop Prot 57:1–7. https://doi.org/10.1016/j.cropro.2013.10.021

Kurabachew H, Stahl F, Wydra K (2013) Global gene expression of rhizobacteria-silicon mediated induced systemic resistance in tomato (*Solanum lycopersicum*) against *Ralstonia solanacearum*. Physiol Mol Plant Pathol 84:44–52. https://doi.org/10.1016/j.pmpp.2013.06.004

Kvedaras OL, An M, Choi YS, Gurr GM (2010) Silicon enhances natural enemy attraction and biological control through induced plant defences. Bull Entomol Res 100:367–371. https://doi.org/10.1017/S0007485309990265

Leroy N, De Tombeur F, Walgraffe Y et al (2019) Silicon and plant natural defenses against insect pests: impact on plant volatile organic compounds and cascade effects on multitrophic interactions. Plants 8:444. https://doi.org/10.3390/plants8110444

Leusch H-J, Buchenauer H (1989) Effect of soil treatments with silica-rich lime fertilizers and sodium trisilicate on the incidence of wheat by *Erysiphe graminis* and *Septoria nodorum* depending on the form of N-fertilizer. Zeitschrift fuer Pflanzenkrankheiten und Pflanzenschutz

Li YC, Bi Y, Ge YH, Sun XJ, Wang Y (2009) Antifungal activity of sodium silicate on Fusarium Sulphureum and its effect on dry rot of potato tubers. J Food Sci 74. https://doi.org/10.1111/j.1750-3841.2009.01154.x

Liang Y, Si J, Römheld V (2005a) Silicon uptake and transport is an active process in *Cucumis sativus*. New Phytol 167:797–804. https://doi.org/10.1111/j.1469-8137.2005.01463.x

Liang YC, Sun WC, Si J, Römheld V (2005b) Effects of foliar- and root-applied silicon on the enhancement of induced resistance to powdery mildew in *Cucumis sativus*. Plant Pathol 54:678–685. https://doi.org/10.1111/j.1365-3059.2005.01246.x

Lin Y, Sun Z, Li Z et al (2019) Deficiency in silicon transporter Lsi1 compromises inducibility of anti-herbivore defense in rice plants. Front Plant Sci 10:0652. https://doi.org/10.3389/fpls.2019.00652

Lin Y, Lin X, Ding C et al (2022) Priming of rice defense against a sap-sucking insect pest brown planthopper by silicon. J Pest Sci 95:1371–1385. https://doi.org/10.1007/s10340-021-01462-4

Liu L, Guo Y, Bi Y et al (2009) Inhabited mechanisms of silicon compounds against Fusarium rot (Fusarium Spp.) of postharvest Chinese cantaloupe. J Food Process Preserv 33:187–202. https://doi.org/10.1111/j.1745-4549.2008.00292.x

Liu M, Cai K, Chen Y et al (2014) Proteomic analysis of silicon-mediated resistance to *Magnaporthe oryzae* in rice (*Oryza sativa* L.). Eur J Plant Pathol 139:579–592. https://doi.org/10.1007/s10658-014-0414-9

Liu J, Zhu J, Zhang P et al (2017) Silicon supplementation alters the composition of herbivore induced plant volatiles and enhances attraction of parasitoids to infested rice plants. Front Plant Sci 8:1265. https://doi.org/10.3389/fpls.2017.01265

Manivannan A, Ahn YK (2017) Silicon regulates potential genes involved in major physiological processes in plants to combat stress. Front Plant Sci 8:1346. https://doi.org/10.3389/fpls.2017.01346

Massey FP, Ennos AR, Hartley SE (2006) Silica in grasses as a defence against insect herbivores: contrasting effects on folivores and a phloem feeder. J Anim Ecol 75:595–603. https://doi.org/10.1111/j.1365-2656.2006.01082.x

Mateos Fernández R, Petek M, Gerasymenko I et al (2022) Insect pest management in the age of synthetic biology. Plant Biotechnol J 20:25–36. https://doi.org/10.1111/pbi.13685

Mburu K, Oduor R, Mgutu A, Tripathi L (2016) Silicon application enhances resistance to xanthomonas wilt disease in banana. Plant Pathol 65:807–818. https://doi.org/10.1111/ppa.12468

McNaughton SJ, Tarrants JL (1983) Grass leaf silicification: natural selection for an inducible defense against herbivores. Proc Natl Acad Sci 80:790–791. https://doi.org/10.1073/pnas.80.3.790

Meena V, Dotaniya ML, Saha JK, Patra AK (2021) Silicon potential to mitigate plant heavy metals stress for sustainable agriculture: a review. Silicon 14:4447–4462. https://doi.org/10.1007/s12633-021-01200-6

Menzies J, Bowen P, Ehret D, Glass ADM (2019) Foliar applications of potassium silicate reduce severity of powdery mildew on cucumber, muskmelon, and zucchini squash. J Am Soc Hortic Sci 117:902–905. https://doi.org/10.21273/jashs.117.6.902

Meyer JH, Keeping MG (2001) Chapter 16 past, present and future research of the role of silicon for sugarcane in southern Africa. Stud Plant Sci 8:257–275. https://doi.org/10.1016/S0928-3420(01)80020-3

Moldes CA, de Lima Filho OF, Merini LJ, Tsai SM, Camiña JM (2016) Occurrence of powdery mildew disease in wheat fertilized with increasing silicon doses: a chemometric analysis of antioxidant response. Acta Physiol Plant 38. https://doi.org/10.1007/s11738-016-2217-4

Moraes JC, Goussain MM, Carvalho GA, Costa RR (2005) Feeding non-preference of the corn leaf aphid Rhopalosiphum maidis (Fitch, 1856) (Hemiptera: Aphididae) to corn plants (Zea mays L.) treated with silicon. Ciência e Agrotecnologia 29:761–766. https://doi.org/10.1590/s1413-70542005000400007

Moraes SRG, Pozza EA, Alves E et al (2006) Effects of silicon sources on the incidence and severity of the common beans anthracnose. Fitopatol Bras 31:69–75. https://doi.org/10.1590/s0100-41582006000100012

Murali-Baskaran RK, Senthil-Nathan S, Hunter WB (2021) Anti-herbivore activity of soluble silicon for crop protection in agriculture: a review. Environ Sci Pollut Res 28:2626–2637. https://doi.org/10.1007/s11356-020-11453-0

Najihah NI, Hanafi MM, Idris AS, Hakim MA (2015) Silicon treatment in oil palms confers resistance to basal stem rot disease caused by Ganoderma boninense. Crop Prot 67:151–159. https://doi.org/10.1016/j.cropro.2014.10.004

Nazaralian S, Majd A, Irian S et al (2017) Comparison of silicon nanoparticles and silicate treatments in fenugreek. Plant Physiol Biochem 115:25–33. https://doi.org/10.1016/j.plaphy.2017.03.009

Ng LC, Hafiz ES, Sariam O, Razi RIM (2019) The effect of calcium silicate as foliar application on aerobic rice blast disease development. Eur J Plant Pathol 153:533–543. https://doi.org/10.1007/s10658-018-1580-y

Noman A, Aqeel M, Khalid N et al (2019) Zinc finger protein transcription factors: integrated line of action for plant antimicrobial activity. Microb Pathog 132:141–149. https://doi.org/10.1016/j.micpath.2019.04.042

Noman A, Aqeel M, Islam W et al (2021) Insects–plants-pathogens: toxicity, dependence and defense dynamics. Toxicon 197:87–98. https://doi.org/10.1016/j.toxicon.2021.04.004

Oliveira JC, Albuquerque GMR, Mariano RLR et al (2012) Reduction of the severity of angular leaf spot of cotton mediated by silicon. J Plant Pathol 94:297–304

Pang Z, Tayyab M, Islam W et al (2019) Silicon-mediated improvement in tolerance of economically important crops under drought stress. Appl Ecol Environ Res 17:6151–6170. https://doi.org/10.15666/aeer/1703_61516170

Pazdiora PC, da Rosa Dorneles K, Forcelini CA et al (2018) Silicon suppresses tan spot development on wheat infected by *Pyrenophora tritici*-repentis. Eur J Plant Pathol 150:49–56. https://doi.org/10.1007/s10658-017-1251-4

Pirrello J, Prasad BCN, Zhang W et al (2012) Functional analysis and binding affinity of tomato ethylene response factors provide insight on the molecular bases of plant differential responses to ethylene. BMC Plant Biol 12:190. https://doi.org/10.1186/1471-2229-12-190

Polanco LR, Rodrigues FA, Nascimento KJT et al (2014) Photosynthetic gas exchange and antioxidative system in common bean plants infected by *Colletotrichum lindemuthianum* and supplied with silicon. Trop Plant Pathol 39:35–42. https://doi.org/10.1590/S1982-56762014000100005

Pozza AAA, Alves E, Pozza EA et al (2004) Effect of silicon on the control of brown eye spot in three coffee cultivars. Fitopatol Bras 29:185–188. https://doi.org/10.1590/s0100-41582004000200010

Prabhu AS, Barbosa Filho MP, Datnoff LE, Snyder GH, Berni RF, Rodrigues FA, Dallagnol LJ (2012) Silicon reduces brown spot severity and grain discoloration on several rice genotypes. Trop Plant Pathol 37:409–414. https://doi.org/10.1590/S1982-56762012000600005

Prasannath K (2017) Plant defense-related enzymes against pathogens: a review. AGRIEAST J Agric Sci 11:38. https://doi.org/10.4038/agrieast.v11i1.33

Qasim M, Islam W, Ashraf HJ et al (2020) Saponins in insect pest control. In: Reference series in phytochemistry. Springer Nature Switzerland AG, Switzerland, pp 897–924

Qin GZ, Tian SP (2005) Enhancement of biocontrol activity of *Cryptococcus laurentii* by silicon and the possible mechanisms involved. Phytopathology 95:69–75. https://doi.org/10.1094/PHYTO-95-0069

Rahman A, Wallis CM, Uddin W (2015) Silicon-induced systemic defense responses in perennial ryegrass against infection by *Magnaporthe oryzae*. Phytopathology 105:748–757. https://doi.org/10.1094/PHYTO-12-14-0378-R

Ramírez-Godoy A, Puentes-Pérez G, Restrepo-Díaz H (2018) An evaluation of the use of calcium, potassium and silicon for the management of Diaphorina citri populations in Tahiti lime trees. Not Bot Horti Agrobot Cluj-Napoca 46:546–552. https://doi.org/10.15835/nbha46211152

Ramouthar PV, Caldwell PM, McFarlane SA (2016) Effect of silicon on the severity of brown rust of sugarcane in South Africa. Eur J Plant Pathol 145:53–60. https://doi.org/10.1007/s10658-015-0812-7

Rasoolizadeh A, Labbé C, Sonah H, Deshmukh RK, Belzile F, Menzies JG, Bélanger RR (2018) Silicon protects soybean plants against Phytophthora sojae by interfering with effector-receptor expression. BMC Plant Biol 18. https://doi.org/10.1186/s12870-018-1312-7

Rastogi A, Tripathi DK, Yadav S et al (2019) Application of silicon nanoparticles in agriculture. 3 Biotech 9:1626. https://doi.org/10.1007/s13205-019-1626-7

Ratnayake RMRNK, Daundasekera WAM, Ariyarathne HM, Ganehenege MYU (2016) Some biochemical defense responses enhanced by soluble silicon in bitter gourdpowdery mildew pathosystem. Australas Plant Pathol 45:425–433. https://doi.org/10.1007/s13313-016-0429-0

Resende RS, Rodrigues FÁ, Costa RV, Silva DD (2013a) Silicon and fungicide effects on anthracnose in moderately resistant and susceptible sorghum lines. J Phytopathol 161:11–17. https://doi.org/10.1111/jph.12020

Resende RS, Rodrigues FA, Gomes RJ, Nascimento KJT (2013b) Microscopic and biochemical aspects of sorghum resistance to anthracnose mediated by silicon. Ann Appl Biol 163:114–123. https://doi.org/10.1111/aab.12040

Richmond KE, Sussman M (2003) Got silicon? The non-essential beneficial plant nutrient. Curr Opin Plant Biol 6:268–272

Rodgers-Gray BS, Shaw MW (2004) Effects of straw and silicon soil amendments on some foliar and stem-base diseases in pot-grown winter wheat. Plant Pathol 53:733–740. https://doi.org/10.1111/j.1365-3059.2004.01102.x

Rodrigues FA, Datnoff LE (2005) Silicon and rice disease management. Fitopatol Bras 30:457–469. https://doi.org/10.1590/s0100-41582005000500001

Rodrigues FÁ, Jurick WM, Datnoff LE et al (2005) Silicon influences cytological and molecular events in compatible and incompatible rice-Magnaporthe grisea interactions. Physiol Mol Plant Pathol 66:144–159. https://doi.org/10.1016/j.pmpp.2005.06.002

Rodrigues FÁ, Duarte HSS, Rezende DC et al (2010) Foliar spray of potassium silicate on the control of angular leaf spot on beans. J Plant Nutr 33:2082–2093. https://doi.org/10.1080/01904167.2010.519082

Rodrigues FA, Carré-Missio V, Jham GN et al (2011) Chlorogenic acid levels in leaves of coffee plants supplied with silicon and infected by *Hemileia vastatrix*. Trop Plant Pathol 36:404–408. https://doi.org/10.1590/S1982-56762011000600010

Rodrigues FA, Polanco LR, Duarte HSS, Resende RS, do Vale FXR (2015) Photosynthetic gas exchange in common bean submitted to foliar sprays of potassium silicate, sodium Molybdate and fungicide and infected with Colletotrichum lindemuthianum. J Phytopathol 163:554–559. https://doi.org/10.1111/jph.12353

Sahebi M, Hanafi MM, Akmar ASN et al (2015) Importance of silicon and mechanisms of biosilica formation in plants. Biomed Res Int 2015:6010. https://doi.org/10.1155/2015/396010

Sakr N (2021) Soluble silicon controls fusarium head blight in bread and durum wheat plants. Gesunde Pflanz 73:479–493. https://doi.org/10.1007/s10343-021-00568-0

Savvas D, Giotis D, Chatzieustratiou E, Bakea M, Patakioutas G (2009) Silicon supply in soilless cultivations of zucchini alleviates stress induced by salinity and powdery mildew infections. Environ Exp Bot 65:11–17. https://doi.org/10.1016/j.envexpbot.2008.07.004

Schurt DA, Cruz MFA, Nascimento KJT, Filippi MCC, Rodrigues FA (2014) Silicon potentiates the activities of defense enzymes in the leaf sheaths of rice plants infected by Rhizoctonia solani. Trop Plant Pathol 39:457–463. https://doi.org/10.1590/S1982-56762014000600007

Seal P, Das P, Biswas AK (2018) Versatile potentiality of silicon in mitigation of biotic and abiotic stresses in plants: a review. Am J Plant Sci 9:1433–1454. https://doi.org/10.4236/ajps.2018.97105

Seebold KW, Datnoff LE, Correa-Victoria FJ et al (2004) Effects of silicon and fungicides on the control of leaf and neck blast in upland rice. Plant Dis 88:253–258. https://doi.org/10.1094/PDIS.2004.88.3.253

Shetty R, Jensen B, Shetty NP et al (2012) Silicon induced resistance against powdery mildew of roses caused by *Podosphaera pannosa*. Plant Pathol 61:120–131. https://doi.org/10.1111/j.1365-3059.2011.02493.x

Shivaraj SM, Mandlik R, Bhat JA et al (2022) Outstanding questions on the beneficial role of silicon in crop plants. Plant Cell Physiol 63:4–18. https://doi.org/10.1093/pcp/pcab145

Shoaib A, Elabasy A, Waqas M et al (2018) Entomotoxic effect of silicon dioxide nanoparticles on *Plutella xylostella* (L.) (Lepidoptera: Plutellidae) under laboratory conditions. Toxicol Environ Chem 100:80–91. https://doi.org/10.1080/02772248.2017.1387786

Siddiqui JA, Zhang Y, Luo Y et al (2022) Comprehensive detoxification mechanism assessment of red imported fire ant (*Solenopsis invicta*) against Indoxacarb. Molecules 27:870. https://doi.org/10.3390/molecules27030870

Silva IT, Rodrigues FÁ, Oliveira JR et al (2010a) Wheat resistance to bacterial leaf streak mediated by silicon. J Phytopathol 158:253–262. https://doi.org/10.1111/j.1439-0434.2009.01610.x

Silva RV, Oliveira RDL, Nascimento KJT, Rodrigues FA (2010b) Biochemical responses of coffee resistance against *Meloidogyne exigua* mediated by silicon. Plant Pathol 59:586–593. https://doi.org/10.1111/j.1365-3059.2009.02228.x

Song A, Xue G, Cui P et al (2016) The role of silicon in enhancing resistance to bacterial blight of hydroponic- and soil-cultured rice. Sci Rep 6. https://doi.org/10.1038/srep24640

Tatagiba SD, Damatta FM, Rodrigues FA (2016) Silicon partially preserves the photosynthetic performance of rice plants infected by Monographella albescens. Ann Appl Biol 168:111–121. https://doi.org/10.1111/aab.12246

Tayyab M, Islam W, Zhang H (2018) Promising role of silicon to enhance drought resistance in wheat. Commun Soil Sci Plant Anal 49:2932–2941. https://doi.org/10.1080/00103624.2018.1547394

Telles Nascimento KJ, Debona D, Silveira PR et al (2016) Silicon-induced changes in the antioxidant system reduce soybean resistance to frogeye leaf spot. J Phytopathol 164:768–778. https://doi.org/10.1111/jph.12497

Tripathi DK, Vishwakarma K, Singh VP et al (2021) Silicon crosstalk with reactive oxygen species, phytohormones and other signaling molecules. J Hazard Mater 408. https://doi.org/10.1016/j.jhazmat.2020.124820

Tuan LNA, Du BD, Ha LDT, Dzung LTK, Van Phu D, Hien NQ (2019) Induction of Chitinase and Brown spot disease resistance by Oligochitosan and Nanosilica– Oligochitosan in dragon fruit plants. Agric Res 8:184–190. https://doi.org/10.1007/s40003-018-0384-9

Van Bockhaven J, Spíchal L, Novák O et al (2015a) Silicon induces resistance to the brown spot fungus *Cochliobolus miyabeanus* by preventing the pathogen from hijacking the rice ethylene pathway. New Phytol 206:761–773. https://doi.org/10.1111/nph.13270

Van Bockhaven J, Steppe K, Bauweraerts I et al (2015b) Primary metabolism plays a central role in moulding silicon-inducible brown spot resistance in rice. Mol Plant Pathol 16:811–824. https://doi.org/10.1111/mpp.12236

Verma KK, Song XP, Tian DD et al (2021) Influence of silicon on biocontrol strategies to manage biotic stress for crop protection, performance, and improvement. Plants 10:2163. https://doi.org/10.3390/plants10102163

Vivancos J, Labbé C, Menzies JG, Bélanger RR (2015) Silicon-mediated resistance of Arabidopsis against powdery mildew involves mechanisms other than the salicylic acid (SA)-dependent defence pathway. Mol Plant Pathol 16:572–582. https://doi.org/10.1111/mpp.12213

Wang L, Cai K, Chen Y, Wang G (2013) Silicon-mediated tomato resistance against *Ralstonia solanacearum* is associated with modification of soil microbial community structure and activity. Biol Trace Elem Res 152:275–283. https://doi.org/10.1007/s12011-013-9611-1

Wang M, Gao L, Dong S et al (2017) Role of silicon on plant–pathogen interactions. Front Plant Sci 8:701. https://doi.org/10.3389/fpls.2017.00701

Wang M, Wang R, Mur LAJ et al (2021) Functions of silicon in plant drought stress responses. Hortic Res 8:254. https://doi.org/10.1038/s41438-021-00681-1

Whan JA, Dann EK, Aitken EAB (2016) Effects of silicon treatment and inoculation with Fusarium oxysporum f. sp. vasinfectum on cellular defences in root tissues of two cotton cultivars. Ann Bot 118:219–226. https://doi.org/10.1093/aob/mcw095

Wiese J, Wiese H, Schwartz J, Schubert S (2005) Osmotic stress and silicon act additively in enhancing pathogen resistance in barley against barley powdery mildew. J Plant Nutr Soil Sci 168:269–274. https://doi.org/10.1002/jpln.200420490

Xavier Filha MS, Rodrigues FA, Domiciano GP et al (2011) Wheat resistance to leaf blast mediated by silicon. Australas Plant Pathol 40:28–38. https://doi.org/10.1007/s13313-010-0010-1

Xue X, Geng T, Liu H et al (2021) Foliar application of silicon enhances resistance against phytophthora infestans through the ET/JA- and NPR1- dependent signaling pathways in potato. Front Plant Sci 12:9870. https://doi.org/10.3389/fpls.2021.609870

Yang L, Han Y, Li P et al (2017) Silicon amendment to rice plants impairs sucking behaviors and population growth in the phloem feeder *Nilaparvata lugens* (Hemiptera: Delphacidae). Sci Rep 7:1060. https://doi.org/10.1038/s41598-017-01060-4

Yang L, Li P, Li F et al (2018) Silicon amendment to rice plants contributes to reduced feeding in a phloem-sucking insect through modulation of callose deposition. Ecol Evol 8:631–637. https://doi.org/10.1002/ece3.3653

Yang J, Song J, Jeong BR (2022) Drenched silicon suppresses disease and insect pests in coffee plant grown in controlled environment by improving physiology and upregulating defense genes. Int J Mol Sci 23:3543. https://doi.org/10.3390/ijms23073543

Ye M, Song Y, Long J et al (2013) Priming of jasmonate-mediated antiherbivore defense responses in rice by silicon. Proc Natl Acad Sci U S A 110:E3631–E3639. https://doi.org/10.1073/pnas.1305848110

Zargar SM, Mahajan R, Bhat JA et al (2019) Role of silicon in plant stress tolerance: opportunities to achieve a sustainable cropping system. 3 Biotech 9:73. https://doi.org/10.1007/s13205-019-1613-z

Zellner W, Datnoff L (2020) Silicon as a biostimulant in agriculture. Burleigh Dodds Science Publishing, Milton, pp 149–196

Zellner W, Frantz J, Leisner S (2011) Silicon delays Tobacco ringspot virus systemic symptoms in *Nicotiana tabacum*. J Plant Physiol 168:1866–1869. https://doi.org/10.1016/j.jplph.2011.04.002

Zhan LP, Peng DL, Wang XL, Kong LA, Peng H, Liu SM, Liu Y, Huang WK (2018) Priming effect of root-applied silicon on the enhancement of induced resistance to the rootknot nematode Meloidogyne graminicola in rice. BMC Plant Biol 18. https://doi.org/10.1186/s12870-018-1266-9

Zhang H, Sonnewald U (2017) Differences and commonalities of plant responses to single and combined stresses. Plant J 90:839–855. https://doi.org/10.1111/tpj.13557

Zhang Q, Fry J, Lowe K, Tisserat N (2006) Evaluation of calcium silicate for brown patch and dollar spot suppression on turfgrasses. Crop Sci 46:1635–1643. https://doi.org/10.2135/cropsci2005.04-0002

Zhang W, Xie Z, Lang D et al (2017) Beneficial effects of silicon on abiotic stress tolerance in legumes. J Plant Nutr 40:2224–2236. https://doi.org/10.1080/01904167.2017.1346127

Ziv C, Zhao Z, Gao YG, Xia Y (2018) Multifunctional roles of plant cuticle during plant-pathogen interactions. Front Plant Sci 9:1088. https://doi.org/10.3389/fpls.2018.01088

MIX
Papier aus verantwortungsvollen Quellen
Paper from responsible sources
FSC® C105338

If you have any concerns about our products,
you can contact us on
ProductSafety@springernature.com

In case Publisher is established outside the EU,
the EU authorized representative is:
Springer Nature Customer Service Center GmbH
Europaplatz 3, 69115 Heidelberg, Germany

Printed by Libri Plureos GmbH
in Hamburg, Germany